KU-505-859

Fifth Edition

Measurement and Evaluation — *in* — Human Performance

James R. Morrow, Jr., PhD
University of North Texas

Dale P. Mood, PhD
University of Colorado

James G. Disch, PED
Rice University

Minsoo Kang, PhD
Middle Tennessee State University

HUMAN KINETICS

Library of Congress Cataloging-in-Publication Data

Morrow, James R., Jr., 1947- , author.

Measurement and evaluation in human performance / James R. Morrow, Jr., Dale P. Mood, James G. Disch, Minsoo Kang. -- Fifth edition.

p. ; cm.

Preceded by Measurement and evaluation in human performance / James R. Morrow Jr. ... [et al.]. 4th ed. c2011.

Includes bibliographical references and index.

I. Mood, Dale, author. II. Disch, James G., 1947- , author. III. Kang, Minsoo (Physical education teacher), author. IV. Title.

[DNLM: 1. Athletic Performance. 2. Physical Fitness. 3. Monitoring, Physiologic--methods. QT 255]

QP301

613.7--dc23

2014042571

ISBN: 978-1-4504-7043-8 (print)

The web addresses cited in this text were current as of May 2015, unless otherwise noted.

Acquisitions Editor: Amy N. Tocco; **Developmental Editor:** Judy Park; **Managing Editor:** Carly S. O'Connor; **Copyeditor:** Ann Prisland; **Indexer:** Andrea Hepner; **Permissions Manager:** Dalene Reeder; **Graphic Designer:** Joe Buck; **Cover Designer:** Keith Blomberg; **Photograph (cover):** © Human Kinetics; **Photographs (interior):** © Human Kinetics, unless otherwise noted; **Photo Asset Manager:** Laura Fitch; **Photo Production Manager:** Jason Allen; **Art Manager:** Kelly Hendren; **Associate Art Manager:** Alan L. Wilborn; **Illustrations:** © Human Kinetics, unless otherwise noted; **Printer:** Thomson-Shore, Inc.

Printed in the United States of America

10 9 8 7 6 5 4 3 2 1

The paper in this book is certified under a sustainable forestry program.

Human Kinetics
Website: www.HumanKinetics.com

United States: Human Kinetics
P.O. Box 5076
Champaign, IL 61825-5076
800-747-4457
e-mail: humank@hkusa.com

Canada: Human Kinetics
475 Devonshire Road Unit 100
Windsor, ON N8Y 2L5
800-465-7301 (in Canada only)
e-mail: info@hkcanada.com

Europe: Human Kinetics
107 Bradford Road
Stanningley
Leeds LS28 6AT, United Kingdom
+44 (0) 113 255 5665
e-mail: hk@hkeurope.com

Australia: Human Kinetics
57A Price Avenue
Lower Mitcham, South Australia 5062
08 8372 0999
e-mail: info@hkaustralia.com

New Zealand: Human Kinetics
P.O. Box 80
Mitcham Shopping Centre, South Australia 5062
0800 222 062
e-mail: info@hknewzealand.com

E6200

Contents

Preface

In *Measurement and Evaluation in Human Performance, Fifth Edition,* our objective remains what we originally developed nearly two decades ago. We want an effective, interactive textbook that is understandable to undergraduate students in human performance, kinesiology, exercise science, or physical education.

Students, you are about to study concepts in human performance, kinesiology, exercise science, and physical education that your textbook authors believe are the most important you will learn about as an undergraduate. We understand that your professors of motor behavior, exercise physiology, and biomechanics and even your professors of English, mathematics, and history feel the same about their courses. We are not suggesting that those courses are unimportant or even a great deal less important. However, the concepts of reliability, objectivity, and validity that serve as the focus of this textbook and the course you are completing transcend all courses and all phases of life. We are confident that you want to make good decisions in all aspects of your professional and personal life. In the truest sense, this means you want to make evidence-based decisions that are accurate, trustworthy, truthful, and specific to making a good decision. That is exactly what this textbook and the course in which you are enrolled are all about.

Whether your desired career path is physical therapy, another allied health profession, teaching, athletic training, general kinesiology or exercise science, health and fitness, or sport studies or sport management, in this textbook you will find concepts important to your daily work once you become employed. It is common knowledge that physical activity is related to quality of life and a wide variety of morbidities and mortality. The ability to measure physical activity, physical fitness, and physical skills is important to you in your career, regardless of the specific career path you enter. It is important that you understand and use sound measurement procedures whether testing in the psychomotor, cognitive, or affective domains.

When you obtain data as a result of some type of testing or survey process, you want to be confident that the data you have for decision making are trustworthy. That is, can you count on these data? Are they truthful? Consider whether you are testing the fitness level of a student or client, or receiving test results from your personal physician. You want the results to be valid because you will be making important life decisions based on the results. In this text we will help you learn how to gather data, analyze them, interpret the results, and feel confident that what you have learned is valuable enough to make good decisions.

The authors of *Measurement and Evaluation in Human Performance* have been university professors for more than 150 combined years. We have taught hundreds of sections and thousands of students in the content area of your textbook. We have received much feedback and commentary from these students. We have attempted to incorporate all of this valuable information into the textbook so that you will find the course and materials interesting, entertaining, informative, and useful. We understand that some (or many) students are intimidated by the content presented in a measurement and evaluation course. We believe that we can solve the mysteries of such a course and help you appreciate and learn this material. Know early on that the level of mathematics involved in this textbook is no more than that in an introductory algebra course. We use these introductory concepts and combine them with reliability and validity theory in a way that we hope will affect your personal and professional life so when you have a decision facing you, you will immediately think about how you might gather data, analyze them, and use your analyses to make a good decision. That does not mean that the course is necessarily easy. It is certainly not as hard as some students think. It does, however, require study, review, and logical thinking to help you put all of the concepts together. We believe we do this well in the textbook.

UPDATES TO THE FIFTH EDITION

As we present the fifth edition of our text, we have maintained elements that have worked well, added some new points, and moved information around a bit to permit instructors and students to have a better understanding of the textbook and apply the information to their specific courses and career goals. The key changes in the fifth edition are the following:

- The focus in our text is data collection and analysis so that you can make appropriate decisions based on evidence. The concept of evidence-based decision making is important regardless of your chosen career path.
- While we continue with major use of SPSS (IBM SPSS Statistics), we also include use of Microsoft (MS) Excel with examples in an appendix that includes instructions, screen captures, and templates for calculating some statistics. Excel templates are located in the web study guide (WSG) in chapters 5 and 7. A PowerPoint presentation illustrating Excel use is available in the WSG in chapter 2.
- The amount of information on physical activity assessment has been increased across all chapters.
- We continue to use large datasets in each chapter that result in your reviewing previously learned concepts and methods to illustrate how valuable computer skills are in measurement and evaluation. Note that these large datasets are for illustrative purposes and may not include actual data from any specific research study. Although students often overlook or ignore these ancillaries, they are presented to help you learn and refresh content. **Students should definitely take advantage of these many resources!**
- The textbook's website includes a wide range of study and practice activities. Mastery Items and homework problems (and their answers) are available to students. We have actually reduced the number of homework problems themselves (so they do not appear overwhelming) but increased the types of resources you can use to help master the course content.
- Given the increased interest in prediction in sport and the application of data collection, analysis, and prediction, we have included a section in chapter 11 about sports analytics.
- We also provide brief video segments from professionals in the field who comment on and illustrate the importance of measurement and evaluation, data collection, and decision making in their respective careers.

For instructors, the ancillary items are updated, including the presentation package; a new image bank that includes most of the figures, content photos, and tables from the book; new and expanded chapter review questions in the instructor guide; new and revised questions in the test package; a new learning management system- compatible chapter quiz ancillary with 140 new questions, and 52 video clips with experts in the field discussing chapter topics.

TEXTBOOK ORGANIZATION

The textbook is presented in four parts. Part I, Introduction to Tests and Measurements in Human Performance, consists of two chapters. Chapter 1 introduces you to concepts in measurement and evaluation in human performance and the domains of interest in which you will use measurement and evaluation concepts and tools. Chapter 2, Using Technology in Measurement and Evaluation, is important. We use the information in the textbook every day of our lives. However, to do all of this work by hand would be tedious and susceptible to errors. The use of statistical packages does not eliminate all errors (you might still enter the wrong data, analyze the data the wrong way, or interpret the results incorrectly), but it does save you a great deal of time. We use SPSS and Excel to help you complete analyses. You are introduced to these tools in chapter 2, and examples are then used throughout the remainder of the textbook. Is it important to learn about tests and measurements? How can you learn about these areas, and can you become skilled at using them? We think so, and we hope you will agree after completing part I.

Part II, Basic Statistical Concepts, consists of three chapters and provides the statistical background for many of the decisions and interpretations that you will encounter in the remainder of the textbook. You will expand your SPSS or Excel experiences (or both) in chapters 3, 4, and 5 and continue to do so throughout the remainder of the textbook. The concepts that you learn in part II are the reason some students refer to this course content as statistics. Although the basic statistics presented are important, we prefer to refer to these as foundational elements for reliability and validity decisions. Again, the level of mathematics knowledge required in these areas is high school or college algebra. Importantly, we are not mathematicians and do not expect you to be. We are, however, practical users of the information in your textbook and we actually use this material *every* day of our lives. We try to keep the mathematics to a minimum and focus on concepts. Although having a stronger mathematics background can be an advantage, we present material in such a way that minimal mathematical expertise is expected or required. Moreover, the use of SPSS and Excel helps you complete time-consuming and difficult tasks with large numbers of observations quickly and accurately. Chapter 3, Descriptive Statistics and the Normal Distribution, illustrates how to describe test results and interpret them with graphs and charts. Chapter 4, Correlation and Prediction, helps you understand relations between variables and how knowledge about one variable tells you something about other variables. The content of chapter 5, Inferential Statistics, is used daily by human performance researchers to help make decisions from research study results and determine how these results can be generalized. It is common knowledge that physical activity, physical fitness, and energy expenditure are important to quality of life, health, and disease and death risk. Much of what we know about the relations between these variables and quality of life has been generated as a result of the types of analyses you will learn in part II. Would you like to have your test results presented in interesting and meaningful ways? Would you like to be able to read, interpret, understand, and use the scientific research literature related to your profession? You will be able to do so after completing part II.

Part III, Reliability and Validity Theory, presents the most important concepts of the textbook. Everything that is done in all of the textbook chapters can be directed toward or emanate from the concepts of reliability and validity. Chapter 6 presents reliability and validity from a norm-referenced perspective, while chapter 7 does so from a criterion-referenced perspective. Chapters 6 and 7 have a great deal in common. The key difference is the level of measurement involved. In chapter 6, test results are continuous in nature (like body fat or oxygen consumption), and in chapter 7 they are categorical (like pass/fail or alive/dead). Beyond that, the concepts of reliability and validity cross both chapters. Consider the most recent test you completed. It could have been in school, in a physician's office, at work, or anywhere else. How do you know the test results were reliable and accurately reflected your true results? You will be better able to interpret these results once you have completed part III.

Part IV, Human Performance Applications, consists of seven application chapters. The intent of part IV is to illustrate practical settings where you will use the kinds of knowledge you have gained to this point in the textbook. Chapter 8, Developing Written Tests and Surveys, shows you how to create tests that discriminate among people with differing levels of knowledge. Although it is important to be able to discriminate among levels of knowledge, this is difficult to do. Another important ability in this area is the development of survey and questionnaire materials that accurately reflect a person's knowledge or attitude or a group's general knowledge or attitude. Chapters 9 and 10 focus on the psychomotor domain and illustrate the assessment of physical fitness and physical activity in adults (chapter 9) and children and youth (chapter 10). Much has been said about the fitness and physical activity levels of adults and youth in the United States and around the world. What measures might be taken to assess physical fitness, physical activity levels, or both? Chapters 9 and 10 address these issues. Chapter 11 illustrates reliability and validity issues when assessing sport skills and motor abilities. Many textbooks use this type of chapter to list tests that can or should be used to assess a particular skill (e.g., tennis or golf) or ability (e.g., jumping or throwing). Rather than present which tests you should use, we prefer to identify the important concepts to be considered when choosing a test and then let you decide if a particular test appropriately meets your needs. Assume that you want to measure the physical skills, abilities, or status of students, clients, or athletes with whom you work. How might you best assess their abilities and skills, and how might you interpret and use these results? For example, the National Football League does this with what are called their combine camps each year, where attempts are made to determine which player should be drafted by a team. You will be able to use measurement techniques to answer many of these questions after completing chapter 11.

Chapter 12, Psychological Measurements in Sports and Exercise, presents scales that can be used in the affective domain. Scales discussed here assess attitudes, beliefs, and concepts that are largely unobservable, yet theoretically (and realistically) existent. For example, it is easy to think about those who have an attitude toward physical activity, but how would you measure this attitude? You will be better able to interpret results obtained from the affective domain after completing chapter 12.

The final two chapters, Classroom Grading: A Summative Evaluation (chapter 13) and Performance-Based Assessment: Alternative Ways to Assess Student Learning (chapter 14), are most appropriate for those with career goals in public or private school instruction. Concepts are presented that are important to those evaluating performance in the cognitive and psychomotor domains, with a focus on the important points of fairness, freedom from bias, and truthfulness of measurement. Even students whose career options are not directed toward evaluating student achievement will learn much from these chapters that they can apply to the classes they are currently enrolled in or will enroll in during their undergraduate career. Are you well informed about the grading procedures used in the classes in which you are enrolled? What might you encourage the professor to do to make grading fairer? Chapters 13 and 14 will help you better answer these questions.

Appendix: Microsoft Excel Applications provides Excel support for those without access to SPSS. Directions for calculating measurement and evaluation statistics with Excel are presented. The results are similar to those obtained with SPSS. Templates are provided in some cases (chi-square and epidemiological statistics) that can be used rather than the many involved steps necessary in these cases with Excel. Some Excel resources are available on the textbook's Internet site.

STUDENT RESOURCES

Key to the fifth edition of our textbook is the variety of student resources that help you learn the material and see how it applies to your career and everyday decision making in human performance. In each chapter you'll find many items that will help you understand and retain information:

- Chapter outlines are presented at the start of the chapter, introducing you to the organization of the chapter and helping you locate key information.
- Chapter objectives tell you the main points you should take away after having finished studying a chapter.
- Key terms are highlighted in the text, and their definitions are provided for you in the glossary.
- Measurement and Evaluation Challenge sidebars both introduce and close the chapters. The opening sidebar presents a scenario in which a person is faced with a situation; the closing sidebar shows how the concepts covered in the chapter can help that person deal with the issues presented in the opening scenario.
- Mastery Items test whether or not you have mastered a particular concept. These items include problems and student activities that will help you confirm that you have mastered the information at certain points throughout a chapter. Some Mastery Items require you to complete the task on a computer; a computer icon identifies which Mastery Items need to be completed using computer software.
- Dataset Applications give you an opportunity to practice many of the techniques presented in chapters 2 through 14. Large datasets are available in each chapter's section of the WSG. (See the next section for more information on the WSG.) You will gain valuable experience in using statistical software by following the instructions in your textbook and using the datasets in the WSG.

We encourage you to complete all of the resources provided with each textbook chapter. Completion of these resources will help you better understand the concepts and apply them to your career. Equally important, they will help prepare you for examinations.

WEB STUDY GUIDE

The WSG is a great resource for you. It includes outlines, interactive quiz questions, homework problems, data tables from the textbook, video clips, and student learning activities that will enhance understanding and application of the concepts presented in the text. There are callouts throughout the chapters that guide you to the WSG. When you see one of these callouts, go to the appropriate location in the WSG and download the information or complete the activity.

Student study activities ask you to think about a certain issue or complete a task that will help you better understand chapter content. Each chapter ends by directing you to the WSG, where you will find homework problems and quizzes to ensure you've fully grasped a chapter's content.

Students who ignore the many opportunities provided in the WSG miss great opportunities to better understand the course content. Frankly, better understanding and application is what you will be tested on. **Students, *beware!* Ignore the WSG at your own peril!** If you want to do well in the course, apply yourself and use the many resources available to you. You can access the WSG at www.HumanKinetics.com/MeasurementAndEvaluationInHumanPerformance. If your instructor has a personal or class Internet site, he or she might even have a link to the WSG from there.

HELPFUL STUDY HINTS

Here are a few hints that should help you learn and use the contents of your measurement and evaluation textbook. Frankly, most of these suggestions apply to any course in which you are enrolled and are common knowledge for most students. The hard part is actually doing what you know you need to do. These are our suggestions:

1. Download the chapter outlines from the WSG.
2. Read the Measurement and Evaluation Challenge at the start of the chapter. Think about it.
3. Keeping the Measurement and Evaluation Challenge in mind, scan the chapter before you read it. Highlight key things that pique your interest.
4. Read the chapter. Highlight key points in more detail.
5. Try to do the homework problems.
6. Attend class every day. Do not sit near your friends. Take notes in class. Ask questions.
7. Study together in groups.
8. Rework the homework problems.
9. Reread the chapter.
10. Return to the Measurement and Evaluation Challenge that started the chapter to see if you can determine how the chapter information has helped you address the challenge.
11. Work with the large dataset available for each chapter beginning with chapter 2. Conduct the analyses as directed, but also conduct analyses of your own that are related to the dataset.
12. If you have questions, consider how you might learn the answer. Look through the chapter, review your notes, look at the homework, and go to the WSG. Some instructors suggest, "Ask three and then me." The idea is for you to ask three of your classmates (or other sources) about your questions. The interaction will help all of you learn the material better. If after asking three people you still have questions, go ask the instructor.

We believe that we have prepared a textbook that is understandable, interesting, informative, and easy to read. Yet it provides important information and concepts that will help you in your decision making. You are encouraged to use the textbook and the various resources to their fullest extent. We have learned that students who apply themselves, spend the time necessary for learning, attend class, are prepared, and follow the preceding suggestions have a greater appreciation for reliability and validity theory, understand the concepts better, and apply the concepts more often in their careers than students who do not use these techniques, methods, and strategies. We hope you will do what we want you to do: read, study, learn, understand, and most importantly *use* measurement and evaluation concepts throughout your academic, personal, and professional lives.

INSTRUCTOR RESOURCES

Several ancillaries are available to help instructors teach the material presented in this textbook:

- ***Instructor guide.*** The instructor guide provides sample course syllabi, chapter review questions and answers, homework problems and answers, and answers to all of the Mastery Items, and Student Activities. While we hope instructors will use these materials, we also encourage instructors to create their own additions and extensions of these materials so that the course reflects the instructor personally and addresses current important topics.
- ***Instructor videos.*** New to the fifth edition is a library of 52 online video clips, featuring interviews with leading experts in the field that instructors can use during their in-person class lectures or online courses. This video is the same as the one in the student web study guide, but it is in an easy-to-access list to quickly supplement lectures.
- ***Test package.*** The test package contains hundreds of test questions. Here again, we encourage instructors to use and modify these items so that they reflect current information from the scientific literature. The literature about measurement and evaluation in human performance is expanding daily. Take advantage of the advances in our field and develop test items around these scientific advances.
- ***Chapter quizzes.*** New to the fifth edition are chapter quizzes. These learning management system (LMS)-compatible, ready-made quizzes can be used to measure student learning of the most important concepts for each chapter. One hundred forty unique questions (ten questions per chapter) are included in multiple choice format.
- ***Presentation package.*** The presentation package contains slides, as well as real-life examples, that match key figures, tables, and concepts from the book. Existing slides can be modified, and the presentation can be easily adapted to accommodate additional slides.
- ***Image bank.*** The image bank includes most of the figures, content photos, and tables from the text. You can use these items to add to the presentation package or to create your own PowerPoint presentation covering the textbook material. These items can also be used in handouts for students.
- ***Web study guide.*** The web study guide is a valuable resource for your students. Here, they can find specific information for each chapter, including a printable outline of the chapter, an interactive chapter quiz, homework problems plus answers to the odd-numbered problems, student activities plus answers to the odd-numbered activities, answers to the Mastery Items, and large datasets that are needed to complete the Dataset Application activities in the chapters. The study guide also contains 52 video clips with interviews of top researchers who offer greater insight into the field.

You can access all of the ancillaries at www.HumanKinetics.com/MeasurementAndEvaluationInHumanPerformance.

Acknowledgments

The fifth edition of our textbook has brought us near the end of our academic careers. As such, we have begun a transitioning authorship. Allen W. Jackson has agreed to move from textbook coauthor to chapter author (see chapters 9 and 10). Allen has been a key player on our team and we could not have gotten to a fifth edition without his insight, writing, and research activities. While his name is no longer on the front cover, he is a major contributor to our textbook. Minsoo Kang of Middle Tennessee State University has joined us as an author. Minsoo was a student of Weimo Zhu in the Kinesmetrics Laboratory at the University of Illinois. Weimo is perhaps the most knowledgeable measurement expert in our field. Minsoo's role in the fifth edition has largely been the updating of the textbook ancillaries. We are pleased that Minsoo has joined us.

We acknowledge the significant contributions by Robert S. Weinberg and Joseph F. Kerns of Miami University in Oxford, Ohio, for the content in chapter 12. Likewise, Jacalyn L. Lund of Georgia State University made a major contribution with the authoring of chapter 14.

Importantly, this text could not have been completed without much guidance, many suggestions, and a great deal of encouragement from numerous professionals. Our association with the professionals at Human Kinetics has been most rewarding. We particularly acknowledge Amy Tocco and Judy Park. Joe Buck spent much time and effort designing the layout and presentation, and Carly S. O'Connor helped us tie things together at the end. Their efforts are greatly appreciated. Doug Fink and Lisa Morgan were instrumental in finalizing the numerous online components for this edition. We value the measurement and evaluation professionals from whom we have learned much. These (our mentors, friends, and students) include ASJ, ATS, BAM, CHS, DJH, GVG, HHM, HRB, JAS, JEF, JLW, JMP, KDH, LDH, LRO, LSF, MAL, MEC, MJL, MJS, MJS, MSB, MTM, RGB, RGF, RWS, SSS, SNB, TAB, TMW, VWS, WBE, and WZ. As we have in the past, we want to acknowledge our families, who tolerated the many hours we spent with our computers. Our endeavors to influence measurement and evaluation thinking and activities are truly a reflection of our professional colleagues and personal friends. We thank you.

Part I

Introduction to Tests and Measurements in Human Performance

We all want to make good decisions. In part I, we introduce you to concepts of measurement and evaluation and their importance in decision making. These concepts are the foundation for your study of measurement in human performance throughout the remainder of the book. Chapter 1 presents a rationale for and an overview of the scope and use of measurement in human performance. Chapter 2 describes computer applications in human performance, with specific attention to applications of measurement, testing, and evaluation; this chapter introduces advanced technology for conducting many of the exercises in the remainder of the book. Specifically, you will use the Internet and statistical software (SPSS [IBM SPSS Statistics] and Microsoft Excel) to help solve measurement and evaluation problems.

Part I will provide you with much of the background and computer skills necessary for making valid measurement decisions. For example, you'll learn to create data tables in SPSS, read Excel files into SPSS, and analyze a dataset using the appropriate procedures. These procedures will be used throughout the remainder of the text in Mastery Items and in other activities available to you through your textbook and through the web study guide (WSG). You should go to the text's website, www.HumanKinetics.com/MeasurementAndEvaluationInHumanPerformance, where you will find many helpful resources as you learn more about measurement and evaluation.

CHAPTER

1

Concepts in Tests and Measurements

OUTLINE

OBJECTIVES

After studying this chapter, you will be able to

- define the terms test, measurement, and evaluation;
- differentiate between norm-referenced and criterion-referenced standards;
- differentiate between formative and summative evaluation;
- discuss the importance of measurement and evaluation processes;
- identify the purposes of measurement and evaluation;
- identify the importance of objectives in the decision-making process; and
- differentiate among the cognitive, affective, and psychomotor domains as they relate to human performance.

 The lecture outline in the WSG will help you identify the major concepts of the chapter.

MEASUREMENT AND EVALUATION CHALLENGE

As you begin learning about measurement and evaluation processes in human performance, we present an overview of what you will study throughout this textbook. This first measurement and evaluation challenge presents a scenario that relates to most of the chapters and concepts you will study. We first describe a scenario, and then at the end of the chapter we explain how you might answer the questions that arise in the scenario.

Imagine that your father talks with you about his recent physical examination. It has been a number of years since he had a medical examination. His physician conducted a battery of tests and asked your father about his lifestyle. As a result, the physician told your father that he is at risk for developing cardiovascular disease. Your father was told that his weight, blood pressure, physical activity level, cholesterol, nutritional habits, and stress levels have increased his chances of developing cardiovascular disease. Your father tells you that he feels great, was physically active throughout high school and college, looks better than most people his age, and cannot imagine that he is truly at an elevated risk. Because he knows you are aware of cardiovascular disease risk factors, he asks you the following questions:

1. How does one know if the measures taken are accurate?
2. What evidence suggests that these characteristics are truly related to developing cardiovascular disease?
3. How likely is it that the physician's evaluation of the tests is correct?
4. What aspect of the obtained values places one at increased risk? For example, how was a systolic blood pressure of 140 mmHg originally identified as the point at which one is at increased risk? Why not 130 mmHg or 150 mmHg? Why has the blood pressure risk been decreased from 140 to 130 and even 120 mmHg? Your father reports being physically active, but what does that mean? Is he engaging in sufficient physical activity to be at increased health or reduced risk for negative health outcomes? Similar questions could be asked about each of the measurements obtained.
5. What evidence exists that changing any of these factors will reduce risk?

Your father is concerned because he doesn't know what these numbers mean. Likewise, he and you are concerned about the accuracy of these measurements. You would like to explain to him how to interpret these results and encourage him to make the necessary lifestyle changes to reduce his cardiovascular risk.

Interpreting measurement results and determining the quality of the information one receives are what this course is all about. Information gained from this course will help you make informed decisions about the accuracy and truthfulness of obtained measures and decisions based on human performance measurements. In general, good measurement and subsequent evaluation should lead to good decisions, such as changing one's lifestyle to improve health. We focus on measurements obtained in the cognitive, psychomotor, and affective domains.

Why is testing important? Is it really necessary for knowing many statistical concepts? What decisions are involved in the measurement process? How can one know the amount and influence of measurement errors on decisions? How you answer these questions is important to your development as a competent professional in human performance.

Decision making is important in every phase of life whether it is related to professional or personal decisions. The quality of the data collected and the way in which one approaches decision making will affect the quality of one's decisions. The statistical and measurement concepts presented in this text provide a framework for making accurate, truthful decisions based on sound data.

We all gather data before making decisions, whether the decision-making process occurs in research, in education, or in other pursuits. For example, you might gather information for student grades, research projects, or fitness evaluation. Researchers gather data on fitness characteristics to examine the relationships among fitness, physical activity, mortality, morbidity, and quality of life. Examples of variables measured might include the amount and type of physical activity, blood pressure, and cholesterol levels. Weight loss and weight control are major health concerns, so you might be interested in measuring energy expenditure to estimate caloric balance. Likewise, you gather data about the weather before venturing out for a morning run, and you adjust your behavior based on the data you obtain (e.g., rain, warm, dark, cold). Before purchasing a stock for investment, you gather data on the company's history, leadership, earnings, and goals. All of these are examples of testing and measuring. In each case, making the best possible decision is based on collecting relevant data and using them to make an accurate decision.

The course you are embarking on was historically called Tests and Measurements. Although some students refer to this as a statistics book, that does not accurately describe what this text and most courses using it are about. Some basic statistical concepts are presented in part II (chapters 3, 4, and 5); however, the statistical and mathematical knowledge necessary for testing and measurement is not extensive, requiring only basic algebraic concepts. On the other hand, every chapter in this text focuses in some way on the important issues of reliability and validity. In kinesiology (a subdomain of human movement and performance), measurement and evaluation is now becoming known as kinesmetrics (i.e., movement measurement) (Mahar and Rowe, 2014). To make good decisions, you must measure and evaluate accurately. *Making effective decisions depends on first obtaining relevant information.* Consider youth physical activity levels in the United States. How is physical activity measured, reported, and tracked? The *2008 Physical Activity Guidelines for Americans* (U.S. Department of Health and Human Services [USDHHS] 2008) recommend 60 minutes of moderate-to-vigorous physical activity (MVPA) daily for children and youth. However, the Centers for Disease and Control and Prevention (CDC) reports that in 2013, 15.2% of high school youth did not engage in at least 60 minutes of physical activity on ANY day in the previous 7 days and that 72.9% of students engaged in 60 minutes per day fewer than 7 days out of the previous 7 days. The data are from the Centers for Disease Control and Prevention's Youth Risk Behavior Surveillance System (YRBSS; Kann et al., 2014). Are these reported values reliable and valid? Can intervention intended to affect lifestyle behavior changes be based on these data? This is where testing and measurement (and reliability and validity) enter the picture.

NATURE OF MEASUREMENT AND EVALUATION

The terms we use in measurement and evaluation have specific meanings. *Measurement, test,* and *evaluation* refer to specific elements of the decision-making process. Although the three terms are related, each has a distinct meaning and should be used correctly. **Measurement** is the act of assessing. Usually this results in assigning a number to quantify the amount of the characteristic that is being assessed. For example, people might be asked to self-report the number of days per week they engage in MVPA (as was done with the YRBSS results presented earlier). Alternatively, they could be asked to report the

YOUR TEXTBOOK AND THE WEB STUDY GUIDE

In writing this textbook, we have taken advantage of the Internet to help you better learn, practice, and use the information presented in your textbook. A web study guide (WSG) has been developed for this textbook; you can access it by going to www.HumanKinetics.com/MeasurementAndEvaluationInHumanPerformance. If your instructor has a personal or class website, he or she might even have a link to the WSG from there. The key things that you can obtain at the WSG include

- outlines associated with each chapter in the textbook,
- answers to mastery items,
- additional student activities that are similar to the textbook's mastery items,
- practice homework problems with selected answers, and
- quizzes.

The WSG is organized so that you go to the site, choose the chapter you would like to obtain information about, and click on the link to that chapter. Once you reach the chapter itself, there are various links that get you to resources that help you prepare for class, review course material, take practice quizzes, complete practice problems, and prepare for examinations. You should go to the site, look around, and examine the resources there.

number of minutes of MVPA per week. Their physical activity behaviors could be tracked with an electronic device (e.g., a pedometer or accelerometer). A **test** is an instrument or tool used to make the particular measurement. This tool can be written, oral, physiological, or psychological, or it can be a mechanical device (such as a treadmill). To determine the amount of MVPA conducted in a week, one might use self-report, direct observation, pedometers, or a motion sensor (i.e., pedometer or accelerometer). **Evaluation** is a statement of quality, goodness, merit, value, or worthiness about what has been assessed. Evaluation implies comparison and decision making. Once we determine how physically active a person is, we can compare that result to national standards based on the U.S. Health and Human Services guidelines for physical activity (USDHHS, 2008) to see if the person is sufficiently active for health benefits. You could evaluate someone's MVPA per week and compare it to the national adult recommendation of 150 minutes per week. Achieving (or failing to achieve) is where evaluation comes into play. You are comparing a measured value, obtained with a test, to some criterion or other measure. Evaluation enters into the discussion when you tell the person whether or not he or she meets the physical activity guidelines. Alternatively, you may report how many steps the person averages per day and compare that to how many steps people of that age generally take per day.

Consider another example where you want to measure cardiorespiratory fitness, often reported as $\dot{V}O_2max$, a measure of aerobic capacity. You can measure a person's maximal oxygen consumption or uptake in several ways. You might have someone perform a maximal run on a treadmill while you collect and analyze expired gases. You might collect expired gases from a maximal cycle ergometer protocol. You might have a participant perform either a submaximal treadmill exercise or a cycle exercise, and then you predict $\dot{V}O_2max$ from heart rate or workload. You might measure the distance a person runs in 12 minutes or the time it takes to complete a 1.5-mile (2.4 km) run. Each of these tools results in a number, such as percent O_2 and CO_2, heart rate, minutes, or yards. Having assessed $\dot{V}O_2max$ with one of these tools does not mean that you have evaluated it. Obtaining and reporting data have little meaning unless you reference the data to something. This is where evaluation enters the process.

Assume that you test someone's $\dot{V}O_2max$. Furthermore, assume that this participant has no knowledge of what the $\dot{V}O_2max$ value means. Certainly, the participant might be aware that the treadmill test is used to measure cardiorespiratory fitness. However, the first question most people ask after completing some measurement is *How did I do?* or *How does it look?* To simply report, "Your $\dot{V}O_2max$ was 30 ml · kg^{-1} · min^{-1}" says little. You need to provide an evaluation. An evaluative statement about the performance introduces the element of merit, or quality, often based on knowledge of normative data. For example, a $\dot{V}O_2max$ of 30 might be considered quite good for a 70-year-old female but quite poor for a healthy 25-year-old male.

Go to the WSG to complete Student Activity 1.1 and view Videos 1.1 and 1.2.

Norm-Referenced and Criterion-Referenced Standards

To make an evaluative decision, you must have a reference perspective. *You can make evaluative decisions from either norm-referenced (normative) or criterion-referenced standards.* Using an evaluative decision based on a norm-referenced standard means that you report how a performance compares with that of others (perhaps of people of the same gender, age, or class). Thus, as in the earlier example, you might report that a $\dot{V}O_2max$ of 30 is relatively good or poor for someone's particular age and gender. Conversely, you might simply report a person's performance relative to a criterion that you would like him or her to achieve. Assume that the $\dot{V}O_2max$ of 30 was measured on someone who previously had a heart attack. A physician may be interested in whether the patient achieved a $\dot{V}O_2max$ of at least 25 ml · kg^{-1} · min^{-1}, which would indicate that the patient had achieved a functional level of cardiorespiratory fitness. This is an example of a criterion-referenced standard. You are not interested in how someone compares with others; the comparison is with the standard, or criterion. The criterion often is initially based on norm-referenced data and the best judgment of experts in the content area.

Consider assessing physical activity behaviors with a pedometer. How many steps are taken could be an indication of general physical activity behavior and be used to determine if one is sufficiently active for health benefits. The fact that one person takes more steps than another does not mean that the more active person is sufficiently active. Comparing the number of steps to another person's is a norm-referenced comparison. Comparing the number of steps to a specific minimum is a criterion-referenced comparison. For example, Tudor-Locke and Bassett (2004) suggested pedometer criteria for adult public health. Someone could be interested in how many steps he or she takes compared to others (a norm-referenced comparison), but the more important measure might be if the person is taking sufficient steps for health benefits (a criterion-referenced comparison). Note that no national criterion has been developed for steps per day to be healthy. Setting such a criterion (and many others) is a difficult challenge for measurement and evaluation experts.

You will learn much more about setting standards and the validity of standards in chapter 7.

Changes in youth fitness evaluation processes over the last 40+ years provide a good comparison of norm-referenced versus criterion-referenced standards. Fitness scores used to be evaluated using a norm-referenced system, that is, relative to a child's classmates, by age and gender. A child completed a fitness test and then was informed what percentage of children he or she did better than. This is a norm-referenced comparison. Many youth fitness tests now are criterion referenced. Consider a 12-year-old boy who completes a 1-mile

HOW MANY STEPS ARE ENOUGH?

Steps per day	Public health index
≤5000	Sedentary lifestyle
5001-7499	Low active
7500-9999	Somewhat active
≥10,000	Active
≥12,500	Highly active

(2.4 km) run and his predicted $\dot{V}O_2max$ is 37.5. This does not meet the minimum criterion for the Fitnessgram Healthy Fitness Zone (40.3). The Healthy Fitness Zone represents a criterion-referenced standard. The Fitnessgram uses criterion-referenced standards (called Healthy Fitness Zones) for assessing **health-related physical fitness**. It is not important how one compares with peers on the Fitnessgram. What is important is whether or not one achieves the Healthy Fitness Zone (the criterion-referenced standard).

Mastery Item 1.1

Are the following measures usually evaluated from a norm-referenced or a criterion-referenced perspective?

- Blood pressure
- Fitness level
- Blood cholesterol
- A written driver's license examination
- Performance in a college class

Go to the WSG to complete Student Activity 1.2.

Formative and Summative Evaluation

Evaluations occur in two perspectives, formative and summative. **Formative evaluations** are initial or intermediate evaluations, such as the administration of a pretest and the subsequent evaluation of its results. Formative evaluation should occur throughout an instructional, training, or research process. Ongoing measurement, evaluation, and feedback are essential to the achievement of the goals in a program in human performance. For example, after shoulder surgery, your goal may be to regain range of motion (ROM) in the shoulder joint. Your physical therapist could assess your ROM and suggest activities to improve it. These ongoing evaluations need not involve formal testing; simple observation and feedback sequences between the therapist and the patient are fine. Formative evaluation processes are obviously important to track changes (in this course, for example). **Summative evaluations** are final evaluations that typically come at the end of an instructional or training unit. You, as a student in this course, are interested in the summative evaluation—the grade—that you will receive at the end of the semester.

The difference between formative and summative evaluations might seem to be merely the difference in timing of their data collection; however, the actual use of the data collected distinguishes the evaluation as formative or summative. Thus, in some situations the same data can be used for formative and summative evaluations.

A weight-loss or weight-control program provides a simple and useful example for applying both formative and summative evaluations. Assume that you have measured a participant's body weight and percent body fat. Your formative evaluation indicates that this participant has a percent body fat of 30% and needs to lose 10 pounds (4.5 kg) to achieve a desired percent fat of 25%. You establish a diet and exercise program designed to produce a weight loss of 1 pound (.5 kg) per week for 10 weeks. Each week you weigh the participant, measure the percent body fat, and give the person feedback on the formative evaluations you are conducting. The participant knows the amount of progress or lack of progress that is occurring each week. At the end of the 10-week program, you measure body weight and percent body fat again and conduct a simple summative evaluation. Were the weight-loss and percent-fat goals achieved at the end of the program?

Consider pedometer steps again. One might have a long-term goal of 10,000 steps per day but a short-term goal of increasing the steps taken by 500 a day over a period of a few weeks. This goal can then be adjusted as one becomes increasingly physically active. The short-term goals can be viewed as formative, and the long-term, final goal can be viewed as the summative evaluation. Can you see how you might use formative and summative evaluations in many of your daily decisions, regardless of your career goal?

Go to the WSG to complete Student Activity 1.3 and view Video 1.3.

PURPOSES OF MEASUREMENT, TESTING, AND EVALUATION

Prospective professionals in kinesiology, human performance, physical activity, health promotion, and the fitness industry must understand measurement, testing, and evaluation because they will be making evaluative decisions daily. Our patients, students, athletes, clients, and colleagues ask us what tools are best and how to interpret and evaluate performance and measurements. Regardless of your specific area of interest, the best tools to use and how to interpret data may be the most important concepts that you will study. Related evaluation concepts are objectivity (rater consistency), reliability (consistency), relevance (relatedness), and validity (truthfulness). These terms are discussed in greater detail in chapters 6 and 7.

The evaluative process in human performance may be used in many ways. For instance, consider the issue of accountability. Your employer might hold you accountable for a project; that is, you might be responsible for obtaining a particular outcome for a person or program. Tests, measurement, and the evaluation processes are used to show whether you have met the goals. Obviously, you want the evaluation to accurately reflect the results of your work—assuming that you did a good job! Certainly, if you enter the teaching profession, you will hold your students accountable for learning and retaining the content of the courses you teach. Likewise, your students should hold you accountable for preparing the best possible tests for evaluating their class performance.

Go to the WSG to complete Student Activity 1.4 and view Video 1.4.

As you will discover during your course of study, you need considerable knowledge and skill to conduct correct and effective measurement and evaluation. *As with any academic or professional effort, having a thorough understanding of the purposes of executing a measurement and evaluation process is critical.* There are six general purposes of measurement and evaluation: placement, diagnosis, prediction, motivation, achievement, and program evaluation.

Placement

Placement is an initial test and evaluation allowing a professional to group students into instructional or training groups according to their abilities. In some cases, instruction, training, and learning in human performance can be facilitated by grouping participants according to their abilities. All participants in a group can then have a similar starting point and can improve at a fairly consistent rate. It is difficult to teach a swimming class if half the students are non-swimmers and the others are members of the swim team, but even less extreme differences can affect learning.

Diagnosis

Evaluation of test results is often used to determine weaknesses or deficiencies in students, medical patients, athletes, and fitness program participants. Cardiologists may administer treadmill stress tests to obtain exercise electrocardiograms of cardiac patients to diagnose the possible presence and magnitude of cardiovascular disease. Recall the measurement and evaluation challenge highlighted at the beginning of this chapter. The doctor made a diagnosis based on a number of physiological and behavioral measures. This was possible because of the known relationships between the measures and the incidence of heart disease. While there is currently much interest in MVPA, there is also much interest in sedentary behaviors. The amount of MVPA and sedentary time in which one engages can be diagnostic of physically active lifestyle behaviors.

Athletic teams are created by placing players of similar skill together, and those players are motivated to reach a certain level of achievement. *Placement, motivation,* and *achievement* are all important factors in measuring and evaluating human performance.

Prediction

One of the goals of scientific research is to predict future events or results from present or past data. This is also one of the most difficult research goals to achieve. You probably took the Scholastic Aptitude Test (SAT) or the American College Test (ACT) during your junior or senior year of high school. Your test scores can be viewed as predictors of your future success in college and perhaps were part of the admissions process used by your college or university. The exercise epidemiologist may use physical activity patterns, cardiovascular endurance measures, blood pressure, body fat, or other factors to predict your risk of developing cardiovascular disease.

Motivation

The measurement and evaluation process is necessary for motivating your students and program participants. People need the challenge and stimulation they get from an evaluation of their achievement. There would not be any athletes if there were only practices and no games or competitions. When are you going to be most motivated to learn the material for this course? What would motivate you to study and learn the material for this course or any other if you knew you would not be tested and evaluated? Can you see how simply measuring your weight can be motivational? Likewise, knowing the number of steps one takes per day might provide impetus for increasing physical activity behaviors.

Achievement

In a program of instruction or training, a set of objectives must be established by which participants' achievement levels can be evaluated. For instance, in this course, your final achievement level will be evaluated and a grade will be assigned on the basis of how well you met objectives set forth by the instructor. Developing the knowledge and skills needed for proper grading is one objective of this book; chapters 13 and 14 are devoted to the topics of assessment and grading. Improvement in human performance is an important goal in instruction and training programs, but it is difficult to evaluate fairly and accurately. Is the final achievement level going to be judged with criterion-referenced standards on a pass/fail basis or with norm-referenced standards and grades (e.g., grading on the curve)? Assessment of achievement is a summative evaluation task that requires measurement and evaluation.

Program Evaluation

You may have to conduct program evaluations in the future to justify your treatment, instruction, and training programs. The goal of program evaluation is to demonstrate (with sound evidence) the successful achievement of program objectives. Perhaps you have a program objective of increasing MVPA in your community. You desire to measure behavior changes with self-reported MVPA and pedometer steps. You would measure the physical activity behaviors and then make decisions based on the data you obtain. Alternatively, if you are a physical education teacher, you may be asked to demonstrate that your students are receiving appropriate physical activity experiences. You might compare your students' fitness test results with the test results of students in your school district or with national test results. You might gather student and parent evaluations of your program. Professionals in community, corporate, or commercial fitness centers can evaluate their programs in terms of membership levels and participation, participant test results, participant evaluations, and physiological assessments. Your job and professional future could depend on your being able to conduct a comprehensive and effective program evaluation.

 Go to the WSG to complete Student Activity 1.5 and view Video 1.5.

RELIABILITY AND VALIDITY

The terms reliability and validity are seen throughout this textbook. They are the most important concepts in this textbook. Regardless of the purpose for which you are using measurement, you want your results to be reliable and valid. **Reliability** refers to consistency of measurement. **Validity** refers to truthfulness of measurement. If a test or measuring process is valid, it must first be reliable. If a test process is reliable and it is relevant to what is being measured, then it is valid. Consider the conundrum where you have a valid test but it is not reliable. How can this be? How can it be that something truthfully measures something but does not do it consistently? It can't! Testing processes must first be reliable before they can be valid. When obtaining any measure and using it for any purpose, your first thoughts should be about reliability and validity. Essentially, you can think of reliability + relevance → validity. A word closely aligned with reliability is objectivity. **Objectivity** refers to reliability between raters (interrater reliability). The terms objectivity, reliability, and validity are presented in detail in chapters 6 and 7; these terms are key to measurement in human performance.

DOMAINS OF HUMAN PERFORMANCE

The purposes we've just discussed are related to the objectives of your program. Objectives are specific outcomes that you hope to achieve with your program. To be accurately measured and truthfully evaluated, these outcomes need to be measurable. Objectives in the area of human performance fall into three areas: the cognitive domain, the affective domain, and the psychomotor domain. Students of measurement and evaluation in education or psychology are concerned with objectives in the first two areas. *For students of human performance, the distinctive objectives are those in the psychomotor domain.*

Bloom (1956) presented a **taxonomy** (classification system) of cognitive objectives (see table 1.1). A hierarchical list of Bloom's levels includes knowledge, comprehension, application, analysis, synthesis, and evaluation. Anderson and Krathwohl (2001) have modified Bloom's original taxonomy and included *creating* as the highest level of cognitive endeavors. Objectives in the **cognitive domain** deal with knowledge-based information. Objectives in the **affective domain** concern psychological and emotional attributes. A taxonomy of these objectives, from Krathwohl, Bloom, and Masia (1964), is as follows: receiving, responding, valuing, organizing, and characterizing by a value complex. Affective objectives, which concern, for example, how people feel about their performance, are important but often difficult to measure. Affective objectives are not normally measured for grading purposes. The third domain of objectives is the **psychomotor domain** (Harrow 1972); these are reflexive movements, basic locomotor movements, perceptual motor abilities, physical abilities, skilled movements, and nondiscursive movements. The measurement techniques and concepts associated with the psychomotor domain differentiate human performance students from students in other areas. Other taxonomies are used for the cognitive, affective, and psychomotor domains; those in table 1.1 are only examples.

When you are measuring and evaluating participants with a specific test, you must take into account the level of the domain your participants have achieved. Each taxonomy is a hierarchy; each level is based on the earlier levels having been achieved. For example, it

Table 1.1 Taxonomies in Domains of Human Performance

Taxonomy of the cognitive domain (Bloom 1956)	Taxonomy of the affective domain (Krathwohl, Bloom, and Masia 1964)	Taxonomy of the psychomotor domain (Harrow 1972)
Knowledge • Of specifics • Of ways and means of dealing with specifics • Of the universals and abstractions in a field **Comprehension** • Translation • Interpretation • Extrapolation **Application** **Analysis** • Of elements • Of relationships • Of organizational principles **Synthesis** • Production of unique communications • Production of a plan for operations • Derivation of a set of abstract relations **Evaluation** • Judgments in terms of internal evidence • Judgments in terms of external evidence	**Receiving** • Awareness • Willingness to receive • Controlled or selected attention **Responding** • Acquiescence in responding • Willingness to respond • Satisfaction in response **Valuing** • Acceptance of a value • Preference for a value • Commitment **Organizing** • Conceptualization of a value • Organization of a value system **Characterizing by a value complex** • Generalized set • Characterization	**Reflex movements** • Segmental reflexes • Intersegmental reflexes • Suprasegmental reflexes **Fundamental movements** • Locomotor movement • Nonlocomotor movement • Manipulative movement **Perceptual abilities** • Kinesthetic discrimination • Visual discrimination • Auditory discrimination • Tactile discrimination • Coordinated discrimination **Physical abilities** • Endurance • Strength • Flexibility • Agility **Skilled movements** • Simple adaptive skill • Compound adaptive skill • Complex adaptive skill **Nondiscursive movements** • Expressive movement • Interpretive movement

would be inappropriate for you to attempt to measure complex motor skills in 7-year-old children because most of them have not achieved prior levels of the taxonomic structure. Likewise, it is difficult, if not impossible, for younger participants to achieve the higher-level cognitive objectives of a written test.

Physical activity is a behavior defined as bodily movement. Physical activities can range from those that substantially increase energy expenditure to light activities, strengthening activities, and even to physical inactivity (i.e., sedentary behaviors). These physical activity behaviors can take place during leisure, transportation, occupational, and household activities. Physical fitness, on the other hand, is a set of attributes that people have or achieve that relates to the ability to perform physical activity. Physical activity is something people do (i.e., a behavior), whereas physical fitness is something people have or achieve. Heredity plays an important role in both factors but is probably more important in physical fitness. Physical activity is more difficult to measure reliably and validly than physical fitness. Measuring a behavior is generally more difficult than measuring an attribute.

Pettee-Gabriel, Morrow, and Woolsey's (2012) diagram (figure 1.1) illustrates the differences between physical activity (the behavior) and physical fitness (the attribute). Measurement considerations are key to assessing physical activity behaviors and physical attributes.

Just as there are a variety of physical activity surveys (e.g., BRFSS, YRBSS), fitness tests, and test protocols for measuring specific fitness attributes, there are a variety of techniques to measure physical activity and fitness. These techniques include use of motion sensors,

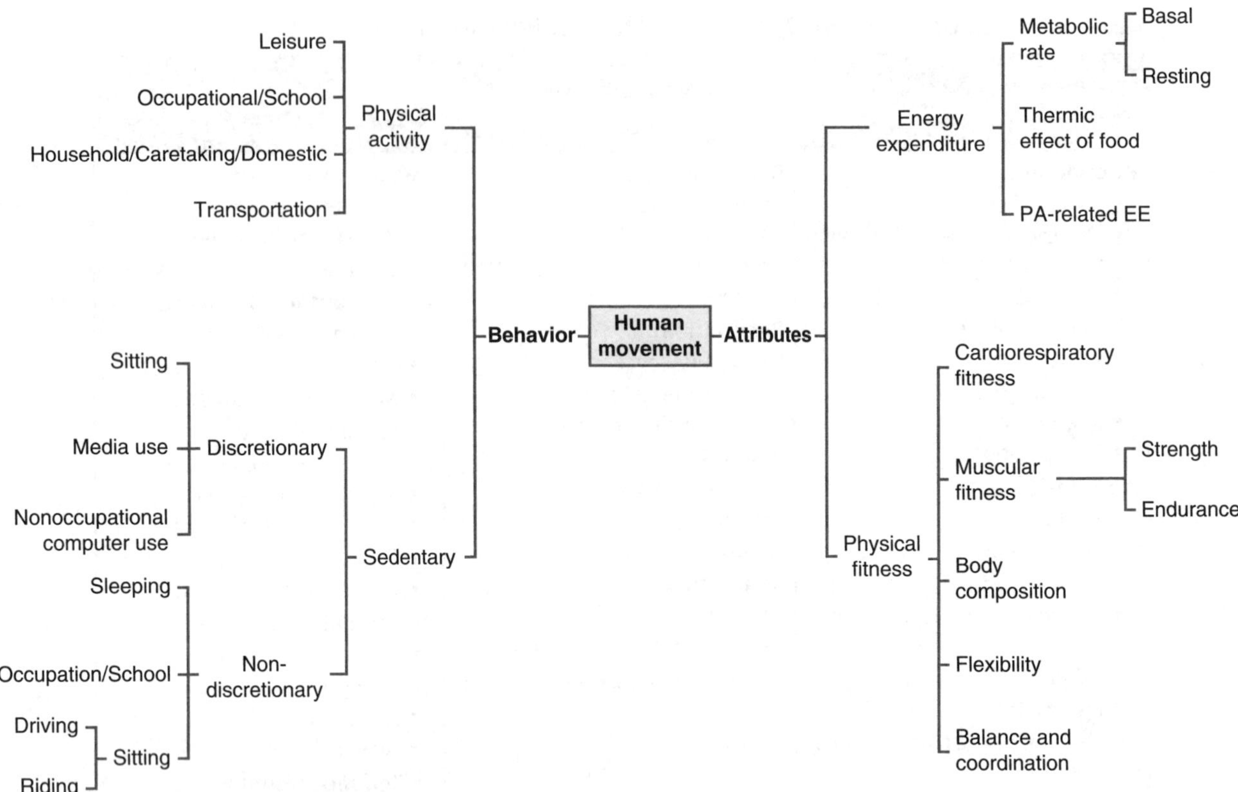

Figure 1.1 A framework for physical activity.

Adapted from Pettee-Gabriel, Morrow, and Woolsey 2012.

PSYCHOMOTOR DOMAIN: PHYSICAL ACTIVITY AND PHYSICAL FITNESS

For many years, physical educators, exercise scientists, personal trainers, athletic coaches, and public health leaders have been concerned about the definition, the reliable and valid measurement, and the evaluation of physical fitness in people of all ages. This concern led to a growing number of fitness tests and protocols for both mass and individual testing. For example, the Cooper Institute's Fitnessgram, the President's Council on Fitness, Sports & Nutrition, the President's Challenge, and the European test Eurofit are all youth fitness test batteries. Each of these test batteries consisted of different test items, but all have served the purpose of assessing levels of physical fitness. You'll learn more about these test batteries in chapter 10. A large number of research studies have been conducted to demonstrate the feasibility, reliability, and validity of such fitness tests. Normative surveys have been conducted to establish the fitness levels of various populations. In this book we have devoted two chapters to physical fitness testing (chapter 9 for adults and chapter 10 for youth).

In the latter part of the 20th century, the health-related aspects of physical activity became a dominant concern of public health officials. The culmination of this concern was presented in the release of *Physical Activity and Health: A Report of the Surgeon General* (USDHHS 1996). This publication, led by senior scientific editor Steven N. Blair, presented a detailed case for the health benefits of a lifestyle that includes regular and consistent participation in moderate-to-vigorous levels of physical activity. Unfortunately, 2011 data (Harris et al., 2013) from the Centers for Disease Control and Prevention's Behavioral Risk Factor Surveillance System (BRFSS) suggests that only about half of the adult U.S. population engages in the recommended 150 minutes of MVPA per week and less than 30% meet the muscle-strengthening guideline of 2 or more days per week.

MEASUREMENT AND EVALUATION CHALLENGE

The issues just presented about standards, evaluation, purposes, and domains of human performance directly relate to the scenario about your father. For example, throughout the remainder of this course, you will learn about the tools available to answer the questions your father raised. In chapter 2 you will learn about accessing the Internet to obtain information and calculate health risks. You will also be introduced to powerful computer programs to help you analyze data and make decisions; you will use these computer programs throughout the remainder of the book. Chapters 3 through 5 will help you understand the statistical procedures necessary for making evidence-based decisions. Chapter 3 presents information about descriptive statistics and measurement distributions. Chapter 4 presents information about quantifying the relationship of one variable to another (e.g., relating decreased physical activity to increased cardiovascular risk). Chapter 5 presents an overview of research methods to help you decide if an intervention makes a significant difference in a specific outcome of interest (e.g., does moderate exercise reduce body fat?). Chapter 6 illustrates how to determine the accuracy (reliability) of measurement. You will learn how to determine the best measure to use, how to interpret the measurement, and what influences the errors of measurement that are always present when measures are taken. Chapter 7 illustrates how specific health standards are set and whether these standards affect the odds or risk of developing a specific disease.

Chapter 8 provides tools with which to make an accurate assessment of knowledge. Knowledge is required but not sufficient for behavior change. For example, does your father know how to be physically active for a health benefit? A simple knowledge test might tell us this; however, knowledge tests must be accurate. You will learn how to determine if a knowledge test is accurate and truly reflects learning. Chapter 9 illustrates how to measure risk factors associated with cardiovascular disease. As an example, you will learn about some simple tests of aerobic capacity, body composition, physical fitness, and physical activity. Chapter 10 illustrates physical fitness and physical activity assessment in youth, values that could be predictive of future health risk. Chapter 11 presents strategies for measuring and evaluating sport skills and physical abilities. Physical abilities, such as strength and flexibility, might be related to your father's overall health risks. Chapter 12 illustrates methods of measuring psychological stress levels, which might be pertinent in your father's case. Chapter 13, although primarily concerned with grading, contains important information on how to properly add various measurements together to obtain a composite score. Chapter 14 provides examples of how to obtain alternative measures. For example, one could measure physical activity by asking for a self-report, by using a pedometer, or by directly observing a participant's daily behaviors.

Your interest and that of your father in the measurements taken are right on target. How do you know they were accurate? How do you know they are truly predictive? How do you know if he is really at increased risk? How do you know if specific interventions help reduce risk? These and other questions are what your future learning in this book is all about.

written recalls, self-reporting, and heart rate biotelemetry. Techniques that work for adults are not always appropriate for children. An increasing body of scientific literature is appearing about the reliability and validity of physical activity measurement in a variety of situations and populations. In chapters 9 and 10 we explore the measurement and evaluation of physical activity and physical fitness, elements in the psychomotor domain.

 Go to the WSG to complete Student Activity 1.6 and view Video 1.6.

SUMMARY

As a student, you are aware that nearly all educational decisions rely greatly on the processes of measurement and evaluation. Figure 1.2 illustrates the relationships among testing, measurement, and evaluation and indicates that as a human performance professional you will have to make a variety of decisions regarding the methods of collecting and interpreting data in the measurement process. A wide range of instruments (tests) are used to assess abilities in the cognitive, psychomotor, and affective domains. You will have to determine the domains in which you wish to have objectives and then develop specific objectives and select tests that produce objective, reliable, relevant, and valid measurements of your objectives. Once you collect the data, evaluative decisions can be either norm referenced or criterion referenced. In norm-referenced comparisons, a participant's performance is compared with that of others who were tested. Criterion-referenced standards are used to compare a participant's performance with a predetermined standard that is referenced to a particular behavior or trait. Evaluations can be formative (judged during the program or at intervals of the program) or summative (judged at the end of the program).

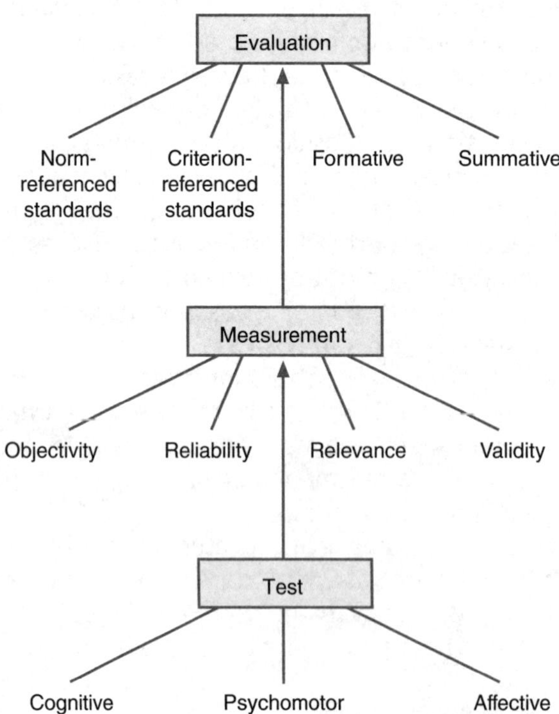

Figure 1.2 Relationships among test, measurement, and evaluation.

 Go to the WSG for homework assignments and quizzes that will help you master this chapter's content.

CHAPTER 2

Using Technology in Measurement and Evaluation

OUTLINE

OBJECTIVES

After studying this chapter, you will be able to

- identify the potential uses of computers in your field;
- identify sources of computer software and hardware for use in exercise science and physical education;
- present examples of computer use in exercise science, physical education, kinesiology, and clinical health and fitness settings and describe how various testing procedures can be facilitated with computers;
- use SPSS to create and save data files;
- use Microsoft Excel to create a data file to be used in SPSS; and
- use SPSS and Excel to analyze data and set the groundwork for reliability and validity analyses that will follow in subsequent chapters.

 The lecture outline in the WSG will help you identify the major concepts of the chapter.

MEASUREMENT AND EVALUATION CHALLENGE

Computers are ubiquitous. They are at homes, at work, and even fit into your hand. Computer usage makes measurement and evaluation decisions manageable. Jessica works for a research company that conducts interventions intended to increase the amount of moderate-to-vigorous physical activity (MVPA) in which people engage. The company's intent is to encourage people to engage in sufficient MVPA to meet the *2008 Physical Activity Guidelines for Americans*. Their current project involves strategies to get people to walk more. As a means of determining how much MVPA each person gets, the study participants are asked to wear motion sensors. The question facing the study team is *How many consecutive days of monitoring are necessary for obtaining a reliable measure of MVPA?* Additionally, they wonder if a weekend day or two is necessary. Their study participants come from throughout a large metropolitan area and represent the public at large. Handling such data by hand would be cumbersome and time consuming. The study team plans to store their data on a computer and have it readily available for analyses throughout their study.

Computers have become commonplace in schools. Tasks that only a few years ago were tedious and time consuming are now completed in a matter of seconds by computers, and computers can now easily fit in a briefcase or even in your hand. Wireless communication is available nearly everywhere, which permits those gathering data to record measures in nearly any location for real-time or later analyses. Clearly, your ability to use a computer for decision making will affect you in your career and leisure activities.

This runner's running watch provides precise speed, distance, and pace data. Accompanying software allows the runner to download these data for detailed analyses. You can use technology, too, to more easily track and analyze your research data.

Undoubtedly, the development of computers is one of the most important technological advances of the past 30 years. Originally, mainframe computers were large machines that took up entire rooms (or even floors) in buildings. Users typically had to be connected via telephone or other electronic line to the mainframe, and mainframes were inaccessible to most people. However, the development of the microprocessor has resulted in small, powerful, relatively inexpensive personal computers. Indeed, it has been suggested that if the change occurring in computer technology over the past 50 years had also occurred in the automobile industry, cars would now be able to travel much farther on a gallon or liter of gas! However, even though computers are now widely available, many students and professionals who measure and evaluate human performance have not taken full advantage of the power that computers provide.

Additionally, the development and worldwide use of the Internet have had a significant impact on how people obtain information and communicate with one another. In 2013 the Pew Research Center (Pew Internet and American Life Project) reported that more than 84% of Americans used the Internet. Usage declines from a high of 98% for those aged 18 to 29 to 83% of those between age 50 and 64, and to only 56% of those age 65 or older. A positive relationship exists between Internet usage and household income and amount of education. Bottom line: It is nearly impossible to think of a career in human performance that does not use computers to analyze or report data. Clearly, regardless of your age or profession, you will have Internet resources and computer applications that help you analyze and report data in meaningful, interesting ways.

Numerous websites are of specific interest and value to the human performance student. Professional organizations such as the American College of Sports Medicine (ACSM; www.acsm.org), the Society of Health and Physical Educators (SHAPE America; www.shapeAmerica.org), and the American Heart Association (AHA; www.americanheart.org) use the Internet to get their messages out, gather data, and attempt to influence healthy lifestyle behaviors. Government agencies such as the U.S. Centers for Disease Control and Prevention (www.cdc.gov) and the National Heart, Lung, and Blood Institute (www.nhlbi.nih.gov) provide important health information. The Behavioral Risk Factor Surveillance System (BRFSS) (www.cdc.gov/brfss) and the Youth Risk Behavior Surveillance System (YRBSS) (www.cdc.gov/HealthyYouth/yrbs) introduced in chapter 1 have excellent resources for reporting data. Others provide scientific and content-based information that may be related to your job responsibilities (e.g., courses you may teach, fitness and health information, athlete-training information).

Mastery Item 2.1

Go to www.pubmed.gov (the U.S. National Library of Medicine) and enter the words "pedometer reliability validity" (without the quotes) to see some of the research that has been conducted with pedometers.

Hardware and software need to be differentiated. Hardware consists of the physical machines that make up your computer and its accessories. Software is the computer code, generated by a computer programmer, by which you interact with the computer as you enter data and conduct analyses, create text, and draw graphs. Applications (apps) are available for handheld and tablet devices. These apps, often inexpensive (or free!), provide the means for completing sophisticated analyses and data collection and reporting strategies. You don't have to be a software programmer to be a competent computer user—the vast majority of expert computer users do not do programming.

USING COMPUTERS TO ANALYZE DATA

Computer technology is now pervasive in schools and businesses. Many schools and businesses require students and employees to be computer literate—able to interact with computers daily for work and pleasure. Some universities require students to have personal computers, whereas others actually give them to students when they pay tuition. Computers have such a big influence in our daily lives (they're involved in everything from grocery shopping and banking to using the telephone) that we have to be able to use them. Computer literacy does not require one to be a computer programmer; one simply needs to be able to use computers in daily life, for example, to conduct daily tasks or for enjoyment (e.g., surfing the Internet).

Mastery Item 2.2

Go to an Internet search engine and enter a particular kinesiology topic (e.g., physical activity prevalence, obesity changes). See how many sources you can identify related to this topic. Consider how you might use this information in this or another class or in your career.

Exercise scientists and physical educators must make many measurement and evaluation decisions that involve numbers, which computers are particularly adept at handling. Because the exercise and human performance professions require daily use of computers, you must familiarize yourself with their features and uses specific to your field so that you can understand and use the concepts presented in this text. Many of the decisions that you will make in your field require data analysis. Thus, we will introduce you to SPSS (IBM SPSS Statistics), a powerful data analysis program that will help you save, retrieve, and analyze much of the measurement and evaluation data that you will encounter daily.

SPSS makes number crunching fast, efficient, and almost painless. For example, the most important characteristics of any test are its reliability and validity. As you will learn in chapters 3, 4, 6, and 7, computers can generate data related to reliability and validity in a matter of seconds. This will be illustrated throughout the textbook. Many statistics can help you make valid decisions. Chapters 3 through 14 provide you with many opportunities to practice using SPSS in scenarios similar to those you will encounter in your profession.

Additionally, we present information about how to create databases with Microsoft Excel. These Excel databases can be easily read with SPSS. The benefit of creating your database with Excel is that it is readily available on computers. Thus, you could create your database while at home and then conduct the analysis with SPSS. We provide more information for Excel users in the appendix. You will learn more about that as you read through chapter 2.

Measurement uses for a computer in human performance, kinesiology, and physical education include the following:

- ***Accessing the Internet to obtain information relative to your specific job responsibilities.*** Whether seeking normative strength measures for your personal trainer clients, researching reports regarding the most effective modality for treating patients in your PT clinic, or accessing health and fitness data from large scale populations, you will use the Internet on a nearly daily basis once you have completed your training and entered your professional career.
- ***Determining test reliability and validity.*** Statistics learned in chapters 3, 4, and 5 can be used to estimate the reliability (consistency) and validity (truthfulness) of test results in the cognitive, affective, and psychomotor domains. SPSS examples are provided in chapters 3 through 14.
- ***Evaluating cognitive test results and physiological test performance.*** Computers can help evaluate and report individual test results. Likewise, you can quickly retrieve, analyze, and return test results to study participants. You can estimate your risk for development of diabetes from the American Diabetes Association (www.diabetes.org) and your risk of cardiovascular disease from the American Heart Association (www.americanheart.org). Consider the physical therapist who wants to track patient improvement. Using the computer to track and display data serves as an excellent source of formative and summative evaluation.
- ***Conducting program evaluation.*** Computers can calculate changes in overall student performance and learning across teaching units or track individual changes in a student's performance.

• ***Conducting research activities.*** You can compare an experimental group of study participants with a control group to determine if your new intervention program has a significant effect on cognitive or physiological performance.

• ***Developing presentations.*** Specialized software can be used to create powerful presentations you can make before students, potential clients, patients, and professional peers. The presentations can include text, pictures, video, graphics, animations, and sound to effectively present your message. Perhaps your instructor is using the presentation package that accompanies this textbook to illustrate specific points.

• ***Assessing student performance.*** Students and clients are always interested in how they perform on tests, whether the tests are cognitive, psychomotor, or physiological. Students, teachers, and clinicians are interested in what their individual scores are, how they are interpreted, what they mean, and what effect they have. Computers make it easy to provide the answers to all these questions.

• ***Storing test items.*** Teachers always have to keep records of student grades. Programs that permit entry and manipulation of student data records are called spreadsheets. Spreadsheets are essentially computer versions of a data matrix with rows and columns of information. Students' names are often found in the first column, and data from course assignments fill the remaining columns. Thus, each row represents a student and each column holds scores from tests and other assignments. If the instructor keeps a daily record of class grades, then average grades, final grades, printed reports, and so on can be generated with a few computer keystrokes. Likewise, health and fitness professionals can keep records of workouts and changes in weight, strength, aerobic capacity, and so forth.

• ***Creating written tests.*** Computers can serve as a bank for written test items. Rather than having to develop a new test each time you teach a unit, you can store test items on your computer and generate a different test each time you teach the unit. Test banks can be built using word-processing or test-development software. Some test-development programs are quite sophisticated and permit you to choose an item not only by content area but also by type of item, degree of difficulty, or date created.

• ***Calculating numerous statistics.*** Physiological measurements often involve equations for estimating values. For example, skinfolds are used to estimate percent body fat, and distance runs and heart rate measurements are used to estimate maximal oxygen consumption. The computer can greatly assist in calculating these values. Rather than substituting each number into an equation and going through the steps to complete the calculation, you can enter the formula into the computer once and automatically calculate the desired value for each person. For example, go to the National Heart, Lung, and Blood Institute (NHLBI) of the National Institutes of Health (NIH) website (www.nhlbi.nih.gov/guidelines/obesity/BMI/bmicalc.htm) and calculate your body mass index (BMI).

Go to the WSG to complete Student Activity 2.1 and view Videos 2.1 and 2.2.

Mastery Item 2.3

Think of some time-consuming tasks that you need to do regularly. How would a computer help you complete them more efficiently? What kinds of tasks can kinesiology, exercise science, or physical education majors complete using computers?

Physical fitness testing is an important component in most physical education, kinesiology, and exercise science programs. Not too many years ago, fitness test results were reported orally, on poorly prepared reports, or on purple mimeographed copies. Today, youth fitness programs use computer software to analyze student results. The Cooper Institute's Fitnessgram is an example of an excellent software program. Figure 2.1 illustrates the type of computer output that the teacher can give students (and their parents) to better inform them of their physical fitness status and/or progress.

 Go to the WSG to complete Student Activity 2.2 and view Videos 2.3 and 2.4.

USING SPSS

Many of the decisions that you will have to make about reliability and validity, whether you are a kinesiologist, physical therapist, clinician, coach, instructor, or educator, are based on statistical evidence. Now don't get scared! This is not a statistics text. It is a measurement and evaluation text. However, the statistics presented in chapters 3, 4, and 5 provide the framework for much of your decision making in measuring and evaluating human performance. Although we will use SPSS (a sophisticated statistical package that is widely available on many university campuses) and Microsoft Excel throughout this text to assist you in calculating statistics used in reliability and validity decisions, your instructor may choose for you to conduct the analyses with another type of software. Regardless, the calculations will be nearly identical (within rounding error) and the interpretation of the results will be exactly the same, regardless of the particular statistical package you use. Nearly all of what we present in this text is available in both the full and student versions of SPSS. SPSS is continually updated, and new versions are available. You might need to be a little flexible when you access SPSS because SPSS is continually updated and the specific method that you use to access SPSS at your location may not be the same as at other locations.

SPSS is software developed to analyze numbers (e.g., calculate the mean or draw a graph). However, SPSS must have a database on which to conduct the analysis. Thus, each analysis conducted by SPSS is run on a set of data created and saved through the SPSS data editor. The SPSS data editor lets you create a database (also called a data matrix) consisting of N rows of people with p columns of variables. Table 2.1 illustrates a data matrix consisting of 10 rows of people (represented by "id" numbers 1 through 10) with five variables ("gender," "age," "weightkg," "heightcm," "stepsperday"). Weight is measured in kilograms (kg) and height in centimeters (cm).

Table 2.1 Sample Database (Data Matrix)

id	gender	age	weightkg	heightcm	stepsperday
1	0	20	50	165	5000
2	0	24	51	160	6000
3	0	21	62	173	7000
4	0	19	59	178	6500
5	0	23	43	145	4500
6	1	22	86	193	4800
7	1	25	65	183	4000
8	1	24	61	178	4200
9	1	28	75	173	3900
10	1	20	70	178	3500

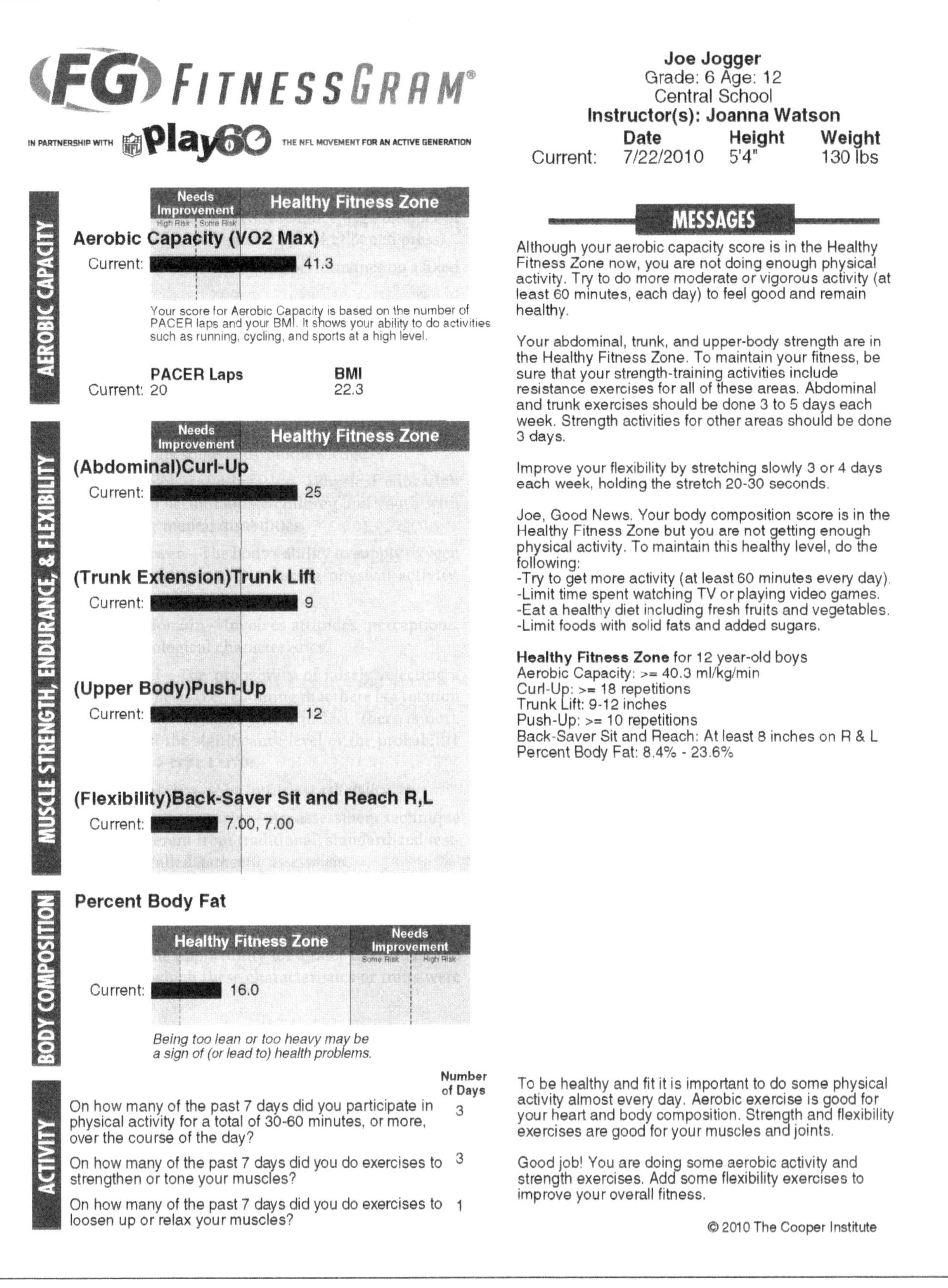

Figure 2.1 Fitnessgram reports such as this one can be given to parents and students to illustrate a student's improvement.

Reprinted with permission from The Cooper Institute, 2010, *FITNESSGRAM/ACTIVITYGRAM test administration manual,* updated 4th ed. (Champaign, IL: Human Kinetics), 106.

SPSS permits you to enter and manipulate data and conduct analyses that result in a variety of numbers, charts, and graphs. Each of the data tables used in your textbook is located on the website for each chapter. You can download them in SPSS or Excel formats. You will learn more about this in the following paragraphs.

MICROSOFT EXCEL PROCEDURES

The Appendix of your textbook contains Excel steps for each of the procedures illustrated in chapter 2. Students without access to SPSS can use Excel to conduct the statistical procedures. SPSS will be used and illustrated throughout the remainder of the text. However, each of the procedures is also illustrated in the appendix by the specific chapters. We have provided templates on the textbook's website for some of the Excel procedures (particularly for chapters 5 and 7) because the number of steps necessary for calculating these statistics is large and the process is rather cumbersome. Excel users will be well served by reviewing the appendix at this time and then throughout the remainder of the textbook when conducting analyses. Each time that SPSS is presented in a chapter, Excel users should turn to the appendix and use the appropriate steps. You should be aware that the statistical procedures learned in early chapters generalize to later chapters, so the same procedures are used repeatedly.

Getting Started

Locate and double-click on the icon for SPSS on your computer's desktop. Alternatively, you may have to go through the Start menu on the bottom left of your computer to locate SPSS. It will be different depending on your university or computer. Once you locate and begin SPSS, note that a blank data matrix appears. The upper left corner has the name Untitled1 [DataSet0]—SPSS Statistics Data Editor. The data editor lets you define and enter data (figure 2.2).

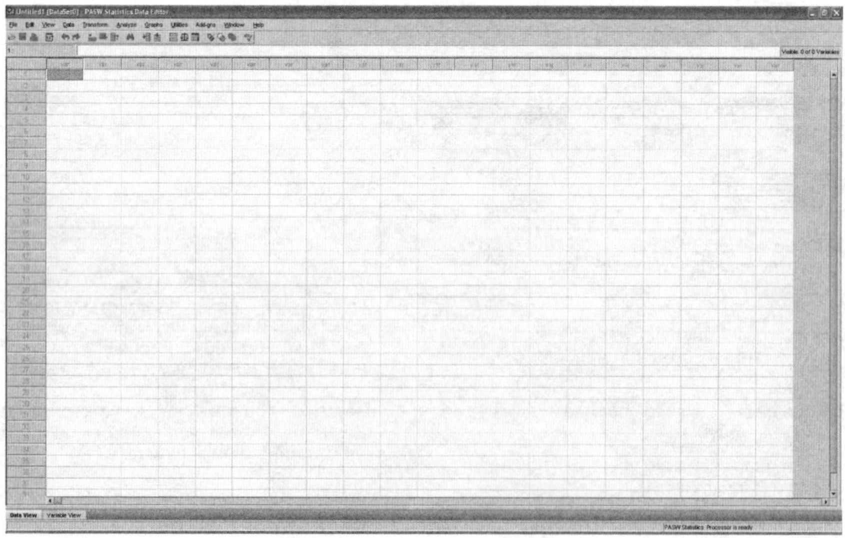

Figure 2.2 Screen capture of SPSS Data Editor.

Note there are two tabs near the bottom left of the SPSS window. One of them says Data View and the other says Variable View. The Data View window presents the data you have entered or provides a spreadsheet that permits you to enter data. The Variable View window permits you to define and name the variables themselves. It also permits you to identify variable labels, value labels, and missing values. These are illustrated in subsequent paragraphs. Note also that there are several pull-down menus across the top of the data matrix. These generally provide you with the following functions:

- *File.* Among other functions, permits you to create a new data matrix, open a data matrix that was previously saved, save the current data matrix, print the current data matrix or analysis results, and exit the program.
- *Edit.* Permits you to undo a previous command; cut, copy, or paste something from the window; insert variables or cases; or find a specific piece of data.
- *View.* Permits you to change the font in which your data appear, change the appearance of the data matrix window, and so on.
- *Data.* Among other functions, permits you to sort the data and select specific cases.
- *Transform.* Permits you to modify your variables in a number of ways. You will use the compute function under the transform tab often.
- *Analyze.* Lists the statistical procedures that you will use. You will become familiar with the Analyze menu as you use this textbook. Don't worry—there are many listed here but we will not be using all of them. Note that each of the options in this menu has an arrow next to it. The arrow indicates that additional submenus are available to you under this particular statistical procedure. You will become familiar with these submenus as you work through the textbook.
- *Direct Marketing.* Permits the user to break data into groups and look at group tendencies. We will not use the Direct Marketing menu in your textbook.
- *Graphs.* Lists the various types of graphs you might use to present your data. We will use a limited number of these options with your textbook.
- *Utilities.* Permits you to modify your data matrix in a number of ways. We will not be using this menu in your textbook.
- *Add-ons.* Includes a number of advanced procedures for data analysis. We will not use Add-ons with your textbook.
- *Window.* Lets you split the Data View into four quadrants. This is helpful when you have a large number of cases or variables and you want to look through the data. The Window tab also lets you hide the data window when you are running several programs at a time and switch from the data matrix window to the SPSS output window once you have conducted an analysis. You can split the data window to facilitate ease of reading columns.
- *Help.* Provides you with a variety of help resources when you are running SPSS and have a question. You might find the Topics (and then Index) submenus helpful.

Each of the pull-down menus has additional functions that you might enjoy investigating, but we provide you with sufficient information to conduct the SPSS processes you will use with your textbook. You are encouraged to investigate these various menus as you learn SPSS and use the Help windows provided within SPSS. The more you interact with SPSS, the better you will understand its capabilities and use it to make your work much easier. SPSS instructions are based on version 20.0. These instructions may change slightly as SPSS updates its software. We have provided SPSS sample datasets in the WSG for each of the remaining chapters of your textbook.

Creating and Saving Data Files

Use the steps in Mastery Item 2.4 to create and save your first SPSS data file named table 2-1. Once you have the idea, you will be able to go through these steps quite rapidly. For the time being, trust us and follow the steps closely. Note that you will need to save the data matrix as table 2-1 (with a hyphen and not a period). This is because the computer could interpret the period as a file extension and may cause you difficulty when you attempt to access the table at some later point. Thus, when you create and save your tables, name them with the following style: chapter number-table number. For example, in chapter 2, the second table would be table 2-2, the third table in chapter 2 would be 2-3, and so on.

SPSS variable names must begin with a letter and be no longer than 64 characters in length. You cannot use spaces in the name, and you should avoid special characters when naming a variable. We have used a mnemonic name to identify our variables. Thus, "weightkg" is actually weight in kilograms (kg). The mnemonic helps you remember exactly what the variable is and the units in which it was measured (figure 2.3).

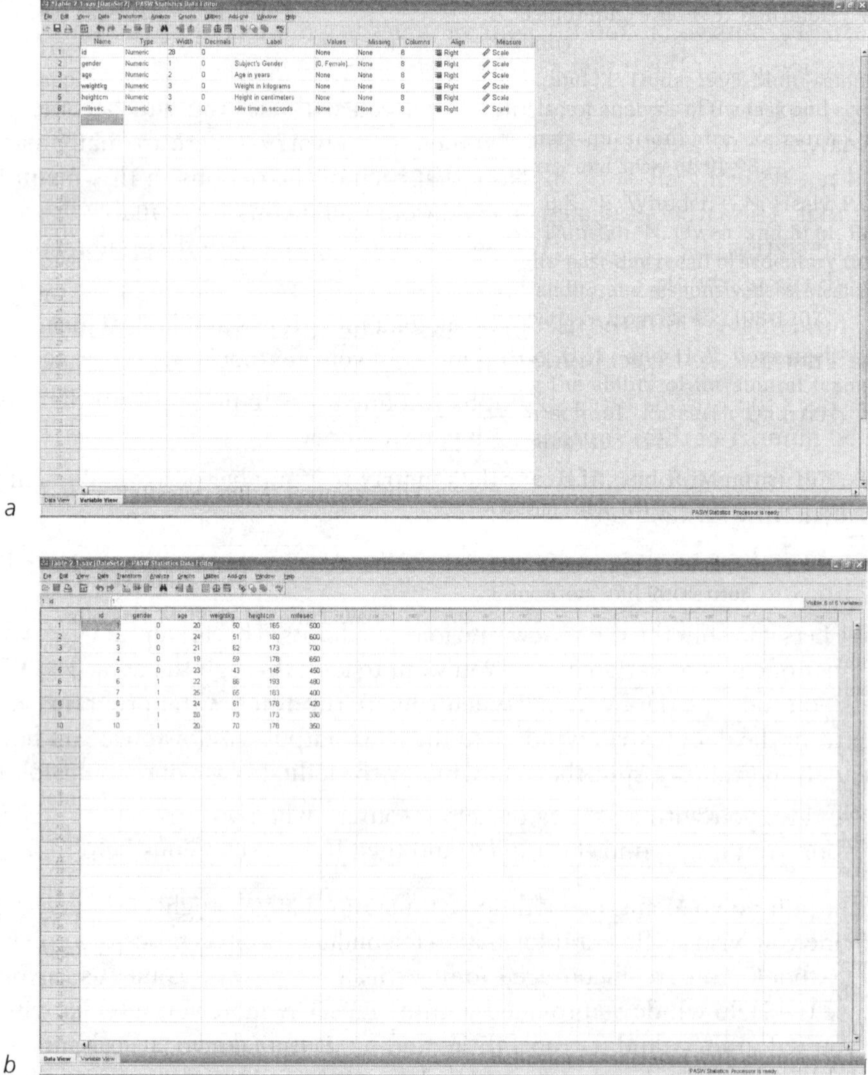

Figure 2.3 Screen capture of SPSS *(a)* variable view and *(b)* data view windows.

 Go to the WSG to complete Student Activity 2.3 and view Video 2.5.

Mastery Item 2.4

Follow these step-by-step procedures to create a SPSS file named table 2-1 and save it to a data storage device (DSD) (e.g., flashdrive, disk, the cloud, etc.). Excel users turn to the appendix for instructions.

Create and Save an SPSS Data File

1. Be certain to have a DSD with you before you start this assignment. In some systems you may have to save your data to an electronic account.
2. Place the DSD in the machine and note the drive location.
3. Locate the SPSS icon and click on it. (Alternatively, you might have to go to the Start button on the bottom left of your computer and locate SPSS among the programs listed in the Start menu.)
4. First you will name the variables, define the variables, and essentially build a codebook that helps you remember what the variables are.
5. Click on the Variable View tab on the bottom left and note that the window now looks like that illustrated in figure 2.3*a* but without the information in it.
6. Name each of the variables in the first column. Note the variable name must start with a letter, have no spaces, contain no special characters, and be no more than 64 characters in length.
7. For the time being, skip over the Type, Width, and Decimals columns.
8. You can expand on the variable names in the Label column.
9. Click on the right-hand edge of the Values column for the second variable (i.e., gender). Notice that you get a box that helps you define the values associated with numbers for gender. In our case, we have coded females as 0 and males as 1. Enter these values, click on the Add button each time, and then click on OK. You have defined your variables and you are ready to begin entering data.
10. Click on the Data View tab to get to the Data View window.
11. Enter the data from table 2.1 into SPSS. Your results should look like those in figure 2.3*b*.
12. You are now ready to save the data on the DSD that you brought with you. Be certain the DSD is inserted properly. Go to the File menu and scroll down to Save As (see figure 2.4).
13. In the File Name box, enter "table 2-1" (without the quotes).
14. Save the data to your DSD (and not to the hard drive). Go to the Save In box at the top of the screen and click on the downward pointing arrow. Scroll down to the location where you just placed your DSD and click.
15. Now click Save. Your table 2.1 data are now saved to your DSD.
16. Go to the File menu, scroll down to Exit SPSS, and click. Doing so will exit you from SPSS.

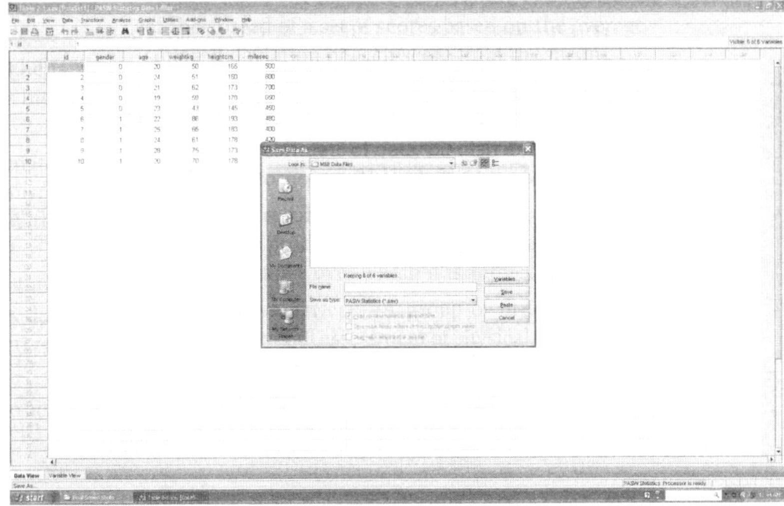

Figure 2.4 Screen capture of SPSS showing how to save a file.

Mastery Item 2.5

You may not always have SPSS available on the computer with which you are working. Thus, you can use Excel to enter your data and then have SPSS read the Excel data file. We will give you an example of how to do this with the data from table 2.1. Follow these steps to create an Excel database of table 2.1 and read the data into SPSS:

Create an Excel Database

1. Open Excel on your computer. You will see a blank data sheet.
2. Enter the variable names in the first row. Continue to use the SPSS restrictions on variable names. (Each variable name must start with a letter, have no spaces, contain no special characters, and be no longer than 64 characters.)
3. Place your cursor in cell a2 and begin to enter the data from table 2.1. Once you have entered all of the data, your Excel data file should look like that presented in figure 2.5.
4. Go to the File menu, scroll down to Save As, and save the Excel version of table 2.1 (remember to name it table 2-1) to your DSD just as you were instructed to do with the SPSS version of table 2.1.

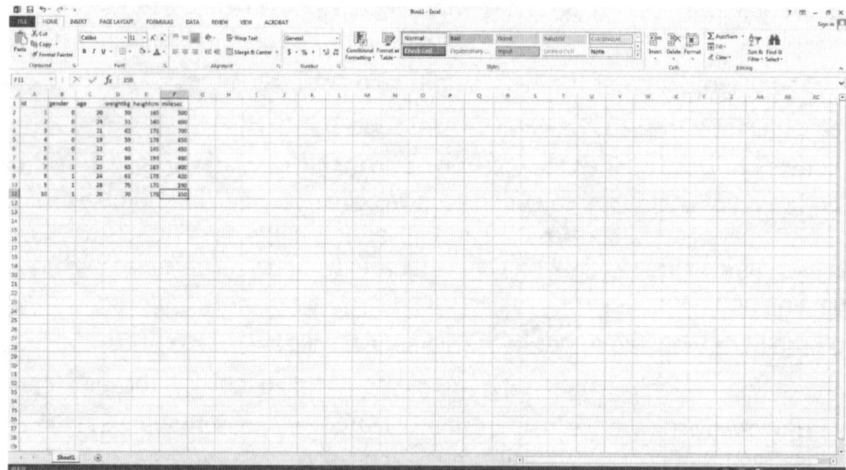

Figure 2.5 **Excel data file with data from table 2.1 inserted.**

You are now ready to access your Excel data with SPSS. SPSS is able to read data from Excel and place it into SPSS for data analysis. Do the following to read the Excel data file into SPSS:

Access an Excel Database with SPSS

1. Open SPSS as you have previously been instructed.
2. Go to the File menu and scroll down to Open and over to Data . . . ; you will see the screen presented in figure 2.6.
3. You will be presented with an Open File window. Go to the Files of type near the bottom of the window and click on the downward arrow so that you can then highlight the Excel (*.xls) indicator. This will indicate that you want to import an Excel data file into SPSS. This is illustrated in figure 2.7.
4. Locate the Excel file that you want to read into SPSS. Click on the file name and it will then appear in the file name box. Click on Open.
5. You will see an Opening Excel Data Source window. Click on the square that has "Read variable names from the first row of data." Recall that you placed the variable names in the first row of your Excel data file. Click OK. This screen is presented in figure 2.8.

6. Your data will automatically be placed into SPSS. Compare the results that you have just imported with those presented in figure 2.3*b*, where you entered the data directly into SPSS.
7. Note that only the variable names and data have been imported into SPSS. You will need to go to the Variable View tab and enter labels and values as you did originally with SPSS. That is one disadvantage of reading your data from an Excel file—you lose the codebook information and must create it.
8. However, you have the variable names and the data now in SPSS.

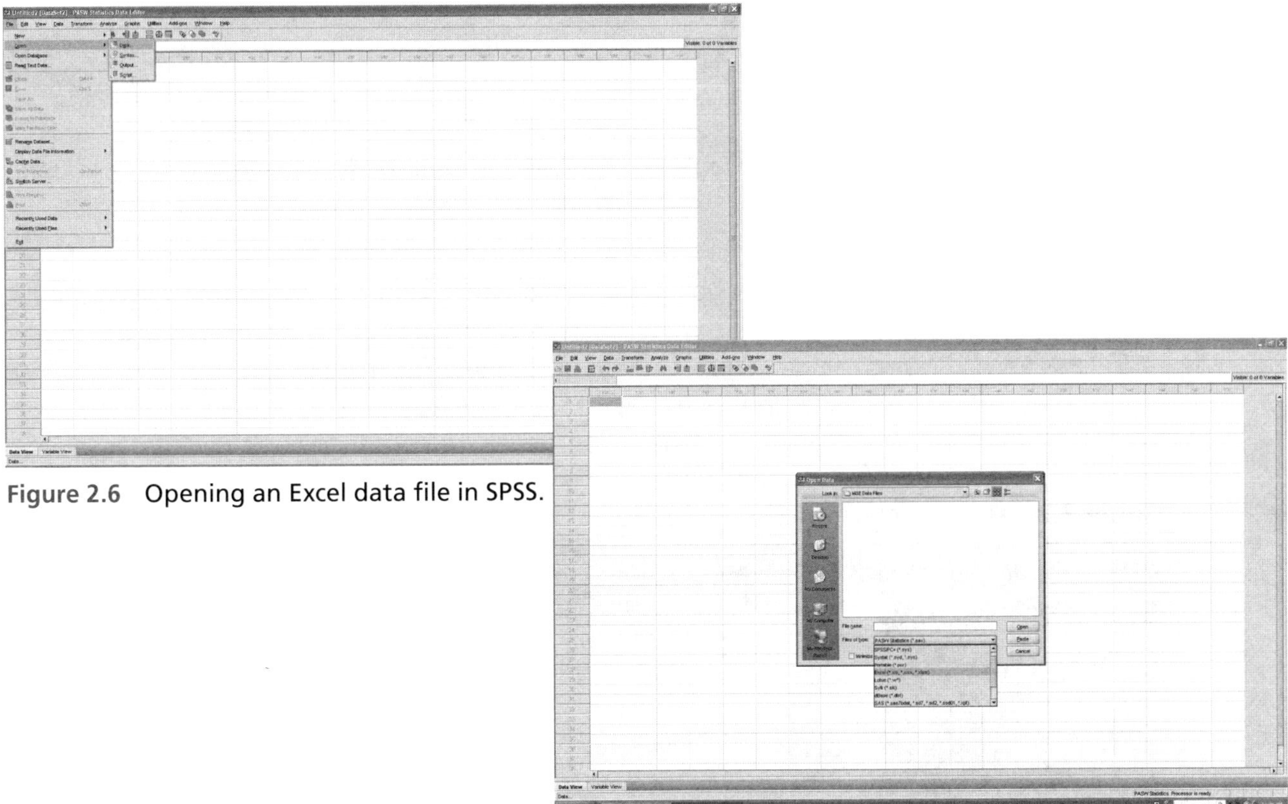

Figure 2.6 Opening an Excel data file in SPSS.

Figure 2.7 Accessing an Excel database to read into SPSS.

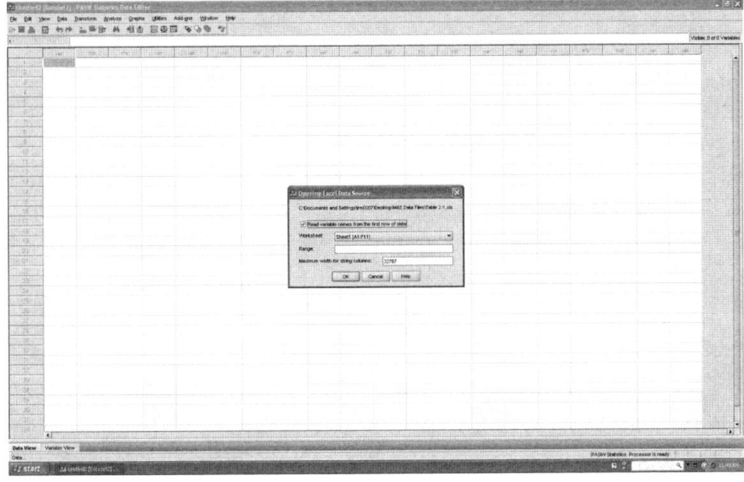

Figure 2.8 Opening an Excel file into SPSS.

Recalling and Analyzing Data

Now that you have created and saved the data, let's recall it and conduct an analysis with it using the procedures in Mastery Item 2.6. Mastery Item 2.7 will demonstrate one of the most powerful functions of SPSS—the ability to manipulate data easily. BMI will be discussed further in chapter 9; we use it here because it provides an excellent example of SPSS data modification.

Mastery Item 2.6

Use table 2-1 that you have created with SPSS to obtain some descriptive statistics on the data.

Recall and Analyze Data

1. First go to your DSD and locate table 2-1. Double-click on it to begin SPSS.
2. Go to the Analyze menu.
3. Scroll down to Descriptive Statistics, over to Descriptives, and click.
4. When the Descriptives window appears, use the arrow to move "age," "weightkg," "heightcm," and "stepsperday" into the Variable(s) box.
5. Click OK and then compare your results with those presented in table 2.2.
6. If your results are different, go back to table 2.1 and compare the data in the table with what you have in the SPSS data editor.

Table 2.2 Descriptive Statistics

	N	Minimum	Maximum	Mean	Standard deviation
Age in years	10	19	28	22.60	2.757
Weight in kilograms	10	43	86	62.20	12.709
Height in centimeters	10	145	193	172.60	13.293
Steps per day	10	3500	7000	4940.00	1183.404
Valid N (listwise)	10				

Mastery Item 2.7

We'll use the following instructions to create some new variables for the 10 participants listed in table 2.1. We'll use weight and height to calculate each person's BMI.

Use Compute Statements in SPSS

1. Access your table 2-1 data as you did in Mastery Item 2.6.
2. Go to the Transform menu, scroll down to Compute, and click—a new Compute Variable window will appear.
3. Type "weightlb" in the Target Variable box.
4. Put "weightkg" in the Numeric Expression box by using the arrow to move it.
5. Go to the keypad in the window and click on the "*" for multiplication.

6. Place the cursor next to the "*" in the Numeric Expression box and enter 2.2 (recall that to calculate weight in pounds from weight in kilograms, you simply multiply weight in kilograms by 2.2).
7. Click on OK.
8. Note that a new variable, "weightlb," has been created and added in a column to the right of "stepsperday."
9. Do the same thing to change height in centimeters to height in inches. Note that you divide height in centimeters by 2.54 to obtain height in inches.

Calculating BMI is a bit more involved. BMI is weight in kilograms divided by height in meters squared. BMI can be calculated from kilograms and centimeters or from inches and pounds. We'll take you step by step.

Calculate BMI in SPSS

1. Use the Compute submenu (under Transform) to create a variable called BMI.
2. Use the Compute statement to create BMI from "weightlb" and "heightin." The formula is weightlb/(heightin * heightin) * 703. Put this formula in the Numeric Expression box on the right and then click OK.
3. Save the revised version of table 2-1 to your DSD with the Save command under the File menu.
4. Calculate the mean for the BMI you just created and confirm that the mean value is 20.6524.
5. If you do not get this number, check the original numbers you entered and recheck how you created the variables at the various steps.
6. If you find a variable that you created incorrectly, simply highlight the column for this variable and press Delete. The column will be removed from the dataset and you can re-create it.

DOWNLOADING DATA MATRICES

As previously indicated, selected data tables from many of the chapters in this textbook are available at the textbook website. We will now illustrate how you can download the data for use in class assignments, practice, and learning. Let's begin by having you log in to your textbook's website for chapter 3. Once you reach the location where you find the data matrices, you will notice that there are two columns with essentially the same names. The differences are in the file extensions (one contains data in SPSS format and the other contains the same data in Excel data files). To download a file, simply click on the file name. Depending on the settings on your computer, either the file will be automatically opened for you in the specific file format (i.e., SPSS or Excel) or you will be able to save it to your DSD.

- ***SPSS download.*** When downloading a SPSS file, you will get all of the data in the table as well as everything in the Variable View window that defines and describes the variables. If the computer that you are using does not have SPSS on it, you will not be able to view the SPSS data matrix. Don't worry. Simply take your DSD to a computer that does have SPSS on it and then double-click on the SPSS data file to open the file in SPSS.

- ***Excel download.*** When you are downloading Excel files, the process is the same as that with SPSS. However, recall that the Excel file has only the variable names and data and nothing that represents the codebook found in SPSS's Variable View.

OPEN SOURCE STATISTICAL SOFTWARE

An open source program called PSPP mimics some, but not all, of the SPSS analyses used in your textbook. The program is publicly available and looks similar to SPSS in many ways. Users without access to SPSS can use PSPP to conduct many of the analyses illustrated in this textbook. Open source means that users can continually recommend updates to the computer source code. Thus, it is possible that PSPP will be expanded to conduct additional SPSS analyses. Google "PSPP" to locate the download site and documentation.

MEASUREMENT AND EVALUATION CHALLENGE

Jessica can use SPSS, Excel, or some other statistical package to help her make decisions about how much MVPA study participants are doing. Use the chapter 2 large dataset available on the text website. Assume these are pedometer step counts for a small sample (n = 100) of Jessica's study participants. Use SPSS to determine the average number of steps that are recorded on Monday, Wednesday, Friday, and Saturday. Go to Analyze → Descriptive Statistics → Descriptives and move the days to the right and then click on OK. How might these results influence Jessica's decision about how many days of steps should be recorded? Imagine that Jessica needs to analyze her data from the full study. Statistical packages and the computer will make her tasks much easier.

SUMMARY

Faster and more capable computers are continuing to change every aspect of our lives. Tasks that previously took hours now take only seconds. Whether used for research, testing, evaluating, teaching, or grading, computers—in conjunction with statistics software—can greatly help measurement and evaluation users to develop data on which to make decisions. Specialized software packages are available for developing written tests and for assessing adult and youth fitness tests.

Computing skills, although perhaps difficult to learn at first, are some of the most valuable skills any professional can have. Development of the Internet has had implications for the gathering and transmission of knowledge that affect every educator, allied health professional, and fitness instructor.

An excellent resource to help you learn more about the statistical methods that you will be studying in chapters 3, 4, and 5 is found on the Internet at the Rice Virtual Lab in Statistics (http://onlinestatbook.com/rvls.html). The various examples and simulations available at this site are excellent learning tools.

Go to the WSG for homework assignments and quizzes that will help you master this chapter's content.

Part II

Basic Statistical Concepts

We have stressed in the preface, and in chapter 1, the importance of making good decisions based on accurate and valid measurements. In this section we emphasize the importance of statistics as a tool for helping to make these decisions. In general, statistics help us to determine the probability of an event's occurrence. Knowing the probability can be an important factor in making a decision. In chapter 3 you will learn how to describe distributions, using measures of central tendency and variability. You will also use the normal distribution to describe measured variables and calculate various numbers based on the normal distribution. In chapter 4 you will learn about the relations between variables and the possibility of predicting one variable from one or more variables. In chapter 5 you will learn how to test scientific hypotheses of differences between groups. For example, consider measures of physical activity. In chapter 3 you will learn how to describe the distribution of physical activity behaviors (e.g., minutes of moderate-to-vigorous physical activity [MVPA] per week). In chapter 4 you will learn how different measures of physical activity might be related (e.g., are weekly self-reported MVPA minutes related to steps taken per week?). Then, in chapter 5 you will learn how to test differences in physical activity behaviors following an intervention developed to increase physical activity. For example, what is the probability that the mean weekly minutes of MVPA involvement for two groups (150 and 140 MVPA minutes) are *true* differences? That is, how likely is it that this difference would occur simply by chance? If this probability is extremely small, it might lead us to conclude that the first group was involved in some form of activity that resulted in increased MVPA minutes per week.

Although statistics can be quite involved, the level of mathematical skills necessary for chapters 3, 4, and 5 is only college algebra. It is not the math that is hard; it is the concepts that are more difficult. The key to using statistics for measurement decisions is understanding the underlying reasoning and concepts and then applying the appropriate statistical procedures. Your research question and the scale of measurement of the variables involved will lead you to the appropriate statistical technique to analyze your data. Thus, careful reading and practicing (including the use of SPSS) of the concepts presented in part II will help you throughout the remainder of the textbook.

CHAPTER

3

Descriptive Statistics and the Normal Distribution

OUTLINE

OBJECTIVES

After studying this chapter, you will be able to

- illustrate types of data and associated measurement scales;
- calculate descriptive statistics on data;
- graph and depict data; and
- use SPSS computer software in data analysis.

 The lecture outline in the WSG will help you identify the major concepts of the chapter.

MEASUREMENT AND EVALUATION CHALLENGE

James, a college student, recently had a complete health and fitness evaluation conducted at The Cooper Clinic in Dallas, Texas. Part of the evaluation required him to run to exhaustion on a treadmill. James ran for 24 minutes and 15 seconds. Using his treadmill time, the technician estimated James' $\dot{V}O_2max$ to be 50 $ml \cdot kg^{-1} \cdot min^{-1}$. How does James interpret this value? Is this result high, average, or low? How does it compare with the results of others his age and gender? The concepts in this chapter will help James better interpret statistical results from any type of test he might complete.

Researchers and teachers often work with large amounts of data. Data can consist of ordinary alphabetical characters (such as student names or gender), but data are typically numerical. In this chapter we cover the basics of data analysis to help you develop the skills necessary for measurement and evaluation. Understanding fundamental statistical analysis is required to accomplish this goal. If you can add, subtract, multiply, divide, and (with a calculator or computer) take a square root, you have the mathematical skills necessary for completing much work in measurement theory. In fact, with the SPSS program introduced in chapter 2, the computer does most of the work for you. However, you must understand the concepts of statistical analysis, when to use them, and how to interpret the results.

Descriptive statistics provide you with mathematical summaries of performance (e.g., the best score) and performance characteristics (e.g., central tendency, variability). They can also describe characteristics of the distributions, such as symmetry or amplitude.

SCALES OF MEASUREMENT

Taking a measurement often results in assigning a number to represent the measurement, such as the weight, height, distance, time, or amount of moderate-to-vigorous physical activity (MVPA). However, not all numbers are the same. Some types of numbers can be added and subtracted, and the results are meaningful. With other types of numbers, the results have little or no meaning. One method of classifying numbers is using scales of measurement, as presented here. Note that scales—or levels—of measurement represent a taxonomy from lowest level (nominal) to highest (ratio).

- *Nominal.* Naming or classifying, such as a football position (quarterback or tight end), gender (male or female), type of car (sports car, truck, SUV), or group into which you belong (experimental or control). A nominal scale is categorical in nature, simply identifying mutually exclusive things on some characteristic. There is no notion of order, magnitude, or size. Everyone within the group is assumed to have the same degree of the trait that determines their group. You are either in a group or not.

- *Ordinal.* Ranking, such as the finishing place in a race. Things are ranked in order, but the differences between ranked positions are not comparable (e.g., the difference between rank number 1 and 2 may be quite small, but the difference between rank number 4 and 5 may be quite large). Consider football rankings. The difference in teams ranked 1 and 2 might be quite small and the difference in teams ranked 10 and 11 might be considerable. The person finishing first might be 0.2 seconds in front of second place. Yet, third place might be 2 seconds later and then fourth place 6 seconds later. Regardless, 1 is ranked higher than 2, 3 is higher than 4, and so on.

What nominal, ordinal, and continuous scales of measurement can you identify in this medal ceremony?

- *Continuous* Numbers are said to be continuous in nature if they can be added, subtracted, multiplied, or divided and the results have meaning. The numbers occupy a distinct place on a number line. Continuous numbers can be either interval or ratio in form.
    - *Interval.* Using an equal or common unit of measurement, such as temperature (°F or °C) or IQ. The zero point is arbitrarily chosen: A value of zero simply represents a point on a number line. It does not mean that something doesn't exist. For example, in the centigrade temperature scale, 0 °C does not indicate the absence of heat but rather the temperature at which water freezes. It is possible to have lower temperatures, referred to as below zero.
    - *Ratio.* Same as interval, except having an absolute (true) zero, such as temperature on the Kelvin scale. Weight, times, or shot-put distance are examples of ratio measures. With a true zero, ratios are possible. For example, if one person is 6 feet (1.82 m) tall and another is 3 feet (0.91 m) tall, the first person is twice as tall as the second.

To help put these scales into perspective, consider physical activity. You might be interested in steps taken by men and women (gender is a nominal variable). You might be interested in moderate versus vigorous physical activity (an ordinal variable because vigorous is higher intensity than moderate but not all moderate or vigorous physical activity is the same). You might be interested in steps taken per day (a ratio variable because if you take 5,000 steps per week and your sister takes 10,000 steps, then she has taken twice as many steps as you).

An important concept to remember is that certain characteristics must exist before mathematical operations can be conducted. *Numbers can be perceived within a scale (i.e., taxonomy) from nominal to ordinal to interval to ratio.* The scales of measurement are hierarchical in that each builds on the previous level or levels. That is, if a number is ordinal, it can also be categorized (nominal); if the number is intervally scaled, it also conveys ordinal and nominal information; and if the number is ratio in nature, it also conveys all three lower levels of information—nominal, ordinal, interval. *Only interval and ratio numbers can be subjected to mathematical operation (e.g., added, divided).* People sometimes use lower scales of measurement—ordinal and nominal—as if they were the higher-scale interval or ratio. Consider if you have males coded 0 and females coded 1. Does it make any sense to talk about the average gender being 0.40? Of course not! Gender is a nominal variable. Note also, it is inappropriate to calculate an average from ordinal data. The average will always be the middle rank. This is another reason it is important to distinguish the level of measurement of the data before applying statistical tests.

Mastery Item 3.1

Diving and gymnastic scores are on what scale of measurement?

SUMMATION NOTATION

To represent what they want to accomplish mathematically, mathematicians have developed a shorthand system called summation notation. Although summation notation can become quite complex, for present purposes you need learn only a few concepts. Three points are important for you to remember: N is the number of cases, X is any observed variable that you might measure (e.g., height, weight, distance, MVPA minutes), and Σ (the capital Greek letter sigma) means "the sum of." In summation notation, $\Sigma X = (X_1 + X_2 + \ldots X_{n)}$, where n represents the nth (or last) observation. This reads, "The sum of all X values is equal to (X_1, plus X_2, plus . . . $X_{n)}$."

Be certain to recall the order of operations when using summation notation. The major rules to remember concern parentheses and exponents. Recall that you do all operations within parentheses before moving outside them. If there are no parentheses, the precedence rules of mathematical operations hold: First, conduct any exponentiation; follow this with multiplication and division and then addition and subtraction. For example, ΣX^2 is read as "the sum of the squared X scores," whereas $(\Sigma X)^2$ is "the sum of X, the quantity squared." The distinction is important because the two terms represent different values. With ΣX^2, one squares each X and then sums them up. With $(\Sigma X)^2$, one sums all of the X values and then squares that sum. You can confirm that these values will not be the same with the following values of X: 1, 2, 2, 3, 3. $\Sigma X^2 = 27$ and $(\Sigma X)^2 = 121$.

Mastery Item 3.2

Use the following numbers to calculate the summation notation values indicated:
3, 1, 2, 2, 4, 5, 1, 4, 3, 5.

Confirm that ΣX is equal to 30.

Confirm that ΣX^2 is equal to 110.

Confirm that the following equation resolves to 2.22.

$$\frac{\Sigma X^2 - \frac{(\Sigma X)^2}{n}}{n-1}$$

REPORTING DATA

You might be interested in how many steps you take per week (or day). Or, after you measure your students or subjects on a variable, you may want to know how they performed. Don't you usually want to know how well you did on a test? If your teacher told you only your score, you would know little about how well you performed. If you were only informed about the number of steps you take per week or day but had no comparison (either norm referenced or criterion referenced), having the score alone would be rather meaningless. You need some additional information. Often people want to compare themselves with others who have completed a similar test. Norm-referenced measurement allows just this. It tells you how well you performed relative to everyone else who completed the test.

One method for deciding how your performance compares with others is to develop a **frequency distribution** of the test results. A frequency distribution is a method of organizing data that involves noting the frequency with which various scores occur. The results from $\dot{V}O_2$max testing for 65 students are presented in table 3.1. Look at them and assume your measured $\dot{V}O_2$max was 46 ml · kg^{-1} · min^{-1}. How well did you do? This is difficult to determine when numbers are presented as they are in the table. But a frequency distribution can clarify how your score compares to those of others (i.e., a norm-referenced comparison). Alternatively, you might want to make a criterion-referenced decision based on your $\dot{V}O_2$max. You will learn more about these interpretations later.

Table 3.1 65 $\dot{V}O_2$max Values

48	45	50	49	46	47	47	49	50	50
45	51	51	48	49	46	44	44	52	53
48	43	48	41	48	49	47	49	51	54
51	43	53	45	48	47	51	46	49	50
48	48	45	46	49	48	46	48	52	54
52	50	51	47	45	47	43	47	49	50
44	55	48	50	53					

Mastery Item 3.3

Use SPSS to obtain a frequency distribution and percentiles for the 65 values presented in table 3.1. Confirm your analysis with the results presented in figure 3.1. SPSS commands for obtaining a frequency distribution and percentiles follow:

1. Start SPSS.
2. Open table 3.1 data from the WSG (or enter the values from table 3.1).
3. Click on the Analyze menu.
4. Scroll to Descriptive Statistics and across to Frequencies and click.
5. Highlight "$\dot{V}O_2$max" and place it in the Variable(s) box by clicking the arrow key.
6. Click OK.

Statistics

$\dot{V}O_2$

N	Valid	65
	Missing	0

$\dot{V}O_2$max

		Frequency	Percent	Valid percent	Cumulative percent
Valid	41	1	1.5	1.5	1.5
	43	3	4.6	4.6	6.2
	44	3	4.6	4.6	10.8
	45	5	7.7	7.7	18.5
	46	5	7.7	7.7	26.2
	47	7	10.8	10.8	36.9
	48	11	16.9	16.9	53.8
	49	8	12.3	12.3	66.2
	50	7	10.8	10.8	76.9
	51	6	9.2	9.2	86.2
	52	3	4.6	4.6	90.8
	53	3	4.6	4.6	95.4
	54	2	3.1	3.1	98.5
	55	1	1.5	1.5	100.0
	Total	65	100.0	100.0	

Figure 3.1 $\dot{V}O_2$max for 65 students.

You may be asking yourself, Why do this? Recall that we were interested in determining how well you performed relative to the rest of the class. Your $\dot{V}O_2$max of 46 appears in the lower half of the distribution. The fifth column, cumulative percentage, of the frequency distribution is a percentile. The percentile is obtained by summing the percentage (percent, third column) of scores that fall at and below the percentile you are calculating. *A percentile represents the percent of observations at or below a given score.* This is a norm-referenced comparison. This concept is extremely important because test results, such as those from standardized college admissions tests, are often reported in percentiles. If you achieve at the 90th percentile (P_{90}), this simply means that you have a $\dot{V}O_2$max higher than 90% of the people who were tested. Conversely, if you scored at the 10th percentile (P_{10}), 90% of the people taking the test had a higher $\dot{V}O_2$max than you did. Your $\dot{V}O_2$max of 46 is at the 26.2nd percentile. That is, 26.2% of the people achieved at or lower than you did, and thus 73.8% had a higher $\dot{V}O_2$max. Conversely, you might be interested in determining if your $\dot{V}O_2$max is high enough to put you at reduced risk for cardiovascular disease. Research suggests that men with a $\dot{V}O_2$max ≥35 are sufficiently fit for health benefits. The $\dot{V}O_2$max of 46 exceeds this criterion.

The presentation in figure 3.1 assumes that a higher score is better than a lower score. This is not always the case. If lower scores are better—for example, in golf or a timed event where a faster time is better—you need to reflect this in your interpretation of the scores. For example, if the data presented in table 3.1 were golf scores for nine holes, your lower score of 46 represents the 73.8th percentile (100 – 26.2 = 73.8).

Dataset Application

Now it is time to try this with a larger dataset. Go to the WSG and use SPSS to calculate percentiles for the mean number of steps taken per week for the 250 people in the chapter 3 large dataset. When you run the Frequencies you will see that the output is much longer (and unwieldy for you). To reduce this, click on the Statistics button in the Frequencies window; then click on the box next to Cutpoints and you will get 10 equal groups on your output. Click on the Continue button and then OK. Notice that the first part of your output is in deciles—10 groups. You'll see the steps per week associated with 10, 20, 30 percentiles and so on.

CENTRAL TENDENCY

Now you know where you fit into the distribution. You are somewhere near the lower 30% of the people tested. Let's further consider how you can interpret your score. One way is to determine how you compare with the so-called typical person who took the test. To do so, you look at where the scores tend to center—their central tendency. We will briefly describe three measures of central tendency:

- ***Mean.*** The arithmetic average; the sum of the scores divided by the number of scores. From summation notation, this definition may be represented as

$$M = \frac{\Sigma X}{n} \tag{3.1}$$

where M is the mean, X is the value of each observation, and N is the number of observations. Every score in the data is used to determine the mean—it is the most stable measure of central tendency. Using the four scores (4, 3, 2, 5), the mean is (4 + 3 + 2 + 5) / 4 = 3.5.

- ***Median.*** The middle score; the 50th percentile. To obtain the median, order the scores from high to low and find the middle one. The median value for the data presented in table 3.1 is 48, which is the middle score. Note that the median is a specific percentile: P_{50}. The median is the most typical score in the distribution. Half of the people score higher and half score lower.

- ***Mode.*** The most frequently observed score. The mode is the most unstable measure of central tendency but the most easily obtained one. You should confirm from figure 3.1 that the mode is 48 (occurring 11 times).

Dataset Application

Use the chapter 3 large dataset available with the WSG and learn what the mean, median, and mode are for each of the variables in the dataset. Hint: Use Analyze → Descriptive Statistics → Frequencies → Options and click on the buttons in the Central Tendency box.

DISTRIBUTION SHAPES

Simply knowing a distribution's measures of central tendency does not tell you everything about the scores. Not all distributions have the same shape. Figure 3.2 illustrates several shapes that distributions may assume. The statistical term for the shape (or symmetry) of a distribution is skewness. Skewness values typically range from +1 to –1. A positively skewed distribution has a tail toward the positive end (right) of the number line, and a negatively skewed distribution has a tail toward the negative end (left) of the number line. Distributions with little skewness are characterized by values close to 0. If you have a skewed distribution, the measures of central tendency generally fall in alphabetical order from the tail to the highest point (i.e., the mode) of the frequency distribution. Thus, the median is typically between the mean and the mode when the distribution is skewed and the mean is generally closest to the tail. Notice how the mode is the most easily identified measure of central tendency.

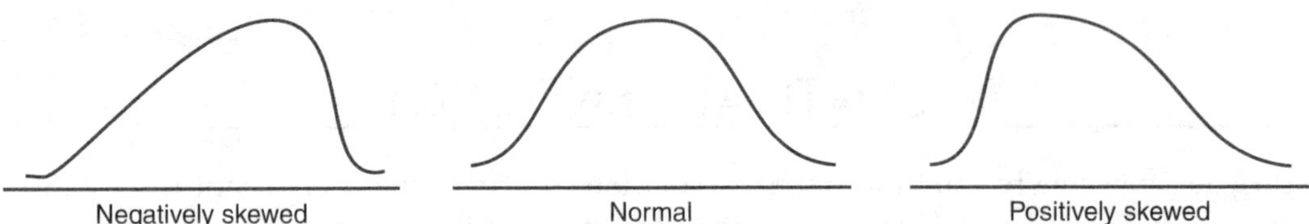

Figure 3.2 Distribution shapes.

Another property of shape is associated with a distribution. In figure 3.3, all three distributions are symmetrical curves with identical means, medians, and modes. However, there is an obvious difference in the amplitudes (or peakedness) of the distributions. The peakedness of a curve is referred to as its **kurtosis**. The top curve (normal) is said to be mesokurtic (i.e., an average amount). The flatter curve is said to be platykurtic (flat) and the steep curve leptokurtic (peaked). Leptokurtic curves have a positive kurtosis value and platykurtic curves have a negative kurtosis value. A normal curve is a distribution with no skewness that is mesokurtic.

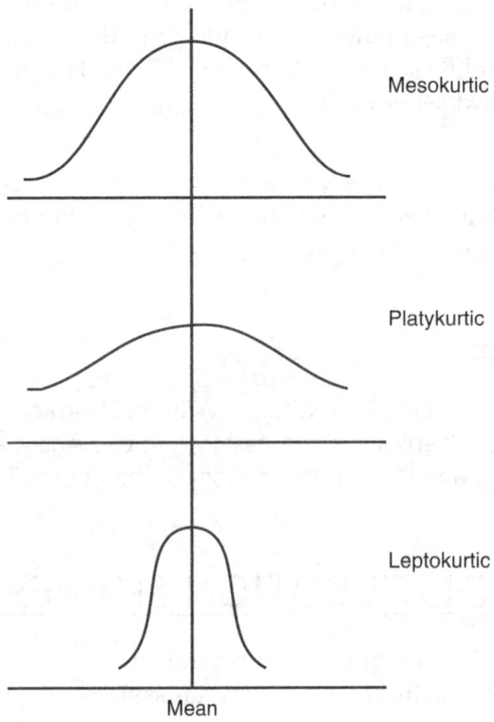

Figure 3.3 Three symmetrical curves.

Dataset Application

Use the chapter 3 large dataset available with the WSG and learn what the skewness and kurtosis are for each of the variables in the dataset. Hint: Use Analyze → Descriptive Statistics → Frequencies → Statistics and click on the buttons in the Distribution box for Skewness and Kurtosis. Can you identify which is the most positive skewed, most leptokurtic, and so on?

A good way to determine the shape of the distribution you are working with is to develop a histogram. A **histogram** is a graph that consists of columns to represent the frequencies with which various scores are observed in the data. It lists the score values on the horizontal axis and the frequency on the vertical axis.

Go to the WSG to complete Student Activity 3.1 and view Video 3.1.

Mastery Item 3.4

Use the data in table 3.1 to create a histogram of the 65 $\dot{V}O_2$max values (see figure 3.4). SPSS commands for obtaining a histogram are the following:

1. Start SPSS.
2. Open table 3.1 data.
3. Click on the Analyze menu.
4. Scroll to Descriptive Statistics and across to Frequencies and click.
5. Highlight "$\dot{V}O_2$max" and place it in the Variable(s) box by clicking the arrow key.
6. Click the Charts button.
7. Click the Histogram button.
8. Click Continue.
9. Click OK.

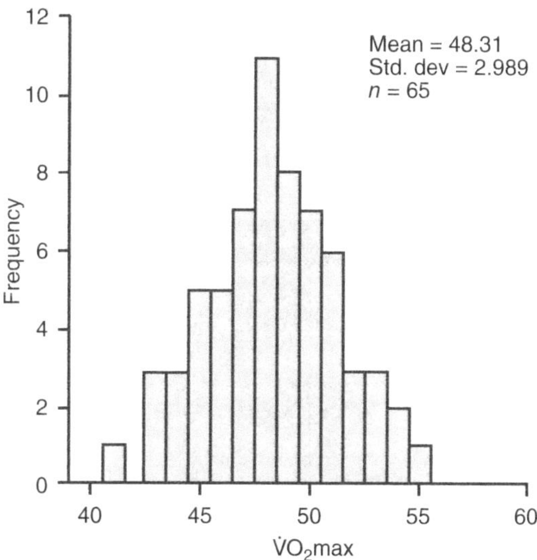

Figure 3.4 Histogram output for 65 values.

Note that you could also obtain a histogram directly by going to Graphs → Legacy Dialog → Histogram and placing the variable of interest in the variable box on the right. You can also have the normal curve displayed by clicking the box Display normal curve.

Dataset Application

Earlier you used SPSS to determine which variables in the chapter 3 large dataset were most skewed. Create some histograms to visually confirm what these distributions look like. In SPSS go to Graphs→ Legacy Dialogs → Histogram and place the variable of interest in the Variables box. You should click on the Display normal curve box. Can you confirm your earlier results with these histograms?

VARIABILITY

A curve's kurtosis leads directly to the next important descriptive measure of a set of scores. The platykurtic curve in figure 3.3 is said to contain more heterogeneous (dissimilar) data, whereas the leptokurtic curve is said to have observations that are more homogeneous (similar). The spread of a distribution of scores is reflected in various measures of variability. We present three measures of variability here: range, variance, and standard deviation.

Range

The range is the high score minus the low score. It is the least stable measure of variability because it depends on only two scores and does not reflect how the remaining scores are distributed.

Variance

The variance (s^2) is a measure of the spread of a set of scores based on the squared deviation of each score from the mean. It is used much more frequently than the range in reporting the heterogeneity of scores. The variance is the most stable measure of variability. Two sets of scores that have vastly different spreads will have vastly different variances. *Many types of variance (such as observed variance, true variance, error variance, sample variance, between-subject variance, and within-subject variance) will become important as measurement and evaluation issues are presented throughout this textbook.*

As an illustration of how to calculate the variance, consider the following scores: 3, 1, 2, 2, 4, 5, 1, 4, 3, 5 (table 3.2).

Table 3.2 Calculating the Variance

X (observed score)	–	*M* (mean)	=	*x* (deviation score)	x^2 (squared deviation score)
3	–	3	=	0	0
1	–	3	=	–2	4
2	–	3	=	–1	1
2	–	3	=	–1	1
4	–	3	=	1	1
5	–	3	=	2	4
1	–	3	=	–2	4
4	–	3	=	1	1
3	–	3	=	0	0
5	–	3	=	2	4
Total				0	20

The steps to calculate the variance by hand are as follows:

1. Calculate the mean.
2. Subtract the mean from each score.
3. Take each difference (deviation) and square it.
4. Add the results together and divide by the number of scores minus 1.

Equation 3.2 (the didactic formula) is used to illustrate the variance.

$$s^2 = \frac{\Sigma(X-M)^2}{n-1} = \frac{\Sigma x^2}{n-1} = \frac{20}{9} = 2.22 \qquad (3.2)$$

That is, the variance is the average of the squared deviations from the mean (hence, the term *mean square* is sometimes used for a variance). Note that you divide by $n - 1$ and not n, so it isn't exactly the mean. You should use the following calculation formula to obtain the variance from a set of scores because it is generally easier to use.

$$s^2 = \frac{\Sigma X^2 - \frac{(\Sigma X)^2}{n}}{n-1} \qquad (3.3)$$

You should confirm that using the calculation formula will result in the same value for the variance as the didactic formula. (Hint: $\Sigma X^2 = 110$; $\Sigma X = 30$.) *Note that when all of the scores are identical, the variance is zero. Typically, this is not something you want in measurement.* Variation among scores is preferable. The reasons for this will be illustrated throughout this textbook.

You should turn to table 3.2 and see that you have really already learned to calculate the variance from the data that were provided there. You have simply used summation notation that you learned there.

Variance in general can be illustrated with a square, as in figure 3.5. The square (total variance) is divided into two types of variance: *true variance* and *error variance*. You might think of the scores on a recent test as having the three types of variance. Not everyone scored the same, so there is total or observed score variance. Not everyone has the same true knowledge about the content of the test, so there is true score variance. Last, there

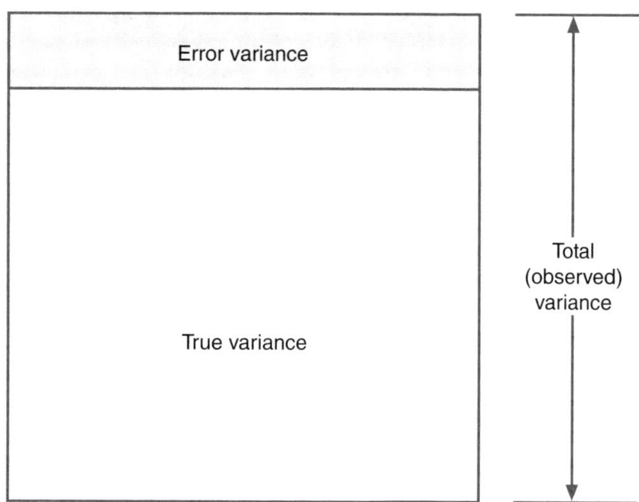

Figure 3.5 Three types of variance.

is some error in the test, but not everyone has the same amount of error reflected in their scores, so there is **error score variance** (error could result from several sources, such as student guessing or teacher incorrectly scoring the test). These important concepts will be further discussed in chapter 6 when you study reliability theory.

The types of variance will become increasingly important throughout the remainder of this textbook, but it is important that you understand variance at this point. Although it is a calculated *number,* we will often speak of variance in conceptual terms (i.e., variability). It may be helpful for you to remember the square in figure 3.5 and the fact that not all scores are identical.

Standard Deviation

Whereas the variance is important, a related number—the **standard deviation**—is often used in descriptive statistics to illustrate the variability of a set of scores. *The standard deviation is the square root of the variance.* It is helpful to think of the standard deviation as a linear measure of variability. Consider the square (figure 3.5) used to illustrate the variance. When you take the square root of a square, you get a linear measure. The same concept holds true for the standard deviation. The standard deviation is important because it is used as a measure of linear variability for a set of scores. Knowing the standard deviation of a set of scores can tell us a great deal about the heterogeneity or homogeneity of the scores.

You are encouraged to use the calculation formula (equation 3.3) to calculate the standard deviation. Simply calculate s^2 and take its square root. You should confirm that the standard deviation from the data in table 3.2 is 1.49 (i.e., $\sqrt{2.22}$).

Dataset Application

Use the chapter 3 large dataset available with the WSG and determine the variance and standard deviation for each of the variables in the dataset. Hint: Use Analyze → Descriptive Statistics → Frequencies → Options and click on the buttons in the Dispersion box. Can you identify which are the most heterogeneous and homogeneous variables? You should create histograms to see the variabilities visually.

Why is the standard deviation so important, and what exactly does it mean? For example, in a **normal distribution**, a bell-shaped, symmetric probability distribution (figure 3.6), knowing the standard deviation tells you a great deal about the distribution. In any normal distribution if you add to and subtract from the mean the value of 1 standard deviation, you will obtain a range that encompasses approximately 68.26% of the observations. If you add and subtract the value of 2 standard deviations, you will obtain a range that encompasses approximately 95.44% of the observations, and adding and subtracting the value of 3 standard deviations will give a range encompassing 99.74% of the observations. This is true regardless of what the mean and standard deviation are, as long as the distribution is normal. To summarize:

$$M \pm 1s \rightarrow 68.26\% \textit{ of observations}$$

$$M \pm 2s \rightarrow 95.44\% \textit{ of observations}$$

$$M \pm 3s \rightarrow 99.74\% \textit{ of observations}$$

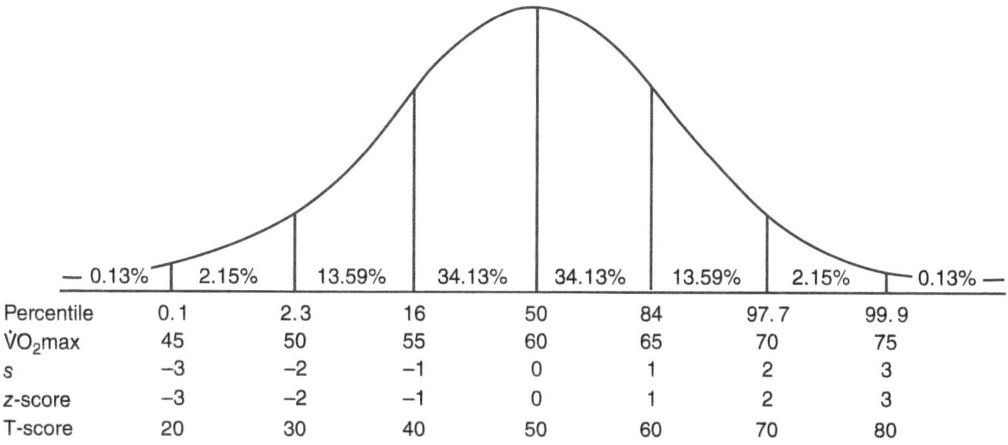

Figure 3.6 Normal distribution of scores.

Using the information obtained about the mean and standard deviation for a set of observations, you can approximate the percentile for any observation. Looking again at figure 3.6, assume that it illustrates the observations on a recent maximal oxygen consumption ($\dot{V}O_2max$) test that had a mean of 60 ml · kg^{-1} · min^{-1} and a standard deviation of 5 ml · kg^{-1} · min^{-1} and that the values are normally distributed. Use the figure to approximate percentiles for $\dot{V}O_2max$ values of 50, 55, 60, 65, and 70. How likely would it be to obtain a $\dot{V}O_2max$ in excess of 70 ml · kg^{-1} · min^{-1} on similar subjects? Figure 3.6 also presents information about two standard scores. You will learn about standard scores next.

STANDARD SCORES

Knowing the mean and standard deviation makes it easy to calculate a standard score. *A standard score is a set of observations that have been standardized around a given mean and standard deviation.* The most fundamental standard score is the *z*-score. It is calculated as follows:

$$z = \frac{X - M}{s} \tag{3.4}$$

In other words, you obtain the *z*-score for any observation by subtracting the mean, *M*, from the observed score, *X*, and dividing this difference by the standard deviation, s. If you do this for all observations in a set of data, you get a set of scores that themselves have a mean of 0 and a standard deviation of 1 (i.e., they have been standardized). Note that we have included *z*-scores in figure 3.6 and that *z*-scores always have a mean of 0 and a standard deviation of 1.

Another commonly used standard score is the T-score, which is calculated as follows:

$$T = 50 + \frac{10(X - M)}{s} \quad \text{or} \quad T = 50 + 10z \tag{3.5}$$

The mean of the T-scores for all the observations in a set of data is always 50, and the standard deviation is always 10. Note that in a normal distribution, 99.74% of the scores will fall between a *z*-score of −3 and +3 and between T-scores of 20 and 80. Can you see why this is the case? (Hint: See figure 3.6.)

You are probably asking yourself, What is the point of standard scores? To answer this, let's assume you are a physical education teacher teaching a basketball unit in which you are going to grade students on only two skill items (we're not recommending this!). Assume that you think shooting and dribbling are two skills important to basketball and that you can validly measure them. For shooting, you measure the number of baskets a student makes in 1 minute; for dribbling, you measure the amount of time a student takes to dribble through an obstacle course. You assume dribbling and shooting are equally important in basketball (we probably don't agree with you!), so you wish to weight them equally. A quick look at the scores illustrates that you cannot simply add these two scores together for each student to determine who is the best basketball player. High scores are better on the shooting test, but low scores are better on the dribbling test. In addition, each score is measured in different units (i.e., number of baskets made and number of seconds in which dribbling was completed).

This is exactly where standard scores can help you. To weight these two skills equally on this test, you must first convert each person's scores to a standard form, such as z-scores or T-scores. Note that the dribble test is a timed test, so you must correct the obtained z-score so that a faster time results in a higher z-score. You do this by changing the sign of the z-score (e.g., 2 becomes –2; –1.5 becomes 1.5). You can then add each student's two z-scores together and obtain a total z-score that is based on the fact that each test now has the same weight. You could not do this using the original raw scores, nor could you do it with any sets of scores having unequal variances. Generally, the set of scores with the larger variance would be weighted more heavily in the total if you simply added two raw scores together for each student. But if you convert them to a standard score, they have the same weight because they have the same standard deviation (remember that z-scores always have a standard deviation equal to 1.0).

You can take this example one more step and decide to give the shooting test twice the weight of the dribbling test. All you need to do is multiply the z-score for the shooting test by 2 and add it to the z-score for the dribbling test. The key concept is that whichever test is most variable will carry the greater weight. In fact, if there is no variability on an examination (i.e., if everyone scores the same), the examination contributes absolutely nothing toward differentiating student performance. This concept is further illustrated for you in chapter 13 on grading.

Table 3.3 summarizes important information about the mean and standard deviation of z- and T-scores. Note the means and standard deviations of these standard scores never change. You can also see this by looking at figure 3.6.

Table 3.3 Standard Score Means and Standard Deviations

	Mean	Standard deviation
z-score	0	1
T-score	50	10

NORMAL-CURVE AREAS (z-TABLE)

If you examine equation 3.4, you'll see that converting a score to a z-score actually expresses the distance a score is from its own mean in standard deviation units. That is, a z-score indicates the number of standard deviations that a score lies below or above the mean. *In addition to weighting performance for grades, z-scores can be used for a number of other purposes, mainly determination of (a) percentiles and (b) the percentage of observations that fall within a particular area under the normal distribution.* Consider again the normal distribution presented in figure 3.6. The area under the distribution is represented as a total of 100%. A student who scores at the mean has achieved the 50th percentile (i.e., has scored better than 50% of the group) and has a *z*-score of 0 (and a T-score of 50).

Observed scores alone may tell you little about performance (and, in fact, tell you nothing about relative performance). However, if a teacher reports your score in *z*-score form and you have a positive *z*-score, you know immediately that you have achieved better than average; a negative *z*-score indicates you scored below the mean. Using the *z*-score and the table of normal-curve areas (table 3.4), you can determine the percentile associated with any *z*-score for data that are normally distributed.

Table 3.4 Normal-Curve Areas

z	0.00	0.01	0.02	0.03	0.04	0.05	0.06	0.07	0.08	0.09
0.0	00.00	00.40	00.80	01.20	01.60	01.99	02.39	02.79	03.19	03.59
0.1	03.98	04.38	04.78	05.17	05.57	05.96	06.36	06.75	07.14	07.53
0.2	07.93	08.32	08.71	09.10	09.48	09.87	10.26	10.64	11.03	11.41
0.3	11.79	12.17	12.55	12.95	13.31	13.68	14.06	14.43	14.80	15.17
0.4	15.54	15.91	16.28	16.64	17.00	17.36	17.72	18.08	18.44	18.79
0.5	19.15	19.50	19.85	20.19	20.54	20.88	21.23	21.57	21.90	22.24
0.6	22.57	22.91	23.24	23.57	23.89	24.22	24.54	24.86	25.17	25.49
0.7	25.80	26.11	26.42	26.73	27.04	27.34	27.64	27.94	28.23	28.52
0.8	28.81	29.10	29.39	29.67	29.95	30.23	30.51	30.78	31.06	31.33
0.9	31.59	31.86	32.12	32.38	32.64	32.90	33.15	33.40	33.65	33.89
1.0	34.13	34.38	34.61	34.85	35.08	35.31	35.54	35.77	35.99	36.21
1.1	36.43	36.65	36.86	37.08	37.29	37.49	37.70	37.90	38.10	38.30
1.2	38.49	38.69	38.88	39.07	39.25	39.44	39.62	39.80	39.97	40.15
1.3	40.32	40.49	40.60	40.82	40.99	41.15	41.31	41.47	41.62	41.77
1.4	41.92	42.07	42.22	42.36	42.51	42.65	42.79	42.92	43.06	43.19
1.5	43.32	43.45	43.57	43.70	43.83	43.94	44.06	44.18	44.29	44.41
1.6	44.52	44.63	44.74	44.84	44.95	45.05	45.15	45.25	45.35	45.45
1.7	45.54	45.64	45.73	45.82	45.91	45.99	46.08	46.16	46.25	46.33
1.8	46.41	46.49	46.56	46.64	46.71	46.78	46.86	46.93	46.99	47.06
1.9	47.13	47.19	47.26	47.32	47.38	47.44	47.50	47.56	47.61	47.67
2.0	47.72	47.78	47.83	47.88	47.93	47.98	48.03	48.08	48.12	48.17
2.1	48.21	48.26	48.30	48.34	48.38	48.42	48.46	48.50	48.54	48.57
2.2	48.61	48.64	48.68	48.71	48.75	48.78	48.81	48.84	48.87	48.90
2.3	48.93	48.96	48.98	49.01	49.04	49.06	49.09	49.11	49.13	49.16
2.4	49.18	49.20	49.22	49.25	49.27	49.29	49.31	49.32	49.34	49.36
2.5	49.38	49.40	49.41	49.43	49.45	49.46	49.48	49.49	49.51	49.52
2.6	49.53	49.55	49.56	49.57	49.59	49.60	49.61	49.62	49.63	49.64
2.7	49.65	49.66	49.67	49.68	49.69	49.70	49.71	49.72	49.73	49.74
2.8	49.74	49.75	49.76	49.77	49.77	49.78	49.79	49.79	49.80	49.81
2.9	49.81	49.82	49.82	49.83	49.84	49.84	49.85	49.85	49.86	49.86
3.5	49.98									
4.0	49.997									
5.0	49.99997									

Based on Lindquist 1942.

The values down the left side of table 3.4 are *z*-scores in whole numbers and tenths, and the numbers across the top represent *z*-scores to the hundredths place. (For example, for a *z*-score of 1.53, you find 1.5 on the left and read across to the 0.03 column.) The numbers in the body of the table are percentages of observations that lie between the mean and any given standard deviation distance from the mean. Verify that 34.13% of the scores lie between the mean and 1 standard deviation above the mean (i.e., a *z*-score of 1.00). Note that no negative *z*-scores are shown in the table: They are not necessary because the normal distribution is symmetrical. Therefore, 34.13% of the observations also fall between the mean and 1 standard deviation below the mean (i.e., a *z*-score of –1.00). Note that ±1 standard deviation is 68.26% (34.13 + 34.13). See if you can determine the percentage of scores that fall between 1 and 1.5 standard deviations above the mean. Figure 3.7 provides an illustration of this area. Use table 3.4 and figure 3.7 to help you determine the answer (43.32 – 34.13 = 9.19%).

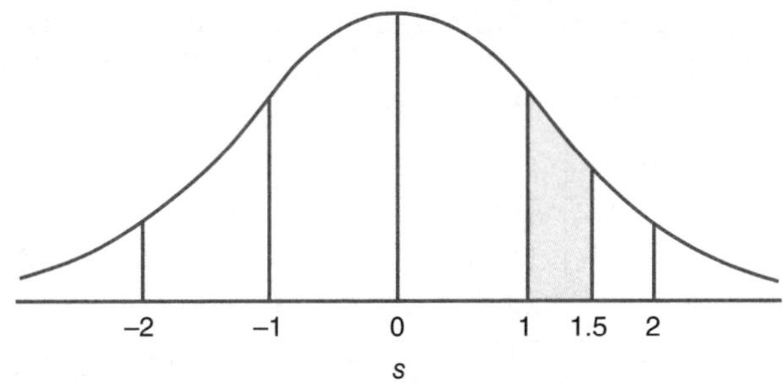

Figure 3.7 Area of normal curve between 1 and 1.5 standard deviations above the mean.

In summary, the important points to remember when using the table of normal-curve areas are that

- the reference point is the mean,
- *z*-scores are presented to the nearest one-hundredth, and
- the numbers in the body of the table are percentages.

Use table 3.4 to confirm that a *z*-score of 1.23 is (approximately) the 89th percentile. Remember that the reference point is the mean and that a percentile represents all people who score at or below a given point. (Hint: This includes those scoring below the mean if your observed score is above the mean or the percentile is greater than 50 or the *z*-score is greater than 0.00.)

Use table 3.4 to confirm that a *z*-score of –1.23 represents (approximately) the 11th percentile. You can do this using the following method:

1. Find 1.2 in the left column of the table and 0.03 at the top. They intersect at the percent value of 39.07.
2. Recall what this number means: 39.07% of the scores lie between the mean and a *z*-score of either +1.23 or –1.23.
3. Recall that 50% of the scores fall below the mean.
4. Thus, 50 – 39.07 gives the leftover area below a *z*-score of –1.23; the answer is 10.93 (or approximately the 11th percentile).

Use table 3.4 to confirm that 14.98% of a set of normally distributed observations fall between a *z*-score of 0.50 and 1.00. (Draw a picture to help you see this.) Some hints for completing these types of items are the following:

- Draw a picture of the question (starting with the normal distribution).
- Consider whether the *z*-score is helpful. (It *often* is.)
- Can the *z*-table (table 3.4) help you? (It *often* can.)

If a score is reported in T-score form, simply transform it to *z*-score form and then estimate the percentile. To change from a T-score to a *z*-score, simply substitute the T-score mean (50) and standard deviation (10) into the *z*-score formula. If your T-score is 30, your *z*-score would be

$$(30 - 50) / 10 = -2.00$$

Although you can use table 3.4 to convert from *z*-scores to percentiles, you can also use the table to convert from percentiles to *z*-scores for normally distributed data. This might be helpful if you wanted to create a summation of *z*-scores to determine the best overall performance, as in our earlier basketball example. Assume you were told that you performed at the 69th percentile on a test. You can determine your *z*-score on the test as follows:

1. Calculate 69% – 50% = 19% to get the area between your percentile and the mean.
2. Find 19 in the body of the table—the closest value is 19.15.
3. Find what *z*-score that 19.15 corresponds to by going over to the left margin and up to the top row. The *z*-score is 0.50. (It is positive, because you were above the mean.)

If you wanted to confirm that this is a T-score of 55, you would simply substitute 0.50 into the T-score formula (equation 3.5).

Go to the WSG to complete Student Activity 3.2 and view Video 3.2.

Mastery Item 3.5

Use table 3.4 to confirm that if the cutoff to achieve an A in this course is a *z*-score of 1.35, 8.85% of the people would be expected to earn a grade of A.

Mastery Item 3.6

Verify that the percentile associated with a T-score of 68 is 96.41.

Mastery Item 3.7

Use SPSS to calculate *z*-scores for the data in table 3.1.

1. Start SPSS.
2. Open table 3.1 data.
3. Click on the Analyze menu.
4. Scroll to Descriptive Statistics and across to Descriptives and click.
5. Highlight "VO_2max" and place it in the Variables box by clicking the arrow key.
6. Click on the Save standardized values as variables box.
7. Click OK.

The scores are converted to *z*-scores, and a new column of *z*-scores (ZVO_2max) is added to your data file. Be certain to save the modified data file because you will need the *z*-scores in the next mastery item.

Mastery Item 3.8

Use SPSS to calculate T-scores.

1. Start SPSS.
2. Open table 3.1 data.
3. Click on the Transform menu.
4. Click on Compute.
5. In the Target Variable box, type "tscore."
6. In the Numeric Expression box, type "50 + 10*ZVO$_2$max."
7. Click OK.
8. Now the T-scores have been added to your data file. Save the data file.

Mastery Item 3.9

Use the data that you just created and SPSS to verify that the mean and standard deviation for the *z*-scores and T-scores are 0 and 1 and 50 and 10, respectively.

MEASUREMENT AND EVALUATION CHALLENGE

What has James learned in this chapter that will help him interpret the results from his treadmill test? James can compare his $\dot{V}O_2$max to that of the average person who is his age and gender. He can determine his percentile based on his score, the mean score, and the variability in $\dot{V}O_2$max. James has learned how to develop and interpret scores based on score distributions and norms. Assuming the average $\dot{V}O_2$max for a male of his age is 45 ml · kg^{-1} · min^{-1} with a standard deviation of 5 ml · kg^{-1} · min^{-1}, James can easily determine that his performance is at the 84th percentile. Only 16% of people his age and gender have $\dot{V}O_2$max values that exceed 50 ml · kg^{-1} · min^{-1}.

SUMMARY

The descriptive statistics presented in this chapter are foundational to the remainder of this textbook. Of particular importance is your ability to understand and use the concepts of central tendency and variability and to use and interpret the normal curve and areas under sections of the normal distribution. If you are more interested in statistical methods, see Glass and Hopkins (1996). Thomas, Nelson, and Silverman (2015) provide excellent examples of research and statistical applications in human performance.

At this point you should be able to accomplish the following tasks:

- Differentiate between the four levels of measurement and provide examples of each.
- Calculate and interpret descriptive statistics.
- Calculate and interpret standard scores.
- Use a table of normal-curve areas (*z*-table) to estimate percentiles.

- Use SPSS to enter data and generate and interpret
    - frequency distributions and the percentiles associated with observed scores,
    - histograms for observed scores,
    - descriptive statistics (mean, standard deviation) on variables, and
    - z- and T-scores.

Go to the WSG for homework assignments and quizzes that will help you master this chapter's content.

CHAPTER

4

Correlation and Prediction

OUTLINE

OBJECTIVES

After studying this chapter, you will be able to

- calculate statistics to determine relationships between variables;
- calculate and interpret the Pearson product-moment correlation coefficient;
- calculate and interpret the standard error of estimate (SEE);
- use scatterplots to interpret the relationship between variables;
- differentiate between simple correlation and multiple correlation; and
- use SPSS and Excel computer software in data analysis for correlational and regression analyses.

 The lecture outline in the WSG will help you identify the major concepts of the chapter.

MEASUREMENT AND EVALUATION CHALLENGE

Now that you know how to calculate descriptive statistics, you are ready to learn other statistical procedures that are important. In chapter 3, a technician reported James' $\dot{V}O_2max$ as 50 ml · kg^{-1} · min^{-1}. However, James had recently read that $\dot{V}O_2max$ should actually be measured by collecting expired air and analyzing it for oxygen and carbon dioxide content. James had not conducted any of these measurements during his treadmill run. He recalls the technician saying, "James, your treadmill time predicts a $\dot{V}O_2max$ of 50 ml · kg^{-1} · min^{-1}." James now realizes that his treadmill time was related to his actual $\dot{V}O_2max$. Correlational procedures are used to determine the relationship between and among variables. Indeed, predictions and estimations of one variable from one or more other variables are common in kinesiology, human performance, and sport and exercise science. For example, in James' case, his treadmill time was used to estimate his $\dot{V}O_2max$ without having to actually collect expired oxygen and carbon dioxide.

How are two variables related? As performance on one variable increases, does performance on the other variable change? The relationship between variables is examined statistically by the **correlation coefficient**. **Correlations** help you describe relations and in some cases predict outcomes; they are useful in measurement theory.

CORRELATION COEFFICIENT

In chapter 3 you learned to describe data using measures of central tendency and variability, and we discussed one variable or measure at a time. However, teachers, clinicians, and researchers often measure more than one variable. They are then interested in describing and reporting the **relationship**—the statistical association between these variables. This is a fundamental statistical task that data analysts, kinesiologists, teachers, clinicians, and researchers should be able to perform. An example would be this: *What is the relationship between the bench press and the leg press strength tests?* That is, do these strength measures have anything in common? To measure this relationship, we would calculate the correlation coefficient, specifically the **Pearson product-moment correlation coefficient (PPM)**, symbolized by *r*.

The correlation coefficient is an index of the **linear relationship** (an association that can best be depicted by a straight line) between two variables. It indicates the magnitude, or amount of relationship, and the direction of the relationship. Figure 4.1 illustrates these aspects of the correlation coefficient. *As you can see, the correlation coefficient can have either a positive or negative direction and a magnitude between –1.00 and +1.00.*

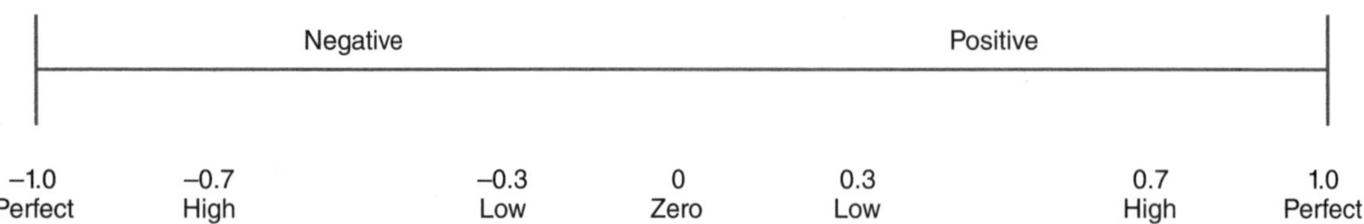

Figure 4.1 Attributes of *r*.

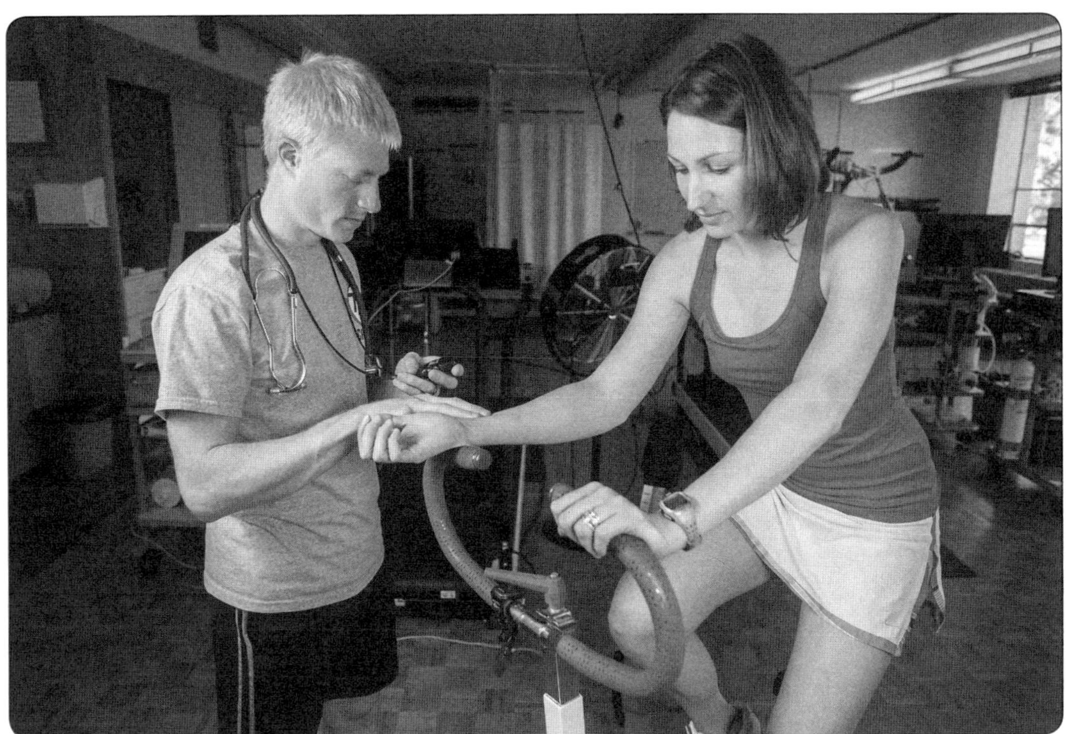

Correlation coefficients are used to look at the relationship between variables, such as heart rate during submaximal exercise on the cycle ergometer and $\dot{V}O_2max$.

The terms *high* and *low* are subjective and are affected by how the correlation was obtained, by the participant measured, by the variability in the data, and by how the correlation will be used. There is nothing bad about a negative *r*. The sign of *r* simply indicates how the two variables covary (i.e., go together). A positive *r* indicates that participants scoring above the mean on one variable, *X*, will usually be above the mean on the second variable, *Y*. A negative *r* indicates that those scoring above the mean on *X* will generally be below the mean on *Y*. Thus, a correlation of –.5 is not less than one of +.5. In fact, they are equal in strength but opposite in direction.

Mastery Item 4.1

If you were calculating a correlation coefficient and computed it to be +1.5, what must have happened?

Let's examine the factors of direction and magnitude of the correlation coefficient. Table 4.1 provides the scores for 10 students on three measures: body weight, chin-ups, and pull-ups. As you can see by examining the data, low values for chin-ups are generally paired with low values for pull-ups. If you were to calculate the correlation between these values (described later), you would find *r* to be indicating a **direct relationship** (positive relationship). In contrast, if you examine body weight versus pull-ups, you can see that high body weights are generally paired with low pull-up scores. These measures have an **indirect relationship** (or negative or **inverse relationship**). Figures 4.2 and 4.3 provide illustrations in the form of scatterplots of these relationships. A **scatterplot** is a graphic representation of the correlation between two variables. (You can create a scatterplot by labeling the axes with names and units of measurement. Then plot each pair of scores for each participant.)

Table 4.1 Sample Correlation Data

Participant	Body weight	Chin-ups	Pull-ups
1	130	10	8
2	130	9	7
3	140	15	12
4	150	9	10
5	150	7	6
6	160	5	3
7	160	3	4
8	160	8	7
9	170	4	5
10	170	6	3

Note: During a pull-up, palms face away from the person; during a chin-up, palms face toward the person.

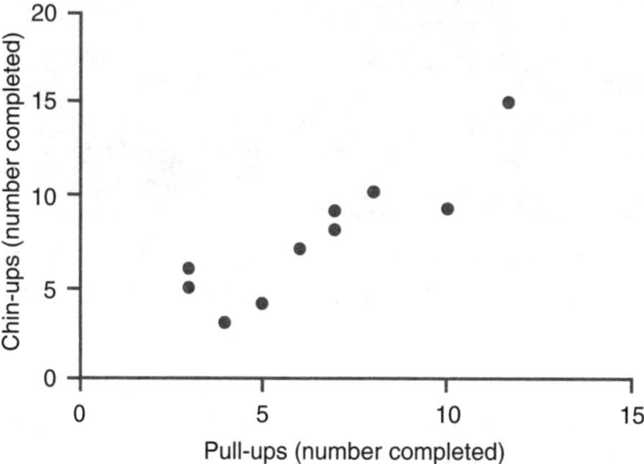

Figure 4.2 Scatterplot of correlation between pull-ups and chin-ups.

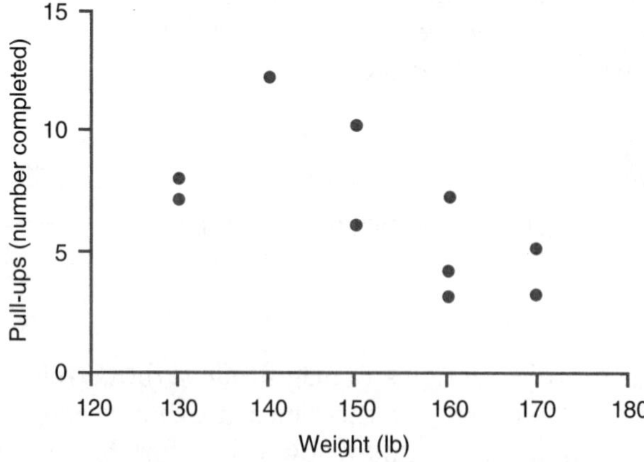

Figure 4.3 Scatterplot of correlation between body weight and pull-ups.

Keep in mind that the correlation coefficient is an index of linear relationship. All paired points must be on a straight line for a correlation to be perfect, –1 or +1. If two variables have a correlation coefficient of 0 (or **zero correlation**), their scatterplot would demonstrate nothing even resembling a straight line; in fact, it would look more like a circle of dots. Figure 4.4 illustrates a sample scatterplot for a zero correlation.

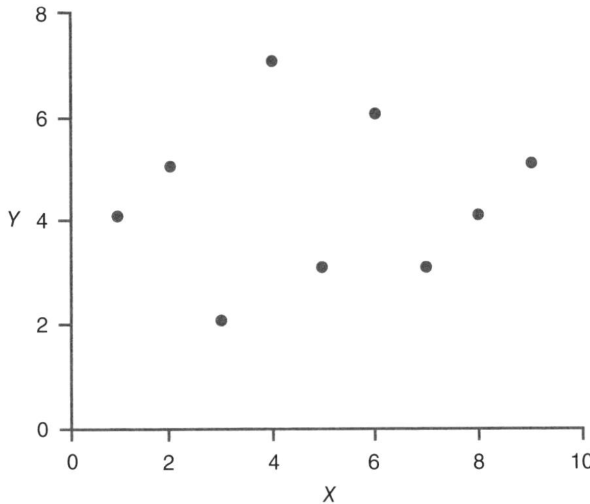

Figure 4.4 Scatterplot of zero correlation.

Go to the WSG to complete Student Activity 4.1 and view Video 4.1.

CALCULATING *r*

Having developed a fundamental understanding of the correlation coefficient and the ability to illustrate correlations with simple scatterplots, let us turn to the calculation of r.

Steps in Calculating r

1. Arrange your data in paired columns (here, X is number of chin-ups, and Y is the number of pull-ups).
2. Square each X value and each Y value and place the results in two additional columns, X^2 and Y^2.
3. Multiply each X by its corresponding Y and place the results in a new column (called cross product, or XY).
4. Determine the sum for each column (X, Y, X^2, Y^2, XY).
5. Use the following formula to calculate r:

$$r = \frac{n(\Sigma XY) - (\Sigma X)(\Sigma Y)}{\sqrt{[n(\Sigma X^2) - (\Sigma X)^2][n(\Sigma Y^2) - (\Sigma Y)^2]}} \tag{4.1}$$

Note that everyone must have two scores. If a student completes only one test, his or her score cannot be used in the calculation of r.

Using the data in table 4.2, follow the steps here to calculate the correlation coefficient between chin-ups and pull-ups.

$$r = \frac{10(576) - (76)(65)}{\sqrt{[(10 \times 686 - (76)^2)(10 \times 501 - (65)^2)]}} = .89$$

Table 4.2 Calculating the Correlation Coefficient

	CHIN-UPS		PULL-UPS		
Participant	X	X^2	Y	Y^2	XY
1	10	100	8	64	80
2	9	81	7	49	63
3	15	225	12	144	180
4	9	81	10	100	90
5	7	49	6	36	42
6	5	25	3	9	15
7	3	9	4	16	12
8	8	64	7	49	56
9	4	16	5	25	20
10	6	36	3	9	18
Σ	**76**	**686**	**65**	**501**	**576**

Mastery Item 4.2

How would you describe the magnitude of the correlation computed between chin-ups and pull-ups? (Hint: Examine figures 4.1 and 4.2.)

Go to the WSG to complete Student Activities 4.2, 4.3, and 4.4.

Coefficient of Determination

An additional statistic that provides further information about the relationship between two measures is r^2. This square of the correlation coefficient is called the **coefficient of determination**. This value represents the proportion of shared variance between the two measures in question. To understand shared variance, let us examine a specific example. If the correlation between a distance-run test and $\dot{V}O_2max$ is $r = 0.90$, then r^2 would be 0.81; the percentage of shared variance would be 0.81 × 100 = 81%. This means that performance in the distance run accounts for 81% of the variation in $\dot{V}O_2max$ values. Nineteen percent of the variance in $\dot{V}O_2max$ values (100%–81%) is the nonpredicted variance in $\dot{V}O_2max$ that is unexplained by performance in the distance run. Thus, 19% is error or residual variance. That is, it is variance remaining after you have used a predictor (*X*) to account for variation in the criterion (*Y*) variable. The coefficient of determination is important in statistics and measurement because it reflects the amount of variation found in one variable that can be predicted from another variable. Figure 4.5 illustrates the variation in $\dot{V}O_2max$ that can be predicted from distance run. Note the similarity here to that presented with variance in chapter 3.

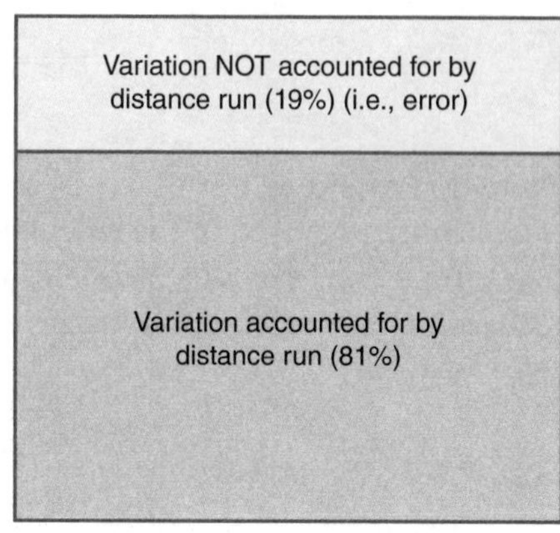

Figure 4.5 Variation in $\dot{V}O_2max$ accounted for by distance run when $r = 0.90$.

@ Go to the WSG to complete Student Activity 4.5.

Negative Correlations

Two measures can have a **negative correlation** coefficient for one of two reasons. First, a negative correlation coefficient can result from two measures having opposite scoring scales. For example, the distance covered in a 12-minute run and the time required to run 1.5 miles (2.4 km) would be negatively correlated. Runners with more endurance will cover more distance and have *greater* distance values on the 12-minute run; the same runners will run the 1.5 mile (2.4 km) faster and have *lower* time values. Runners with less endurance will have opposite results. Notice that this negative correlation has a positive connotation because the so-called best and worst scores for each test are associated with each other.

A second reason for a negative correlation coefficient is that two measures can have a true negative relationship. A good example is provided by the measures of body weight and pull-ups in table 4.1. In general, heavier people have a more difficult time lifting or moving their body weight than do lighter people.

Mastery Item 4.3

What would be the reason for a negative correlation between marathon run times and $\dot{V}O_2$max?

Limitations of *r*

The correlation coefficient is an index of the linear relationship between two variables. If two variables happen to have a **curvilinear relationship**, such as the one depicted in figure 4.6 between arousal level and performance, the PPM correlation coefficient would be near 0, indicating no *linear* relationship between the two variables. However, to say that there is no relationship at all between the variables would obviously be incorrect. Both low- and high-arousal scores are related to lower performance, and midrange arousal scores are associated with higher performance. This limitation of *r* is one reason for graphing the relationships between variables using the scatterplot technique.

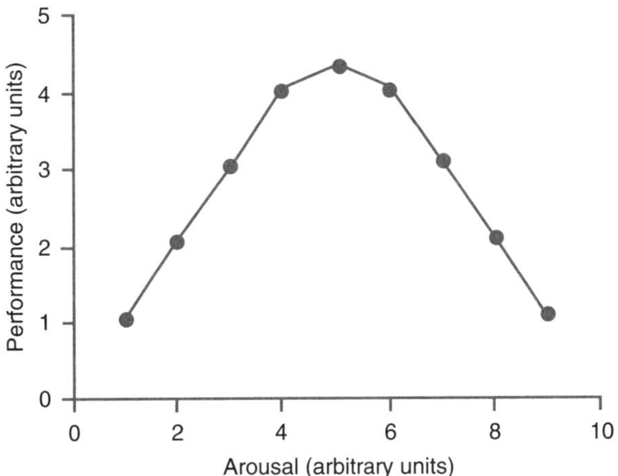

Figure 4.6 Graph of curvilinear relationship.

A second limitation of—or disclaimer about—*r* is that correlation is not necessarily an indication of a cause-and-effect relationship. Even if the correlation coefficient for two variables is +1 or −1, a conclusion, on the basis of *r* alone, that one variable is a cause of a measurable effect in another is incorrect. Some third variable may be the cause of the relationship detected by a high *r* value. For example, if we found the correlation coefficient between body weight and power (the amount of work performed in a fixed amount of time) to be +1.00 and assumed a cause-and-effect relationship, we might be tempted to use the following logic: (a) Higher body weights cause higher power; therefore (b) all athletes who need power should gain weight. Athletes might then become so heavy that they would suffer decreases in power. In a second example, the PPM correlation coefficient between shoe size and grade level for elementary school children would be high. However, having big feet does not cause a child to be in a higher grade nor does being in fifth grade cause a child to have big feet. Obviously, both variables are highly related to age, and this is the variable responsible for the high *r* between shoe size and grade level. Although the lack of a cause-and-effect relationship is relatively easy to explain in this example, in other situations the third variable may not be so apparent.

Go to the WSG to complete Student Activity 4.6.

Another limitation of *r* is the effect of the variance or range of the data on the magnitude of *r*. Variables with larger variances or ranges tend to have higher *r* values than variables with restricted variances or ranges. Figure 4.7 demonstrates this phenomenon. The value of *r* in either the solid or open dots in the figure is much smaller than the value of *r* for the combined set of data. This is because the variance is greater in the complete set of data than it is in either subset. The take-home message is that the interpretation of the value of *r* goes beyond just noting its magnitude. Consideration must also be given to other issues, such as linearity, other related variables, and the variability of the data involved.

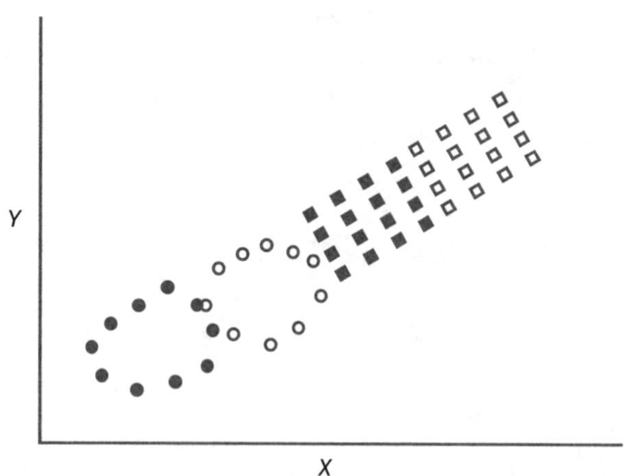

Figure 4.7 Correlation example of range restriction.

@ Go to the WSG to complete Student Activity 4.7.

PREDICTION

In science, one of the most meaningful research results is the successful forecast. A valuable use of correlations is in **prediction**—that is, estimating the value of one variable from those of one or more other variables. From a mathematical standpoint, if there is a relationship between X and Y, then X can predict Y *to some degree* and vice versa. This still does not mean that X and Y are causally related, however. To establish a cause-and-effect relationship between X and Y, another type of study and analysis would be necessary (i.e., establishing and testing a hypothesis with an experimental study).

Straight Line

As you may recall from high school geometry, any point on a plane marked with an x-axis and y-axis can be identified by its coordinates (X, Y) on the plane, and a straight line can be defined by the equation $\hat{Y} = bX + c$, where b is the slope of the line and c is where the line intercepts the Y-axis. The slope indicates how much Y changes for a unit change in X. The Y-intercept represents the value of Y when $X = 0$. In figure 4.8 we have marked five coordinate points: (0,1), (1,2), (2,3), (3,4), and (4,5); these points fall on a straight line. In figure 4.8, the Y-intercept is 1 and the slope of the line is 1. When all of the paired points do not fall in a straight line, a line of best fit can be plotted through the set of points—this is called the regression line (also called the line of best fit or prediction line).

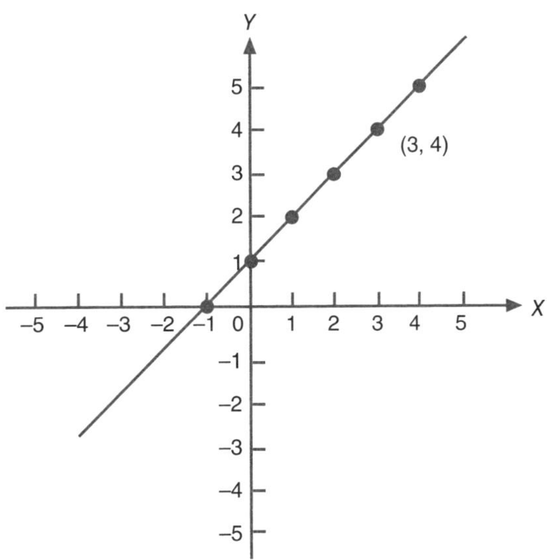

Figure 4.8 Plot of a straight line.

Go to the WSG to complete Student Activity 4.8.

Simple Linear Prediction

Simple linear prediction, also called *regression*, is a statistical method used to predict the criterion, outcome, or **dependent variable**, Y, from a single predictor or **independent variable**, X. If the two variables are correlated, which indicates that they have some amount of linear relationship, we can compute a prediction equation. The prediction equation has the same form as the equation for a straight line in plane geometry:

$$\hat{Y} = \text{b}X + \text{c} \tag{4.2}$$

We must think in terms of $\hat{Y}$ because, unless the correlation between X and Y is –1.00 or +1.00, $\hat{Y}$ is only an estimate of Y. Generally, $\hat{Y}$ will not equal Y. The following formulas are used for calculating *b* and *c*:

$$\text{b} = \frac{n(\Sigma XY) - (\Sigma X)(\Sigma Y)}{\text{n}(\Sigma X^2) - (\Sigma X)^2} \tag{4.3}$$

$$\text{c} = \bar{Y} - \text{b}\bar{X} \tag{4.4}$$

For example, using the data in table 4.1, we can calculate *b* and *c* and present the prediction equation for predicting pull-ups from body weight. Note the prediction line presented in figure 4.9 represents the prediction equation and also illustrates the error of prediction.

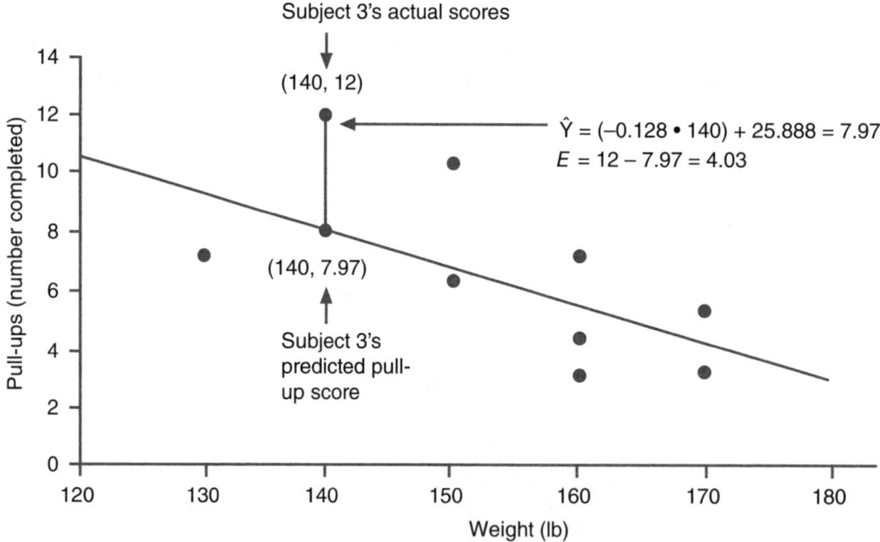

Figure 4.9 Error (residual) scores.

Vema, Sajwan, and Debnath (2009) provide an example of the use of simple linear prediction in human performance. They found the correlation coefficient between $\dot{V}O_2max$ (estimated from a 1-mile [1.6 km] run or walk test) and heart rate after a 3-minute step test for a sample of athletes attending the Lakshmibai National Institute of Physical Education to be –0.70. They calculated a linear prediction equation to predict $\dot{V}O_2max$ from heart rate for the athletes. Their equation is

$$\text{Predicted } \dot{V}O_2\text{max} = 114.38 - 0.24(X) \text{ (where X is heart rate)}$$

where –0.24 is the slope, 114.38 is the *Y*-intercept, and *X* is the athlete's heart rate after a 3-minute step test. $\dot{V}O_2max$ is estimated in ml · kg^{-1} · min^{-1}. The 1-minute heart rate is estimated from a 15-second pulse immediately after completing the step test.

Mastery Item 4.4

Using the equation provided in figure 4.9, find the value of Y if X is 160.

Mastery Item 4.5

Use the data in table 4.1 to hand calculate ***b*** and ***c*** and provide the equation for predicting the number of chin-ups from body weight.

 Go to the WSG to complete Student Activity 4.9.

Errors in Prediction

Unless the correlation coefficient is −1.00 or +1.00, $\hat{Y}$ will not necessarily equal Y. The following equation summarizes this:

$$\text{E} = Y - \hat{Y} \tag{4.5}$$

E (error) represents the inaccuracy of our predictions of Y based on the prediction equation. Figure 4.9 provides a demonstration of the error, or residual score. Participant 3 actually did 12 pull-ups, but the regression equation predicts 7.97 pull-ups for a participant weighing 140 pounds (63.6 kg). Thus, for this participant the residual score (or error) is 4.03. Residual scores are important for several reasons. First, they represent pure error of estimation or prediction. If these errors can be minimized, prediction can be improved. Second, residual scores can represent lack of fit, which means that the independent variable does not predict a portion of the criterion (Y). The predictor should then be examined to reduce this problem. Perhaps more predictors could be added to reduce the error (more about that soon). Finally, residual scores could represent a pure measure of a trait with the predictor statistically removed. In our earlier example of the correlation between pull-ups and body weight, pull-ups could be predicted from body weight. The resulting residual score is interpreted as a person's ability to do pull-ups with body weight statistically controlled. Or, in other words, if everyone was the same weight, how many pull-ups could you do? Positive residual scores would indicate that you did more pull-ups per unit weight than predicted. An important point to remember is that E has a zero correlation with X. This means that the prediction equation is equally accurate (or inaccurate) at any place along the X score scale.

The **standard error of estimate** (SEE), also called the *standard error of prediction (SEP)* or simply, the *standard error (SE),* is a statistic that reflects the average amount of error in the process of predicting Y from X. Technically, it is the standard deviation of the error, or residual scores. The following formula is used for its calculation:

$$s_e = s_y\sqrt{1 - r^2} \tag{4.6}$$

s_y is the standard deviation of the Y scores. If we use the data in table 4.1, the SEE for predicting pull-ups (Y) from body weight (X) would be as follows:

$$s_e = 2.953\sqrt{1 - (.637)^2}$$
$$s_e = 2.953\sqrt{1 - (.4058)}$$
$$s_e = 2.276$$

Because this is a standard deviation of the error or residual scores, it could be used as follows: If we predicted 7.97 pull-ups for someone weighing 140 pounds (63.6 kg), about 95% of the people who weigh 140 pounds will have pull-up scores between approximately 12 and 4. Remember from Chapter 3 that approximately 95% of the time a score is located ±2 standard deviations from the mean of a normal distribution and error scores are assumed to be normally distributed. Thus,

$$7.97 \pm 2 \times 2.276 = 12.5 \text{ to } 3.4$$

The standard deviation here is actually the SEE because SEE is the standard deviation of the errors of prediction. Note also that the 7.97 predicted from 140 pounds (63.6 kg) is the mean value of pull-ups for those who weigh 140 pounds (63.6 kg).

Figure 4.10 illustrates all of these concepts, the line of best fit, the predicted value, and the SEE when estimating DXA-determined percentage of body fat from skinfolds.

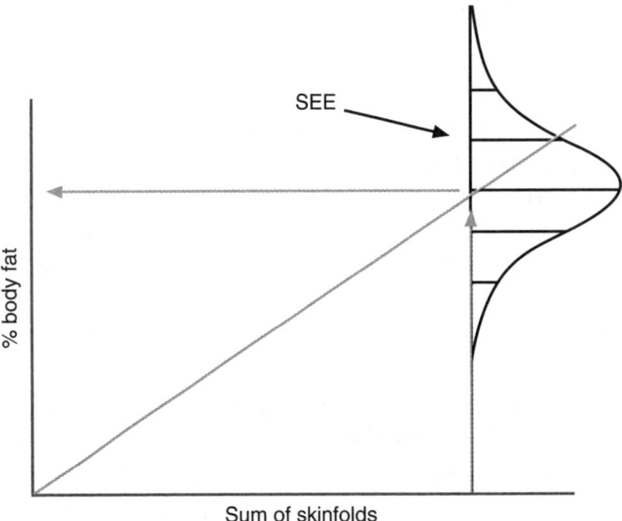

Figure 4.10. **Predicting DXA-determined percentage of body fat from skinfolds.**

 Go to the WSG to complete Student Activity 4.10 and view Video 4.2.

MULTIPLE CORRELATION OR MULTIPLE REGRESSION

Correlation and prediction are two interrelated topics that are based on the assumption that two variables have a linear relationship. We have examined the notion of simple linear prediction that has one predictor, *X*, of the criterion, *Y*. A more complex prediction of *Y* can be developed based on more than one predictor: X_1, X_2, and so on. This is called **multiple correlation**, multiple prediction, or multiple regression. The mathematics of this approach are much more complicated. If *X* and *Y* have a curvilinear relationship, then nonlinear regression can be used to predict *Y* from *X*.

The general form of the multiple correlation equation is presented below:

$$\hat{Y} = b_1X_1 + b_2X_2 + b_3X_3 + \ldots b_pX_p + c \quad (4.7)$$

Where $\hat{Y}$ is the predicted value, b_1 represents the regression coefficient for variable X_1, b_2 represents the regression coefficient for variable X_2, and so on, and c represents the constant.

Although the mathematics involved in these techniques are beyond the scope of this book and will not be discussed here, we provide two examples of multiple regression equations. Jackson and Pollock (1978) published prediction equations that combined both multiple and nonlinear prediction. Their equation predicted hydrostatically measured body density (BD) (dependent variable) of men from age (A), sum of skinfolds (SK), and sum of skinfolds squared (SK^2). The three predictors (independent variables) are the multiple predictors, and the sum of skinfolds squared is the nonlinear component in the prediction.

$$BD = 1.10938 - 0.0008267(\Sigma SK) + 0.0000016(\Sigma SK^2) - 0.0002574(A)$$

Using the BD, one can then estimate percentage of body fat.

The second example is based on Kline et al. (1987). Their intent was to estimate $\dot{V}O_2$max from weight, age, sex, time to complete a 1-mile (1.6 km) walk and heart rate at the end of the walk. The correlation between measured and estimated $\dot{V}O_2$max, was .88 and the SEE was 5.0 ml · kg^{-1} · min^{-1}.

Their equation was as follows:

$$\text{Predicted } \dot{V}O_2\text{max} = (-0.0769)\ (\text{weight in lb}) + (-0.3877)\ (\text{age in yr}) + (6.3150)\ (\text{gender}) + (-3.2649)\ (\text{time in min}) + (-0.1565)\ (\text{heart rate})$$

Males are coded 1 and females are coded 0.

Can you see how the regression coefficients reflect the correlations between the predictors and the predicted value? That is, for increases in weight, age, time, and heart rate, there is a decrease in the predicted $\dot{V}O_2$max. Males, on average, have a $\dot{V}O_2$max 6.3150 units higher than females.

Dataset Application

Return to figure 4.5. This figure illustrates the relationship between $\dot{V}O_2$max and distance run. Consider what would happen to the variation explained (and the resultant error variation) if additional variables were added to the regression equation. Assume that you added gender and age. What would happen to the amount of error illustrated in the figure? Why would this happen? What other variables might you include in the model to reduce the amount of error?

Use the chapter 4 large dataset in the WSG to illustrate correlations. Calculate the PPM (Analyze → Correlate → Bivariate and move all of the variables to the right) to determine which variables are most related to weekly step counts. Interpret the resulting correlation matrix. Create scatterplots to illustrate the relationships. (Graphs → Legacy Dialogs → Scatter/Dot and click on Simple Scatter and then Define. Select your X-axis and Y-axis variables and click on OK.) Create a multiple correlation (Analyze → Regression → Linear) and predict weekly step counts from gender and body mass index (BMI). Put "Weekly Steps" in the Dependent cell and "Gender" and "BMI" in the Independent(s) cell and click on OK. What happens to the correlation (and the ability to predict) and the SEE when you use additional predictors? Explain why. Now add weight to the prediction equation. Explain what happens to the ability to predict beyond that with gender and BMI only. Would you need or want to add weight to the prediction equation? Why?

MEASUREMENT AND EVALUATION CHALLENGE

James now understands the statistical methods used to estimate his $\dot{V}O_2max$ from treadmill time. The type of treadmill test that James completed is called a Balke protocol. Treadmill protocols are further described in chapter 9. Research has shown the correlation between treadmill time with the Balke protocol and measure $\dot{V}O_2max$ to be >.90. The prediction equation is $\dot{V}O_2max = 14.99 + 1.444$ X, where X equals treadmill minutes in decimal form. Note that $\dot{V}O_2max$ is actually $\hat{Y}$ (i.e., the predicted value of Y based on X). Thus, James, who ran for 24 minutes and 15 seconds (i.e., 24.25 min), has a predicted $\dot{V}O_2max$ of $14.99 + 1.444\ X\ (24.25) = 50$ ml · kg^{-1} · min^{-1}. However, James also realizes that there is some error in the prediction equation because the correlation is not perfect (i.e., ±1.00). The standard error of estimate (SEE) reflects the amount of error in the prediction equation. With this equation, the SEE is about 3 ml · kg^{-1} · min^{-1}. Thus, James can be 68% confident that his actual $\dot{V}O_2max$ is between 47 and 53 ml · kg^{-1} · min^{-1} (his predicted score ±1.00 SEE).

SUMMARY

The correlation and prediction statistics presented in this chapter lay the foundation and provide the many necessary skills for the measurement and evaluation process. As you will see, these skills are necessary for generating measurement theory, as well as applying reliability and validity concepts to practical problems in exercise and human performance.

At this point you should be able to accomplish the following tasks:

- Calculate and interpret measures of correlation.
- Calculate and interpret a prediction equation.
- Calculate the SEE.
- Use SPSS or Excel to enter data and to generate and interpret
    - correlation coefficients,
    - scatterplots of variables, and
    - simple linear prediction equations.

Go to the WSG for homework assignments and quizzes that will help you master this chapter's content.

CHAPTER

5

Inferential Statistics

OUTLINE

OBJECTIVES

After studying this chapter, you will be able to

- understand the scientific method and the hypotheses associated with it;
- perform an inferential statistical analysis to test a hypothesis; and
- use selected programs from the SPSS or Excel computer software in data analysis.

 The lecture outline in the WSG will help you identify the major concepts of the chapter.

MEASUREMENT AND EVALUATION CHALLENGE

Logan is taking a course in measurement and evaluation in human performance. He is also enrolled in a course titled Physiological Basis of Human Performance. His instructor has assigned a research article for the class to read. In the article the researcher hypothesized that drinking a carbohydrate solution in water would improve cycling endurance performance beyond what would result from drinking water only. The research study compared two groups of endurance cyclists. One of the groups drank water only. The other group drank water that contained a 4% solution of carbohydrate. The cyclists were then tested to see how long they could cycle at a given workload.

A t-test indicated that the carbohydrate-drinking group cycled significantly ($p < .05$) longer than the water-only group. Logan wants to understand what a t-test is and what *significance* means in this context. He wonders what $p < .05$ means. Does it mean the researchers proved that the carbohydrate drink was better than drinking only water if you want to increase your cycling endurance? You will find out how to interpret these and other results and will learn about other statistical methods in this chapter.

The descriptive statistical techniques that we have presented thus far are those that you will most commonly use for measurement problems; however, there are a number of other statistics that you may need to use in various measurement situations. The most common of these examines group differences. When these techniques are used to relate the characteristics of a small group (sample) to those of a larger group (population), they are referred to as **inferential statistics**. Much human performance research is conducted using inferential statistics. Inferential statistics is an extension of the correlational examples presented in chapter 4. In the example just described, the researcher is interested in determining if there is a relationship between the type of drink used and endurance performance: Does the drink produce a difference in endurance performance? Is endurance performance a function of what the athlete drinks? Draw a figure illustrating this relation in two ways, once where there is no effect and once where there is an effect. Put drink type on the horizontal axis and cycling performance on the vertical axis. Draw distributions of cycling performance for the two possible results: (1) there is no effect (difference) and (2) there is an effect (a difference). Can you see how this is similar to examining the relation between variables?

HYPOTHESIS TESTING

The scientific method uses inferential statistics for obtaining knowledge. The **scientific method** requires both the development of a scientific hypothesis and an inferential statistical test of that hypothesis versus another competing hypothesis. A **hypothesis** is a statement of a presumed relation between at least two variables in a population. A **population** is the entire group of people or observations in question (e.g., college seniors). A measure of interest in the population is called a **parameter**. Inevitably, because entire populations are so large and unwieldy (imagine surveying all U.S. college seniors on a certain matter), you study hypotheses about a population by using a subgroup of the population, called a **sample**. The measure of the variable of interest in the sample is called a **statistic**. Using various techniques, you can make an inference—but not an absolute statement—about the whole population from your work with a sample. Table 5.1 contains the symbols that are commonly used to distinguish sample statistics from population parameters.

Table 5.1 Statistical Symbols

Measure	Population parameter	Sample statistics
Mean	μ	M
Standard deviation	σ	s
Correlation	ρ	r

Consider the following examples. A teacher was interested in the minutes of physical activity conducted in a typical physical education class (parameter) of fifth graders in the school. There were 200 fifth graders (population). The teacher randomly selected 50 students (sample) and had them wear pedometers that indicated minutes of moderate-to-vigorous physical activity (MVPA). The MVPA minutes were analyzed, and the sample values (statistic, e.g., the mean) were considered to be representative of the population parameter. Surveys taken before presidential elections use samples to estimate the percentage of people preferring a particular candidate. Note, however, that there is *always* some error in these techniques.

Mastery Item 5.1

Create a research problem related to something of interest to you. Identify the following: (a) population, (b) sample, (c) parameter, and (d) statistic.

Hypotheses are the tools that allow research questions to be explored. A hypothesis may be one of several types:

- ***Research hypothesis.*** What the researcher actually believes will occur. For example, assume you believe that training method is related to oxygen uptake. Your research hypothesis is this: *There will be differences in oxygen uptake based on the type of aerobic training one uses.* Or perhaps you are a physical therapist who wants to determine if treatment modality A is more or less effective than treatment modality B. You can investigate these hypotheses with a t-test or analysis of variance (ANOVA).
- ***Null hypothesis (H_0).*** A statement that there is no relation (association, relationship, or difference) between variables ($\mu_1 = \mu_2$). In the earlier examples, your null hypothesis is that the mean oxygen uptake is not different for training groups that use different training methods. Or, the physical therapist has a null hypothesis that there will be no difference in the effectiveness of the two treatment modalities. The null hypothesis is the one that you will actually test (and hope to discredit) using the techniques of inferential statistics.
- ***Alternative hypothesis (H_1).*** A statement that there is a relation (association, relationship, or difference), typically the converse of H_0. Here, your alternative hypothesis is $\mu_1 \neq \mu_2$ where μ_1 is the population mean for group 1 and μ_2 is the population mean for group 2. Remember that you actually obtain data on samples only and then infer your results to the population. In this example, the research hypothesis is H_1.

Before you perform the appropriate statistical test, select a probability (*p*) level beyond which the results are considered to be statistically significant. This probability value is called the **significance**, or **alpha level** (α), and allows you to test the probability of the actual occurrence of your result. The alpha level is conventionally set at .05 or .01 (i.e., 5% or 1%). For example, if the researcher sets the alpha level at .05, he or she is saying that the probability of obtaining the statistic just by chance must be less than 5 times out of 100 before

Photo courtesy of Judy Park.

Students' physical activity levels can be measured with inexpensive pedometers. Data can then be used to make hypotheses about a larger group's level of activity.

he or she will decide the null hypothesis is not tenable. You may recall from chapter 3 that 5% is in the extreme tails (2.5% on each side) of the normal curve. In effect, you assume no relation between variables until you have evidence to the contrary. The statistical data may provide the contrary evidence.

Recall the normal curve and distribution presented in chapter 3. Pay particular attention to figure 3.6 and recall how about 2.5% of the distribution is outside about ±2 standard deviations. However, the researcher might reach an incorrect conclusion (i.e., say there is a relationship or difference when in fact there is not). The probability of making such an error is the alpha level. This error is referred to as a **type I error**. The alpha level is set at .05 or .01 to make the probability of a type I error extremely small. You can also make a second type of error, a **type II error**, by concluding that there is no relation between the variables in the population when in fact there truly is.

The SPSS computer program (or other statistical software) will calculate the actual alpha level for you. If the probability is less than the preset alpha level of .05 or .01, you conclude that there is a significant relation between the variables. Thus, H_0 is rejected and H_1 is accepted. Essentially, you assume that there is no relationship between the variables of interest (i.e., the null hypothesis is true). Then you gather data. If the data you obtain is unlikely (e.g., $p < .05$) to occur if the null hypothesis is true, then you reject the null hypothesis and conclude that the alternative hypothesis is the true state of the relationship in the population. Figure 5.1 illustrates the types of decisions and errors you might make. You can never know the true state of the null hypothesis in the population, so you always run the risk of making a type I or type II error. You cannot make both a type I and type II error in the same research study. Can you look at figure 5.1 and see why this is the case?

		True state in population	
		H_0 is true H_1 is false	H_0 is false H_1 is true
Your decision	Reject H_0, accept H_1	Type I error (alpha)	Correct decision
	Retain H_0, reject H_1	Correct decision	Type II error (beta)

Figure 5.1 Type I and type II errors.

Mastery Item 5.2

A conditioning coach wants to study the best way to develop strength in older adults. He randomly divides a sample into three groups based on the following methods: free-weight strength training, machine strength training, and elastic band strength training. Write the appropriate null and alternative hypotheses for this problem.

Selection of the appropriate statistical technique is based on your research question and the level of measurement of the variables. The number of groups and the level of measurement of the data determine the appropriate statistics to use. Some of the most common are these:

- ***chi-square test*** **(χ^2).** Used to examine associations in nominal data.
- ***t-test.*** Used to examine a difference in a continuous (interval or ratio) dependent variable between two and only two groups.
- ***ANOVA (analysis of variance).*** Used to examine differences in a continuous (interval or ratio) dependent variable among two or more groups.

INDEPENDENT AND DEPENDENT VARIABLES

The differences between independent and dependent variables are important. The dependent variable is the criterion variable; its existence is the reason you are conducting the research study. The independent variable exists solely to determine if it is related to (or influences) the dependent variable. Independent and dependent variables can be characterized in a number of ways; these are presented in table 5.2.

If the dependent variable is nominally scaled, the differences between groups (or cells) are measured by frequencies or proportions. If you are dealing with continuous (interval or ratio) data, then differences in mean values are often examined. For example, consider the strength training example described in Mastery Item 5.2. The variable you select to

Table 5.2 Variable Classification

Independent	Dependent
Presumed cause	Presumed effect
The antecedent	The consequence
Manipulated or measured by researcher	Outcome (measured)
Predicted from	Predicted to
Predictor	Criterion
X	Y

measure the training effect is strength. This is the dependent variable. The independent variable is method of training and has three levels: free weight, machines, and elastic bands. Can you see why ANOVA would be used in this situation?

 Go to the WSG to complete Student Activity 5.1 and view Video 5.1.

OVERVIEW OF HYPOTHESES TESTING AND INFERENTIAL STATISTICS

All inferential statistical tests follow the same reasoning and processes:

1. Develop a research hypothesis about the relation between variables (e.g., There is a relation between the type of exercise in which you engage [e.g., moderate or vigorous] and $\dot{V}O_2$max). Alternatively, this could be stated that there is a difference in $\dot{V}O_2$max that depends on whether you engage in moderate or vigorous physical activity.
2. State a null (H_0) hypothesis, reflecting no relation (or difference) (e.g., There is *no* relation [i.e., difference] between the type of exercise in which one engages and $\dot{V}O_2$max).
3. State an alternative hypothesis (H_1), which is the opposite of the null hypothesis. It is a direct reflection of the research hypothesis in #1. Note: Hypotheses come in pairs and must cover all possible outcomes!
4. Gather data and analyze them based on the research question and the types of variables.
5. Make a decision based on the probability of the null hypothesis being correct given the data you have collected.

Note that if the null hypothesis is true, then the mean $\dot{V}O_2$max for moderate and vigorous groups would be the same. That is, there is *no* relation between type of exercise and $\dot{V}O_2$max; further, if the null is true, the difference between the two means would be zero! Recall the normal curve where zero is in the center. However, if the null hypothesis is *not* true, then the difference between the moderate and vigorous $\dot{V}O_2$max means will be nonzero. Consider if the nonzero value was in an extreme tail of the normal distribution. This would suggest that this nonzero finding would be extremely rare *if the null hypothesis is true.* Thus, you conclude that the null hypothesis is *not* true and believe that the alternative hypothesis is true. By doing so, you reject the null hypothesis and conclude that the alternative hypothesis is a more accurate reflection of the data.

This same logic is used regardless of the statistical test that you conduct (i.e., chi-square, t-test, or ANOVA). The chi-square (χ^2), t-test (t), and ANOVA (F) use different distributions but they are akin and related to the normal distribution. Thus, you can think of the χ^2, *t*, and F as if they are *z*-scores. They are *not*, but they are closely related. When you have a large (positive or negative) *z*-score, this is a rare occurrence (way out in the tail of the distribution). Thus, in hypothesis testing, if you obtain a χ^2, *t*, or F that would be rare (e.g., <5 times in 100), you conclude that the null hypothesis is not true. Computer programs typically report the probability associated with the χ^2, *t*, or F. This is interpreted as the likelihood of obtaining a value this extreme *if the null hypothesis is true*. The researcher would reject the null and conclude that there is a significant relation (or difference) between the levels of the independent variable and the dependent variable if this probability is low (often < .05 or < .01).

This logic can be extended to the most sophisticated statistical inference. In fact, many research studies in human performance use this logic. You will typically see probabilities reported in research reports. Effectively, the researcher sets up a falsifiable hypothesis (the null) and then collects and analyzes data and makes a decision about the truth of the null (or its alternative) based on the sample data.

SELECTED STATISTICAL TESTS

The following are statistical tests that examine associations or differences between groups. The techniques selected represent common basic inferential tests.

Chi-square (χ^2)

Purpose: To determine if there is an association between levels (cells) of two nominally scaled variables.

Example: An aerobics instructor is teaching two classes, aerobic dance and circuit weight training. The instructor wants to know if the proportion of males and females is the same in each class. The null hypothesis is that there is no association (relation) between gender and type of class in which students are enrolled. The alternative hypothesis is that there is an association. One can only reject the null hypothesis and believe that the alternative hypothesis is the true state of circumstances in the population when the probability of the null hypothesis being true is small (i.e., $p < .05$) based on the sample data. The data are found in table 5.3. Use them in conjunction with the following SPSS commands to calculate the chi-square test of association and verify the results with those presented in figure 5.2. (An Excel template is in the WSG in chapters 5 and 7.)

1. Start SPSS.
2. Open table 5.3.
3. Click on the Analyze menu.
4. Scroll down to Descriptive Statistics and across to Crosstabs and click.
5. Put "Class" in the rows and "Gender" in the columns by clicking the arrow keys.
6. Click Statistics.
7. Check the Chi-Square box.
8. Click Continue.
9. Click OK.

The resulting SPSS printout is presented as figure 5.2. Although a number of statistics are calculated for us, the test statistic in which we are interested is the Pearson chi-square. The observed chi-square value is 22.5. Think of this chi-square value as if it is a *z*-score (it

Table 5.3 Data Entry for χ^2 Example

Id	Gender	Class	Id	Gender	Class
1	1	1	16	2	2
2	1	1	17	2	2
3	1	1	18	2	2
4	1	1	19	2	2
5	1	1	20	2	2
6	1	1	21	2	2
7	1	1	22	2	2
8	1	1	23	2	2
9	1	1	24	2	2
10	1	1	25	2	2
11	1	2	26	2	2
12	1	2	27	2	2
13	2	2	28	2	2
14	2	2	29	2	2
15	2	2	30	2	2

Note: Gender code: 1 = male, 2 = female. Class code: 1 = circuit, 2 = dance.

Case processing summary

	Cases					
	Valid		Missing		Total	
	N	Percent	*N*	Percent	*N*	Percent
Class enrollment * Gender of the subject	30	100.0%	0	.0%	30	100.0%

Class enrollment * Gender of the subject crosstabulation

Count

		Gender of the subject		
		Male	Female	Total
Class enrollment	Circuit weight training	10		10
	Aerobic dance	2	18	20
Total		12	18	30

Chi-square tests

	Value	*df*	Asymp. sig. (2-sided)	Exact sig. (2-sided)	Exact sig. (1-sided)
Pearson chi-square	22.500[b]	1	.000		
Continuity correction[a]	18.906	1	.000		
Likelihood ratio	27.377	1	.000		
Fisher's exact test				.000	.000
Linear-by-linear association	21.750	1	.000		
N of valid cases	30				

a. Computed only for a 2 x 2 table.
b. 1 cells (25.0%) have expected count less than 5. The minimum expected count is 4.00.

Figure 5.2 SPSS crosstabs output.

is *not,* but it is related). Where is a *z*-score of 22.5? It is way out on the extreme end of the normal distribution. This is an unlikely occurrence, particularly if the null hypothesis is true and there is no relation between gender and class. The probability associated with this test is reported to be .000 (labeled as Asymp. Sig [2-sided]) on the SPSS crosstabs output. This is the probability of the cell distribution occurring as they did *if* the null hypothesis is true. In actuality, however, you can never have a probability of zero. It is simply the case that the SPSS computer program calculates the probability (i.e., significance) to three decimal places. In any case, you should interpret this as .001. Thus, it is exceedingly rare to find the cell frequencies being distributed as 5.3they are when, in fact, there is no association between type of class and gender (the H_0). Because of this exceedingly small probability, the aerobics instructor can conclude that there is an association between gender and type of class. The null hypothesis (H_0) of no association is rejected, and the instructor concludes that there is an association between gender and type of class in which a person enrolls. Figure 5.2 shows that 10 of the 12 males registered for circuit weight training, whereas all 18 of the 18 females registered for aerobic dance. This association might help the instructor in planning the types of activities for the classes.

t-test for Two Independent Groups

Purpose: To examine the difference in one continuous dependent variable between two (and only two) independent groups. Independent groups are groups that are not in any way related (e.g., boys and girls; experimental and control groups).
Example: A high school volleyball coach is selecting players for the varsity team and is using serving accuracy as a selection factor. After the team is selected, the coach wants to quantify the differences in serving accuracy between varsity and subvarsity players. The serving scores are presented in table 5.4.

Table 5.4 Serving Scores

Varsity	20, 18, 17, 19, 20, 16, 18, 19
Subvarsity	16, 15, 17, 14, 15, 13, 14, 12

The research hypothesis for this study is *There will be a difference in serving accuracy between varsity (v) and subvarsity (sv) players.* The null hypothesis to be tested is that the mean volleyball serving score for the varsity players will be equal to the mean for the subvarsity players (i.e., there is no difference between the two):

$$H_0: \mu_v = \mu_{sv} \quad (5.1)$$

The alternative hypothesis is that the mean for the varsity players will not be equal to the mean of the subvarsity players:

$$H_1: \mu_v \neq \mu_{sv} \quad (5.2)$$

For the coach's purpose, the alpha level is set at .05. The SPSS procedure for the t-test may be used to analyze the data. Use the data in table 5.4 to calculate an independent group's t-test and confirm your results with those presented in figure 5.3.

1. Start SPSS.
2. Open table 5.4.

3. Click on the Analyze menu.
4. Scroll down to Compare Means and across to Independent-Samples T-Test and click.
5. Put "score" in the Test Variable(s) box by clicking the arrow key.
6. Put "group" in the Grouping Variable box by clicking the arrow key.
7. Click the Define Groups button.
8. Put "1" in the Group 1 box and "2" in the Group 2 box.
9. Click Continue.
10. Click OK.

Group statistics

	Team level	*N*	Mean	Std. deviation	Std. error mean
Serving score	Varsity	8	18.38	1.408	.498
	Subvarsity	8	14.50	1.604	.567

Independent samples test

		Levene's Test for Equality of Variances		*t*-test for Equality of Means						
									95% confidence interval of the difference	
		F	Sig.	*t*	*df*	Sig. (2–tailed)	Mean difference	Std. error difference	Lower	Upper
Serving score	Equal variances assumed	.095	.763	5.136	14	.000	3.88	.754	2.257	5.493
	Equal variances not assumed			5.136	13.769	.000	3.88	.754	2.254	5.496

Figure 5.3 Sample t-test output: group statistics and independent samples test.

SPSS output is displayed in figure 5.3. Inspection of the means indicates that the varsity players (group 1; mean = 18.38) served significantly (Sig. [2-tailed] = .000) better than the subvarsity players (group 2; mean = 14.50).

A number of statistics in the output from the t-test are beyond the scope of this text. For our purposes, you can ignore the results under Levene's Test for Equality of Variances. Focus your attention on the areas below t-test for Equality of Means. Notice the *t* presented with a value of 5.136 (actually presented twice). Think of this *t* as if it is a *z*-score that you learned about in chapter 3. Again, it is not a *z*-score but it is much like a *z*-score. If the *z*-score was large (e.g., greater than 3 in absolute value), you know that the probability of finding a value this large is small. The same can be said of the calculated t-value. Thus, you can see that the *t* is relatively far into the tail of the distribution. This *t* is generally a rare occurrence. Of most importance to you is the box titled Sig. (2-tailed). This is the probability that the null hypothesis is true given the data from the sample. Because the probability is less than .05, the coach rejects the null hypothesis and retains the alternative hypothesis. This is an example of using a t-test with independent groups. Suppose the coach wanted to examine the serving accuracy of the varsity during preseason as opposed to the end of the season. A paired t-test would be used because the same group is being measured at two points in time. This dependent t-test is illustrated next.

Excel Commands for the independent t-test are found in the appendix.

Dependent t-test for Paired Groups

Purpose: To compare two related (paired) groups on one dependent variable. Groups can be paired by matching them on some external characteristic (e.g., siblings) or by measuring the same group twice (i.e., pre- and postperformance).

Example: Let's extend the previous independent t-test example. The coach is interested in serving accuracy for the varsity squad at preseason and the end of the season. The coach tests the players at the beginning and the end of the season. The research hypothesis is that there will be a difference in preseason and postseason serving accuracy. The null hypothesis is that there will be no difference in the players' serving accuracy over the course of the season. To test the null hypothesis, the SPSS t-test procedure is used again. However, the data are entered differently from the previous example because each person is tested twice (compare tables 5.4 and 5.5). This allows SPSS to properly pair the data so that the correct result is calculated.

Table 5.5 Data Format for Paired t-test

Preseason	Postseason
18	20
20	24
17	20
16	19
15	20
18	22
19	21
17	21

Use the data in table 5.5 to calculate a paired (dependent) t-test, and confirm your results with those presented in figure 5.4.

1. Start SPSS.
2. Open table 5.5.
3. Click on the Analyze menu.
4. Scroll down to Compare Means and across to Paired-Samples T-Test and click.
5. Put "preseason" and "postseason" in the Paired Variables box by clicking the arrow key.
6. Click OK.

The mean difference between the posttest and the pretest was found to be 3.38. The observed t-value was found to be 9.000, with an associated probability (alpha level) that approaches zero (Sig. [2-tailed]). Again, think of this t-value as if it was a *z*-score. A *z*-score of 9 is way out in the tail, an unlikely occurrence. Thus, the null hypothesis is rejected and the alternative hypothesis is adopted as the true state of differences. The coach can conclude that the differences in serving accuracy from the beginning of the season to the end can be attributed to something other than chance.

Excel Commands for the paired t-test are found in the appendix.

Paired samples statistics

		Mean	*N*	Std. deviation	Std. error mean
Pair 1	Postseason performance	20.88	8	1.553	.549
	Preseason performance	17.50	8	1.604	.567

Paired samples correlations

		N	Correlation	Sig.
Pair 1	Postseason performance and preseason performance	8	.775	.024

Paired samples test

		Paired differences							
					95% confidence interval of the difference				
		Mean	Std. deviation	Std. error mean	Lower	Upper	*t*	*df*	Sig. (2–tailed)
Pair 1	Postseason performance and preseason performance	3.38	1.061	.375	2.49	4.26	9.000	7	.000

Figure 5.4 Sample t-test output: paired samples statistics, paired samples correlations, and paired samples test.

Note that the differences in serving accuracy could have been caused by a number of factors. Because a time lag occurred between the first and second measurements, the differences could have been attributable to growth, maturation, or some other factor not under the control of the researcher. In an actual experiment, these factors would have to be controlled.

Go to the WSG to complete Student Activity 5.2.

One-Way ANOVA

Purpose: To examine group differences between one continuous (interval-scaled or ratio-scaled) dependent variable and one nominal-scaled independent variable. Unlike the t-test, ANOVA can handle independent variables with more than two levels (groups) of data.

Example: The data for this example are based on the strength-training example previously presented with three types of training methodologies, free weights, machines, and elastic bands. The study participants engage in one (and only one) of the three methods of strength training for 8 weeks. In this example, the independent variable is strength training method (coded 1, 2, and 3), and the dependent variable is one-repetition maximum (1RM) bench press (we realize that the researcher might also assess other strength measures). The problem to be examined is whether there are differences in bench press strength across the three training groups.

The null hypothesis is that means for the 1RM bench press for the three training method groups will be equivalent:

$$H_0: \mu_1 = \mu_2 = \mu_3 \tag{5.3}$$

The alternative hypothesis is that the means will not be equivalent (for at least one of the means):

$$H_1: \mu_1 \neq \mu_2 \neq \mu_3 \tag{5.4}$$

The alpha level was set at .01 rather than .05, indicating that the researcher wanted to reduce the probability of making a type I error and increase confidence that if differences were found among the means they were not attributable to chance. The data for this problem are presented in table 5.6.

Table 5.6 Input Format for One-Way ANOVA

Id	Group	1RM bench press
1	1	193
2	1	190
3	1	195
4	1	175
5	1	188
6	2	148
7	2	170
8	2	172
9	2	168
10	2	165
11	3	170
12	3	157
13	3	140
14	3	148
15	3	150

Use the following SPSS procedures to conduct ANOVA on these data:

1. Start SPSS.
2. Input the data from table 5.5 in this chapter or open table 5.5 from the online study guide.
3. Click on the Analyze menu.
4. Scroll down to Compare Means and across to One-Way ANOVA and click.
5. Put "1RM Bench Press" in the Dependent List by clicking the arrow key.
6. Put "Group" in the Factor box.
7. Click Options.
8. Click Descriptives.
9. Click Continue.
10. Click OK.

 Go to the WSG to complete Student Activity 5.3.

The key information presented in figure 5.5 is the Sig. (significance or probability). The other information is used to obtain the significance. For ANOVA, the significance test is an F or F-ratio (17.145). Again, think of this F-ratio like the *z*-score you learned about in chapter 3. An F-ratio is not a *z*-score but it is somewhat like a *z*-score. High values are rare, and the probability of obtaining a high value is reduced when the groups do not differ much from each other. Because the probability level for the observed event is less than .01 (the computer gives it as .000), the null hypothesis is rejected and the alternative hypothesis is adopted. What type of statistical error is possible with this conclusion? Type I or Type II?

Descriptives

1-RM bench press

	N	Mean	Std. deviation	Std. error mean	95% confidence interval for mean Lower bound	Upper bound	Minimum	Maximum
1.00	5	188.2000	7.8549	3.5128	178.4468	197.9532	175.00	195.00
2.00	5	164.6000	9.6333	4.3081	152.6387	176.5613	148.00	172.00
3.00	5	153.0000	11.2694	5.0398	139.0072	166.9928	140.00	170.00
Total	15	168.6000	17.6141	4.5479	158.8456	178.3544	140.00	195.00

ANOVA

1-RM bench press

	Sum of squares	*df*	Mean square	*F*	Sig.
Between groups	3217.600	2	1608.800	17.145	.000
Within groups	1126.000	12	93.833		
Total	4343.600	14			

Figure 5.5 One-way ANOVA results.

Examination of the group means in figure 5.5 indicates that the participants in the free-weight group (group 1) had the highest 1RM bench press (*M* = 188.2 lb [85.5 kg]); participants from the elastic band group (group 3) had the lowest 1RM bench press (*M* = 153 lb [69.5 kg]); and participants in the machine-training group (group 2) were between (*M* = 164.6 lb [74.8 kg]) the other two training methods. Thus, the researcher concludes that there is a differential effect of each of the three training methods on the 1RM bench press. Note that the F value indicates there are differences among the three groups, but it does not specify which groups are different from the others! Statistical tests called multiple-comparison tests exist to compare the specific groups to one another; however, they are beyond the scope of this text.

Excel Commands for one-way ANOVA are found in the appendix.

Two-Way ANOVA

Purpose: To examine group differences between one interval- or ratio-scaled dependent variable and two nominal independent variables.

Example: The data were taken from a hypothetical test on fielding coverages administered to Minor League baseball players. The dependent variable was their score on the coverages test. The two independent variables were position played (pitchers or position players) and instructional method (drills or scrimmages). There is a research and null hypothesis for each independent variable (called main effects) and a research or null hypothesis for the interaction between the two independent variables. The first null hypothesis is that no difference will exist between pitchers and position players. The second null hypothesis is that no difference will exist between instructional methods. The interaction hypothesis tests whether or not the instructional effect will be constant between the positions played.

The data for this example are presented in Table 5.7. To analyze this problem, use the SPSS commands that follow. The output from SPSS is summarized in Figures 5.6 and 5.7.

Table 5.7 Data for Two-Way ANOVA

Method	Position	Score	Method	Position	Score
1	1	29	2	1	29
1	1	30	2	1	34
1	1	24	2	1	34
1	1	23	2	1	38
1	1	28	2	1	28
1	1	29	2	1	29
1	2	34	2	2	28
1	2	34	2	2	34
1	2	31	2	2	31
1	2	30	2	2	26
1	2	32	2	2	27
1	2	28	2	2	32

Note: For Method, group 1 is Scrimmage, group 2 is Drills; for Position, group 1 is Pitchers, group 2 is Position Players.

SPSS Commands for Two-Way ANOVA

1. Start SPSS.
2. Click on the Analyze menu.
3. Select General Linear Model-Univariate.
4. Select Score as the dependent variable.
5. Next, select Method and Position as fixed factors.
6. Click Options.
7. Check the Descriptive statistics box.
8. Click Continue.
9. Click OK.

Figure 5.6 presents the labels associated with the analysis along with the descriptives. Figure 5.7 presents the actual ANOVA analysis. The important information in the Tests of Between-Subjects Effects are the F values for method, position, and method * position (the interaction of the two independent variables). Note that neither test for method or position is significant; however, the interaction between method and position is! Inspection of the cell means in Figure 5.6 shows that the pitchers (mean = 32) learned the coverages better through drills, whereas the position players (mean = 31.5) performed better through scrimmaging. A more detailed explanation of interaction is beyond the scope of this text. This example does point out that you may have a significant interaction even if your main effects were not significant.

Excel Commands for two-way ANOVA are found in the appendix.

Descriptive statistics

Dependent variable: Defensive positioning test

Method	Position	Mean	Std. deviation	N
Scrimmage	Pitchers	27.17	2.927	6
	Position players	31.50	2.345	6
	Total	29.33	3.393	12
Drills	Pitchers	32.00	3.950	6
	Position players	29.67	3.141	6
	Total	30.83	3.614	12
Total	Pitchers	29.58	4.166	12
	Position players	30.58	2.811	12
	Total	30.08	3.513	24

Between-subject factors

		Value label	*N*
Method	1	Scrimmage	12
	2	Drills	12
Position	1	Pitchers	12
	2	Position players	12

Figure 5.6 SPSS two-way variable labels and descriptives.

Tests of between-subjects effects

Dependent variable: Defensive positioning test

Source	Type III sum of squares	*df*	Mean square	F	Sig.
Corrected model	86.167[a]	3	28.722	2.906	.060
Intercept	21720.167	1	21720.167	2197.656	.000
Method	13.500	1	13.500	1.366	.256
Position	6.000	1	6.000	.607	.445
Method * position	66.667	1	66.667	6.745	.017
Error	197.667	20	9.883		
Total	22004.000	24			
Corrected total	283.833	23			

a. R^2 = .304 (adjusted R^2 = .199)

Figure 5.7 SPSS two-way ANOVA summary output.

Dataset Application

Use the chapter 5 large dataset in the WSG to examine differences in pedometer steps for the people listed there. Sample data are included for steps from school-age boys and girls from North America, Europe, and Asia. Answer the following questions based on the data. Calculate the means and standard deviations across both genders and then do so for boys and girls separately (chapter 3). Do boys and girls differ (use an independent t-test)? Do results from the three regions of the world differ (use ANOVA)? Actually, the better way to analyze these data would be with a two-way ANOVA (gender by region). Use SPSS to conduct a two-way ANOVA (gender by region). Lastly, use gender and the Does8500steps variable to see if there is an association between gender and physical activity based on whether one averaged at least 8500 steps per day (chi-square).

MEASUREMENT AND EVALUATION CHALLENGE

Based on the research article, Logan has learned that a t-test was conducted because there was a single independent variable (drink type); the control group drank water only and the experimental group drank the carbohydrate solution. He also learned that the researcher hypothesized that cycling duration (the dependent variable) was a function of which drink was consumed. Logan now knows that $p < .05$ means that a null hypothesis was rejected and an alternative hypothesis was accepted. He realizes that the researcher may have made a type I error, but the probability of such an error is less than 5 times in 100.

Thus, based on this sample, the researchers concluded that carbohydrate drinks will improve cycling endurance for most cyclists similar to those from whom this sample of cyclists was drawn; however, this conclusion is not a certainty because a type I error could have been made. Because the null hypothesis was rejected, it is impossible for the researcher to have made a type II error. Inferential tests provide evidence only to support or not support the hypotheses; therefore, Logan has learned that the researcher's hypothesis about the influence of carbohydrate drinks can never be fully proven with hypothesis testing.

SUMMARY

This chapter provided a brief overview of the tests used in inferential statistics; however, many assumptions regarding these techniques were not discussed. Statistical tests of significance often obscure practical differences. There is no substitute for blending statistical findings with intuitive logic. In-depth treatment of statistical methods can be found in Glass and Hopkins (1996). Thomas, Nelson, and Silverman (2015) provide excellent examples of research in human performance.

At this point you should be able to accomplish the following tasks:

- Understand and interpret the scientific method.
- Write and interpret research, null, and alternative hypotheses.
- Use SPSS to enter data and to generate and interpret
    - chi-square tests of association,
    - independent and dependent t-tests,
    - one-way ANOVA, and
    - two-way ANOVA.

Go to the WSG for homework assignments and quizzes that will help you master this chapter's content.

Part III

Reliability and Validity Theory

In this part, what you learned about basic statistics and computer applications in part II will be extended and applied to issues related to valid decision making. We all make decisions in life, and each of us attempts to make the best decisions possible. A decision you make in the field of human performance might be about a person's aerobic capacity, muscular strength, or amount of daily physical activity. You may also need to make valid decisions about cognitive knowledge and report grades or levels of achievement to students, clients, or program participants. Alternatively, you may need to evaluate a program that you direct. Good decisions are based on sound data, which in turn reflect the characteristics of reliability, validity, and objectivity. You will use the SPSS skills that you gained in parts I and II to accomplish specific tasks related to these characteristics. Each chapter provides you the opportunity to use SPSS procedures to illustrate and analyze measurement problems.

Chapter 6 provides the important steps you will need to follow to judge the quality of norm-referenced data. To make accurate decisions about individuals or groups, you must use data that are sufficiently reliable, valid, and objective. For example, when reporting someone's aerobic capacity, you will need to be certain that the value is truthful. Invalid data can result in inappropriate decisions. Chapters 6 and 7 help you analyze data so that you can report them in such a way as to make valid interpretations and decisions. No measurement technique is perfectly reliable or valid, but you need to know how to interpret the amount of reliability and validity reflected in your measurement protocol so that you can make appropriate decisions. Chapter 7 addresses these issues from a criterion-referenced perspective. Actually, chapters 6 and 7 have a great deal in common. A key difference is in the way measurements are taken and reported. In chapter 6, the measures are continuous; in chapter 7, data are nominal (i.e., categorical) in nature. Otherwise the concepts of reliability (consistency) and validity (truthfulness) are quite consistent between the two chapters.

CHAPTER

6

Norm-Referenced Reliability and Validity

OUTLINE

OBJECTIVES

After studying this chapter, you will be able to

- discuss the concepts of reliability and validity;
- differentiate among the types of reliability and how to calculate them;
- identify the types of validity evidence that can be used to provide information about a measure's truthfulness and calculate the appropriate statistics;
- describe the relationship between reliability and validity and comment on why these concepts are important to measurement;
- evaluate the evidence for reliability and validity typically presented in the measurement of human performance; and
- use SPSS and Excel to calculate reliability and validity statistics.

 The lecture outline in the WSG will help you identify the major concepts of the chapter.

MEASUREMENT AND EVALUATION CHALLENGE

Kelly, a YMCA fitness director, wants to assess the cardiorespiratory fitness of young adult members. Kelly is interested in determining the $\dot{V}O_2max$ of her members. However, she has heard that the best measure of cardiorespiratory fitness is to have a person run on a treadmill until he or she is exhausted. This test requires collecting the air that participants expire while on the treadmill and, therefore, requires considerable and expensive equipment. Because of these difficulties, Kelly is interested in using a surrogate (field test) measure such as the YMCA 3-Minute Step Test (see chapter 9) or a distance run (e.g., 1-mile or 1.6-kilometer) in place of treadmill performance. She has even heard that nonexercise models can be used to estimate one's $\dot{V}O_2max$. Kelly wonders whether she should use that method to save time, money, and equipment, and reduce the risk to participants. However, Kelly is concerned that a field test might lack the accuracy that a treadmill test provides. Kelly's concern is a very real one. How can she determine if the measure that is being obtained with the field test is reliable (i.e., consistent) and valid (i.e., truthful)?

Regardless of the area of human performance in which you work, you will need to make decisions based on the data you collect. Often, these decisions require you to make comparisons among people or report test results to a specific person. For example, Kelly may need to report the results of her testing to the programming director or the Y's board of directors to keep a particular fitness program funded. Your decisions and reports need to be accurate. The accuracy of your decisions relates to the norm-referenced characteristics of your variables. As you learned in chapter 1, the most important measurement characteristics are reliability, objectivity, and validity. (Recall from chapter 1 that a norm-referenced standard is a level of achievement relative to a clearly defined subgroup.)

Reliability and validity are the most important concepts presented in this textbook. The many computational, theoretical, and practical examples presented throughout the textbook can be traced to these concepts. *Reliability relates to the consistency or repeatability of an observation; it is the degree to which repeated measurements of the same trait are reproducible under the same conditions.* Reliability can also be described as accuracy, consistency, dependability, stability, and precision. A measure is said to be reliable if one obtains the same (or nearly the same) score each time the instrument is administered to the same person under the same conditions. As you can see from this terminology, test reliability will be extremely important as Kelly determines which field test to administer.

Validity is the degree of truthfulness of a test score. That is, does the test score, once found to be reliable, accurately measure what it is intended to measure? *Validity is dependent on two characteristics: reliability and relevance.* **Relevance** is the degree to which a test pertains to its objectives. Thus, for a measure to be valid, it must measure the particular trait, characteristic, or ability consistently, and it must be relevant. That is, the instrument or test must be related to the characteristic reportedly being measured.

Thus, you can see that reliability and validity are important concerns for Kelly. She must be certain that the field test produces results that are consistent from trial to trial yet also accurately estimates what the $\dot{V}O_2max$ would be if it were measured on the treadmill. Can you see that a measure might be reliable yet not valid, but if it is valid, it *must* be reliable? That is, a measure may be reliable but unrelated to what you want to assess. However, if a measure is valid, it must certainly be reliable. How could something be valid if it does not measure the characteristic reliably?

Mastery Item 6.1

What variables might Kelly obtain if she considered a nonexercise model to estimate $\dot{V}O_2max$? Be sure to consider the reliability and validity of the variables.

A test may be valid in one set of circumstances but not in another. There are many tests that have sufficient reliability but poor validity. For example, assessment of total body weight is typically a reliable measure. It changes little from day to day, and two evaluators would likely report the same or nearly the same value when measuring it. However, total body weight is not a valid measure of total body fatness because total body weight is made up of fat, bone, and lean tissue. Thus, one's weight depends on the relative proportions of these body components.

Objectivity is a special kind of reliability. *Objectivity is interrater reliability.* You have probably taken objective tests (such as those with multiple-choice items) and subjective tests (such as those with essay items). These tests are classified as such because of the type of scoring system used when grading them. Multiple-choice, true–false, and matching test items are said to be objective because they have a high amount of interrater reliability. That is, regardless of who the grader is, the scores on these types of items are consistent from one grader to another because there is a well-defined scoring system for the correct (or best) answer. However, a test can be objective in nature yet not accurate or reliable. If the questions are poorly written, a multiple-choice examination may be an unreliable and invalid measure of student knowledge. The scoring of essay tests tends to be subjective—different readers will give different scores to an answer—but there are ways to increase the objectivity of scoring essay test items (see chapter 8).

Go to the WSG to complete Student Activity 6.1 and view Video 6.1.

RELIABILITY

Many of the basic statistical concepts presented in chapters 3, 4, and 5 help us determine whether a test is reliable and valid. Teachers and researchers generally need specific evidence about a measure's reliability and validity—and not general statements suggesting that it is reliable and valid. Numerous statistics are used to provide evidence of reliability and validity. *The variance (presented in chapter 3) and the Pearson product-moment (PPM) correlation coefficient (presented in chapter 4) are used to provide evidence of a test's reliability and validity and thus need to be understood fully.* Before getting into the number crunching associated with reliability and validity, however, we need to consider these concepts from theoretical perspectives so you are clear about exactly what these constructs are. With a deeper understanding, you will be better able to determine which statistical procedure you should use and how to interpret the results.

Observed, Error, and True Scores

Consider the scores obtained at a recent blood pressure screening (table 6.1). Each of the 10 participants has an observed recorded blood pressure; however, it is possible that errors of measurement may have entered into the recording system so that the observed score is not the person's true blood pressure value. For example, the observed score may be in error because of the amount of experience the tester has, how the actual measurement was taken, when it was taken, where it was taken, the type of instrument used, the time of day, what events took place before testing, and so forth.

Table 6.1 Systolic Blood Pressure Recordings for 10 Participants

Participant	Observed blood pressure	= True blood pressure	+ Error
1	103	105	−2
2	117	115	+2
3	116	120	−4
4	123	125	−2
5	127	125	+2
6	125	125	0
7	135	125	+10
8	126	130	−4
9	133	135	−2
10	145	145	0
Sum (Σ)	1250	1250	0
Mean *(M)*	125.0	125.0	0
Standard deviation *(s)*	11.6	10.8	4.1
Variance (s^2)	133.6 =	116.7 +	16.9

Note: Units are mmHg.

Data based on an example from Sax 1980.

Although it is unlikely that we can ever know exactly (without any error) what a person's true blood pressure is, imagine that we can develop a method by which to measure it more accurately than is typically done in a laboratory or clinical setting. For example, we might place a pressure-sensitive apparatus directly in an artery to determine the pressure exerted during systole. (Obviously, we would have to ignore the fact that even thinking about such a procedure would undoubtedly alter a person's blood pressure reading.) Assume that we have done this for the participants whose scores are reported in table 6.1. You will note that only 2 of the 10 have observed blood pressures that are equal to their true blood pressures. The other blood pressure readings have various amounts of error associated with them. Some of the errors result in overestimating the true blood pressure, whereas others result in underestimating the true blood pressure.

A few key points can be seen in table 6.1:

- Each person's observed score is the sum of the true score and the error score. Your true score theoretically exists but is impossible to measure and can be thought of as what you actually know or how well you actually perform; it is without error. You can think of it as the average of an infinite number of administrations of the test in which you don't get any better because of practice nor do you get any worse because of fatigue. In a sense, your true score never changes for a specific point in time, and it is perfectly reliable. Your error score results from anything that causes your observed score to be different from your true score; it is a value that theoretically exists but is impossible to measure. Sources of error include individual variability, instrument inaccuracy, cheating, testing conditions, and so forth.
- There is variation in the observed, true, and error scores (the standard deviations and variances are calculated for you).
- Error can be positive (increase the observed score) or negative (decrease the observed score).
- Error scores contribute relatively little to the observed variation.
- The error score mean is zero.

- The observed score variance (133.6) is equal to the sum of the true score variance (116.7) plus the error score variance (16.9).

Using observed (total), true, and error score variances, the reliability ($r_{xx'}$) is defined as that proportion of observed score variance that is true score variance (i.e., true score variance divided by observed [total] score variance):

$$r_{xx'} = \frac{s_t^2}{s_o^2} = \frac{(s_o^2 - s_e^2)}{s_o^2} \tag{6.1}$$

where s^2_t is the true score variance, s^2_o is the observed (total) score variance, and s_e^2 is the error score variance. In table 6.1, the reliability is 116.7/133.6 = .87.

In theory, the true score is perfectly reliable, with $r_{xx'}$ = 1.00. (Certainly, the true score changes if there is a change in the phenomenon being measured, but at any time, the true score is viewed as perfectly reliable and thus contains no error.) Thus, a test is said to be reliable to the extent that the observed score variation is made up of true score variation.

Importantly, you can see that the observed score variance is equal to the sum of the true and error score variances; knowing any two of these variances results in your ability to calculate (or estimate) the third. Using equation 6.1, we see that the limits on reliability are 0 and 1.00. If the observed score is made up of no true score variation, the reliability is 0.00, and if the observed score variation (s^2_o) is made up of only true score variation (s^2_t), the reliability is 1.00. Neither of these two cases generally arises; however, for a test to be valid, it must be reliable, so a test's reliability must be reported. *Depending on the nature of the decisions made from the test results, you generally want reliability to be .80 or higher. Although .80 is a goal, other results might be acceptable with reliabilities less than this, and others would require reliabilities higher than .80. Consider a police officer's radar gun. Is it reliable? More importantly, is it a valid reading of your speed?*

Return now to our original measurement and evaluation challenge. Kelly is interested in learning the reliability of the scores from the field test she will use because that measurement will tell her whether the results she obtains are consistent from one testing period to another. If the measure is to be of any value to Kelly, the results should vary little from one testing session to another. Additionally, the observed differences between people assessed for $\dot{V}O_2$max obtained from the field test should reflect true differences in $\dot{V}O_2$max and not be simply a function of measurement errors.

The following practical implications derive from this presentation:

- There should be observed score variance. (If there is none, the reliability is undefined—because of division by zero.)
- Error score variance should be relatively small in relation to the total variance.
- Generally, longer tests are more reliable than shorter tests. This is true because as tests increase in length, there is an increase in observed score variance that is more likely a function of an increase in true score variance than in error score variance.

 Go to the WSG to complete Student Activity 6.2.

You may be saying to yourself, *This is all fine, but how does one ever know a person's true score?* This is absolutely correct: One can never know the person's true score. However, the observed score is readily available, and there are ways of estimating the error score variation for a set of scores. Therefore, as noted on the right side of equation 6.1, we can estimate the reliability using the observed score variance and error score variance.

 Go to the WSG to complete Student Activity 6.3.

Calculating the Reliability Coefficient

Let's turn to the actual calculation of a reliability coefficient. Reliability coefficients are classified into one of two broad types: interclass reliability coefficients (based on the PPM correlation coefficient, presented in chapter 4) and intraclass reliability coefficients (based on analysis of variance, or ANOVA models, presented in chapter 5).

Interclass Reliability

Let us first look at three **interclass reliability** methods: test–retest reliability, equivalence reliability, and split-halves reliability.

Test–Retest Reliability Consider the simplest way of determining if a measure is reliable, or consistent. We could simply administer the test participants on two occasions (e.g., on the same day) and then correlate the two sets of observations using the PPM correlation coefficient to see if the correlation is high. This is exactly what is done with the test–retest reliability coefficient. Look at the two trials of sit-up performance presented in table 6.2. The PPM correlation coefficient is calculated to be .927, a correlation between scores from the two test administrations that is certainly high enough to call the measure reliable. The coefficient suggests that 92.7% of the observed score variance is true score variance. If the time between testing occasions is longer (e.g., days or weeks), the test–retest reliability coefficient may be called **stability reliability**. That is, the measure is consistent or stable across time. Importantly, the interclass reliability method should only be used when you are quite certain that there is no change in the mean from one trial to the second trial. You can test for mean differences in two trials with the dependent t-test presented in chapter 5. If there are multiple trials, an extension of ANOVA presented in chapter 5, called repeated-measures ANOVA, is used to test for mean differences.

Table 6.2 Sit-Up Performance for 10 Participants

Participant	Trial 1	Trial 2
1	45	49
2	38	36
3	54	50
4	38	38
5	47	49
6	39	38
7	39	43
8	42	43
9	29	30
10	42	42
Sum (Σ)	413	418
Mean (*M*)	41.3	41.8
Standard deviation (*s*)	6.6	6.5
Variance (s^2)	43.6	41.7
	$r_{xx'} = .927$	

Mastery Item 6.2

Use SPSS to confirm the reliability of .927 reported in table 6.2. (Hint: Use SPSS to calculate a correlation coefficient as illustrated in chapter 4.) For a graphic representation of the data, create a scatterplot of the two trials. Can you see the high relationship demonstrated and why this can be interpreted as a reliability coefficient?

Equivalence Reliability A second way to determine interclass reliability is by using an equivalence reliability coefficient. Consider a teacher who is concerned about students cheating during a written test. The teacher develops two parallel or equivalent forms of an exam and distributes the exams in the class so that no two adjacent students receive the same examination. However, how should this instructor now determine grades? Should there be two grading procedures for the same class? Does student performance depend on which exam was completed? This teacher should first determine the equivalence of the two examinations. To do so, a test group would be asked to take each of the examinations (both forms) under nearly identical conditions. Half of the participants should take form A first and half should take form B first, so that no order effect causes scores on the test to be affected. An assumption must be made that the tests are parallel and that taking the first test neither hinders nor helps a student on the second test. The results from the two administrations are then correlated to determine if there is reliability, or consistency, between the two test forms. Note again that this is simply a PPM calculation in which the two variables correlated are the test scores from the respective forms. This is an equivalence reliability coefficient. As with test–retest reliability, you could use SPSS to display a scatterplot illustrating the relation between scores on Form A and those on Form B.

You may be thinking that both of the interclass reliability examples are a bit far-fetched, because it is unlikely that teachers will administer tests on more than one occasion (which is a requisite for determining the reliability of a test). You are correct! Teachers will typically administer a test only once because of time constraints and because examinee fatigue might otherwise come into play and negatively affect scores on subsequent trials. Additionally, practice can also affect the subsequent scores and thus the calculated reliability. However, there are ways to make minor adjustments in the equivalence methods and still reach a conclusion regarding a test's reliability. Consider how a teacher might create two equivalent forms of a single test. The equivalent forms can be created after the test is administered by assigning each person a score on two halves of the test (e.g., a score for the odd items and a score for the even items). Thus, the odd and even portions of the test can be perceived of as equivalent forms.

Go to the WSG to complete Student Activity 6.4.

Split-Halves Reliability The PPM correlation coefficient can be calculated between the scores on the halves of the test and used as an estimate of the reliability of the measure. Table 6.3 has sample data for calculating split-halves reliability. The split-halves reliability using the PPM correlation coefficient is .639. Because it was suggested earlier that a reliability of .80 or higher is desirable, you might be tempted to dismiss as unreliable the test whose scores are shown in table 6.3. However, an additional aspect of the data presented in the table needs to be considered. The reliability coefficient calculated from the data in table 6.3 is the correlation between two halves of the test (let's assume that each half consists of 13 items, giving a total test length of 26 items). We indicated that longer tests are generally more reliable than shorter ones. Because the .639 calculated in table 6.3 was based on a test of 13 items in length, we must now estimate the reliability of the original 26-item test.

You might think that you just multiply the obtained reliability by 2 since the actual test is twice as long. That would be incorrect. Notice that multiplying .639 × 2 results in 1.278. By definition (see equation 6.1), reliability cannot exceed 1.0, so multiplying by 2 is incorrect. Rather, the Spearman–Brown prophecy formula, equation 6.2, is used to estimate the reliability of a measure when the test length is changed:

$$r_{kk} = \frac{k \times r_{11}}{1 + r_{11}(k-1)} \tag{6.2}$$

where r_{kk} is the predicted (prophesied) reliability coefficient when the test length is changed k times, k is the number of times the test length is changed and is defined as

$$\frac{\text{the number of items for which an estimate of reliability is desired}}{\text{the number of items for which the reliability has been calculated}}$$

and r_{11} is the reliability that has been previously calculated. Thus, to estimate the reliability for the 26-item test, we substitute into the Spearman–Brown prophecy formula and obtain the predicted (estimated) reliability of the measure across all 26 items. Note that $k = 2$ (i.e., 26 / 13) and r_{11} = .639 (our original reliability estimate).

$$r_{kk} = \frac{(26 \div 13) \times .639}{1 + .639([26 \div 13] - 1)}$$

$$r_{kk} = \frac{1.278}{1.639}$$

$$r_{kk} = .78$$

Table 6.3 Odd–Even Scores for 10 Participants

Participant	Odd score	Even score
1	12	13
2	9	11
3	10	8
4	9	6
5	11	8
6	7	10
7	9	9
8	12	10
9	5	4
10	8	7
Sum (Σ)	92	86
Mean (*M*)	9.2	8.6
Standard deviation (*s*)	2.2	2.6
Variance (s^2)	4.8	6.7
		$r_{xx'}$ = .639

Thus, the estimated reliability for the original 26-item measure is .78. The reliability is said to have been adjusted with the Spearman–Brown prophecy formula. You should also note that it makes no difference how many items you started with. When the number of items or trials is doubled (i.e., $k = 2$), the predicted reliability will be the same. Had the .639 reliability been obtained on 50 items and you doubled it to 100 items, the predicted r_{kk} would still be .78.

This number can also be estimated from table 6.4, which shows values of r_{kk} calculated from equation 6.2 using numerous values of r_{11} (left column) and k (column headings). The number of times you want to change the test length (k) is listed across the top of table 6.4 (0.25-5.0). You determine the predicted reliability (r_{kk}) by finding where the appropriate row and column intersect. For example, if your calculated reliability (r_{11}) is .40 and you increase the test length by a factor of five, the estimated reliability is .77.

You will note that there are values of k listed in table 6.4 that are less than 1. This indicates that the instructor can estimate the reliability for a test of shortened length. For example, assume that an instructor has a written test of 100 items with a reliability of .92. Randomly splitting the test into equal parts with 50 items each would result in two tests with predicted reliabilities of .85. This would reduce the time to administer and score the examination (and make students happier!) and give a teacher the opportunity to have two forms of the test. The Spearman–Brown prophecy formula can be used to estimate the reliability of a measure when the test length is changed. The formula can be used with the interclass reliability estimate or with the intraclass reliability, which is discussed next.

 Go to the WSG to complete Student Activity 6.5.

Intraclass Reliability

The interclass reliability, based on the correlation between two measures, is different from the **intraclass reliability**, based on ANOVA. Assume that you measured a certain skinfold on one group of participants three times. You might want to estimate the reliability of the three measures. However, the interclass model permits you to correlate only two trials of skinfolds reputedly measuring the same thing because the PPM is used to correlate only two things at a time. The intraclass model, on the other hand, allows you to estimate reliability for more than two trials. You might want to estimate the reliability for more than two trials, given that reliability generally increases as the number of trials increases.

Additionally, if there is a constant difference between two trials (i.e., each person's score goes up or down by the same amount), the interclass reliability would be 1.00, but from a theoretical perspective the results would not be consistent (something does seem wrong). For example, in testing skinfold fat, the measures could become smaller with each measurement if the subcutaneous fat is still compressed from the previous measure. Another example of constant change is demonstrated in table 6.5, where the PPM correlation is perfect ($r_{xx'}$ = 1.00) yet the reliability (i.e., consistency of measurement) is lacking because each person's score increased by 10 on the second trial. The intraclass reliability model can address this issue. Significant mean differences across trials necessitate a deeper look into the changes across trials. It may be that participant learning or fatigue is affecting the reliability.

The most common names used for the intraclass reliability models are Cronbach's alpha coefficient, Kuder–Richardson formula 20 (KR_{20}), and ANOVA reliabilities. Each of these is calculated in a similar manner. The total variance in scores is partitioned into three sources of variation: people, trials, and people-by-trials. People variance is the observed score (total) variance between the participants (s^2_o). Trial variance is based on the variance across the trials. Think about it—if a measure is reliable, you would expect little variation across trials. Thus, variation across trials can be considered error (s^2_e). People-by-trials variation is based on the fact that not all participants perform equally differently across the trials. People variance is considered total variance. People-by-trials variance and trial variance are considered sources of error variance. Reliability is estimated by subtracting error variance from total (observed) variance and dividing the result by the total (observed) variance.

Table 6.4 Values of r_{kk} from Spearman–Brown Prophecy Formula

	k (CHANGE IN TEST LENGTH)							
r_{11}	.25	.33	.50	1.50	2.00	3.00	4.00	5.00
.10	.03	.04	.05	.14	.18	.25	.31	.36
.12	.03	.04	.06	.17	.21	.29	.35	.41
.14	.04	.05	.08	.20	.25	.33	.39	.45
.16	.05	.06	.09	.22	.28	.36	.43	.49
.18	.05	.07	.10	.25	.31	.40	.47	.52
.20	.06	.08	.11	.27	.33	.43	.50	.56
.22	.07	.09	.12	.30	.36	.46	.53	.59
.24	.07	.09	.14	.32	.39	.49	.56	.61
.26	.08	.10	.15	.35	.41	.51	.58	.64
.28	.09	.11	.16	.37	.44	.54	.61	.66
.30	.10	.12	.18	.39	.46	.56	.63	.68
.32	.11	.13	.19	.41	.48	.59	.65	.70
.34	.11	.15	.20	.44	.51	.61	.67	.72
.36	.12	.16	.22	.46	.53	.63	.69	.74
.38	.13	.17	.23	.48	.55	.65	.71	.75
.40	.14	.18	.25	.50	.57	.67	.73	.77
.42	.15	.19	.27	.52	.59	.68	.74	.78
.44	.16	.21	.28	.54	.61	.70	.76	.80
.46	.18	.22	.30	.56	.63	.72	.77	.81
.48	.19	.23	.32	.58	.65	.73	.79	.82
.50	.20	.25	.33	.60	.67	.75	.80	.83
.52	.21	.26	.35	.62	.68	.76	.81	.84
.54	.23	.28	.37	.64	.70	.78	.82	.85
.56	.24	.30	.39	.66	.72	.79	.84	.86
.58	.26	.31	.41	.67	.73	.81	.85	.87
.60	.27	.33	.43	.69	.75	.82	.86	.88
.62	.29	.35	.45	.71	.77	.83	.87	.89
.64	.31	.37	.47	.73	.78	.84	.88	.90
.66	.33	.39	.49	.74	.80	.85	.89	.91
.68	.35	.41	.52	.76	.81	.86	.89	.91
.70	.37	.44	.54	.78	.82	.88	.90	.92
.72	.39	.46	.56	.79	.84	.89	.91	.93
.74	.42	.48	.59	.81	.85	.90	.92	.93
.76	.44	.51	.61	.83	.86	.90	.93	.94
.78	.47	.54	.64	.84	.88	.91	.93	.95
.80	.50	.57	.67	.86	.89	.92	.94	.95
.82	.53	.60	.69	.87	.90	.93	.95	.96
.84	.57	.63	.72	.89	.91	.94	.95	.96
.86	.61	.67	.75	.90	.92	.95	.96	.97
.88	.65	.71	.79	.92	.94	.96	.97	.97
.90	.69	.75	.82	.93	.95	.96	.97	.98
.92	.74	.79	.85	.95	.96	.97	.98	.98
.94	.80	.84	.89	.96	.97	.98	.98	.99
.96	.86	.89	.92	.97	.98	.99	.99	.99

Table 6.5 Effect of a Constant Change in Measures

Participant	Trial 1	Trial 2
1	15	25
2	17	27
3	10	20
4	20	30
5	23	33
6	26	36
7	27	37
8	30	40
9	32	42
10	33	43
Sum (Σ)	233	333
Mean (*M*)	23.3	33.3
Standard deviation (*s*)	7.7	7.7
Variance (s^2)	59.1	59.1
	$r_{xx'}$ = 1.00	

Consider equation 6.1, in which reliability can be estimated from observed score and error score variance. People variance is observed variance (s^2_o). People-by-trials variance can be perceived as error variance, or all variance not attributable to people (i.e., trials and people-by-trials variances) can be perceived as error (s^2_e). Having estimates of observed and error variance permits you to use equation 6.1 to estimate the reliability of the scores.

The alpha coefficient is calculated as follows:

$$r_{xx'} = \text{alpha coefficient} = \left(\frac{k}{k-1}\right)\left(1 - \frac{\Sigma s^2_{\text{trials}}}{s^2_{\text{total}}}\right) \tag{6.3}$$

where k is the number of trials, $\Sigma s^2_{\text{trials}}$ is the sum of the variance of each trial, and s^2_{total} is the variance for the sum across all trials.

Table 6.6 presents an example of a calculation of the alpha coefficient. The variance calculations are identical to those you learned in chapter 3. Note that the **alpha reliability** (i.e., intraclass reliability) estimates the reliability for the total score (i.e., the sum across all trials). You can then use this result in the Spearman–Brown prophecy formula (equation 6.2) to estimate the change in the reliability coefficient if the number of trials is increased or reduced.

Table 6.6 Calculating the Alpha Coefficient

Participant	Trial 1	Trial 2	Trial 3	Total
1	3	5	3	11
2	2	2	2	6
3	6	5	3	14
4	5	3	5	13
5	3	4	4	11
ΣX	19	19	17	55
ΣX^2	83	79	63	643
s^2	2.70	1.70	1.30	9.50

$$k/(k-1) \times (1 - (\Sigma s^2_{\text{trials}} / s^2_{\text{total}}))$$

$$3/(3-1) \times (1 - (2.70 + 1.70 + 1.30) / 9.50)$$

$$3/2 \times (1 - 5.7/9.50)$$

$$1.5 \times (1 - .60)$$

$$1.5 \times .40 = .60 = \text{alpha coefficient}$$

 Go to the WSG to complete Student Activity 6.6.

Mastery Item 6.3

Use the Spearman–Brown prophecy formula to estimate the reliability of six trials from the data in table 6.6. Note that k is = 2 (6 / 3) and r_{11} is that obtained with the alpha coefficient (.60).

Mastery Item 6.4

Use SPSS to confirm the reliability estimate reported for the data in table 6.6. Let's calculate the alpha coefficient in two ways with SPSS. The first way uses the variances and equation 6.3. The second way takes advantage of a SPSS program to calculate alpha directly.

1. Download table 6.6 data from the WSG.
2. Go to Analyze → Descriptive Statistics → Descriptives.
3. Put all three trials and the Total in the Variable(s) box.
4. Click on Options.
5. Click *only* the box for Variance in the Dispersion box.
6. Click OK.
7. Your output has the four variances necessary for substitution into equation 6.3.

The second way takes advantage of SPSS commands to calculate the alpha coefficient.

1. Download table 6.6 from the WSG.
2. Start SPSS.
3. Click on the Analyze menu.
4. Scroll down to Scale and over to Reliability Analysis and click.
5. Highlight "trial1," "trial2," and "trial3" and use the arrow to place them in the Item box. Note: *Do not* include the total in this listing. SPSS will calculate the total for you.
6. Click on OK.

The alpha coefficient can also be used when data are scored as correct (1) or incorrect (0). In this case, the alpha coefficient is referred to as the Kuder–Richardson formula 20 (KR_{20}). The alpha coefficient and KR_{20} are mathematically equivalent. You will learn more about this in chapter 8. Jackson, Jackson, and Bell (1980) provided an excellent discussion of the alpha coefficient.

 Go to the WSG to complete Student Activity 6.7.

Mastery Item 6.5

Here are some activities that illustrate interclass, intraclass, and the Spearman–Brown prophecy formula. Go to the WSG and download the chapter 6 large dataset–reliability. These data represent four consecutive weeks of pedometer counts from a large sample. Do the following:

1. Analyze → Correlate → Bivariate and put all of the variables in the box at the right. Note these will be interclass correlations illustrating consistency of steps across paired weeks. Review the correlations and see that they range from .55 to .76 with a median of about .70. Notice also that week 1 correlates lower with the other weeks.
2. Analyze → Scale → Reliability Analysis and put all 4 weeks in the Items box. Notice the alpha coefficient for all 4 weeks is .885. This is the reliability of the entire 4-week period.
3. How would you estimate the reliability for a single week from the .885? Use table 6.4 to estimate this value.
4. Use the estimated value .70 obtained for a single week (from step 1) and substitute into the Spearman–Brown prophecy formula (or use table 6.4). Notice that the prophesized value (.90) is similar to that obtained with the alpha coefficient (.885).

Index of Reliability

Another statistic important to the interpretation of the reliability coefficient is the **index of reliability**. *The index of reliability is the theoretical correlation between observed scores and true scores and is calculated as the square root of the reliability coefficient (equation 6.4):*

$$\text{index of reliability} = \sqrt{r_{xx'}} \tag{6.4}$$

Thus, if the reliability of a test is .81, the theoretical correlation between observed and true scores is .90. Note that if reliability is 1.0, there is perfect correlation between observed and true scores. If the reliability is 0.00 then the correlation between observed and true scores is 0.0!

Standard Error of Measurement

Reliability obviously deals with a person's true score. Although true score can never be actually determined, as suggested earlier in this chapter, it can be thought of as the average of an infinite number of administrations of the test (where neither fatigue nor practice affects participants' scores). *Thus, for any given test administration, your best estimate of a person's true score is the obtained score.*

If the test is administered twice, you would average the test scores to get a better estimate of the true score. The theory is that random positive errors and negative errors will balance out in the long run. Regardless of a person's score, there will obviously be some error associated with it. In other words, it is unlikely in a real-life setting to have a score that is totally without error. Thus, a person's score is expected to change from test administration to test administration. *The standard error of measurement (SEM) reflects the degree to which a person's observed score fluctuates as a result of errors of measurement.* You should not confuse the **standard error of measurement** (SEM) with the standard error of estimate (SEE) presented in chapter 4. Although the two have similar interpretations (and look quite similar), they are different: The SEM relates to the reliability of measurement, whereas the SEE concerns the validity of an estimate.

The SEM is calculated as follows:

$$\text{SEM} = s\sqrt{1 - r_{xx'}} \tag{6.5}$$

where s is the standard deviation of the test results and $r_{xx'}$ is the estimated reliability.

Assume a test has a standard deviation of 100 and a reliability of .84. The SEM is calculated to be

$$\text{SEM} = 100\sqrt{1 - .84}$$
$$\text{SEM} = 100(.4)$$
$$\text{SEM} = 40$$

If a person scored 500 on a test whose SEM was 40, one can place confidence limits around that person's observed score in an attempt to estimate what the true score is. *The SEM, just like the standard error of estimate, is interpreted as a standard deviation: The SEM is the standard deviation of the errors of measurement around an observed score.* It reflects how much a person's observed score would be expected to change from test administration to test administration as a result of measurement error. Because error scores are expected to be normally distributed, 68% of the scores would be expected to fall within ±1 SEM of the observed score. In our example, therefore, there is a 68% chance that the person's true score is between 460 and 540 (i.e., 500 ± 40). Note that you could use table 3.4 to place confidence intervals around a particular observed score. You should be able to see that if you take the observed score and add and subtract 2 SEMs to the observed score, you have placed an approximately 95% confidence interval around the observed score. This is because, as you learned in chapter 3, the mean score plus and minus 2 standard deviations results in capturing approximately 95% of the scores in a normal distribution. Remember that the SEM is a standard deviation. As such, it can be used along with table 3.4 to provide evidence of the accuracy of a measure obtained with a test that has specific variability (the test standard deviation) and reliability.

Mastery Item 6.6

You should verify that approximately 95% of true scores are within the range of 420 and 580 when the observed score is 500 and the SEM is 40 points (i.e., ±2 SEM).

Go to the WSG to complete Student Activity 6.8.

A test or measure does not necessarily have reliability in all settings. Said another way, the reliability of a measure is situation specific. *Measures (i.e., test scores) are reliable under particular circumstances, administered in a particular way, and with a specific group of people.* It is not appropriate to assume that simply because measures are reliable for one group of participants (e.g., females) they are automatically reliable for another group (e.g., males). The following list provides factors that can affect a test's reliability.

- ***Fatigue.*** Fatigue generally decreases reliability.
- ***Practice.*** Practice generally increases reliability. Thus, practice trials during teaching and training should be encouraged.
- ***Participant variability.*** Greater variability in participants being tested results in greater reliability.

- ***Time between testing.*** Reliability generally decreases as the time between test administrations increases.
- ***Circumstances surrounding the testing periods.*** Reliability generally increases with a greater similarity between the testing periods.
- ***Appropriate level of difficulty for testing participants.*** The test should be neither too difficult nor too easy.
- ***Precision of measurement.*** Adequate accuracy must be assured with the measuring instrument. For example, one should measure the 50-yard (45.7 m) dash not to the nearest second but to the nearest hundredth of a second.
- ***Environmental conditions.*** Such factors as noise, excessive heat, and poor lighting can negatively affect the measurement process.

Test users need to be sensitive to factors that could affect the reliability of any test chosen.

VALIDITY

We have spent time developing procedures for estimating a measure's reliability because of the important role that reliability plays in determining a test's validity. *A measure must first be reliable in order for it to be valid—to truthfully measure what it is intended to measure.* Validity can be subdivided into several types, of which we will discuss three: content-related validity, criterion-related validity, and construct-related validity. These are summarized here and greatly expanded on in the American Psychological Association's *Standards for Educational and Psychological Testing* (1999). Validity can also be broadly classified as logical or statistical in nature. Regardless of the type of validation procedure involved, a criterion of some sort exists. The criterion can be perceived of as the most truthful measure of what you are attempting to measure. For example, the criterion or best measure of $\dot{V}O_2$max is treadmill-determined $\dot{V}O_2$max. The best measure of body fatness is DXA- or hydrostatically determined body fatness.

Content-Related Validity

Content-related validity is evidence of truthfulness based on logical decision making and interpretation. The terms *face validity* and *logical validity* are often used for content-related validity. The universe of interest, or the content universe, for a particular test needs to be well defined. For example, the items that appear on a particular cognitive unit test should reflect the knowledge content presented during the unit. A basketball skills test should theoretically include items that constitute the game of basketball (shooting, dribbling, passing, and jumping). That is, the test should appear to measure the material presented.

The fact that a test reflects content validity, however, does not necessarily make it valid. For example, consider someone who is taking skinfold measures to estimate percent body fat. Certainly skinfold measures have been shown to validly measure body fat. But if the person taking the measures is unqualified to do so (perhaps he or she is not well trained in the use of calipers) or measures at the wrong location (e.g., takes a posterior calf measure instead of a medial calf measure), the measurement may appear to be valid but is not. The criterion for content validity exists in the mind of the interpreter. Content experts, expert judges, colleagues, and textbook writers can serve as sources to content-validate instruments. Teachers developing cognitive tests are writing items that reflect the content of the course (so the items are content valid).

Criterion-Related Validity

Criterion-related validity is based on having a true criterion measure available. Validity is based on determining the systematic relationship between the criterion and other measures used to estimate the criterion. In short, criterion-related validity is evidence that a test has a statistical relationship with the trait being measured. Other terms for criterion-related validity are *statistical validity* and *correlational validity;* these terms are used because criterion-related evidence is based on the PPM correlation (from chapter 4) between a particular test and a criterion. For example, refer back to Kelly's situation in the measurement and evaluation challenge; she wants to measure the maximal oxygen uptake ($\dot{V}O_2$max) for a number of young adult participants. Kelly knows that the best way to do so is to have each person complete a maximal exercise test on a treadmill, cycle ergometer, swimming flume, or other type of ergometer; however, Kelly does not have the equipment and resources for conducting such a maximal test on each participant. Therefore, Kelly is seeking alternative measures that can be used to estimate $\dot{V}O_2$max—submaximal tests, distance runs, and nonexercise models. These alternative measures must first be validated with the criterion measure. To do this, at some point participants must complete both the criterion test and the alternative test (often called a field test) to be used to estimate the criterion. If a strong relationship is found between the criterion and the alternative test, future participants need not complete the criterion measure but can have their value on the criterion estimated from the alternative (i.e., field) measure, or **surrogate measure**. The taking of skinfold measures to estimate body fatness is an excellent example of criterion-related validity.

Criterion-related evidence is often further subdivided into **concurrent validity** and **predictive validity**. Both are based on the PPM correlation coefficient. The main difference between concurrent and predictive validities is the time at which the criterion is measured. For concurrent validity, the criterion is measured at approximately the same time as the alternative field, or surrogate, measure. Using a distance run to estimate $\dot{V}O_2$max is an illustration of concurrent validity. For predictive validity, the criterion is measured in the future. To establish predictive validity, the criterion might be assessed many weeks, months, or even years after the original test is conducted. The prediction of heart disease development in later life is based on predictive validity procedures; the criterion—development of heart disease—is not measured until many years later. However, it has been shown that lack of exercise, high body fat, smoking, high cholesterol, and hypertension are all predictive of future heart disease. (Of course, these same variables can be used to predict if one currently has heart disease. Thus, the time at which the criterion is measured and the interpretation of the correlation help identify whether the criterion-related evidence is concurrent or predictive in nature.) The following list provides some examples of concurrent validity and predictive validity in exercise science, kinesiology, and education. Each criterion is followed by a list of possible predictors.

Concurrent Validity

- $\dot{V}O_2$max (criterion: oxygen consumption)
    - Distance runs (e.g., 1.0 mi [1.6 km], 1.5 mi [2.4 km]; 2 km [1.2 mi]; 9 min, 12 min; 20- m [21.9 yd] shuttle)
    - Submaximal (e.g., cycle, treadmill, swimming)
    - Nonexercise models (e.g., self-reported physical activity)

- Body fat (criterion: dual-energy X-ray absorptiometry, DXA; hydrostatically determined body fat)
  - Skinfolds
  - BMI
  - Anthropometric measures (e.g., girths, widths, and lengths)
- Sport skills (criterion: game performance, expert ratings)
  - Sport skills tests (e.g., wall volley tests, accuracy tests, and total body movement tests)
  - Expert judges' evaluation at skill performance

Predictive Validity

- Heart disease (criterion: heart disease developed in later life)
  - Present diet, exercise behaviors, blood pressure
  - Family history of heart disease or related health issues
- Success in graduate school (criterion: grade-point average or graduation status)
  - Graduate Record Examination scores
  - Undergraduate grade-point average
- Job capabilities (criterion: successful job performance)
  - Physical abilities
  - Cognitive abilities

Sport skills tests are good examples of criterion-related validation procedures. Green, East, and Hensley (1987); Hensley, East, and Stillwell (1979); Hensley (1989); and Hopkins, Schick, and Plack (1984) provide excellent examples of the procedures used to validate sport skills tests. Although these tests are all over 30 years old, they retain their value as robust procedural models. A criterion measure must first be developed and then a variety of skills tests (i.e., a test battery) must be correlated with the criterion measure to determine which of these are most valid and most helpful in estimating the criterion. If a series of tests is used to estimate the criterion, multiple correlational procedures (see chapter 4) are used rather than the simple Pearson correlation coefficient. However, the logic is the same: An attempt is made to account for variation (i.e., increase the coefficient of determination) in the criterion measure from more than one measure. Consider a golf test. The criterion could be the average score from several rounds of golf. Thus, a study could be conducted in which everyone completes several rounds of golf to obtain the criterion measure. Each participant then completes a variety of skills tests (e.g., driving, long irons, short irons, chipping, and putting), which are then correlated with the criterion measure to determine which of the measures or combination of measures best estimates the criterion measure. Note that there will always be some error in all of the measures (criterion and estimators).

Interpretation of the criterion-related validity coefficient depends on its absolute value. Because the criterion-related validity coefficient is simply a PPM correlation coefficient, it must range between –1.00 and +1.00. However, the closer the absolute value of the validity is to 1.00, the greater the validity. For example, Green at al. (1987) reported validities for items on a golf skill test battery ranging from .54 for a pitch shot to .66 for a middle distance shot.

Let's return to the standard error of estimate (SEE) presented in chapter 4. The SEE is often reported with concurrent validity coefficients. For example, assume that you have a submaximal test that estimates $\dot{V}O_2max$ from a timed distance run of 1 mile (1.6 km) in length and the SEE is 4 ml · kg^{-1} · min^{-1}. Assume someone has a $\dot{V}O_2max$ of 50 ml · kg^{-1} · min^{-1}. You can place confidence limits around the predicted score: You can be 68% confident that the actual $\dot{V}O_2max$ is between 46 and 54 (i.e., 50 ± 4) ml · kg^{-1} · min^{-1}. Kline et al. (1987) provide an excellent example of concurrent validity for estimating $\dot{V}O_2max$. Their reported validity is .88 and the SEE is 5 ml · kg^{-1} · min^{-1}.

Note that the SEE reflects the accuracy of estimating your score on the criterion measure; in other words, it is a validity statistic.

Development of the criterion measure is extremely important in criterion-related evidence of validity. Examples of how criterion measures can be obtained include the following:

- ***Actual participation.*** One can actually complete the criterion task (e.g., play golf, shoot archery, conduct job-related activities).
- ***Known valid criterion.*** One can use a criterion (e.g., run on a treadmill, obtain underwater weight) that has been previously shown to be valid.
- ***Expert judges.*** Experts judge the quality of the criterion performance. This is often used with team activities (e.g., volleyball) in which it is difficult or impossible to obtain a number that reflects performance on the task being measured.
- ***Tournament participation.*** Rankings of abilities can be determined when everyone participates with everyone else (best used when the skilled event is an individual sport).
- ***Known valid test.*** Participants can complete a test that has been previously shown to be valid.

Construct-Related Validity

Construct-related procedures are often used to validate measures that are unobservable yet exist in theory. For example, one's intelligence quotient (IQ) exists in theory, but IQ is not something that can be readily measured. The same is true for attitude measures. Certainly, each of us has attitudes about various behaviors (e.g., exercise, dieting, physical activity), but it is difficult to measure these attitudes directly. This is where construct validity comes into play. *Construct-related validity evidence is essentially a marriage between logical (content) and statistical validity procedures.* To provide construct-related evidence for a particular measurement, you gather a variety of statistical information that, when viewed collectively, adds evidence for the existence of the theoretical construct being measured. Constructs are based on theories and theories are validated through research. Mahar and Rowe (2002) and Rowe and Mahar (2006) provide excellent descriptions and examples of the stages involved in providing construct-related validity evidence.

When you are collecting construct-related evidence, your working hypothesis would be worded like this: *If in theory the construct is valid, then such-and-such should occur.* Then test to see if it does occur. The logical part of construct validity is *what should occur.* The statistical part consists of the data that you gather. Continual gathering of information that supports the theory adds to the evidence for the existence of the construct. When what you think should occur is not supported by the data collection, there are two things to consider: It may be that the construct doesn't exist or that the if–then statement was inaccurate. Development of construct validity is highly related to the scientific method presented in chapter 5. A hypothesis is generated, a method is developed, data are collected and analyzed, and a decision is made based on the evidence obtained.

You will sometimes hear the words *convergent* or *discriminant* evidence. Generally, these concepts are based on correlations between variables. Variables that theoretically *should* correlate (either positive or negative, depending on the scales used) provide convergent evidence, and variables that theoretically should have *no* correlation provide discriminant evidence (that is, they measure different things).

Kenyon (1968a, 1968b) developed a multidimensional instrument to measure Attitudes Toward Physical Activity (ATPA). Certainly attitudes toward physical activity exist. Some people like physical activity and others do not. But how can one measure these perceived attitudes? Kenyon provided evidence that there are a variety of reasons people like or dislike (or engage or don't engage in) physical activity (i.e., ATPA is a multidimensional construct). The unobservable, but theoretically existent, dimensions suggested by Kenyon include the following:

- Aesthetic experience
- Catharsis
- Health and fitness
- Social experience
- Pursuit of vertigo (the thrill of victory)
- Ascetic experience

Consider the aesthetic dimension, which indicates that some people like physical activity for the beauty of movement expressed in such activities as dance, ballet, gymnastics, diving, and skating. To provide construct-related evidence that this dimension exists, one could administer the ATPA to groups of people who engage in different types of behaviors. Your working hypothesis would be *If the aesthetic dimension exists, people who attend dance concerts, ballet performances, and gymnastics events should score significantly differently on the aesthetic dimension than people who do not attend such functions.* This is exactly how construct-related evidence is obtained for such unobservable measures.

Construct-related evidence can be used to provide additional evidence for criterion-related validation evidence. Consider the golf test described earlier. A working hypothesis would be *If this is a valid golf test, the following should occur: Students who have never played golf should score poorly on the test, beginning golfers should score better, experienced golfers should score higher, and members of the golf team should score the best.* This is referred to as the known group difference method of construct validation. Conducting such a study and testing for differences in the group means (see ANOVA in chapter 5) could provide construct-related evidence for the golf test.

Dataset Application

Go to the chapter 6 large dataset–validity in the WSG. The dataset contains a number of variables that *might* be related to underwater weighing. Use the information that you learned in chapter 4 to determine which variables are most and least related to hydrostatically determined body fatness. Can you also calculate the SEE for some of the variables?

Go to the WSG to complete Student Activity 6.9.

Figure 6.1 illustrates the relationships among the various aspects of validity that we have introduced. Validity is made up of reliability (the left side) and relevance (the right side). Reliability can be interclass or intraclass. Interclass reliability methods are test–retest reliability, equivalence reliability, and split-halves reliability. Intraclass reliability is based on ANOVA and results in alpha and KR_{20}. Relevance (relatedness) can be content related, criterion related or construct related. Criterion-related evidence can be either concurrent or predictive.

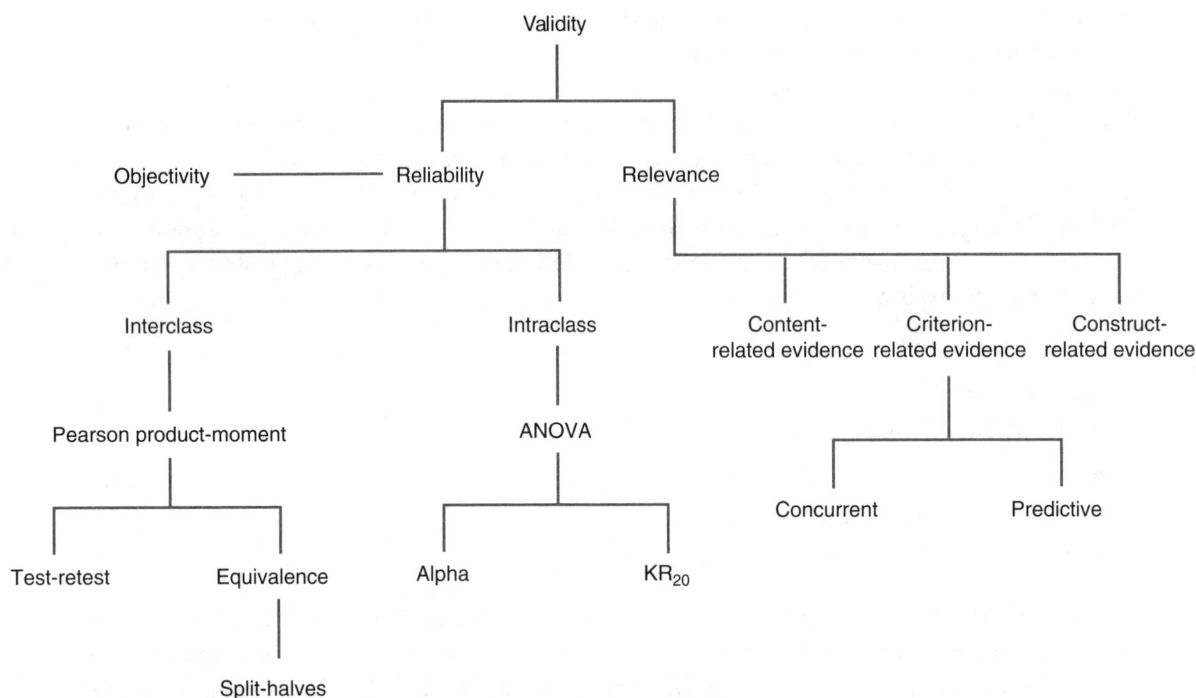

Figure 6.1 Diagram of validity and reliability terms.

Much of the information presented in this chapter has related to the Pearson product-moment (PPM) correlation coefficient, introduced in chapter 4. In some cases the PPM is interpreted as a reliability coefficient. In other cases it could be an objectivity coefficient or a criterion-related (either concurrent or predictive) validity coefficient. In all cases the PPM correlation coefficient is calculated as you learned in chapter 4. The difference in the interpretation depends on what two things are correlated. This is described in figure 6.2. Essentially, if two trials of the same thing are correlated but measured at different times, then this PPM is interpreted as a reliability (stability) coefficient. If two raters are correlated when they scored the same test, this PPM is interpreted as an objectivity coefficient, which would be interrater (between raters) if more than one rater is involved and intrarater (within raters) for the same rater on more than one occasion. If two forms of the same test are being correlated, this is an estimate of equivalence. If one of the measures being correlated is a criterion, then you are working with validity. Whether the PPM calculated is a concurrent or predictive validity coefficient depends on when the criterion was measured. This illustrates the generalized use of the PPM for estimating reliability, objectivity, and validity. Distinguishing between these correlations is important. See Odom and Morrow (2006) for further illustrations of this concept and how to interpret the correlation coefficient.

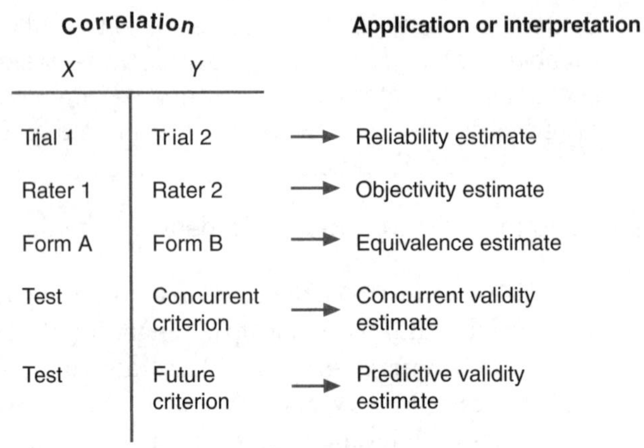

Figure 6.2 Applications of the Pearson product-moment correlation in reliability and validity.

 Go to the WSG to complete Student Activity 6.10.

TYPES OF VALIDITY EVIDENCE

We illustrate here the three types of validity mentioned previously (content-related validity, criterion-related validity, and construct-related validity).

Evidence Based on Content

Evidence based on content is evidence that the test characteristics are representative of the universe of potential items that could have been used. For example, the items on a written test at the end of a unit should reflect the material presented in the unit; the physical tests required for employment should reflect the types of tasks that would be performed on the job.

Evidence Based on Relations to Other Variables

Evidence based on relations to other variables demonstrates that test scores are systematically related to a criterion. A criterion measure is obtained, and test scores are correlated (often using the PPM correlation coefficient) with the criterion. For example, skinfolds validly measure percent fat (the criterion), and distance runs validly estimate $\dot{V}O_2$max (the criterion).

Evidence Based on Response Processes

Evidence based on response processes focuses on the test score as a measure of the unobservable characteristic of interest. Attitudes, **personality** characteristics (the totality of a person's distinctive psychological traits), and unobservable yet theoretically existing traits are often validated with construct-related evidence. For example, attitudes toward physical activity theoretically exist; students can theoretically evaluate the effective teaching that is conducted within a classroom.

APPLIED RELIABILITY AND VALIDITY MEASURES

Let's look at some examples of reliability and validity from exercise, sport science, and kinesiology. Recall that Green and colleagues (1987) used correlation to develop a golf skills test which provided validity information, but their validity matrix contained *no* reliability results. Remember, in order to estimate reliability, you must administer the *same* thing on *at least* two occasions. Green and colleagues further investigated validity through the use of the multiple skills test to predict golf performance and identified the middle-distance shot, pitch shot, and long putt combined to have a concurrent validity of .76. Adding a fourth skills test, the chip shot, only increased the concurrent validity to .77 (Green et al., 1987). Their golf test batteries were determined with multiple regression techniques (see chapter 4) to determine which test items best combine to account for variation in the criterion measure (golfing ability). The test administrator will need to determine if it is worth the time and effort to measure four skills (validity = .77) rather than three (validity = .76). Table 6.7 provides illustrations of reliability estimates from physical performance and a variety of sport skills tests.

Table 6.8 provides concurrent-validity coefficients for estimating $\dot{V}O_2$max from a variety of measures. Some of the authors used a single measure to estimate $\dot{V}O_2$max, whereas others used multiple regression. Look at the results from Murray and colleagues (1993) presented in table 6.8. Can you explain why the correlation increases with additional items? Can you also explain why the SEE decreases with increasing number of items?

Table 6.7 Reliability Measures from Sport Skills and Fitness Tests

Author	Test item	Reliability ($r_{xx'}$)
Engelman & Morrow (1991)	Traditional pull-up (boys) Traditional pull-up (girls) Modified pull-up (boys) Modified pull-up (girls)	.83 to .92 .91 to .92 .68 to .83 .77 to .83
Green, East, & Hensley (1987)	Golf—chip shot (males) Golf—chip shot (females) Golf—long putt (males) Golf—long putt (females) Golf—short putt (males) Golf—short putt (females)	.85 .86 .87 .93 .54 .46
Hensley, East, & Stillwell (1979)	Racquetball—short volley (males) Racquetball—short volley (females) Racquetball—long volley (males) Racquetball—long volley (females)	.77 .86 .85 .82
Hensley (1989)	Tennis—serve (males) Tennis—serve (females) Tennis—volley (males) Tennis—volley (females)	.86 & .95 .79 & .88 .70 & .72 .69 & .79
Hopkins, Schick, & Plack (1984)	Basketball—shot (males) Basketball—shot (females) Basketball—pass (males) Basketball—pass (females)	.84 to .95 .87 to .95 .88 to .96 .82 to .91
Nelson, Yoon, & Nelson (1991)	Modified push-up (boys) Modified push-up (girls)	.78 to .89 .77 to .91
Rikli, Petray, & Baumgartner (1992)	1/2 mile (.8 km) (boys) 1/2 mile (girls) 3/4 mile (1.2 km) (boys) 3/4 mile (1.5 km) (girls) 1 mile (1.6 km) (boys) 1 mile (1.6 km) (girls)	.65 to .82 .32 to .77 .48 to .94 .58 to .83 .44 to .87 .34 to .90
Schick & Berg (1983)	Golf—5 iron	.90

Note: All reliabilities are intraclass.

 Go to the WSG to complete Student Activities 6.11 and 6.12.

ESTIMATING AGREEMENT BETWEEN MEASURES USING THE BLAND-ALTMAN METHOD

Bland and Altman (1986) present a method of estimating the agreement between two clinical or laboratory measures of the same attribute. They argued that using correlation and regression statistical methods had serious drawbacks in demonstrating the degree of agreement between two measures of the same attribute. Specifically, two measures might have a high correlation but still have clinically important differences in terms of the magnitude and the distribution of those differences. We suggested this concern earlier when we moved from the discussion of the interclass to the intraclass reliability method (see table 6.5). The Bland–Altman approach is based on calculating the absolute differences between the two measures and plotting those differences against the averages of the two measures.

In order to understand their approach, examine the data in table 6.9. The data represent two clinical measures of systolic blood pressure. One was taken by a clinician trained in

Table 6.8 Concurrent Validity Measures for $\dot{V}O_2max$

Author	Criterion	Predictor(s)	Validity (r)	Standard error of estimate ($ml \cdot kg^{-1} \cdot min^{-1}$)
Getchell, Kirkendall, & Robbins (1977)	$\dot{V}O_2max$	1.5 mi (2.4 km) run	.92	2.38
Kline et al. (1987)	$\dot{V}O_2max$	1 mi (1.6 km) walk Gender Age Body weight	.88	5.00
Murray et al. (1993)	$\dot{V}O_2peak$	20-min steady-state run	.68	5.32
	$\dot{V}O_2peak$	20-min steady-state run Gender	.73	4.96
	$\dot{V}O_2peak$	20-min steady-state run Gender Weight	.79	4.45
Jurca et al. (2005)	Maximal cardiorespiratory fitness	Gender Age BMI Resting heart rate Self-reported physical activity	.76-.81	6.90-5.08
Wier et al. (2006)	$\dot{V}O_2max$	Gender Activity code Age BMI	.80	4.90

Note: BMI = body mass index.

Table 6.9 Example of the Bland–Altman Method

Participant	Clinician	Automated	Differences	Averages
1	103	105	2	104
2	117	115	−2	116
3	116	120	4	118
4	123	125	2	124
5	127	125	−2	126
6	125	125	0	125
7	135	125	−10	130
8	126	130	4	128
9	133	135	2	134
10	145	145	0	145

blood measurement and the other was taken by an automated device. The differences are calculated by consistently subtracting the clinician's measure from the automated value. The averages represent the average systolic blood pressure for each participant across both measures. To represent the standard error of the agreement between measures, the standard deviation of the differences is calculated. For the example it equals 4.11.

The next step is to develop a scatterplot with the average values on the X-axis and the differences on the Y-axis (figure 6.3). Three horizontal reference lines are added to the scatterplot:

- The first line indicates zero or no differences between the measures. Notice from table 6.3, there are two participants with zero difference scores; those points are located on the line, indicating no difference between the measures.
- The second line is slightly above a difference of +8. This is 0 + 2 × 4.11 (the standard deviation of the differences).
- The third line is slightly below a difference of –8. This is 0 – 2 × 4.11.

Examine the scatterplot with the reference lines to see the number of large errors in agreement. These would be points above the top reference lines or below the lower reference line. Then examine the distribution of the errors across the range of the average blood pressures. Example questions to be answered include these:

- Are the errors consistently above, below, or equally distributed in relation to the zero error reference line?

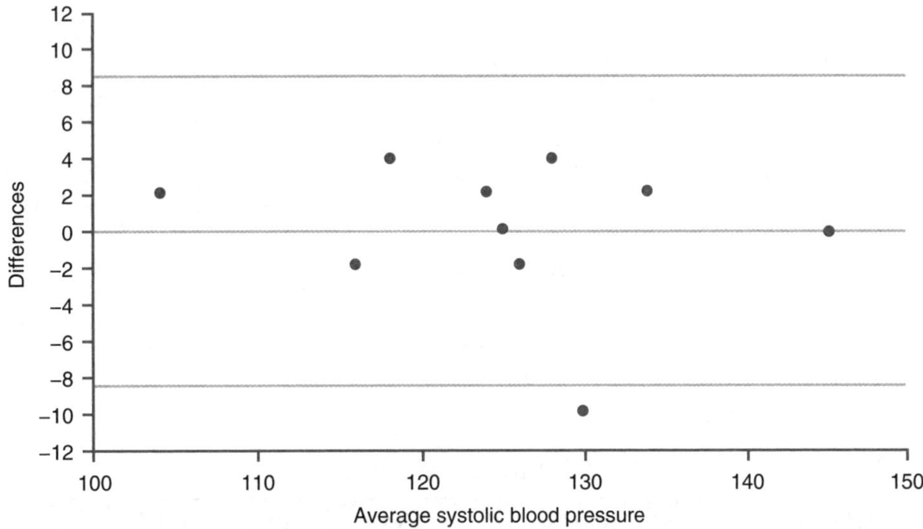

Figure 6.3 Bland–Altman plot of systolic blood pressure (clinician and automated).

MEASUREMENT AND EVALUATION CHALLENGE

You should be able to determine the steps that Kelly must take to select and administer a reliable and valid field test of aerobic capacity to her young adult members. She must first determine if the test she has selected is reliable. That is, are the results consistent from one administration to another, assuming that the participants have not actually changed training or activity levels? She must be sensitive to the standard error of measurement (SEM). Next, she must determine the concurrent validity between the proposed field test and the actual treadmill performance of the participants. Such information might be available in the research literature, or she may need to actually work with a researcher to obtain this vital information. She should concern herself with the types of participants she is testing in comparison to those used in the validation process. If they are similar, she should feel confident that the field test results will provide a fairly accurate estimate of the participants' aerobic capacities. The field test will not provide an exact measure of the participants' aerobic capacities, however. Thus, she must be concerned about the standard error of estimate (SEE) in estimating actual $\dot{V}O_2max$ from the surrogate (i.e., field) measure.

• Are the magnitudes of the errors consistent across the range of blood pressure measures? Are errors larger or smaller at high blood pressures? Are errors larger or smaller at low blood pressures?

Another consideration that Bland and Altman (1986) emphasized was the clinical relevance of the differences. For instance, were the differences (i.e., lack of agreement) between the measures large enough to cause issues in clinical interpretation? In our example, only one of the participants had a systolic blood pressure error below the bottom reference line. Thus, overall, clinical interpretations would be similar between the two measures. However, examine the modified data in Table 6.10. The standard deviation of the differences is now 12.9, which is much larger due to the larger magnitude of differences in the revised data. A systolic blood pressure of 140 indicates a person with hypertension. Only participant 10 is consistently indicated as hypertensive by both methods. The second Bland–Altman plot in Figure 6.4 illustrates that the larger errors are at the higher blood pressure measures.

Table 6.10 Modified Systolic Blood Pressure Data

Participant	Clinician	Automated	Differences	Averages
1	103	105	2	104
2	117	115	−2	116
3	116	120	4	118
4	123	125	2	124
5	147	125	−22	136
6	125	125	0	125
7	151	125	−26	138
8	126	130	4	128
9	133	151	18	142
10	145	145	0	145

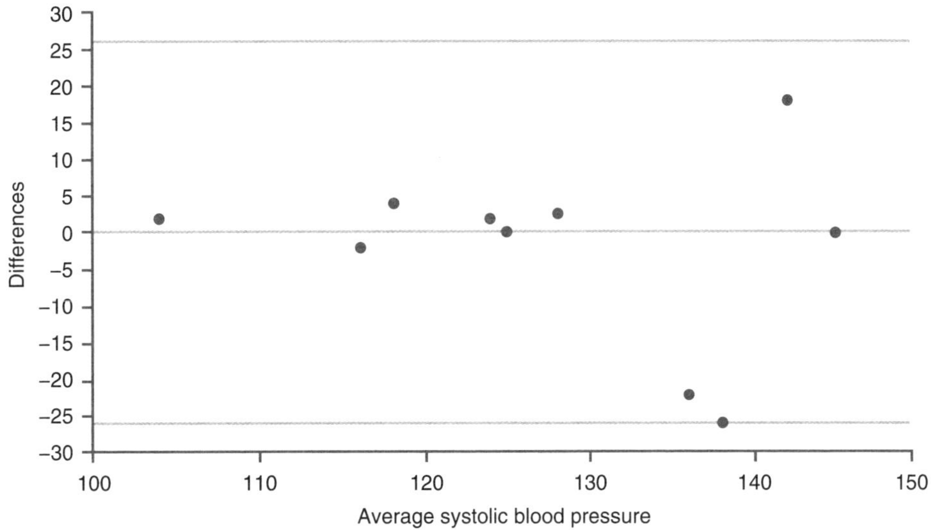

Figure 6.4 Bland–Altman plot of systolic blood pressure.

SUMMARY

The issues of reliability, objectivity, and validity are the most important ones that you will encounter when you test and evaluate human performance, whether the performance is in the cognitive, affective, or psychomotor domain. Reliability coefficients represent the consistency of response and range from 0 (totally unreliable) to 1.00 (perfectly reliable). Likewise, objectivity (interrater reliability) values range from 0 to 1.00. The standard error of measurement (SEM), a reliability statistic, reflects the degree to which a person's score will change as a result of errors of measurement. A validity coefficient represents the degree to which a measure correlates with a criterion. Statistical validity coefficients range from −1.00 to +1.00. The absolute value of the validity coefficient is important. A value of 0 indicates no validity; 1.00 represents perfect correlation with the criterion. The standard error of estimate (SEE), a validity statistic, indicates the degree to which a person's predicted or estimated score will vary from the criterion score.

Last, you should realize that reliability and validity results are not typically generalizable. The reliability or validity obtained is specific to the group tested, the environment of testing, and the testing procedures. You must study whether the reliability and validity results that you obtain can be inferred to another population or setting before making such an inference.

Now that you are familiar with concepts related to reliable and valid assessment, you should be able to better evaluate the instruments that you might use in human performance testing.

Go to the WSG for homework assignments and quizzes that will help you master this chapter's content.

CHAPTER

7

Criterion-Referenced Reliability and Validity

OUTLINE

OBJECTIVES

After studying this chapter, you will be able to

- define a criterion-referenced test;
- explain the approaches for developing criterion-referenced standards;
- explain the advantages and limitations of criterion-referenced measurement;
- select appropriate statistical tests for the analysis of criterion-referenced tests;
- interpret statistics associated with criterion-referenced measurement;
- discuss and interpret epidemiologic statistics; and
- use SPSS and Excel to calculate criterion-referenced statistics.

 The lecture outline in the WSG will help you identify the major concepts of the chapter.

MEASUREMENT AND EVALUATION CHALLENGE

Christina is an athletic trainer working primarily in sport rehabilitation. She recently has been faced with an increasing number of patients with pulled hamstrings. Currently, there is speculation in the sport rehabilitation field that this increase in injuries could be related to creatine use. Christina would like to get more scientific information on the association between hamstring pulls and creatine use. However, when she received her professional preparation, the measurement techniques she learned were associated with only norm-referenced measurement. She is in a quandary about how to approach this question. She is interested in determining if creatine use is related to an increased incidence of hamstring pulls. Christina decides to go to the library and examine the literature to see if she can determine the techniques that will allow her to examine this question.

In the area of human performance, we are fortunate that many variables are measured on the continuous (interval or ratio) scale (see chapter 3). The speed that a person runs, the distance that a person jumps, and the number of pedometer-recorded steps a person takes are common ratio measures. Some variables may not be measurable on this scale and are instead reported as rankings and ratings, or even as pass/fail or sufficient/insufficient. Players can be ranked on overall playing ability or rated on a specific skill. Other variables, such as gender and ethnicity, can be measured only categorically; recall from chapter 3 that these are termed nominal variables. Some variables can be measured in more than one way. For example, height is normally measured on a ratio scale and reported in feet and inches or in centimeters. However, let's assume that a teacher wants to equate teams in a basketball class on height. The teacher could rank students from tallest to shortest and assign teams based on the player's rank. The teacher could likewise cluster students into groups based on height. The taller players could play against one another, whereas the shorter players would be matched against players of similar height.

This last example represents the establishment of **cutoff scores** to create forced categories. Cutoffs are important when comparisons of individual performances are not the primary concern but rather we want to know whether a certain performance (or minimum) level has been achieved. Pass/fail and sufficient/insufficient are two such categorizations that are based on cutoff scores. Conventional basic statistics do not apply when variables are measured in this manner; for example, calculating the mean and standard deviation for data that have been forced into categories would be inappropriate. Therefore, specific techniques applicable to nominal measurement must be used (as you should recall from chapter 5). **Criterion-referenced tests (CRTs)** are appropriate for this situation.

Technically, there is not a great deal of difference between what you learned in chapter 6 and what you will learn in chapter 7. The primary difference is in the level of measurement used to describe performance. In chapter 6, the variables were continuous in nature. The variables presented in chapter 7 are categorical in nature. However, beyond that, the concepts of reliability and validity presented in chapter 6 can be readily adapted to the variables presented in chapter 7. We will present information about criterion-referenced reliability and criterion-referenced validity. Our focus in chapter 6 was on each person's specific score. Our focus in chapter 7 is on the category into which each person is placed.

A CRT is one that is deliberately constructed to yield measurements that are directly interpretable in terms of specific performance standards. Performance standards are generally specified by defining a class or domain of tasks that should be performed by each participant (Nitko 1984, p. 12).

 Go to the WSG to complete Student Activity 7.1 and view Video 7.1.

CRTs are used to make categorical decisions, such as passing or failing or meeting a standard versus not meeting the standard, or to classify participants as masters or nonmasters. When specific, well-defined goals are identifiable, CRTs may best measure the reliability and validity of that particular item. CRTs are not limited to nominal measurement. Often, continuous variables can be used with criterion-referenced testing methods. For example, push-up or sit-up performance can be evaluated using criterion cutoffs rather than norm-referenced methods. Historically, programmed instruction that centered on **behavioral objectives**—specifically written goals with instructions on how they can be obtained—was well suited for this type of measurement approach. Mastery instruments based on behavioral objectives are best exemplified by tests involving licensing, such as the test you took to get a driver's license and, in the human performance area, Red Cross standards for CPR, lifesaving, and swimming certification. It is easy to see in these examples that a minimum standard must be obtained before competency is proclaimed and a license is granted. Another excellent example is the Physical Ability Test used by the Houston Fire Department as part of the selection process for candidates to become fire fighters (http://www.houstontx.gov/fire/employment/applicantprocedure.html). The Houston Fire Department selected seven physical performance tests that are job related and that require balance, coordination, strength, endurance, and cardiorespiratory fitness. In these cases, cutoff scores are referenced to a standard based on a theoretical minimal performance level associated with being a successful firefighter. Achieving the standard (or minimal cut score) provides evidence that a test taker is qualified, passes, or is sufficient in some manner.

Before she could become a lifeguard, this candidate had to pass many tests by demonstrating specific performance standards. These types of tests are known as criterion-referenced tests.

DEVELOPING CRITERION-REFERENCED STANDARDS

Four basic approaches are used to develop criterion-referenced standards for tests of human performance (Safrit, Baumgartner, Jackson, and Stamm 1980):

- The *judgmental* approach is based on the experience or beliefs of experts. It reflects what they believe is an appropriate level based on their background and experience in testing and evaluating human performance. For example, many high school volleyball coaches require players to be able to serve overhand to play on the varsity team. The coach might set a cutoff level, such as placing 8 out of 10 overhand serves in the court.
- The *normative* approach uses norm-referenced data to set standards; some theoretically accepted criterion is chosen. Research from the Cooper Institute in Dallas, Texas, indicates that being above the bottom 20% on fitness levels is associated with reduced risk of negative health outcomes. That is, those who are in the top 80% for their age and gender are less likely to incur negative health outcomes. This criterion is based not only on experts' opinions—and research—but also on the available norms.
- The *empirical* approach relies on the availability of an external measure of the criterion attribute. Cutoff scores are directly established based on the data available on this external attribute. This approach is the least arbitrary of the four. However, it is seldom used because of the lack of a directly measurable external criterion. An example is a firefighter having to scale a 5-foot (1.5 m) wall to perform his or her duties. This is a concrete example of a pass/fail item that is based on the empirical approach. Another excellent example of this approach is the work of Cureton and Warren (1990) on cardiorespiratory fitness, presented later in this chapter. The Fitnessgram Healthy Fitness Zone (see chapter 10) standards were established in this manner.
- The *combination* method involves using all available sources: experts, prior experience, empirical data, and norms. Usually, experts' opinions and norms are the basis for making criterion-referenced decisions in human performance. The combination method is perhaps the best because it employs multiple methods to determine the most appropriate cutoff score.

 Go to the WSG to complete Student Activities 7.2 and 7.3 and view Videos 7.2 and 7.3.

DEVELOPMENT OF CRITERION-REFERENCED TESTING

The specific use of the term *criterion-referenced testing* is generally traced to a 1962 article by Glaser and Klaus. They developed this term because of a number of limitations they believed were inherent in norm-referenced tests; the primary shortcoming is that such tests are constructed to have content validity over a wide range of instructional goals and philosophies. Consequently, the more specific norm-referenced tests are, the less marketable they become. For this reason, norm-referenced tests are not well suited to the assessment of specific objectives. For example, if a norm-referenced approach is used to determine who receives a driver's license, then your ability to pass the test would be based on the group tested and not on your absolute ability to drive a car. The primary goal of a norm-referenced test is to establish a range of behavior to discriminate among levels of knowledge, ability, or performance. If a certain level of performance is necessary, then norm-referenced testing does not provide this information in the most efficient way. CRTs, on the other hand, are usually structured to assess far fewer objectives than a traditional norm-referenced test and therefore can be set up to identify specifically enumerated goals for behavioral items. For example, how many sit-ups should a 10-year-old boy be able to complete to be considered physically fit?

The primary difference between norm-referenced tests and CRTs is that CRTs are evaluated categorically. The traditional statistical techniques used to establish the reliability and validity of norm-referenced tests presented in chapter 6 cannot be used with CRTs. Therefore, you must choose specific techniques that best estimate the reliability and validity of criterion-referenced measures. Indexes of reliability associated with criterion-referenced tests are called indexes of dependability. The indexes allow you to determine not only **proportion of agreement** (P) (which refers to the consistency with which performances are categorized across methods or trials) but also the consistency with which decisions are made. Specific examples of indexes of dependability are presented later in this chapter.

Cureton and Warren (1990) summarized the advantages and limitations of criterion-referenced measurement as follows:

Advantages

- Criterion-referenced standards represent specific, desired performance levels that are explicitly linked to a criterion. In human performance, this criterion is often a health outcome.
- Because criterion-referenced standards are absolute standards, they are independent of the proportion of the population that meets the standard. Regardless of the prevalence of people meeting the standard, the standard is still valid. For example, regardless of the percentage of the population that smokes, you want to reduce the number.
- If standards are not met, then specific diagnostic evaluations can be made to improve performance to the criterion level. If a person does not meet the standard of 150 minutes of moderate-to-vigorous physical activity (MVPA) per week, you can provide specific feedback regarding what he or she should do.
- Because the degree of performance is not important, competition is based on reaching the standard, not on bettering someone else's performance level. For example, has a person stopped smoking or achieved the 150 minutes of MVPA per week?

The following are other key advantages:

- Performance is linked to specific outcomes.
- People know exactly what is expected of them.

Limitations

- Cutoff scores always involve some subjective judgment. Because few criteria are clear cut, philosophical guidelines can drastically affect the selection of the performance criterion. Authorities often disagree on exact levels, so cutoffs are sometimes arbitrarily determined.
- Misclassifications can be severe. Consider a hypothetical situation in which a doctor is prescribing medication based on a criterion-referenced standard. Misclassification of patients could have severe health consequences.
- Because cutoffs must be set at some level, those who attain the cutoff level may not be motivated to continue to improve. Conversely, those who never attain the cutoff could become discouraged and lose interest.

To examine some of these limitations, Cureton and Warren (1990) studied criterion-referenced standards for the 1-mile (1.6 km) run/walk test, for which the Fitnessgram

(Cooper Institute for Aerobics Research 1987) and Physical Best (AAHPERD 1988) both provide criterion-referenced standards. To examine the validity of these standards, these authors developed an external criterion:

> *The criterion was defined as the lowest level of $\dot{V}O_2max$ in children and adolescents consistent with good health, minimized disease risk, and adequate functional capacity for daily living. Because no empirical data specifically identifies [sic] the minimum level, the criterion $\dot{V}O_2max$ was based primarily on indirect evidence relating aerobic capacity to health disease/risk. (p. 10)*

Essentially, Cureton and Warren determined 1-mile (1.6 km) run/walk speeds that corresponded to criterion levels of $\dot{V}O_2max$ and converted these speeds to run times. The authors evaluated data on 581 boys and girls aged 7 to 14 against the Fitnessgram criterion and the Physical Best criterion. Their results are presented in table 7.1. The table indicates that 496 of the 581 cases (85%) were properly classified by the Fitnessgram standards, whereas only 357 (61%) were properly classified by the Physical Best standards. Fifteen percent (11% + 4%) were misclassified by the Fitnessgram standards, and 39% (35% + 4%) were misclassified by the Physical Best standards. This analysis highlights the importance of setting cutoff standards correctly.

Table 7.1 Comparison of Fitnessgram and Physical Best Standards for 1-mile (1.6 km) Run/Walk Times

	FITNESSGRAM	
Run/walk test result	**Below the criterion $\dot{V}O_2$**	**Above the criterion $\dot{V}O_2$**
Did *not* achieve the standard	24 (4%)	21 (4%)
Did achieve the standard	61 (11%)	472 (81%)
	PHYSICAL BEST	
Run/walk test result	**Below the criterion $\dot{V}O_2$**	**Above the criterion $\dot{V}O_2$**
Did *not* achieve the standard	130 (22%)	23 (4%)
Did achieve the standard	201 (35%)	227 (39%)

Another example of criterion-referenced standards is cholesterol levels set by professional associations. The American Heart Association and the National Heart, Lung, and Blood Institute have established cutoff values for blood cholesterol levels related to the risk of coronary heart disease. They are as follows:

- Low risk: <200 mg/dl
- Moderate risk: ≥ 200 mg/dl or ≤ 240 mg/dl
- High risk: >240 mg/dl

A physician who is counseling a patient about the risk of coronary heart disease would use the patient's blood test results and compare them with these standards. The physician might advise the following:

- No need for concern (patient's level = 180 mg/dl).
- Increase physical activity levels and eat a low-fat diet (patient's level = 215 mg/dl).
- Increase physical activity levels, eat a low-fat diet, and take prescription medication (patient's level = 300 mg/dl).

The *2008 Physical Activity Guidelines for Americans* (USDHHS 2008) provide another CRT example. The guidelines call for all adults to engage in 150 minutes of MVPA each week for health benefits. Vigorous activities count twice as much as moderate ones, so if one does 75 minutes of vigorous physical activity, this also meets the guidelines (i.e., 75 × 2 = 150 minutes). Moderate and vigorous physical activities can be combined with vigorous minutes counting twice as much as moderate minutes (e.g., 50 vigorous × 2 + 50 moderate = 150 minutes of MVPA). The important point is to accumulate a total of 150 minutes of MVPA each week. The 150 minutes is the criterion. If someone does not meet the criterion, specific suggestions and prescriptions can be provided to help achieve these guidelines. On the other hand, if a person does 150 minutes, he or she might simply be unmotivated to do additional physical activity because he or she sees the minimum as the goal. The physical activity guidelines specifically state that additional health benefits are achieved with more physical activity (essentially, a dose response). Thus, the criterion serves a good purpose (a goal) but can also be problematic (unmotivated to do beyond the minimum amount).

STATISTICAL ANALYSIS OF CRITERION-REFERENCED TESTS

Not only is the procedure for setting cutoffs critical, but so is the selection of statistical tests for examining the appropriateness of the cutoffs. Selection of the statistical tests to be used to analyze CRTs is based on the same principles as that used for selecting norm-referenced tests. The first factor to consider is the level of measurement of the variables involved. With CRTs, you categorize data into nominal variables; therefore, you must select statistical tests appropriate for this level of measurement. Remember that nominal variables are categorical in nature. For tests that are measured on a continuous scale to be evaluated with criterion-referenced instruments, the scores must first be categorized above and below the cutoff criterion. For criterion-referenced testing, the primary tool for analysis is a statistical technique using a contingency table (a 2 × 2 chi-square; see figure 7.1) for identifying those who score above the cutoff and those who score below the cutoff. Figure 7.1 illustrates the stability (dependability) of the CRT over 2 days. People classified as not meeting the standard (n_1) on both days or meeting the standard (n_4) on both days are consistently classified. Those classified as meeting the standard on one day and not meeting the standard the next

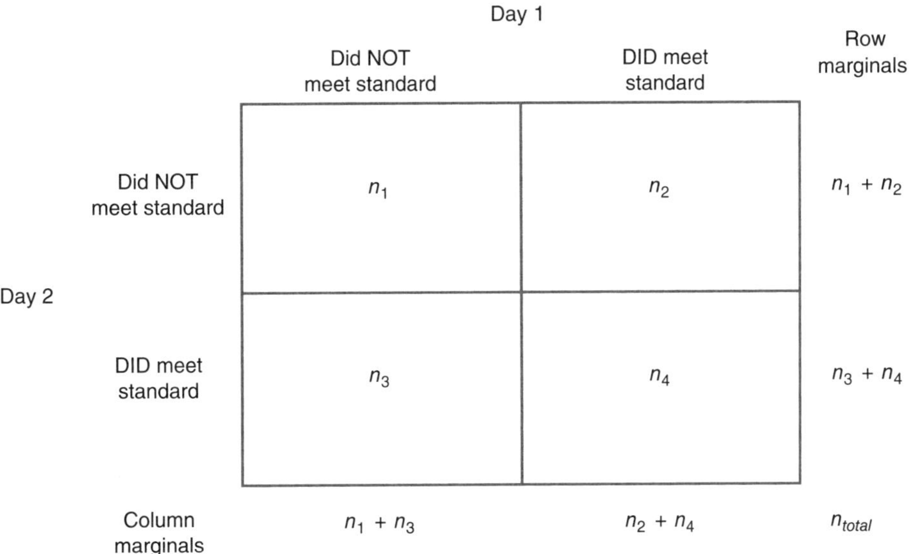

Figure 7.1 A 2 × 2 contingency table for a criterion-referenced test taken over 2 days.

(n_2) or vice versa (n_3) are misclassified. **Marginals** are the sum of observations for a specific row ($n_1 + n_2$ *or* $n_3 + n_4$) or column ($n_1 + n_3$ *or* $n_2 + n_4$) of a contingency table (see figure 7.1).

The next factor to consider in analysis is the specific measurement situation. The measurement situations are the same as those associated with norm-referenced testing. To establish the reliability of a CRT, you first need to determine whether you're concerned with the equivalence or the stability of the test. To measure the validity, you must have a criterion measure. The criterion measure reflects the true state of circumstances regarding the test being investigated. Recall Christina's challenge from the beginning of this chapter. Her interest in the relationship between creatine use and muscle pulls is a validity study. The criterion is whether the person had a muscle pull, and the predictor variable is whether or not the person was taking creatine.

STATISTICAL TECHNIQUES TO USE WITH CRITERION-REFERENCED TESTS

Several statistics are used to estimate the reliability and validity of CRTs. This text presents the techniques of chi-square (chapter 5), proportion of agreement (P), the **phi coefficient** (actually a Pearson product-moment correlation between two dichotomous variables), and Kappa (K). These are techniques that reflect association and agreement and can be used with data measured on a nominal scale.

As illustrated in chapter 5, the chi-square test is a test of association between nominally scaled variables. Logically, you would want there to be an association between how a person does on the first attempt of a CRT and on the second attempt. This is an illustration of CRT stability reliability. Likewise, you would like there to be an association between how a person does on a field test of a measure and how he or she would do on a more truthful measure (i.e., the criterion) of the characteristic being measured. This is an illustration of CRT criterion-related validity. Recall from chapter 5 that the null hypothesis in both of these tests is that there is *no* association (or relation), but rejection of the null hypothesis results in deciding there *is* an association between the variables. Obviously, you would like there to be a relationship between how one is scored multiple times on the same CRT.

Note that the variables are scored 0 or 1 for both of the measures. You can calculate the Pearson product-moment correlation coefficient (chapter 4) between the dichotomously scored variables. This special case of the Pearson product-moment correlation coefficient is called the phi coefficient. The phi coefficient has limits of –1.00 and +1.00 with a value closer to 1.00 in absolute value, indicating increased association, and a value close to zero, indicating no association.

SPSS produces the chi-square and phi coefficient as statistics options within the Crosstabs routine. How this is done will be illustrated later in the chapter.

The proportion of agreement (P) value is established by adding the proportions in the cells that are consistently classified; thus P is equal to the number of agreements ($n_1 + n_4$) divided by the total number ($n_1 + n_2 + n_3 + n_4$). From figure 7.1 it is estimated by the following formula:

$$\mathrm{P} = \frac{(n_1 + n_4)}{(n_1 + n_2 + n_3 + n_4)} \tag{7.1}$$

The P ranges from 0 to 1.00, and the higher the value, the more closely the data are consistently (correctly) assigned to cells. The problem with P is that values up to .50 could happen simply by chance.

The **Kappa (K)** value is a widely used technique that allows for the correction for chance agreements. It is closely associated with the phi (ϕ) coefficient, which is the Pearson product-moment (PPM) correlation calculated on nominal data. K is most appropriately used to assess interobserver agreement but can be used in test–retest situations or to examine the agreement between a predictor and a criterion that are nominally scaled. Although the proportion of agreement is a rough estimate of agreement or association between two nominal variables, the major problem with this statistic is that it does not consider the fact that some of these agreements could be expected purely because of chance. K takes chance agreement into account and therefore gives a more conservative estimate of the association between two nominal variables. The formula for K is

$$K = \frac{(P - P_c)}{(1 - P_c)} \quad (7.2)$$

where P is the proportion of observed agreement and P_c is the proportion of agreement due to chance. Consider the following example. Four hundred elementary students performed the 1-mile (1.6 km) run on each of 2 days. Their instructor wanted to know if the test could consistently measure the students' abilities to achieve the cutoff established in Fitnessgram. Table 7.2 presents these data. Note that this is a reliability example because it is the same test on more than one occasion.

Table 7.2 CRT Test–Retest Reliability Example

DAY 1	DAY 2: Did not achieve the standard	DAY 2: Did achieve the standard	Total
Did not achieve the standard	80	20	100
Did achieve the standard	50	250	300
Total	130	270	400

$\chi^2 = 137.13$, df = 1, $p < .001$; phi coefficient = .586

For this example, P is calculated to be

$$(250 + 80) / 400 = 330 / 400 = .825$$

K is calculated to correct for chance. The P (.825) was estimated previously. The proportion of chance values (P_c) is determined and added as shown here:

$$(130 \times 100) / (400 \times 400) = .081$$

and

$$(270 \times 300) / (400 \times 400) = .506$$

That is, multiply the marginals and divide by n^2. The sum of these properties is .587. Therefore, K = (.825 – .587) / (1 – .587) = .238 / .413 = .576. This value is substantially lower than the P value of .825. Therefore, it is suggested that chi-square, phi, percent agreement, and Kappa values be calculated to give the most information about the association involved.

Thus, given a 2 × 2 table, determine the proportion of observed agreement (P) by summing the number of agreements that appear on the diagonal of the table and dividing by the total number of paired observations. Determine the proportion of chance agreement (P_c). Sum these proportions across all the cells to obtain a total proportion of chance agreement.

Then substitute the proportion of agreement and the proportion of chance agreement into the Kappa formula. The values of K can theoretically range from –1.00 to +1.00; however, a negative value of K implies that the proportions of agreement resulting from chance are greater than those attributable to observed agreements. For that reason, K practically ranges from 0.00 to 1.00. The magnitude of the K is interpreted just as any other reliability or validity coefficient, with the higher the values, the better. However, because of the adjustment for chance agreement, values seldom exceed .75. Kappas of .20 or below indicate slight or poor agreement. Values in the .61 to .80 range are usually considered to be substantial, whereas values ranging from .41 to .60 are often considered to be moderate (Viera and Garrett 2005). K is an extremely useful statistic: It can be used not only to evaluate interobserver agreement, but to measure stability and equivalence in a test–retest situation and also for test validity.

A serious disadvantage of the K coefficient is that it is highly sensitive to low values in the marginals and small contingency tables because chance values are so high. It is also limited to square contingency tables. Again, SPSS can provide the Kappa coefficient as one of the statistics options within the Crosstabs routine.

Criterion-Referenced Reliability

For the most part, the same types of reliability and validity situations exist for CRTs as with norm-referenced data. Equivalence reliability as well as stability reliability can be estimated (see chapter 6).

Equivalence Reliability

Mahar and colleagues (1997) examined the criterion-referenced and norm-referenced reliability of the 1-mile (1.6 km) run/walk and the PACER (Progressive Aerobic Cardiovascular Endurance Run) test (both tests are used on the Fitnessgram). The sample consisted of 266 fourth- and fifth-grade students who were administered two trials of the PACER test and one trial of the 1-mile (1.6 km) run/walk. Equivalence reliability was examined between the 1-mile (1.6 km) run/walk and each trial of the PACER test for the total sample and also by gender. Both P and K values were calculated for all cases. The results are presented in table 7.3.

Inspection of the results indicates that fairly high P values ($.65 \leq P \leq .83$) are associated with varying levels of K ($.30 \leq K \leq .65$). Remember that you expect the K values to be more conservative than the P values. Whereas the equivalence reliability looks to be at least acceptable for the total sample and with boys alone, the values for the girls are much lower (P values of .66 and .65 and K values of .33 and .30). This result points out not only

Table 7.3 Criterion-Referenced Equivalence Reliability Between the 1-Mile (1.6 km) Run/Walk and PACER

Tests	Total sample	Boys	Girls
TRIAL 1			
P	.76	.83	.66
K	.51	.65	.33
TRIAL 2			
P	.71	.76	.65
K	.43	.52	.30

Note: For trial 1, n = 126 boys, n = 95 girls, and total (both) N = 221; for trial 2, n = 122 boys, n = 91 girls, and total (both) N = 213.

the nature of CRT reliability estimates, but also the importance of examining specific reliability situations.

Stability Reliability

Rikli, Petray, and Baumgartner (1992) examined the reliability of distance-run tests for students in kindergarten through grade 4. Test–retest reliability estimates using both the norm-referenced (intraclass reliability) and criterion-referenced techniques were calculated. Data on the 1-mi (1.6 km) and half-mile (.8 km) run/walk tests were gathered in the fall (on 1229 students—621 boys, 608 girls) and in the next spring (1050 students—543 boys, 507 girls). The P values for these data were calculated using the Physical Best and Fitnessgram cutoff values. The results are presented in table 7.4.

Table 7.4 Criterion-Referenced Reliability Estimates

		AGE									
		5		6		7		8		9	
		F	S	F	S	F	S	F	S	F	S
PHYSICAL BEST											
1/2 mi (.8 km)	M	.79	.86	.98	.95	.92	.86	.97	.83	.89	.90
	F	.88	.74	.98	.90	.89	.91	.96	.91	.92	.75
1 mi (1.6 km)	M	.70	.70	.94	.89	.95	.92	.90	.94	.95	.93
	F	.75	.88	.88	.73	.81	.87	.95	.94	.92	.90
FITNESSGRAM											
1 mi (1.6 km)	M	.75	.70	.76	.66	.85	.77	.91	.85	.86	.83
	F	.69	.51	.71	.45	.81	.85	.90	.84	.83	.94

"Criterion-referenced equivalence reliability between the 1-mile run," R.E. Rikli, C. Petray, and T.A. Baumgartner, *Research Quarterly for Exercise and Sport*, Vol. 63. 270-276, 1992 Taylor & Francis, reprinted by permission of Taylor & Francis (Taylor & Francis Ltd, www.tandfonline.com).

Inspection of the results indicates that all reliability estimates fall in the acceptable range (P ≥ .70) except the Fitnessgram standards for 5-year-old girls (fall = .69, spring = .51) and for 6-year-old boys (P = .66) and girls (P = .45) in the spring. These criterion-referenced values are consistently higher than the associated norm-referenced values. This is understandable because P values are not corrected for chance. Rikli and colleagues (1992) also explained this as follows: "The higher values for Physical Best are not surprising because *P* is always larger when there is a large percentage of scores that either meet or do not meet the standard on both the test and retest" (p. 274).

Morrow et al. (2010) used the just described criterion-referenced reliability procedures with school-based teachers to see if repeated administrations of the Fitnessgram test by the same teachers resulted in students achieving or not achieving the Fitnessgram Healthy Fitness Zone consistently on two occasions. Using aerobic fitness, body composition, and muscle strengthening items from the Fitnessgram resulted in percentage of agreements between .74 and .97 (median = .84); modified Kappas between .48 and .94 (median = .68); and phi coefficients between .40 and .92 (median = .65). All chi-square results were significant (P < .001). In total, these results suggested that there was good teacher reliability with Fitnessgram assessments.

Criterion-Referenced Validity

The validity of CRTs is usually established with some type of criterion-related situation, either concurrent or predictive. Construct validity can be demonstrated by examining the overlap of two divergent groups measured on a continuum.

Criterion-Related Validity

An example of the criterion-related validity approach, in this case concurrent validity, can be seen in the work of Cureton and Warren (1990). Remember, Cureton and Warren studied criterion-referenced standards for the 1-mile (1.6 km) run/walk test. The Fitnessgram (Cooper Institute for Aerobics Research 1987) and Physical Best (AAHPERD 1988) tests were used. Both tests provided criterion-referenced standards. The data were presented in table 7.1.

The results from these two CRT examples are presented in table 7.5. These results illustrate some of the problems of interpreting CRT results. Both tests have significant chi-square results, the phi coefficient is higher for the Physical Best standards, and the percent agreement and Kappa coefficient are higher for the Fitnessgram analysis.

Table 7.5 Comparison of Two CRT Validities

	Fitnessgram	Physical Best
Chi-square result	$\chi^2 = 55.35$, $df = 1$, $p < .001$	$\chi^2 = 66.41$, $df = 1$, $p < .001$
Phi coefficient	.309	.338
Percent agreement (P)	.85	.61
Kappa	.288	.277

Look again at table 7.1, which shows that 85% of participants were correctly classified for the Fitnessgram. Eleven percent achieved the standard on the run/walk test but were below the criterion $\dot{V}O_2$. These results are *false* negatives. That is, a participant is said to be okay on the field test (i.e., the run/walk), but in actuality (i.e., the criterion) he or she is below the standard. Notice too that 4% (n = 21) of participants did not meet the standard on the field test but were above the criterion $\dot{V}O_2$max. These participants are referred to as *false positives* because the field test results indicate they do not meet the standard but their performance on the criterion is above the standard. Compare the false negative and false positive results for the Fitnessgram and Physical Best results in table 7.1. The impact of false negatives and false positives can be important in determining which field test you might use. To help you differentiate between false negatives and false positives, consider a field test of cholesterol that involves a simple finger stick to obtain a drop of blood. The criterion method for estimating cholesterol would be from drawing venous blood. The results of your finger stick (i.e., the field test) can be accurate (you have been correctly identified as having a healthy or unhealthy cholesterol level) or inaccurate. If the field test reports that your cholesterol level is healthy when in fact it is not, the results are a false negative. If the field test results indicate that your cholesterol level is too high when it is actually in the healthy range, the result is a false positive.

Construct-Related Validity

Setting cutoff scores is a difficult undertaking. The divergent group method can be used as a construct validation procedure. As illustrated in figure 7.2, the concept is to find two groups that are clearly different from each other. To establish a cutoff using this technique, we plot the distributions of scores for the divergent groups. The point in the curves where the scores overlap is used as the criterion cutoff score. This method was explained in more detail by Plowman (1992). A theoretical application of this approach would be to select two groups of adults (or children). One of these groups would be physically active enough to obtain a health benefit, whereas the other group would not be active enough for a health benefit. Obtaining data on the amount of physical activity for each of these groups and then graphing it should help set a cutoff score for a minimum amount of physical activity needed for a health benefit.

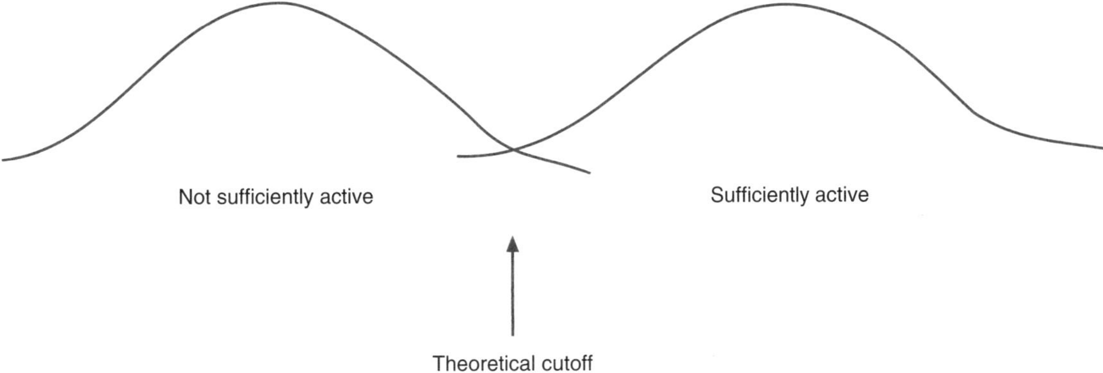

Figure 7.2 A theoretical example of the divergent group method.

Tarter and colleagues (2009) provide an example of construct validity using a construct-related CRT approach called receiver operating curve analysis (ROC). The authors were examining the use of an aggregate performance measure called the composite physical fitness index (CPFI) to predict potential for playing in the National Hockey League (NHL). Playing in the NHL was defined as playing at least five games within a 4-year period after a player was drafted. The ROC technique maximizes the number of discriminations between those who achieved the criterion and those who did not. It was found that setting the cutoff at the 80th percentile of the CPFI for defensemen yielded a 70% probability of success, whereas setting it at the 80th percentile for forwards yielded only a 50% probability of making the NHL. When the CPFI scores were adjusted to the 90th percentile, the probabilities for defensemen and forwards shifted to 72% and 61%, respectively. This study again points to the importance of setting cutoff scores appropriately.

Morrow, Going, and Welk (2011) present eleven manuscripts in a supplement issue of the *American Journal of Preventive Medicine* where similar ROC procedures were used to determine criterion-referenced standards for Fitnessgram aerobic capacity and body composition measures. Effectively using the procedures described above, they were able to determine cutoff scores for the Fitnessgram Healthy Fitness Zone based on whether or not the children and youth had metabolic syndrome. Thus, metabolic syndrome served as the criterion; cutoff scores were determined for aerobic capacity and body composition that differentiated those with and without metabolic syndrome. This example also illustrates a criterion-referenced validity study.

CRITERION-REFERENCED TESTING EXAMPLES

The logic behind the use and interpretation of reliability and validity procedures with CRTs is similar to that with norm-referenced measurement as presented in chapter 6. If two trials of the same measure are administered, then stability reliability is being assessed. With CRTs this is the reliability or dependability of classification. If two tests are being compared that are thought to measure the same thing, equivalence is being assessed. With CRTs, the equivalence is whether the two tests result in equivalent classifications for the people being assessed. There is no analogy to internal consistency reliability (i.e., alpha reliability) with CRT. If one of the measures is a criterion, the matter being investigated is validity. As we have pointed out several times in this chapter, determination of the criterion with CRTs is the most difficult aspect to establish. However, when the analysis is to determine if the measure is significantly associated with a criterion, test validity is being investigated.

The following examples present specific applications of selected techniques for the assessment of reliability and validity using criterion-referenced testing. Try to calculate P and K for the following mastery items.

Mastery Item 7.1

Assume that two criterion-referenced physical fitness tests have been developed to establish the cutoff scores for physical performance on sit-ups. For test 1, participants perform sit-ups with their hands clasped on their chests ("handches"). For test 2, they perform sit-ups with their hands behind their heads ("handhead"). Are the tests equivalent? A sample of participants is administered test 1 and test 2, and a 2× 2 contingency table is developed to determine if there is equivalence in the classification on the two tests. The data are presented in table 7.6. Use the following steps to obtain the chi-square, phi coefficient, and K statistics. SPSS does not calculate P, so you will have to do that by hand from the output you receive. (An Excel template is in the WSG in chapters 5 and 7.)

1. Download table 7.6 from the WSG.
2. Start SPSS.
3. Click on the Analyze menu.
4. Scroll down to Descriptive Statistics and across to Crosstabs and click.
5. Put "handches" in the rows and "handhead" in the columns by clicking the arrow keys.
6. Click Statistics.
7. Check the chi-Square, phi, and Kappa boxes.
8. Click Continue.
9. Click OK.

Table 7.6 Example of Equivalence Reliability

Participant	handches	handhead	Participant	handches	handhead
1	1	1	21	0	0
2	1	1	22	0	0
3	1	1	23	0	0
4	1	1	24	0	0
5	1	1	25	0	0
6	1	1	26	0	0
7	1	1	27	0	0
8	1	1	28	0	0
9	1	1	29	0	0
10	1	1	30	0	0
11	1	1	31	0	0
12	1	1	32	0	0
13	1	1	33	0	0
14	1	0	34	0	0
15	1	0	35	0	1
16	1	0	36	0	1
17	1	0	37	0	1
18	1	0	38	0	1
19	1	0	39	0	1
20	1	0	40	0	1

Note: 0 = fail, 1 = pass

Mastery Item 7.2

Human performance specialists use test–retest reliability—the stability of a test over successive administrations—more frequently than equivalence to determine the reliability of CRTs. Let's assume that we select one test and administer it on Friday (day 1) and then administer it to the same group of students on the following Monday (day 2). We are concerned with the consistency of classification across the two testing periods. The data are presented in table 7.7. Use the SPSS commands from the preceding Mastery Item to calculate the statistics.

Table 7.7 Example of Stability Reliability

Participant	Friday	Monday	Participant	Friday	Monday
1	1	1	21	0	0
2	1	1	22	0	0
3	1	1	23	0	0
4	1	1	24	0	0
5	1	1	25	0	0
6	1	0	26	0	0
7	1	0	27	0	0
8	1	0	28	0	0
9	1	1	29	0	0
10	1	1	30	0	0
11	1	1	31	0	1
12	1	1	32	0	1
13	1	1	33	0	0
14	1	1	34	0	0
15	1	1	35	0	0
16	1	1	36	0	0
17	1	1	37	0	0
18	1	1	38	0	1
19	1	0	39	0	1
20	1	0	40	0	1

Note: 0 = failed to meet criterion, 1 = met the criterion.

Mastery Item 7.3

From a validity standpoint (whether it be predictive validity, concurrent validity, or construct validity), the application of the 2 × 2 contingency table is appropriate. For example, let's assume that we have a standard for body composition and we suspect that if people achieve the Fitnessgram Healthy Fitness Zone for body composition, they are less likely to have metabolic syndrome. Alternatively, if they are in the Needs Improvement Zone, they are more likely to have metabolic syndrome. Therefore, we want to determine if the Fitnessgram body composition cutoff can properly classify those who have metabolic syndrome as opposed to those who do not. The data are presented in table 7.8. Use the same SPSS commands presented for Mastery Item 7.1 and interpret the validity of the cutoff score for body composition to predict metabolic syndrome. Hint: Put the Needs Improvement Zone in the rows and the metabolic syndrome in the columns. Note these data are for illustrative purposes only.

Table 7.8 Example of Statistical Validity

Participant	Metabolic syndrome	Needs Improvement Zone	Participant	Metabolic syndrome	Needs Improvement Zone
1	0	1	31	1	1
2	0	1	32	1	1
3	0	1	33	1	1
4	0	1	34	1	1
5	0	1	35	1	1
6	0	1	36	1	1
7	0	1	37	1	1
8	0	1	38	0	0
9	0	1	39	0	0
10	1	0	40	0	0
11	1	0	41	0	0
12	1	0	42	0	0
13	1	0	43	0	0
14	1	0	44	0	0
15	1	0	45	0	0
16	1	0	46	0	0
17	1	0	47	0	0
18	1	0	48	0	0
19	1	0	49	0	0
20	1	0	50	0	0
21	1	1	51	0	0
22	1	1	52	0	0
23	1	1	53	0	0
24	1	1	54	0	0
25	1	1	55	0	0
26	1	1	56	0	0
27	1	1	57	0	0
28	1	1	58	0	0
29	1	1	59	0	0
30	1	1	60	0	0

Note: For metabolic syndrome 0 = yes, 1 = no; for Needs Improvement Zone 0 = yes, 1 = no (in Healthy Fitness Zone).

APPLYING CRITERION-REFERENCED STANDARDS TO EPIDEMIOLOGY

Epidemiologic research is a tool that is becoming increasingly popular in human performance measurement. It is closely related to CRT because the variables are often nominal in nature and some of the statistics used are those calculated from a 2 × 2 contingency table. The criterion measure is categorical: for example, alive or dead; has a disease or does not have a disease. The predictor variables can be nominal (e.g., gets sufficient physical activity or does not get sufficient physical activity) or continuous (e.g., weight). It is when the predictor and criterion variables are both nominal that epidemiologic statistics are most like those of CRT (and can even be calculated with SPSS Crosstabs or Excel).

Epidemiology is the study of the distribution and determinants of health-related states and events in populations and the applications of this study to the control of health problems (Last 1992). Epidemiology is the fundamental science of public health that uses hypothesis testing, statistics, and research methods to develop an understanding of the frequency and distribution of mortality (death) and morbidity (disease or injury), and more importantly, the risk factors that are causally related to mortality and morbidity (Stone, Armstrong, Macrina, and Pankau 1996). In our fields, modern epidemiologic research has clearly discovered the increased risk for a variety of chronic diseases related to a sedentary or physically inactive lifestyle (Ainsworth and Matthews 2001; Caspersen 1989; USDHHS 1996, 2008).

Descriptive epidemiology seeks to describe the frequency and distribution of mortality and morbidity according to time, place, and person. For instance, what was the rate of breast cancer in adult women in the United States during the 1990s? Epidemiology may help identify risk factors of mortality and morbidity. Analytical epidemiology pursues the causes and prevention of mortality and morbidity. For example, does obesity increase the risk of breast cancer in women? In women who are obese, does moving into a healthy weight range lower the risk of breast cancer?

Epidemiology uses both prospective research approaches, tracking a study group into the future, and retrospective research approaches, looking back at a database of previously collected data. Epidemiologists use a variety of research designs, some of which are depicted in table 7.9. (The Excel template in the WSG calculates these statistics for you.)

Table 7.9 Research Designs in Epidemiology

Type	Description
EXPERIMENTAL	
Randomized clinical trial	Randomly assigns participants to treatments or exposures
Community trial	Randomly assigns whole communities to treatments or exposures
OBSERVATIONAL	
Case series	Notes cases at a particular time or place
Cross-sectional	Takes a snapshot of identifiable groups at one time
Proportionate mortality or morbidity study	Compares results of a study group to the population
Case-control	Compares known cases of mortality or morbidity with matched noncases
Cohort	Longitudinal; generally tracks populations long term

Epidemiology is a science that requires the use of advanced statistics and complicated multivariate models to understand the relationships between risk factors and mortality and morbidity while controlling for confounding factors or extraneous variables. However, the logic is similar to that presented in chapter 5, and the decisions are made about a null hypothesis. Even though the statistics might be different, the logic is identical. Complicated statistical models such as logistic regression and proportional hazards are used to test relationships between disease states and predictors of the disease state. Those types of analyses are beyond the scope of this text and are not necessary for us to know at present. But we do need to know some basic procedures and statistics to understand how criterion-referenced standards play a role in epidemiology. Two basic statistics are the calculations of incidence and prevalence.

- **Incidence.** The number, proportion, rate, or percentage of *new* cases of mortality and morbidity. Incidence could be calculated in a randomized clinical trial or a prospective, longitudinal cohort study.
- **Prevalence.** The number, proportion, rate, or percentage of total cases of mortality and morbidity. Prevalence is calculated in a cross-sectional study.

Values of incidence and prevalence are often expressed as a rate, which is the number of cases per unit of the population. An example would be 10 cases per 1000 in the population or 100 deaths per 100,000 in the population. The value of expressing incidence and prevalence as a rate is that two populations of different sizes can be compared. For example, the rate of mortality in Dallas, Texas, can be compared with the rate of death in New York City.

In analytical epidemiology, we convert measures of incidence or prevalence into estimates of risk:

- **Absolute risk.** The risk (proportion, percentage, rate) of mortality or morbidity in a population that is exposed or not exposed to a risk factor.
- **Relative risk.** The ratio of risks between the exposed or unexposed populations. This statistic is calculated with incidence measures.
- **Odds ratio.** An estimate of relative risk used in prevalence studies.
- **Attributable risk.** The risk of mortality and morbidity directly related to a risk factor. This risk can be thought of as the reduction in risk related to removing a risk factor.

Let's combine criterion-referenced standards with an example of a simple analysis in epidemiology. High cholesterol is defined by the American Heart Association and the National Heart, Lung, and Blood Institute as a value of 240 mg/dl or above. Thus, the criterion-referenced standard for total cholesterol is 240 mg/dl. Let's examine the results of a theoretical epidemiologic study about the relationship of cholesterol and mortality attributable to heart attack. Examine table 7.10, which is a 2 × 2 contingency table. We have conveniently labeled each cell as A, B, C, or D. This will make all descriptive and analytical calculations quite simple. We also conduct our analyses on incidence and prevalence bases. In this study, 56 participants with high cholesterol and 44 with cholesterol below that criterion are compared. All had a genetic history of early coronary heart disease. Note that both variables are categorical in this example.

Table 7.10 Results of a Hypothetical Study Relating Cholesterol and Heart Attack Mortality

	OUTCOME	
Exposure	**Heart attack deaths**	**No heart attack deaths**
High cholesterol	A 25	B 31
No high cholesterol	C 7	D 37

If you examine all the results in figure 7.3 you can observe the following:

- All calculations can be made from the easy-to-follow formulas using the A, B, C, and D cell identifiers.

$$Total = \frac{A + C}{A + B + C + D} = \frac{25 + 7}{25 + 31 + 7 + 37} = \frac{32}{100} = .32 \text{ or } 32\%$$

$$High = \frac{A}{A + B} = \frac{25}{25 + 31} = \frac{25}{56} = .45 \text{ or } 45\%$$

$$Not\ high = \frac{C}{C + D} = \frac{7}{7 + 37} = \frac{7}{44} = .16 \text{ or } 16\%$$

→ Absolute risk

$$RR = \frac{A \div (A + B)}{C \div (C + D)} = \frac{.45}{.16} = 2.81$$

→ Relative risk

$$OR = \frac{AD}{BC} = \frac{25 * 37}{7 * 31} = \frac{925}{217} = 4.26$$

→ Odds ratio

$$AR = \frac{[A \div (A + B)] - [C \div (C + D)]}{A \div (A + B)} = \frac{.45 - .16}{.45} = .64 \text{ or } .64\%$$

→ Attributable risk

Figure 7.3 Statistical analysis of epidemiological data in table 7.10.

- The absolute risk for heart attack death was 32% for all participants, 45% for those with high cholesterol, and 16% for those without high cholesterol.
- If a participant had high cholesterol, the relative risk of 2.81 indicated that high cholesterol elevated the risk of heart attack mortality by a multiplier of 2.81.
- If a participant had high cholesterol, the odds ratio indicated elevated odds of heart attack mortality by a multiplier of 4.26.
- The attributable risk indicated that high cholesterol contributed to 64% of the heart attack mortality. Thus, heart attack mortality could be reduced by 64% if high cholesterol were no longer present in people of this population.

The example used in table 7.10 and figure 7.3 was contrived to serve as a simple demonstration of some basic concepts and analyses in epidemiology. However, research studies using epidemiologic methods have demonstrated strong relationships between levels of physical activity and fitness and a variety of mortality and morbidity outcomes from chronic diseases. Chapter 9 will discuss some of those specific findings in more detail.

Mastery Item 7.4

1. Go to the WSG for chapter 7 and download the data from table 7.10.
2. Confirm that you can calculate the odds ratio and relative risk by using the Crosstabs routine.
3. Do so by running Analyze → Descriptive Statistics → Crosstabs and placing "cholesterol" in the rows and "heart attack" in the columns.
4. Then go to Statistics and click on Risk.
5. When you review the SPSS results, you should see that the odds ratio and relative risk values are presented in the SPSS output.

Note that in Table 7.10 what is viewed as negative exposure is listed in the first row, followed by what is viewed as positive exposure. Likewise, for the outcome, what is viewed as a negative outcome is listed in the first column, followed by what is viewed as a positive outcome. We recommend that you construct the contingency table in this fashion. You can reorder either or both of the variables and arrive at similar conclusions, but setting it up as we suggest will generally make your interpretation more understandable.

Mastery Item 7.5

In table 7.11, a 2×2 contingency table, are results from a study conducted by Bungum, Peaslee, Jackson, and Perez (2000). The study examined the relationship of physical activity during pregnancy and the risk of Cesarean birth compared with normal vaginal birth. Perform the analysis presented in figure 7.3 with these data.

Table 7.11 Results of a Study Relating Physical Activity During Pregnancy and Type of Birth

	OUTCOME	
Exposure	Cesarean section birth	Vaginal birth
Sedentary	A 26	B 67
Active	C 7	D 37

Mastery Item 7.6

Let's now apply these epidemiologic statistics to the data in Table 7.8. Remember that you want to set up the contingency table as suggested previously for the negative and positive exposures and negative and positive outcomes. Run SPSS Crosstabs on the data. Click on Statistics and then check the Risk box. Confirm that the risk of having metabolic syndrome goes up by 294.9% if one is in the Needs Improvement Zone (notice that the Odds ratio = 3.949).

Dataset Application

The chapter 7 large dataset in the WSG consists of data on school-age children who were tested for body composition risk on two Fitnessgram tests. The Fitnessgram permits determination of Healthy Fitness Zone from body mass index (BMI) and from skinfolds (to estimate percentage of fat). Theoretically, it should not make any difference if either of these tests is used. A person who is overweight or at risk should be identified as such with each test. This is an equivalence reliability example. Are the results equivalent regardless of which test you use? Use SPSS to calculate chi-square, the phi coefficient, and Kappa. You will need to calculate percent agreement from the 2 × 2 table that you produce with SPSS. What is your interpretation of the results of these two tests? Are they equivalent? Do you get similar results if you conduct the analyses separately for boys and girls?

MEASUREMENT AND EVALUATION CHALLENGE

When Christina arrived at the library, she perused *Measurement and Evaluation in Human Performance, Fifth Edition,* and found that she needed to select criterion-referenced measurement tools to assess the relationship between pulled hamstrings and creatine use. She decided to ask the athletes she was treating two simple questions:

1. In the past 12 months, have you sustained a hamstring injury?
2. During the past 12 months, have you taken creatine?

The answers would be yes or no. The criterion measure is hamstring pulls, and the predictor is creatine usage. Notice that both variables are nominal (with two categories: yes or no).

From her readings, Christina believes that she could study the validity of creatine use as a predictor of hamstring injuries by examining the proportion of agreement (P) and Kappa (K) values. She would ask all of her patients the two questions (not just the ones who had hamstring injuries) and set up a 2 × 2 contingency table. She would use each of these statistics and epidemiologic statistics to investigate hamstring pull risks associated with use of creatine. She hopes, as a result of this study, to obtain some information that will help her in counseling athletes on creatine use.

SUMMARY

Several specific measurement situations in human performance are well suited for criterion-referenced measurement; moreover, there are specific statistical techniques that must be used with these CRTs. The primary problem associated with criterion-referenced testing in human performance is in establishing a criterion or cutoff score. Because few measurement problems in human performance have concrete criterion scores associated with them, cutoff scores have to be established from experts' opinions or normative data. Cutoff scores can often be arbitrary, thus affecting empirical validity. The establishment of these scores also affects the reliability and the statistical validity of the test. Therefore, criterion scores must be set with a high degree of caution.

In the area of youth fitness testing, criterion-referenced standards have been established by test developers (e.g., Fitnessgram). In other areas of human performance, such as sport skills testing, such standards have typically not been set. In epidemiological research and practice, many cutoffs have been established that are directly related to health risks. See Morrow, Zhu, and Mahar (2013) for a summary of criterion-referenced testing use with the Fitnessgram.

Criterion-referenced statistical techniques are used to analyze data. Criterion-referenced measurement can be a valuable tool for you to examine measurement in human performance. Criterion-referenced testing is the method of choice when variables are categorized and where an obvious level of proficiency must be achieved before proceeding to the next level (e.g., flotation and treading water skills need to be mastered before one enters the deep end of the pool).

The typical statistics used with criterion-referenced testing reliability and validity are chi-square, phi coefficient, proportion of agreement (P), and Kappa (K), which adjusts the proportion of agreement for chance.

Finally, you learned how epidemiologic statistics are closely related to criterion-referenced testing procedures. Epidemiology is a powerful method for identifying risk factors for various disease outcomes.

Go to the WSG for homework assignments and quizzes that will help you master this chapter's content.

Part IV

Human Performance Applications

Part IV of this book is about what you will do after graduation. Some of you will work in school-based instructional settings, others in athletics, and still others in a wide variety of health professional and human performance job settings, such as physical therapy, health clubs, corporate fitness, wellness programs, hospitals, and graduate schools. The particular measurement and evaluation tasks that you will encounter will vary from job to job. However, knowledge of the issues of reliability and validity will serve you well whatever your career.

We introduce part IV with a brief description of the domains that will be illustrated throughout the remainder of the book. The domains reflect the cognitive, psychomotor, and affective learning that you will be evaluating after graduation. Each of these domains is reflected in a taxonomy (review chapter 1). Each level of the taxonomy is built on the levels below it. For example, in the cognitive domain, you must first demonstrate knowledge before you can exhibit comprehension.

Likewise, you would not expect young children to be able to achieve well on a difficult (or higher-order) assignment, and you would expect that all college athletes would score high on a simple motor task. The measurement tasks that you conduct must reflect the appropriate level of learning or performance expected of the people with whom you are working. Thus, measurement protocols for each domain must be carefully considered. A key concept in measurement is to be able to design and use measurement protocols that discriminate among people who are actually at different levels of achievement.

In part I you were introduced to tests and measurement and the use of computers to help you make evaluation decisions. In part II you learned basic statistical concepts, including descriptive statistics (chapter 3), correlation and prediction (chapter 4), and inferential statistics (chapter 5). These two parts provided the background and the tools necessary for making reliable and valid decisions. In part III you learned about reliability and validity theory. You used the foundational knowledge from parts I and II to make these decisions.

Now you have the background, theory, tools, and information with which to make valid decisions. In part IV we turn to the various domains in which you will make these decisions. Chapter 8, on the cognitive domain, provides information about developing valid written tests and surveys. Chapters 9, 10, and 11, on the psychomotor domain, address valid evaluation of physical fitness and physical activity assessment. The medical and scientific literature on the relationships among physical fitness and physical activity and the prevention of disease increases almost daily. Therefore, exercise scientists must fully understand the issues of reliability and validity as they relate to human performance testing. Reliability and validity in the measurement and evaluation of adult physical fitness and physical activity are discussed in chapter 9, whereas chapter 10 deals with measurement and evaluation of youth physical fitness and physical activity. Chapter 11 presents techniques for reliable and valid measurement of ability and skills assessment for sport and human performance. Chapter 12, on the affective domain, provides guidelines for making valid decisions when using psychological measurements in sport and exercise psychology. The last two chapters are specifically targeted toward those students who have career goals of teaching physical education in schools. Chapter 13 addresses making decisions that lead to valid student assessment and reporting of grades. Chapter 14 presents examples of alternative assessment strategies that provide additional ways of assessing student achievement.

CHAPTER

8

Developing Written Tests and Surveys

OUTLINE

OBJECTIVES

After studying this chapter, you will be able to

- plan high-quality written tests;
- construct high-quality written tests;
- score written tests efficiently;
- administer written tests properly;
- analyze written tests; and
- understand concerns associated with planning, constructing, and enhancing the return of questionnaires.

 The lecture outline in the WSG will help you identify the major concepts of the chapter.

MEASUREMENT AND EVALUATION CHALLENGE

Kate, an educational researcher, is conducting an experiment on the effectiveness of using computers to teach basic statistical concepts. She randomly assigns students taking the basic statistics course to one of three sections of the class. One group will be taught by the traditional lecture method. A second group will receive all class lectures and activities online using a newly developed multimedia approach. This group will not attend the general lectures. The third group will complete the course in a blended fashion: They will attend some of the weekly general lectures but use online course materials and activities instead of the other class lectures. This group will receive 50% of their course work in the traditional method and 50% through online instruction and activities. What decisions must Kate consider and what steps does she need to follow to measure how well students in each group learn basic statistical concepts?

One primary objective of a research project or a physical education curriculum often is to increase participant knowledge and understanding of various aspects of physical activity. To determine if this objective is being met, making measurements in the cognitive domain is necessary. *The written test is used for measuring the level of achievement of cognitive objectives. Also, a common objective of research is to assess people's attitudes, opinions, or thoughts about a particular topic. Most often this is accomplished through the use of a questionnaire. The construction and administration of an accurate questionnaire are more complex than you might think.* Although the focus of this chapter is on the process for developing written paper-and-pencil tests and questionnaires, all of the concepts presented apply equally well to online tests, which are the latest innovation in testing.

There are many sources for written tests. In some disciplines, nationally normed standardized tests are available. Textbook publishers (e.g., the one publishing this book) often provide tests or banks of test questions from which you can build your own tests. Some state agencies provide written tests for statewide testing programs, making comparisons possible among schools or districts. In human performance, however, outside sources of written tests are rare. In physical education, the lack of standardized tests is partly due to the great variety of activities embedded in physical education curricula and the fact that there are fewer textbooks available in physical education than in such classroom subjects as English and math. In our discipline, the most common source of written tests and questionnaires is undoubtedly the researcher or teacher interested in measuring cognitive objectives. This is not all bad, because the person making the assessment should be able to construct the most valid measuring instrument (one that measures what it is intended to measure). However, knowing what to measure and knowing how to measure it are different things. There are five requirements for constructing effective written tests:

- You must be knowledgeable in the proper techniques for constructing written tests. Various types of questions have differing efficiencies and uses in certain situations.
- You must have a thorough knowledge of the subject area to be tested. Without this knowledge, it is difficult to construct meaningful test questions.
- You must be skilled at written expression. Test questions devised by people lacking good writing skills are often ambiguous. This ambiguity reduces the validity and reliability of a written test because there is no way of distinguishing whether an incorrect response is chosen due to the participant's lack of knowledge or to an error in interpretation of the question.

In the psychomotor domain, not every student will be able to master every tennis skill; in the cognitive domain, not every student will be able to master every cognitive objective. In both cases, achievement tests are designed to ascertain each student's level of accomplishment.

- You must have an awareness of the level and range of understanding in the group to be tested so that you can construct questions of appropriate difficulty. As will be explained later, not having such awareness can affect the efficiency of the test.
- As the prospective test constructor, you must be willing to spend considerable time and effort on the task. Effective written tests are not put together overnight.

If you examine these five requirements carefully, you will notice that the last four are also qualities of a careful researcher or a dedicated teacher. However, we will limit ourselves in this chapter (and book) to presenting information about the first requirement, proper techniques for constructing a written test. Properly constructed tests can result in reliable and valid decisions about the cognitive ability being assessed. For Kate to determine if there are differences in the manner in which statistical concepts are taught, she must first be able to accurately measure students' levels of understanding of these concepts. Kate can use the information in the following sections as she considers how to construct such a written test.

PLANNING THE TEST

First, consider the differences between mastery tests (criterion-referenced tests) and achievement tests (norm-referenced tests). A **mastery test** is used to determine whether a student has achieved enough knowledge to meet some minimum requirement set by the tester. It is not used to determine the relative ranking of students' cognitive abilities but rather to determine each student's compliance, or lack of compliance, with some previously determined standard or criterion. A familiar example of a mastery test is the type of spelling test in which the expected score is perfect or nearly perfect—all words are spelled correctly. Other examples are the written portion of the test for obtaining a driver's license or other health professional licensures, on which you must get a certain minimum number correct to pass.

The purpose of an **achievement test**, on the other hand, is to discriminate among levels of cognitive accomplishment. Because it is usually not reasonable to expect every student to achieve 100% of every cognitive objective put forth, identifying each student's progress toward meeting the objectives is of great importance.

In human performance, both types of tests have key uses. For example, in a potentially dangerous activity such as gymnastics or swimming, using a mastery test of safety rules is prudent. For the most part, however, this chapter deals with the various phases of constructing and using achievement tests, which are more commonly used for human performance assessments than mastery tests.

 Go to the WSG to complete Student Activity 8.1 and view Video 8.1.

There are two important decisions to make when planning your written test. The first and more significant of these involves determining what is to be measured. One technique for ensuring that a written test measures the desired objectives and that the correct emphasis is given to each objective is to develop a table of specifications. The second fundamental decision in planning a written test involves answering several mechanical questions concerning how the objectives are to be measured, the frequency and timing of testing, the number and type of questions, and the format and scoring procedures that will be used.

What to Measure

The question of what the test will measure should be answered before instruction begins. *The objectives of a course of study, the experiences used to meet these objectives, and the implementation and sequence of these experiences must all be determined in advance if an instructional unit is to be effective.* You can alter these elements as instruction progresses, but radical changes should not be necessary. In any event, testing will allow you to measure the degree to which the objectives of the course are being achieved and to evaluate where problems may exist. When the objectives to be assessed lie in the cognitive domain, the initial step in designing a written test is the development of a table of specifications.

The **table of specifications** is to the test writer what the blueprint is to the home builder. It provides the plans for construction. The table of specifications identifies the relative importance of each content area on the test by assigning it a percentage value. It is a two-way table, with the **content objectives** of the instructional unit along one axis and **educational objectives** along the other. Content objectives are the specific goals determined by an instructor, and educational objectives are generic topics suggested by various experts. A table of specifications helps ensure a test's content validity (the extent to which the items on a test adequately sample the subject matter and abilities that the test is designed to measure).

Let us look at an example that demonstrates the process of formulating a table of specifications for a 60-item test to be used with an instructional badminton unit. The content objectives of the instructional unit and an instructor's decision about their relative importance might be as follows:

History	5%
Values	5%
Equipment	10%
Etiquette	10%

Safety	10%
Rules	20%
Strategy	15%
Techniques of play	25%
Total	**100%**

The educational objectives (see also chapter 1) and the tester's weighting of each might be as follows:

Knowledge	30%
Comprehension	10%
Application	30%
Analysis	20%
Synthesis	0%
Evaluation	10%
Total	**100%**

Once an instructor has determined the content and educational objectives and their relative importance, the table of specifications can be constructed.

For this example, the result is shown in table 8.1. The content objectives and their relative weights are located in the rows, and the educational objectives and their weights are in the columns. The weight associated with a single cell of the table is found by determining the product of the intersection of the appropriate row and column. For example, the weight for knowledge of history is determined by multiplying 5% (the weight of history) by 30% (the weight of knowledge) to get 0.015, or 1.5%. This product for any cell is an expression of the approximate percentage of the test that should be made up of items combining the two types of objectives intersecting in that cell. The actual number of questions of each combination is found by multiplying the obtained percentage by the proposed length of the test. In this case, we wanted a 60-item test, so for knowledge of history we would multiply 0.015 by 60 to get 0.9. In table 8.1, each cell is divided into two halves; the upper number represents the percentage of the test made up of items combining the appropriate combination of objectives, and the bottom number represents the number of questions of this type based on a total test length of 60 items.

Obviously, it is not possible to include on the test 0.9 of a question dealing with knowledge of the history of badminton; the numbers in the table of specifications are to be used as guides, and usually some rounding and adjusting are required. For example, the 0.9 would logically result in one question of the 60 focusing on knowledge of the history of badminton. If the table of specifications is followed closely, the resulting test will contain questions in proportion to the percentages of weighting for each category.

Mastery Item 8.1

Based on the table of specifications in table 8.1, how many questions involving the analysis of techniques of play would be included on a 100-item test?

Various educators and test construction experts have identified educational objectives that may be used in tables of specifications. The educational objectives in table 8.1 come from a list published in *Taxonomy of Educational Objectives* (Bloom 1956). Recall from chapter 1 that the cognitive taxonomy consists of knowledge, comprehension, application,

Table 8.1 Table of Specifications for a 60-Item Written Test on Badminton

			EDUCATIONAL OBJECTIVES						
			Knowledge	Comprehension	Application	Analysis	Synthesis	Evaluation	Totals for content objectives
		Weight	30%	10%	30%	20%	0%	10%	100%
Content objectives	History	5%	1.5% 0.9	0.5% 0.3	1.5% 0.9	1.0% .06	0% 0	0.5% 0.3	3
	Values	5%	1.5% 0.9	0.5% 0.3	1.5% 0.9	1.0% 0.6	0% 0	0.5% 0.3	3
	Equipment	10%	3.0% 1.8	1.0% 0.6	3.0% 1.8	2.0% 1.2	0% 0	1.0% 0.6	6
	Etiquette	10%	3.0% 1.8	1.0% 0.6	3.0% 1.8	2.0% 1.2	0% 0	1.0% 0.6	6
	Safety	10%	3.0% 1.8	1.0% .06	3.0% 1.8	2.0% 1.2	0% 0	1.0% 0.6	6
	Rules	20%	6.0% 3.6	2.0% 1.2	6.0% 3.6	4.0% 2.4	0% 0	2.0% 1.2	12
	Strategy	15%	4.5% 2.7	1.5% 0.9	4.5% 2.7	3.0% 1.8	0% 0	1.5% 0.9	9
	Techniques of play	25%	7.5% 4.5	2.5% 1.5	7.5% 4.5	5.0% 3.0	0% 0	2.5% 1.5	15
	Totals for educational objectives	100%	18	6	18	12	0	6	Test total = 60

Note: The top number in each cell of the table body is the percentage of questions for the combined content and educational objectives for that cell; the bottom number is the actual number of questions (of the 60 in total) that the percentage represents.

analysis, synthesis, and evaluation. Briefly, knowledge is defined as remembering and being able to recall facts; comprehension as the lowest level of understanding; application as the use of abstractions in real situations; analysis as the division of material into its parts to make clear the relationship of the parts and the way they are organized; synthesis as putting together elements and parts to form a whole; and evaluation as judgments about the value of ideas, works, solutions, methods, and materials.

The following list of questions or tasks gives you an idea of how Bloom's taxonomy would apply to a written test for basketball.

- ***Knowledge.*** What is the height of a regulation basketball hoop?
- ***Comprehension.*** What area of the court is the forwards' responsibility in a zone defense?
- ***Application.*** What defense should be used when the opposing team is much faster than your team?
- ***Analysis.*** Prioritize the following basketball skills for each player position: blocking shots, dribbling, passing, shooting.
- ***Synthesis.*** Design a practice schedule for the first 3 weeks of the season for a boys' high school team having four baskets in the gym, 5 days per week to practice, 90 minutes for each practice, and 35 boys out for the team.

• ***Evaluation.*** Present arguments for and against the following statement: Rather than have separate boys' and girls' basketball teams at the junior high school level, the school should have coed teams.

In 2001, Bloom's taxonomy was expanded by Anderson and Krathwohl (2001) to combine the cognitive process with knowledge dimensions. They refer to the highest level as creating, and slight modifications were made to the categories. However, the basic concepts for constructing a table of specifications, as already described, remain valid.

Another list of educational objectives includes the categories of terminology, factual information, generalization, explanation, calculation, prediction, and recommended actions (Ebel 1965). Bloom's and Ebel's examples indicate some of the educational objectives that can be used in constructing the table of specifications. You can also devise your own lists.

 Go to the WSG to complete Student Activity 8.2 and view Video 8.2.

How to Measure

As mentioned previously, determining how to measure usually involves answering several mechanical questions. The answers are often partially resolved by deadlines or by practical considerations, but frequently the answers require understanding the outcomes of various testing procedures.

When to Test

For testing that occurs in a school system, institutional policies may dictate the times testing is done. The type and frequency of grade reporting, a requirement to set aside certain class periods for testing, and various class-scheduling practices may influence the decision of when to test. Most frequently, tests are administered during a regularly scheduled class period at or near the end of each unit of study, and the lengths of the units are designed to coincide with the school's grading periods. These practices are justifiable for the achievement type of test discussed in this chapter. However, there may be valid reasons for administering tests at other times during instruction.

Deadlines, too, usually determine the appropriate time to administer a written test related to a research project a student is completing. Depending on the hypothesis being tested in the research, an investigator may plan cognitive assessments before, at the conclusion of, throughout, or at both ends of an instructional unit embedded in a research project.

Test frequently enough to ensure that you obtain reliable results and yet not so frequently that you needlessly use valuable instructional or student time. *For obvious reasons, there is no set amount of time that you should reserve for measurement purposes, but it is likely that more errors are made by providing too little time for testing than by incorporating too much time.*

How Many Questions

Generally, the reliability of an achievement test increases as its length increases. Reliability increases with increases in test length because the more often an assessment of achievement is made, the less overall effect chance has on the results. Flipping a coin twice and obtaining two heads is meager evidence to support the contention that the coin has two head sides. However, if the coin is tested 50 times and heads occur 50 times, the contention becomes tenable because the chance occurrence of such an event with a normal coin is extremely remote.

The length of a test is a function of other factors in addition to the desire for reliable results. Three other important factors determining the number of questions on a test are

1. the time available for testing,
2. the type of questions used, and
3. the attention span of the students.

In most school situations, the length of the class period is the limiting factor on the length of an achievement test. Often, only the typical 45 to 60 minutes of class time is available. The number of questions that can be answered in this time largely depends on the type of questions used, such as essay, true–false, or multiple-choice. The time required may vary considerably not only by question type but also within one type. For example, few essay questions requiring extensive answers can be completed within one class period, but many more essay questions requiring a one- or two-sentence response can be included. A test composed mainly of factual multiple-choice items can realistically include more questions than one made up of multiple-choice items that require examinees to apply knowledge to novel situations; the factual questions involve mostly recall, whereas the application items require additional thinking and reflection. Finally, differences in the attention spans of examinees influence the decision of how many questions to include on a test. Schools often account for differences in attention spans by adjusting the length of class periods according to the grade level of the students. A researcher has more flexibility than a schoolteacher in varying test length, so the factor of attention span typically becomes the most important limiting factor for any researcher.

Another aspect to consider when determining the number of questions to include on a test is that not all students work at the same rate. What percentage of students should be able to complete the test? In most situations, all or nearly all of those being tested should be able to finish the test. With a few exceptions—such as a sport officiating course or an emergency room diagnosis unit in which an objective is to acquire the ability to make rapid and correct decisions—it is generally true that a measurement of the ability to answer questions correctly is of more value than a measurement of the speed with which correct answers can be produced. Furthermore, to construct a test containing more questions than can be completed by most or all examinees is an inefficient use of your time, because the questions near the end of the test are seldom used.

The numerous combinations of the factors of available time, question type, attention span, and work rate make it inevitable that a certain amount of trial and error will occur in determining the number of questions on a test. However, we suggest some general guidelines that you may adjust to meet your particular situation. Most students of high school age and older should be able to complete three true–false questions, three matching items, one or two completion questions, two recognition-type multiple-choice items, or one application-type multiple-choice item in 1 minute. For younger students, these estimates should be reduced appropriately. Few guidelines can be given regarding the number of essay questions; however, allow enough time for a student to be able to organize an answer and then put in on paper (or online). Also, in general, including many short essay questions measures achievement more effectively than using a few lengthy ones.

Mastery Item 8.2

Approximately how much time should be allotted for a college-aged person to complete a written test containing a combination of 25 true–false questions, 25 recognition-type multiple-choice questions, and 25 application-type multiple-choice questions?

What Type of Test Format

Although achievement tests are most commonly printed, expense, convenience, minimizing opportunities for cheating, and concern for vision- or hearing-impaired examinees all affect the decision of what format to use. *The format selected should maximize the opportunity for every examinee to understand and complete the required task or tasks.*

Oral presentation of test questions is, in general, an unsatisfactory procedure for most types of items, with the possible exception of true–false questions. Although the expense and your preparation time for this format are minimal, all examinees are forced to work at the same set pace, and there is little or no opportunity for examinees to check over answers. Projecting the test by means of slides, PowerPoint, or transparencies on an overhead projector or computer has the same disadvantages as oral presentation. In addition, this format introduces some expense and time-consuming preparation. *Probably the most common, efficient, and preferred method of presenting achievement tests is in written form, in which each examinee receives a copy of the test questions.* Although this method requires advance preparation (in this case, typing, proofreading, duplicating, and possibly compiling), it maximizes convenience for the examinees. Each examinee can work at his or her own rate, the answers can be checked if time permits, and the questions can be answered in any order. You are free to monitor the test. Where a sufficient number of computers are available, a testing format involving online tests is becoming increasingly popular. Although it may slightly increase the time required for test preparation, this format can be extremely efficient, especially in the administration of the test and scoring and analysis of the test results.

Paying attention to the way you lay out your test can help you cut costs and test-preparation time as well as enhance the accuracy of responses. When an examinee actually knows the correct answer to a question but makes an incorrect response because of an illegible test copy, the reliability and validity of the test are reduced. Also, carefully proofreading your test before administering it can eliminate the need to orally correct errors in test items, which wastes valuable testing time. Here are some additional tips to consider:

- Provide students with advance notice about the number and nature of the test items.
- Provide extensive directions for completing the test (and review these the day before the test, if possible).
- If various types of questions are used in one test, group together questions of the same type to reduce fluctuation among the types of mental processes required of examinees.
- Group together questions of similar content (i.e., subject area) on achievement tests.
- Although ordering test questions from easiest to hardest is not generally recommended, including a relatively simple question or two at the beginning of a test may benefit students by reducing anxiety.

Two interesting variations of the typical written test are the open-book–open-notes test and the take-home test. Each has advantages and disadvantages and under certain conditions can be used effectively. The greatest benefit of both is the reduction of student anxiety. In addition, an open-book test typically requires you to ask fewer trivial questions and more application questions; it forces you to invent novel situations rather than present questions based entirely on circumstances presented in the textbook or lectures. Furthermore, open-book tests reduce the possibilities for cheating because students are allowed to use books, notes, and other materials.

One possible disadvantage of an open-book test is that it may reduce a student's incentive to overlearn and reduce the time a student spends preparing for the test. Students tend to rely on being able to obtain answers from their notes and books during the test and may thus spend less time studying. Because examinees can look up answers, you will need to set time limits on open-book tests, or some examinees (usually unprepared and in effect studying while taking the test) will take an inordinate amount of time to finish. If an open-book test is well constructed, most examinees will find that the textbook and notes are of little value except for looking up formulas and tables. Examinees should not be able to answer open-book test items simply by turning to a specific page in the textbook and finding the answer. Online tests should be administered in a supervised setting to avoid the possibility of someone other than your students or subjects completing the test.

Take-home tests can be used in situations in which more time is required to complete a test than is available in a controlled setting. The major problem with them lies in the impossibility of ensuring that each person does his or her own work. Thus, take-home tests should not be used to measure achievement but might be used for illustrating what students should study and as homework assignments.

Go to the WSG to complete Student Activity 8.3.

What Type of Questions

Questions can be classified into three general categories: semiobjective, objective, and essay. Semiobjective questions have characteristics of both of the other categories. There are three types of semiobjective questions: short-answer, completion, and mathematical questions. For these questions, examinees must compose the correct answer; the answer is so short that little or no organization of it is necessary. Some subjectivity may be involved in scoring (e.g., awarding partial credit for the correct procedures but a wrong answer for a mathematical problem, or the incorrect spelling of the correct answer). Scoring procedures are generally similar to those used for objective questions: The response is checked to see whether it matches the previously determined correct answer.

Characteristically, the task of an examinee responding to an *objective question* is to select the correct (or best) answer from a list of two or more possibilities provided. This type of question is considered objective because scoring consists of matching an examinee's response to a previously determined correct answer; that is, this type of scoring is relatively free of any subjective or judgmental decision. Types of questions classified as objective include true–false, matching, multiple-choice, and classification items.

When responding to an *essay question,* an examinee's task is to compose the correct answer. Usually the question or test item provides some direction by including such terms as *compare* or *explain.* Or the item may constrain the answer by including such phrases as *Limit your discussion to . . .* or *Restrict your answer to the year. . . .* Essay questions are considered subjective because scoring usually involves judgmental decisions.

Several differences among the categories—other than objective versus subjective and selecting answers versus composing them—have consequences for either instructors or examinees. For examinees, much of the time available for testing is consumed in writing (essay questions), reading (objective or semiobjective questions), or calculating the answer (mathematical problems). Hence, because reading is less time consuming than writing or calculating, usually a greater number of objective questions can be included on a test than questions from the other two categories. Also, examinees who are weak in one of these areas (writing, reading, or calculating) may be at a disadvantage in taking tests composed mainly of questions requiring the skill in which they are weak. A poor reader, for example, may do worse on an objective test than on an essay test over the same material.

From the test constructor's point of view, essay and semiobjective questions are easier to prepare than objective questions but harder to score. In addition, the quality of an objective test depends almost entirely on your ability as a test constructor rather than as a scorer, whereas the situation is reversed for an essay or semiobjective test. Thus, your decision about what type of test to construct might be influenced, in part, by the time you have available to construct and score it, or whether your abilities lie in constructing or in scoring tests.

It is plausible that students study differently for different types of tests (though evidence for this is inconclusive); for example, some believe that objective tests promote the study of factual and general concepts. However, this belief rests mainly on the mistaken assumption that objective questions cannot measure depth of achievement. Although it is often more difficult to construct, a test composed of objective questions can measure the achievement of almost any objective as well as a test made up of essay questions. *In short, the type of studying promoted by a test is more a function of the* ***quality*** *of the questions than the type of questions.*

It is true, though, that one type of question is more efficient than another in a particular situation. It would be difficult, for example, to conceive how the quality of one's handwriting might be measured efficiently with an objective test, or how the ability to solve mathematical problems might be measured any more validly than by a test composed of mathematical problem questions. However, the fact that it may be more efficient to use objective questions to measure factual knowledge and essay questions to measure the organization and integration of knowledge has stereotyped the way certain questions are used. Also, other testing factors may preclude the use of what appears to be the most efficient type of question. For example, it is often impractical to correct an essay test given to a large number of people. Thus, an objective test may be used even though the measurement involves more than just factual information. Although many nationally standardized tests include some essay type-questions, their heavy reliance on objective-type questions is an example of this situation.

Despite the names of the three categories of questions, remember that subjectivity is a part of every test constructed. Subjective decisions are required in the scoring of essay questions and, to a lesser degree, semiobjective questions. Subjectivity is present in the construction of all types of questions: Your decisions in determining both what questions to ask and how to phrase them are subjective in nature. *To increase the reliability of written tests, reduce the amount of subjectivity involved in their construction and scoring as much as possible.* Practices such as formulating a table of specifications (see the What to Measure section) and consulting with colleagues can ensure that you accomplish this.

Go to the WSG to complete Student Activity 8.4.

Mastery Item 8.3

Review the term *objectivity* discussed in chapter 6. How does the concept of objectivity apply to the administration of a written test? Can you list some of the procedures used in the administration of the ACT or SAT that are meant to increase objectivity?

Regardless of the type or types of questions used on a test, the usefulness of the resulting score depends on its stability (i.e., reliability). A test is designed and constructed to measure the achievement of certain objectives, and the scores resulting from the administration and correction of the test are supposed to express the degree of achievement. If a different construction, administration, or correction of the test by you or a different person was to result in a different set of scores and a corresponding different ordering of the examinees, the stability and thus the usefulness of the score is reduced. The type of questions included on a test affects the stability of the scores in various ways.

For example, if two people were told to construct a test over the same unit of instruction, it is more likely that the two tests would contain similar questions if the two were told to construct an essay test rather than an objective or semiobjective test. On the other hand, if two people each scored an objective test, a semiobjective test, and an essay test, concurrence is much more probable for the objective test than for the semiobjective or essay test.

Understanding the similarities and differences among the types of questions, and being aware of the advantages and disadvantages of each type of question (see the following section), are necessary in selecting the most efficient types of questions for a particular situation. This knowledge, in addition to proficiency in the general requirements of test construction, will allow you to develop valid and reliable written achievement tests.

Kate, from the measurement and evaluation challenge, has decided to develop a table of specifications to ensure the proper emphasis and weighting of the concepts her test will assess. She also will probably choose to have as lengthy a test as possible and to use either multiple-choice questions, mathematical problems, or a combination of both.

CONSTRUCTING AND SCORING THE TEST

Most of the work of a teacher or researcher will be either in the construction or in the scoring of written test items. As we've discussed, essay questions are relatively easy to construct and time consuming to score, whereas multiple-choice questions are the opposite. There are many ways to construct and score the various types of questions to increase their efficiency.

Semiobjective Questions

The three types of semiobjective questions are *short-answer questions, completion questions,* and *mathematical problems.* The short-answer question and the completion question differ only in format: The completion item is presented as an incomplete statement (a fill-in-the-blank), whereas the short-answer item is presented as a question. The task required to answer a mathematical problem is specified by symbols or by words, as in a story problem. We describe the uses, advantages, and limitations and provide construction and scoring suggestions for all three types of questions simultaneously because of their similarities.

Uses and Advantages

Semiobjective questions are especially useful for measuring relatively factual material such as vocabulary words, dates, names, identification of concepts, and mathematical principles. They are also suitable for assessing recall rather than recognition because each examinee supplies the answer. *The advantages of semiobjective questions include relatively simple construction, almost total reduction of the possibility of guessing the correct answer on the part of an examinee, and simple and rapid scoring.*

Limitations

Because of the limited amount of information that can be given in one question or incomplete statement, it is often necessary to include additional material to prevent semiobjective questions from being ambiguous. Even when a situation is explained in fair detail, the danger of ambiguity is not completely removed, especially for completion items. Occasionally, a blank left in a sentence can be filled by a word or phrase that can be defined as being correct even though it is not precisely what the test constructor desired. For example, consider the following completion item: "Basketball was invented by ______________________________." The name *James Naismith,* the phrase *a male,* and the date *1900* are three possibilities that correctly complete the sentence. When this situation occurs, an instructor must decide whether to award credit. With mathematical questions, instructors may have to decide whether to award no credit, partial credit, or total credit if a student followed correct procedures but gave the wrong answer. Similar decisions are necessary when an examinee provides the correct answer but it is unclear how it was derived. These situations introduce some subjectivity and thus the possibility of inconsistency in the scoring procedure. Specific construction techniques can help reduce (but seldom completely eliminate) this problem.

Recommendations for Construction

Of the three types of semiobjective questions, ambiguity is most likely to occur with completion questions. Rephrasing the incomplete sentence into a question—that is, converting it into a short-answer item—often resolves several problems. However, if you prefer a completion question, these suggestions may reduce some ambiguities.

• Avoid or modify indefinite statements for which several answers may be correct and sensible. Do this, in part, by specifying in the incomplete statement what type of answer is required. For example, "Basketball was invented by ______________________________" can be reworded as "The name of the person who invented basketball is ______________________________." A similar method for eliminating ambiguity is to present the item this way: "Basketball was invented by (person's name)."

• Construct the incomplete sentences, when possible, so that the blank occurs near the end of the statement. This technique better identifies the specific type of answer required than when the blank space occurs early in the statement. For example, in the item "The ______________________________ system of team play in doubles badminton is recommended for beginners," the desired correct answer is *side-by-side,* but the blank could logically be filled in with the phrase *least complex* because it is not clear that the name of the system is desired. Rewording the statement so that the blank occurs near the end solves this problem: "The type of team play recommended for beginners in doubles badminton is called the ______________________________ system."

• Do not leave so many blanks in one statement that the item becomes indefinite. Consider this extreme example: "The name of ______________________________ who invented ______________________________ is ______________________________." As the example demonstrates, the more blanks in the statement, the less information given; answering the question becomes a guessing game. Give additional information by either explaining what is required or making several items from the one.

• Do not give inadvertent clues. Occasionally the phrasing of the statement or the use of a particular article (*a* or *an*) or verb reduces the number of possible words or phrases that might complete a statement. Use the following format for the indefinite article: "Basketball was invented by a(n) ______________________________ (nationality)." If more than one blank occurs in a statement, each blank should be the same length to avoid giving students information about the length of the correct response.

• If a numerical answer is required, indicate the units and degree of accuracy desired (e.g., "Round to the nearest one-tenth."). Specifying this information simplifies the scorer's task and eliminates one source of confusion for examinees.

• Use short-answer questions where possible to reduce ambiguity. For example, using a short-answer question such as "An athlete from what country won the gold medal in the pentathlon in the 2012 Olympics?" rather than the completion item, "The gold medal in the pentathlon in the 2012 Olympics was won by ______________________________" increases the probability that the country will be identified rather than other possible information.

Scoring consistency is enhanced because the examinees' task is typically more clearly identified than with completion items. You should phrase short-answer items in such a way that the limits on the length of the response are obvious.

Recommendations for Scoring

If semiobjective questions are well constructed and you encounter no problems (e.g., when two or more answers are plausible for one item), the scoring process is simple, objective, and reliable. The answers can be scored easily by persons other than the test maker.

If the test consists of completion items, you can prepare an answer key by using a copy of the test and cutting out a rectangular area where each blank occurs. Write the correct answer immediately below or adjacent to the rectangular area. When the answer key is superimposed on a completed test, each response can be quickly matched with the keyed answer.

Using separate answer sheets for short-answer items speeds the scoring process. Because only one-word or short-phrase answers are expected, you can distribute, along with the test itself, an answer sheet previously prepared with a numbered blank space corresponding to each test item. Usually, you can place two columns of answers on one side of a standard-sized piece of paper. To score short answers efficiently, construct an answer key by recording the correct responses on a copy of the answer sheet and place this alongside each answer sheet. This procedure eliminates the need to search through the pages of all the tests to locate the answers.

Go to the WSG to complete Student Activity 8.5 and view Video 8.3.

Objective Questions

Questions requiring the selection of one of two or more given responses can be scored with minimal subjective judgment and are thus categorized as *objective questions*. Although there are many similarities among types of objective questions, we give separate consideration to true–false, matching, and multiple-choice questions because of their differences.

True–False Questions

Perhaps unfortunately, *true–false questions* have been widely used by teachers and others, probably because these questions are relatively easy to construct and score. Although there are advantages to true–false questions and situations in which their use is justifiable, they are the least adequate type of objective question because of several weaknesses.

Uses and Advantages Like the various semiobjective questions, true–false items are particularly suited for measuring relatively factual material such as names, dates, and vocabulary words. The advantages of using true–false items include the ease of construction, administration, and scoring and the fact that more true–false items can be answered than any other type of question within a given time.

Limitations Many of the major weaknesses of true–false questions stem from the fact that an unprepared examinee might answer half of the items correctly by chance alone. This makes it difficult to assess a test taker's level of achievement. A correct answer could be an indication of complete understanding of the concept, a correct blind guess, or any shade of understanding between these two extremes. *In addition, the inordinately excessive influence of chance lowers the amount of differentiation among good and poor examinees and consequently the reliability of the test.*

To be fair and avoid ambiguity, a true–false item should be absolutely true or absolutely false. It is difficult to meet this requirement except when factual knowledge is involved. *True–false questions are not well suited for measuring complex mental processes.* Because of this, ill-composed true–false tests can include trivial questions and reward sheer memory rather than understanding.

Recommendations for Construction Generally, writing good true–false questions involves avoiding ambiguity. Here are some specific suggestions. Examples of good and poor true–false questions are given at the end of the section.

• Avoid using an item whose truth or falsity hinges on one insignificant word or phrase. To do so results in measuring alertness rather than knowledge.

• Beware of using indefinite words or phrases. A question whose response depends on the interpretation of such words or phrases as *frequently, many,* or *in most cases* is usually a poor item.

• Include only one main idea in each true–false question. Combining two or more ideas in one statement often leads to ambiguity. If the combination introduces the slightest amount of falsity in an otherwise true statement, the examinee must decide whether to mark true or false on the basis of the amount of truth rather than on the basis of absolute truth.

• Avoid taking statements directly out of textbooks or lecture notes. Out of context, the meaning of the resulting item can be confusing. Few statements made in a text or lecture can stand alone meaningfully. In addition, using textbook sentences as true–false items results in rewarding memorization.

• Use negative statements sparingly and avoid double negatives completely. Inserting the word *not* to make a true statement false borders on trickery and may result in a measurement of vigilance rather than knowledge. Statements containing double negatives, especially if false, are often needlessly confusing and complex.

• Beware of giving clues to the correct choice through **specific determiners** or statement length. Specific determiners are words or phrases that inadvertently provide an indication of the truth or falsity of a statement. For example, true–false items containing words such as *absolutely, all, always, entirely, every, impossible, inevitable, never,* or *none* are more likely to be false than true because an exception can usually be found to any such sweeping generalization. Conversely, such qualifying words as *generally, often, sometimes,* or *usually* are more common in true statements than in false statements. Because it often takes several qualifications to make a statement absolutely true, take care to avoid a pattern of long statements being true and false statements being short.

• Include approximately the same number of true statements and false statements on a test. Having too many of one or the other can bias responses. There is some evidence that false statements are slightly more discriminating, perhaps because an unprepared examinee is more inclined to mark true. For this reason it may be advantageous to include a slightly higher percentage of false statements.

• Avoid arranging a particular pattern of correct responses. Regulate the placement of true and false statements by chance to minimize the possibility that examinees will detect a pattern of responses.

• Arrange for a colleague to review the true–false questions before administering them. Doing so may help you remove ambiguity from questions.

Modifications Test makers have attempted to modify true–false questions with the intent of reducing excessive blind guessing. One method is to require the examinee to identify the portion of a false statement that makes it false. A further modification requires the correction of the inaccurate portion. These modifications are called *corrected true–false*. Although these two modifications partially eliminate the effect of chance on the final score, they simultaneously introduce other problems. Ambiguity may result, as in the following: "James Naismith invented the game of volleyball." The statement is false but may be corrected by replacing the name *James Naismith* with the name *William Morgan*, or by replacing the word *volleyball* with the word *basketball*. These kinds of true–false questions can introduce some subjectivity into the scoring. Furthermore, the advantage of quick scoring is lost.

Another way to modify true–false questions involves changing the answering and scoring procedure to reflect the degree of confidence examinees have in their responses. The intent is to discriminate between those who get an answer wrong because they do not know the correct response and those who know something but not enough to prevent a bad luck choice. Several scoring systems have been devised to accomplish such *confidence weighting* of the response to a true–false item. In the system presented in table 8.2, if the examinee marks A, for example, and the correct answer is "true," the examinee is awarded two points, but if the correct answer is "false," two points are deducted from the examinee's score. This modification, although increasing the discriminatory power of a true–false test, may introduce some undesirable variables. For example, differences in personality traits among examinees (some more willing to gamble than others) and the importance of knowledge of the content being tested as well as an awareness of the nature of one's knowledge become factors influencing the final test results. Thus, these modifications may well increase the reliability and discriminatory power of a true–false test but simultaneously reduce its validity.

In this example, the reason half a point is awarded for choosing C is that if a totally unprepared examinee guessed at every question on a normally scored true–false test, his or her score would be 50% correct (in the long run).

Table 8.2 System for Confidence Weighting Answers to True–False Questions: Scoring Procedure

		POINTS AWARDED OR SUBTRACTED	
Response	**Mark**	**Correct**	**Incorrect**
Definitely true	A	2.0	−2.0
Likely true	B	1.0	0.0
Omit or don't know	C	0.5	0.5
Likely false	D	1.0	0.0
Definitely false	E	2.0	−2.0

Recommendations for Scoring As is true of most semiobjective and objective questions, using a separate answer sheet facilitates the scoring procedure. Because of the similarity between the letters *T* and *F*, it is not a good idea to have the examinees write the letters *T* and *F* on a sheet of paper when completing a true–false test. A previously prepared answer sheet on which examinees block out, circle, or underline the correct response eliminates such scoring problems. Special answer sheets that can be scored by machine are available for most objective questions, including true–false questions. You (or someone not even familiar with the subject matter) can efficiently score the test by hand by matching each response on an answer sheet with a previously prepared correct answer sheet.

 Go to the WSG to complete Student Activity 8.6.

TRUE–FALSE QUESTIONS FOR BASKETBALL

Good Questions

1. Kicking the ball is a team foul. (False)
2. It is generally better to dribble than to pass. (False) (Earlier, *generally* was identified as a specific determiner, and its use was discouraged. However, notice that in this question it is used in a false statement rather than the normally expected true statement.)
3. A double violation occurs when a player commits two fouls at the same time. (False)

Poor Questions

1. Basketball was first introduced in 1901. (False) (Very trivial; it was introduced in 1891.)
2. The overhead pass should always be used by short players. (False) (Use of the specific determiner *always*)
3. The shovel, underhand, and hook passes are made while holding the ball with both hands. (False) (Part of the statement is true and part is false.)
4. In most cases, teams play one-on-one or zone defense. (True) (Use of indefinite phrase *in most cases*)
5. A time-out shouldn't be wasted when the team isn't in trouble. (True) (Double negative)

Matching Questions

Matching questions generally involve a list of questions and a list of possible answers. The examinee's task is to match the correct answer to the proper question. At times, instead of involving a question-and-answer format, this type of question involves matching an item in one list with the item most closely associated with it in the second list.

Uses and Advantages As with true–false questions, matching questions are most efficient for measuring relatively superficial types of knowledge. Measurements of vocabulary, dates, events, and simple relationships, such as authors to books, can be effectively obtained with matching questions. *Matching questions are used to measure who, what, where, and when rather than how or why.* Among the advantages of matching questions are relative ease of construction and the rapidity, accuracy, and objectivity of scoring. Matching questions require developing a cluster of similar questions and similar answers. The most discriminating matching questions often are those used in conjunction with a graph, chart, map, diagram, or similar device for which labels on the illustration are matched with functions, names, or similar categories of answers.

Limitations It is difficult, although not impossible, to construct matching questions that require the examinee to use higher-order mental processes. However, the most limiting aspect of matching questions is that they require similarity within each of the two lists that make up the item. As compliance to this requirement lessens, the discriminating power of the matching item also usually diminishes.

Recommendations for Construction Because it is easier to write questions that measure relatively superficial knowledge than it is to write questions that measure higher-order cognitive processes such as application, analysis, and evaluation, refer often to the table of specifications developed for a test when constructing matching questions. This ensures that you will achieve the desired balance among the areas measured. Unless you exercise caution, a test composed mainly of matching items may concentrate more heavily on factual material than warranted by the table of specifications. Here are some additional suggestions for constructing matching questions. Examples of good and poor matching questions are given at the end of this section.

- Present clear and complete directions. In general, include three details in the instructions:
  1. The basis for matching the item in the two lists
  2. The method to record the answers
  3. Whether a response in the second column may be used more than once
- An instruction such as "Match the statements in column 1 with those in column 2" does not include any of the three points; contrast it with the following complete instruction: "For each type of physical activity listed in column 1, select the physical benefit from column 2 that is most likely to be derived from it. Record your choice on the line preceding the question number. An item in column 2 may be used once, more than once, or not at all."
- Avoid providing clues. Every word or phrase in one column must be a logically and grammatically acceptable answer to every question in the other. Use the same verb tense, either singular or plural words, and the same articles if possible in all questions.
- Avoid including too many questions in one matching item. To be effective, the list of questions and the list of answers in a matching item must be somewhat homogeneous. As the length of the list of questions or answers increases, meeting the requirement of homogeneity becomes increasingly difficult. In most cases five or six questions is the practical limit for each matching item.
- Make sure all questions and answers appear on the same page of the test.
- Include a greater number of answers than questions or allow the repeated use of some answers. This removes the possibility of using the process of elimination to obtain the answer to one question of a matching item.
- Keep the parts of the matching questions as short as possible without sacrificing clarity. Because examinees must completely reread the list of possible answers as they respond to each item, having them read needlessly long answer choices consumes valuable testing time.
- Arrange the two lists of questions and answers in a random fashion. There should not be any particular pattern to the sequence of correct responses.
- Place the answer choices in a logical order (e.g., alphabetical, chronological) if one exists. This allows an examinee who knows the answer to locate it quickly.

Recommendations for Scoring Because matching questions are generally answered on the test itself rather than on a separate answer sheet, arrange the items on the test so that a key can be placed next to the margin for quick scoring. Scoring a matching item can be done by someone not familiar with the subject matter tested.

Go to the WSG to complete Student Activity 8.7.

Multiple-Choice Questions

A multiple-choice question includes two parts: the *stem,* which may be in the form of a question or an incomplete statement, and at least two *responses*, one of which best answers the item or best completes the statement. The task is to select the correct or best response to the item presented in the stem.

Uses and Advantages Multiple-choice questions make up a large portion of almost all nationally standardized written tests for several reasons:

• Questions can be scored and analyzed efficiently, quickly, and reliably.

• Questions of this type often create less ambiguity than other kinds of questions.

• Questions with more than two responses are not as susceptible to chance errors caused by blind guessing.

• Questions can be used to measure the higher-order cognitive processes, such as application, analysis, synthesis, and evaluation.

MATCHING QUESTIONS

Good Question

For each person listed in column 1, select the sport from column 2 for which he or she is most noted. Record your choice on the line preceding the question number. A sport in column 2 may be used once, more than once, or not at all.

_____ 1. Aaron, Hank	a. Baseball
_____ 2. Calipari, John	b. Basketball
_____ 3. Williams, Serena	c. Cycling
_____ 4. Mickelson, Phil	d. Football
_____ 5. Karolyi, Bela	e. Golf
_____ 6. Ruth, Babe	f. Gymnastics
_____ 7. Hamm, Mia	g. Soccer
_____ 8. Bannister, Roger	h. Swimming
	i. Tennis
	j. Track

Poor Question

Match column 1 and column 2.

_____ 1. Sit-and-reach	c. Muscle fibers
_____ 2. 50-yard dash	h. Golf
_____ 3. Pull-up	f. Tennis
_____ 4. Shuttle run	a. $\dot{V}O_2$max
_____ 5. Balke treadmill	e. Agility
_____ 6. Dyer backboard volley	g. Arm strength
_____ 7. Disch putting	d. Speed
_____ 8. Biopsy	b. Flexibility

This is a poor matching question because

- the instructions do not indicate basis for matching, how to record answers, or how many times items in column 2 may be used;
- the items in each column are too heterogeneous, making the answers too obvious;
- both columns contain the same number of items, so the last item could be answered by elimination; and
- some matches are too obvious (e.g., Dyer backboard volley and tennis, Disch putting and golf)

- Questions can measure almost any educational objective.
- Questions can be analyzed to determine their contribution to the test's reliability and validity.

Because multiple-choice questions are capable of measuring all levels of cognitive behavior, are applicable to nearly any subject or grade level, and can be used to measure virtually any educational objective, they can be used in almost any situation. If you are testing a large group of examinees or are planning to reuse a test, multiple-choice tests are most efficient in terms of the time it takes to construct, administer, score, and analyze test items. In the event that fairly rapid feedback is important, multiple-choice tests, because of their quick and accurate scoring characteristics, should be used. Generally, you can include a fairly large number of multiple-choice questions on a test because the time required to answer each item is short. Because of this, and because multiple-choice questions can be constructed to measure most educational objectives, it is less difficult to construct a test fitting the table of specifications by using multiple-choice questions than any other type of question. Finally, scoring is fast and can even be done by someone not familiar with the subject area.

Limitations Multiple-choice questions, because of their versatility, do not have many intrinsic weaknesses. However, the required investment of time makes multiple-choice questions inefficient for small groups or one-time use. A few objectives are not as efficiently measured by multiple-choice questions as by other types of questions. For example, organization of an answer, grammatical construction of sentences, and other writing characteristics are probably best measured by essay questions (although appropriate multiple-choice tests could probably be devised).

Recommendations for Construction Writing good multiple-choice questions requires paying careful attention to many aspects, such as constructing the stem and the responses and avoiding clues. Examples of good and poor multiple-choice questions are given at the end of this section. General considerations include the following:

- As you write the initial draft, realize that each question will probably require revisions.
- Set up a computer file to allow for revision and the addition of information. Record with each question the course objectives and educational objectives it measures so you quickly can determine its place in the table of specifications. Also record the location of the source of the idea around which the question is built, since this information is often lost with the passage of time.
- Base each question on an important, significant, and useful concept. Usually the most successful multiple-choice questions are those based on generalizations and principles rather than on facts and details. For example, a question requiring knowledge of the general organization of Bloom's Taxonomy of Educational Objectives is more valuable than a question requiring the examinee to know that the *third* category of the taxonomy is called application.
- Use novel situations when possible. Generally, effective questions result when you avoid the specific illustrative materials used in the textbook or lectures and use novel situations requiring the application of knowledge instead.
- Phrase each question such that one response can be defended as being the best of the alternatives. It is not always necessary that the response keyed as correct is the best of all possible answers to the question, but it must be able to be defended as the best of the choices listed. Also in this regard, avoid asking a question that requests an opinion, because this results in a no-one-best-answer situation. For example, consider the following stem:

"What do you consider to be the best defense against the fast break in basketball?" Because this asks for an opinion, any choice marked must be regarded as correct, whether or not it agrees with the opinions of basketball authorities or the test constructor.

• Phrase each question clearly and concisely. Ideally, the stem should contain enough information that examinees understand what is being asked and yet should be brief enough that no testing time is wasted reading unnecessary material. Occasionally, it is necessary to include a sentence or two to clarify a situation and avoid ambiguity. However, avoid the practice of teaching on the test, including inserting unnecessary information (called window dressing by some test-construction experts) or using flowery and imaginative language. Flowery language can increase possible interpretations of questions, which then can lead to ambiguity.

• Avoid negatively stated questions as much as possible. When you do use them, capitalize or underline the negative words. The purpose of asking a question is to determine whether the examinee knows the answer, not to see who reads carelessly and who can work through the confusion that sometimes arises with negatively stated questions.

• Do not include a question that all examinees will answer correctly unless you determine that the question must be included to increase the validity of the test. A question that every examinee answers correctly (or incorrectly) is of little value on an achievement test because no discrimination results. In fact, it can be shown mathematically that maximum discrimination can occur only when a question is of middle difficulty—that is, when approximately half the examinees answer the question correctly and half incorrectly. Although it is difficult to estimate the proportion of examinees who will answer a question correctly the first time the question is used, you should attempt to structure multiple-choice questions so they will be of middle difficulty. (Recall that one of the requirements for writing good test questions is to be aware of the level and range of understanding of the group being tested.) The difficulty of a multiple-choice question is most effectively altered by changing the homogeneity of the responses; the more homogeneous the responses, the more difficult the question. A method for obtaining an index describing the difficulty of a multiple-choice question is presented in the Item Analysis section (see equation 8.3).

• Arrange to have the questions reviewed by someone knowledgeable in the same subject. An independent reviewer can often locate ambiguities, grammatical mistakes, idiosyncrasies, and clues—all of which can affect a test negatively. If it is not possible to arrange for another person to review the questions, reread them yourself a few days after you have written them. (An implication of this suggestion is that you should not write the questions the night before the test is to be administered. One of the requirements for writing good test questions is the willingness to spend a considerable amount of time on them.)

• Consider layout issues when formatting and printing out the test. List each response choice on a new line rather than immediately following one after another. Also, unless each response is long (an unlikely event), print the items in two columns instead of across the page. Use letters instead of numbers to identify the responses (this avoids confusion between questions and answers). Keep all the responses to a question on the same page as the question stem. Separate groups of related questions from other questions by a space or dotted line.

• Review the items and responses once you have constructed them to be certain that any correct option does not appear too often in a series of questions and that the correct options are fairly evenly distributed across all questions. Students will begin to second-guess themselves when more than two or three B (or whatever) responses appear as correct answers in a row. Sometimes, students think that the best option if they do not know is to pick C. For all these reasons, spread the correct responses out among the possible choices.

Writing the Stem *If a multiple-choice item is to be meaningful and important, keep in mind a definite concept around which the item is built.* In expressing this concept, the most important part of the multiple-choice item is the stem, and it is the first part constructed.

The stem can take two forms: a direct question or an incomplete sentence. It is usually wise (especially for novice question writers) to use questions rather than incomplete stems so that the examinee's task is clearly defined. *No matter which form is used, by the time an examinee finishes reading the stem, a definite problem needs to have been identified so that the search for the correct response can begin.* A stem such as "Badminton experts agree that . . ." does not provide a specific question or task, because badminton experts agree on many things. The examinee is forced to read through all the responses to determine what exactly is being asked. This stem would not be improved greatly by changing it to the question "On what do badminton experts agree?" If the stem is revised to "On what do badminton experts agree regarding the learning of the rotation strategy by beginning badminton players?" the examinee can begin reading the responses to locate the correct one rather than determining what is being asked. Using incomplete stems is more likely to result in incomplete specification of tasks than is using direct questions. The suggestions provided previously under Recommendations for Construction are especially germane to writing the stem of a multiple-choice question.

Writing the Responses After the stem of a multiple-choice question are usually three, four, or five words, phrases, or sentences known as responses. One of the responses is predetermined to be the correct response (usually called the keyed response). The remaining responses are labeled foils or distractors. When constructing a multiple-choice question, write the keyed response immediately after you write the stem. Following this procedure helps ensure that the question is based on an important concept. On the test, of course, the position of the keyed response among the responses should be determined by some random procedure.

There is no reason that a multiple-choice question must contain any set number of responses or that all the multiple-choice questions on a test have to have the same number of responses. Three, four, or five responses are commonly used because this represents a compromise between the problem of finding several adequate, plausible possibilities and including enough responses so that, as happens with true–false questions, chance does not become an important factor.

There is some evidence (Rodriguez, 2005) that three responses may be the most efficient number for constructing the most accurate test. This is because the reduction of reading time (over four or five responses) permits the addition of an increased number of questions. Recall that longer tests generally tend to be more reliable than shorter tests. However, having only three responses does increase the possibility of answering a question correctly by chance. The most important consideration regarding the best number of responses is whether or not all are reasonable and plausible choices to answer the question. A distractor that is chosen by no examinee is worthless and reduces the discriminating power of the test.

The distractors, the last part of a multiple-choice question developed, should not be constructed for the purpose of tricking the knowledgeable examinee into selecting one of them. However, one should make the distractors attractive to the unprepared examinee. *All the responses should be plausible answers to the item.* Use statements that are true but do not answer the question or use statements that include stereotypical words or phrases. Stereotypical words or phrases are those that are used often in whatever content area the test is assessing. The unprepared examinee may be attracted to a distractor containing such words or phrases simply because they sound familiar. As mentioned, using a ridiculous distractor unlikely to be chosen by any examinee is a waste of testing time.

Take care not to word the keyed response more precisely than the distractors. Recall that the keyed response needs only to be the best of the listed choices, not unequivocally correct under any circumstances. Keep all responses as similar as possible in appearance, length, and grammatical structure to avoid the selection of any response for reasons other than its correctness. As with the stem, keep the responses simple, clear, and concise to avoid ambiguity and to keep reading time to a minimum. If a natural order exists among the responses (such as with dates), list them in that order to remove one possible source of confusion.

Distractors need to appear to be equally correct to an examinee who is not familiar with the content of the item. However, examinees who fully understand the concept being tested should be able to ascertain the correct response. You want an item to appear ambiguous to the ill-prepared student (i.e., to have **extrinsic ambiguity**). If an item appears ambiguous to well-prepared examinees, it has **intrinsic ambiguity**. Extrinsic ambiguity is desirable; intrinsic ambiguity is not. Figure 8.1 depicts the difference between these types of ambiguity.

When sufficient plausible distractors are difficult to invent, it is tempting to use "None of these" as the final response. To avoid confusion, however, do not use this unless the keyed response is absolutely correct (as in a mathematical problem) and not merely the best response. When all the responses are partially correct (even if one of them is more correct than the others), the response "None of these" might be defended as being correct because none of the partially correct answers is absolutely correct. Without the "None of these" response in this situation, the most correct response is defensible as the best and thus the correct answer. A similar problem exists with the response "All of these." When there is not an absolutely correct answer and all responses contain some element of correctness,

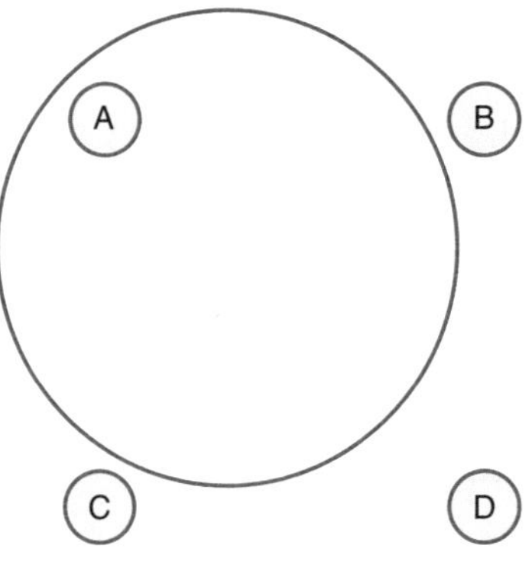

Item that is too easy

Response A (the correct response) is clearly shown to be correct, being inside the realm of the larger circle, and the other responses are clearly incorrect, falling outside of the larger circle. This item will be answered correctly by nearly every examinee and thus will not discriminate among them.

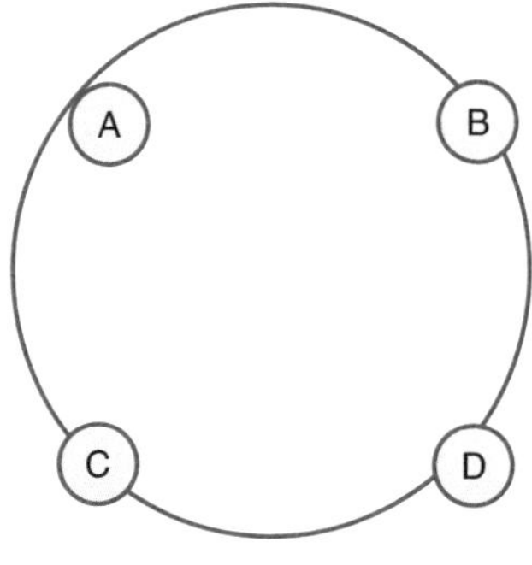

Extrinsic ambiguity

Response A is the best response, but the other responses are reasonable (have some degree of acceptability), and the unprepared student will find it difficult to choose among the possibilities. The prepared student will most likely select A, perceiving it as being better than the other responses.

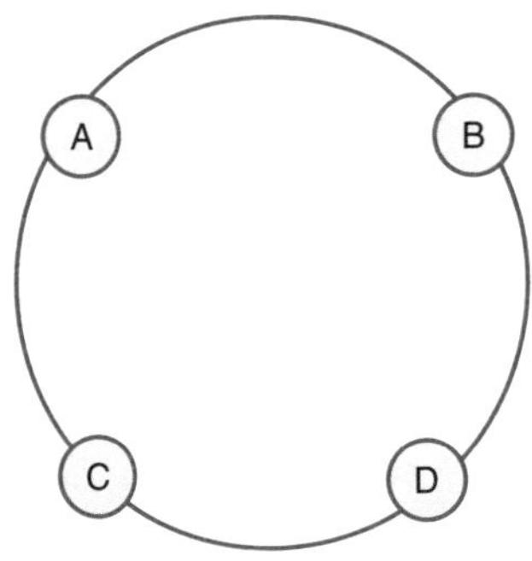

Intrinsic ambiguity

All the responses could be considered to be correct, although response A is depicted as being slightly better than the other responses. This type of item will be ambiguous to well- and ill-prepared students alike and will likely not discriminate among them.

Figure 8.1 The difference between extrinsic and intrinsic ambiguity. The keyed correct response in each example is A.

the response "All of these" could be considered the keyed response, but the examinee is put in a difficult position if one of the responses is a little more correct than the others. If you use these types of responses, make sure they are the keyed response occasionally (especially at the beginning of the test), so that examinees realize that they are to be considered seriously as possible correct answers.

Clues Ideally, an examinee will answer a multiple-choice question correctly only if he or she knows the answer and incorrectly if he or she does not. Two factors, however, can adversely affect this situation. An examinee may blindly guess the correct answer to a question—there is no way to determine whether a correct response indicates a high degree of knowledge or luck. However, in the long run, everyone has an equal chance to be lucky, and the effects of chance can be mathematically accounted for. The second and more serious factor is that of clues included within multiple-choice questions or tests. Because all examinees are not equally adept at spotting clues, the effects are not as predictable as those resulting from chance. The only way to eliminate the problem is to eliminate the clues.

Some clues are rather obvious; others are subtle. For example, it is usually easy to spot the use of a key word in both the stem and the correct response, or a keyed response that is the only one that grammatically agrees with the stem (e.g., stem calls for a plural answer and all but one of the responses are singular). Clang associations, words that sound as if they belong together, such as bats and balls, shoes and socks, up and down, are often relatively difficult for the test constructor to spot but provide immediate clues to test takers. We suggest using stereotypical words or phrases as a method of securing attractive distractors. However, do not use these in the correct response, because an unprepared student may select the keyed response because it sounds good rather than because he or she knows it to be the correct answer.

In the process of asking one question, test constructors may inadvertently give information that answers another item on the test. Such interlocking questions provide clues for a test-wise examinee. This is especially likely to happen if you construct a test by selecting several questions from a file of possible questions, or if you add new questions or revise old questions on a subsequent test. To prevent interlocking items, read the test in its entirety once you have assembled it.

Variations Several variations of multiple-choice questions have been devised to meet the needs of particular situations. For example, the classification item is an efficient form of the multiple-choice format if the same set of responses is applicable to many items. An example of a classification item follows:

For questions 89 through 92, determine the type of test best described by each statement or phrase. For each item, select answer

a. if an essay test is described.
b. if a true–false test is described.
c. if a matching test is described.
d. if a classification test is described.
e. if a multiple-choice test is described.

89. Test limited by difficulty in securing sufficiently similar stimulus words or phrases. (c)
90. Responses generally cover all possible categories. (d)
91. Quality determined by skill of reader of answers. (a)
92. Student can answer most items per minute. (b)

Another variation of the multiple-choice question involves using pictures or diagrams. This is illustrated in figure 8.2.

If the shaded circle represents a top view of a tennis player making a crosscourt forehand stroke, in what location should the ball be when contacted by the racket: A, B, C, or D? (b)

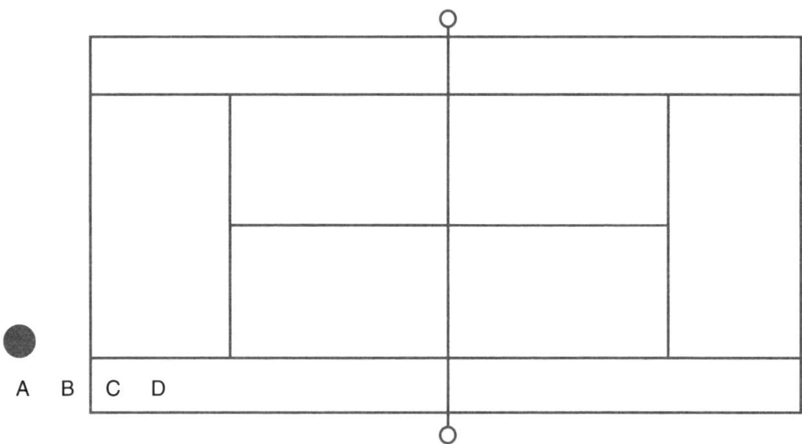

Figure 8.2 Sample of a diagram used on a written test.

You can create other variations to serve particular functions as long as examinees are able to understand their task in answering. Most of the suggestions presented previously will apply to these diverse variations.

Recommendations for Scoring Typically, examinees record the answers to multiple-choice questions on the test itself or on a separate answer sheet. Having students mark directly on the test slightly reduces the chances of mismarking an answer and is convenient when discussing a test after it has been administered. If this procedure is used, you can facilitate the scoring process by arranging the questions so that the answers are recorded along the margins of the test and by using an answer key overlay spaced to match each page.

Although not as convenient for an examinee, recording answers on a separate answer sheet has many advantages for a scorer. You can score the answer sheets quickly and accurately by constructing a key from one of the answer sheets. Punch holes corresponding to the positions of the keyed responses on the answer sheet. When the key is superimposed on an examinee's answer sheet, you can count the number of correct responses. You can also use machine-scorable answer sheets that allow data to be scored and analyzed by the machine and a computer program.

Mastery Item 8.4

Using the suggestions just presented, write five multiple-choice questions on a topic of your choice. Critique your classmates' questions.

MULTIPLE-CHOICE QUESTIONS FOR RACQUETBALL AND HANDBALL

Asterisk denotes correct answer for good questions and poor questions.

Good Questions

Notice that these questions are printed in two columns with each item completed in the column or on the page on which it starts. Notice also that responses are stacked and identified by letters.

1. What is called if a served ball hits the server's partner who is standing in the correct area?
    a. short
    b. fault
    c. handout
    d. *dead ball
2. How many outs are there in the first inning of a doubles game?
    a. one
    b. two
    c. *three
    d. four
3. Which of the following shots is used to get your opponent to move into the backcourt?
    a. kill shot
    b. passing shot
    c. *clear shot
    d. front-wall angle shot
4. What is called if the server stops a served ball that has hit the front wall and has bounced twice in front of the service line?
    a. handout
    b. hinder
    c. fault
    d. *short
5. What is called if a served ball hits the server on the rebound from the front wall?
    a. short
    b. fault
    c. *handout
    d. dead ball

Poor Questions

1. In handball

    a. One short and one hinder put the server out. *b. The fist may be used to hit the ball. c. Receivers can make points. d. The game may be played by only two or four persons.

The stem does not ask a question and thus the examinee must read everything to determine what is being asked. Moreover, the responses have been run together and printed across the page.

2. What is called if a served ball that hits the front wall, side wall, floor, back wall, and the other side wall is not returned by the receiver?

    1. Lucky *2. Point 3. 911 4. Strikeout

All distractors are not plausible answers. Also the question has been printed across the page, numerals have been used to identify responses, and the responses have been run together and printed across the page.

3. What is called if a player gets in the way of his or her opponent?
    a. Point
    b. Short
    c. Kill
    d. *Hinder

"Gets in the way of" and "Hinder" represent a clang association. Most examinees would answer correctly with little or no knowledge of the game.

4. What is called if, in receiving a serve, one receiver causes the ball to hit his or her partner before it touches the front wall or floor?
    a. Hinder
    b. Fault
    c. Handout
    d. *Point, because hitting your partner is your own fault

Wording the keyed response more precisely than the distractors to ensure its correctness might cause examinees to select it even if they are not sure of the answer.

5. What is the best shot to use in racquetball?
    a. Passing shot
    b. *Ceiling shot
    c. Kill shot
    d. None of these

"None of the above" could be defended as being as correct as the keyed response because there is no absolutely "best" shot in every situation.

Essay Questions

To complete an essay question, an examinee must read the question, conceive a response, and write the response. The essay question has many uses, such as requiring the examinee to give definitions, provide interpretations, make evaluations or comparisons, contrast and compare concepts or other topics, and demonstrate knowledge of relationships. For the response to an essay question to be scored accurately, the evaluator must be knowledgeable in the topic being assessed.

Uses and Advantages

Although almost any type of question can effectively measure the ability to organize, analyze, synthesize, and evaluate information, essay questions are easier to construct for this purpose than any other type. The contention that essay questions promote the study of generalizations rather than facts seems reasonable but has not been and probably never will be conclusively substantiated. Essay questions can effectively measure opinions and attitudes and although we are seldom interested in measuring these attributes in an instructional unit, it may be desirable to assess them in a research setting. Questionnaires are often used to measure opinions and attitudes as well. Essay questions on questionnaires are referred to as open-ended questions. Information about questionnaires as measuring instruments is presented at the end of this chapter. In some situations, using essay questions is more efficient or convenient, regardless of the mental processes or subject matter involved. For example, the total time required to construct and correct an essay test is often less than for other types of questions.

You should also consider your personal preferences. If you are confident in your ability to construct and score essay questions but lack confidence in using other types of questions, you should probably use essay tests. However, be aware of the limitations of essay questions and the ways you can eliminate or minimize those limitations. Finally, when scheduling circumstances dictate little time for test construction but ample time for test correction, use essay tests.

Limitations

Even with careful preparation and scoring methods, at least three problems can arise when essay questions are used to measure achievement.

Inability to Obtain a Wide Sample of Achievement Because of the time required to organize and write answers, it is not always possible to include enough essay questions on a test to measure the achievement of each content and educational objective. Consequently, there is some lack of content validity. You can alleviate this problem by constructing a table of specifications, using several essay questions requiring relatively short answers rather than a few questions requiring extended answers, and testing frequently to reduce the amount of material measured by each test.

Inconsistencies in Scoring Procedures *The most serious problem associated with essay questions is the unreliability of the scoring procedures.* Not only does it take a substantial amount of time to correct an essay question properly, but several factors cause inconsistencies in the scores obtained. Because of the freedom an examinee has in constructing the essay answer, it is often necessary for you to decide, sometimes subjectively, whether that examinee has achieved an objective. You can reduce (although not completely eliminate) the subjectivity if you are knowledgeable in the subject matter tested and if you make it clear what task is required of the examinee for each question. Also, prior to the correcting of an essay question, construction of a rubric (see chapter 14) as a scoring guide can help eliminate inconsistent scoring.

Another problem is the halo effect, or generalization—the part of an examinee's score that reflects your overall opinion of him or her. Giving the benefit of the doubt on one question to an examinee who has done well on most of the other questions on a test or to an examinee who has impressed you favorably in the past is an example of this phenomenon. By devising a coding system so that examinees' names do not appear on the answer sheets and by correcting the test question by question rather than paper by paper, you can diminish the consequences of this problem.

Handwriting, spelling, and grammar, for example, can positively or negatively affect the scoring of an essay answer. Unless these are specific objectives of the test, the score should not reflect these elements but should be influenced only by achievement in the area being measured.

Difficulties in Analyzing Test Effectiveness After you have constructed, administered, and corrected a test, you will want to analyze how well the test measured what it was intended to measure, especially if you will use the test again. Analyzing a test generally includes obtaining indications of the overall reliability, validity, and objectivity of the test and the strengths and weaknesses of the test's individual items. Although some of these characteristics can and should be investigated for essay tests, essay questions do not lend themselves to this scrutiny as conveniently as do objective questions.

 Go to the WSG to complete Student Activity 8.8.

Recommendations for Construction

The following eight suggestions on the construction of essay questions will help you overcome some of the weaknesses and problems associated with their scoring.

- Phrase the question such that the mental processes required to respond to it are clearly evident. The objective of a question might be to determine whether mastery of factual material has occurred (e.g., "What are the outside dimensions of a regulation tennis court?"), to ascertain the degree to which a student can apply learned material to novel situations (e.g., "If the rules were changed to allow the shot-put circle to be raised by 2 feet [61 cm], would this increase or decrease the distance the shot would travel if all other factors were kept equal? Why?"), or to evaluate the ability to organize an answer in a logical manner (e.g., "Trace the development of public school physical fitness tests from the Kraus and Weber Low Back Test to the Fitnessgram."). The examinee should be able to recognize the type of answer required by the manner in which a question is stated.
- Use several essay questions requiring relatively short answers rather than a few questions requiring extended answers. This practice usually leads to two positive results: a wider sampling of knowledge and a test made up of relatively specific questions, the answers to which can normally be scored with increased reliability.
- Phrase the question so that the task of the examinee is specifically identified. Avoid asking for opinions when measuring educational achievement. Begin essay questions with such words or phrases as *Explain how, Compare, Contrast,* and *Present arguments for and against.* Do not start essay questions with such words or phrases as *Discuss, What do you think about,* or *Write all you know about.* Also, unless the purpose of a question is to measure the mastery of relatively factual material, do not begin an essay question with such words as *List, Who, Where,* or *When.*
- Set guidelines to indicate the scope of the answer required. Build limiting factors into the question: "Illustrate, through words and figures, how health related physical fitness is associated with academic achievement in school . . ." or "Limiting your answer to team games only, compare . . ." Other methods to indicate how involved the answer should be are to specify the amount of time to be spent on the answer, the number of words needed to provide a best answer, or the number of points the answer is worth in the margin beside the question, or the amount of space in which the answer is to be completed; however, spacing each question differently to provide sufficient room for each may penalize those examinees with large handwriting.
- Prepare for yourself an ideal answer to the question. Because this requires that you identify exactly what the question is intended to measure, ambiguities often become apparent. As mentioned previously, this practice also increases the reliability of the scoring process.
- Avoid giving a choice of essay questions to be answered. If an essay test is designed to measure achievement of the objectives for a group of students all exposed to the same instruction, each student should be required to answer the same questions. The common base of measurement is lost if a choice of questions is given. Optional questions add another variable and increase the possibility of inaccurate assessment.
- Avoid asking students for their opinions. Although the intent is to grade each response based on the substantive support provided for the answer, it is difficult to separate opinions

from truth (and who is to say which opinion is best?—as you might imagine, it is typically an instructor's opinion that is perceived as best).

- Indicate the approximate number of words examinees should write (e.g., 50 words; 150 words; 1 paragraph) or the amount of time they should spend on each item (e.g., 5 minutes; 10 minutes).

Recommendations for Scoring

Certain practices reduce some of the unreliability inherent in the process of scoring an essay answer. Several of these procedures are related to or follow from the previous suggestions for construction.

- Decide in advance what the essay question is intended to measure. If an essay question is designed to measure application of facts, evaluate the answers to the question on that basis and not on the basis of organization, spelling, grammar, neatness, or some other standard. Ignore elements not dealing with the question's objective.
- Use the ideal answer previously prepared as a frame of reference for scoring. This is especially important if you secure an independent rating of the answers (see the last bullet in this list for additional information).
- Determine the method of scoring. Use one of three systems:
    - **Analytic scoring** involves identifying specific facts, points, or ideas in the answer and awarding credit for each one. An answer receiving a perfect score would necessarily include all the specific items occurring in the ideal answer. This type of scoring is especially effective when the question's objective is to measure whether students have acquired the factual material.
    - **Global scoring** consists of reading the answer and converting the general impression obtained into a score. In theory, the general impression is a function of the completeness of the answer in comparison to the ideal answer. Of the three grading methods, this is the most subjective and the one most likely to be affected by extraneous factors.
    - **Relative scoring** consists of reading all students' answers to one question and arranging the papers in order according to their adequacy. You may accomplish this by setting up several categories (such as good, fair, and poor; or excellent, above average, average, and below average) and assigning each answer to one of the categories. Second, third, and possibly more readings may be necessary to arrange the papers within each category, and you may occasionally shift one paper to another category. The end result is an ordering of all the papers with respect to the correctness of the answers to the one question evaluated. After sorting, scores may be assigned to each answer. There is no reason the top paper must be assigned an A or the bottom paper an F; your evaluations should also be influenced by the comparison of each answer to the ideal answer. This ordering of the answers enhances consistency in the scoring procedure and is especially effective when the objective of a question is to measure relatively complex mental processes. Repeat the procedure for each of the remaining questions.
- Develop a system so that you don't know whose paper is being scored. Examinees could sign their names on a piece of paper next to a number corresponding to the number on their test booklet, or they could mark their test copy with unique designs or patterns that only they will recognize. Having each answer recorded on a separate sheet of paper also eliminates the bias caused by noticing the scores given to a previous answer. If several answers do occur on one answer sheet (as would be the case if short answers are required), record

points awarded for each answer on a separate sheet of paper, thus helping to eliminate the halo effect. This procedure is also useful if tests are rescored to check reliability. The second reader, who may or may not be you, will not be influenced by the score previously awarded.

- Score everyone's answers to one question rather than score one complete paper at a time. This process is required if you use global or relative scoring. Although not required for analytic scoring, the process usually leads to the most consistent scoring because it is easier to compare all the answers to one question if answers to other questions do not intervene.
- Arrange for a second scoring of the question. Ensuring the reliability and objectivity of essay test scoring requires that each answer be scored twice and the two scores compared. Ideally, to assess the test's reliability these two scores should be awarded by two scorers to ensure that they are independently obtained. Refer back to chapter 6 regarding how objectivity is a subset of reliability and is actually a measure of interrater reliability. If it is possible to arrange for another person knowledgeable in the area covered by the test to score the responses, provide that scorer with the ideal answers to the questions so that the two scores thus obtained have a common basis. However, if it is not feasible to obtain an independent scorer, score the answers yourself on two occasions, perhaps separated by a week, in an effort to secure some evidence about the consistency of the scoring procedures used.

As should be obvious by this point, the process of constructing and scoring a reliable essay test can be tedious and time consuming. However, to be fair to examinees, the procedures explained here should be followed if an essay test is used to measure cognitive objectives.

Go to the WSG to complete Student Activity 8.9.

ADMINISTERING THE TEST

As we have noted, there are potential problems involved in testing. Before and during a testing session, the anxiety level of some examinees can increase beyond desirable levels, during a testing session cheating can occur, and afterward, when examinees learn their scores, they may experience feelings of humiliation or haughtiness. However, these undesirable circumstances do not *have* to occur. The suggestions presented here should help eliminate or reduce many of the objectionable occurrences that are often associated with test administration. *Although the written test and the scoring procedures used have some influence on these occurrences, the administration of the test itself probably has the greatest impact on whether problems will arise before, during, and after the test.*

Before the Test

- ***Prepare the examinees for the test.*** Generally, less anxiety is associated with a test announced well in advance than with surprise tests, and discussing the content of an upcoming test with students can help reduce their apprehension. It is not logical (or ethical) to include on an achievement test questions on topics that have not been covered or assigned. Items such as the general areas to be measured, the approximate amount of testing time to be devoted to each area, the types of questions that will be asked (e.g., essay, multiple-choice), and the length of the test represent legitimate concerns of examinees. A written test, if properly constructed, should be a precise expression of the objectives of the instructional unit. It is difficult to imagine a situation in which knowledge of these objectives should be withheld from examinees.
- ***Eliminate the test-wise advantage for some examinees.*** Use proper test construction techniques outlined previously in this chapter (avoiding grammatical clues, specific determiners, interlocking items, and the like) and provide examinees with test-taking suggestions. For example, the following recommendations might be made to examinees:

- Realize that all the material measured by a good test cannot be learned the night before the test. Spend this time reviewing, not learning.
- Read the instructions to the test before beginning to answer the questions. Know how the test will be scored. Be aware of (a) whether all questions have the same value; (b) whether neatness, grammar, and organization will be accounted for in the score; and (c) whether a correction-for-guessing formula will be applied.
- Pace yourself.
- Plan an essay answer before starting to write it down.
- Check often to see that you are writing answers in the correct place on the answer sheet.
- Check over your answers if time permits.

- *See the comprehensive list of test-taking skills in the Test-Taking Skills highlight box.*
- *Give any unusual or lengthy instructions before test administration time.* Doing so will save time on the day the test is given and, more important, will enable examinees to begin the test as soon as possible. Providing instructions in advance reduces the time available for anxiety to build, especially for those who feel pressured by time.
- *Proofread the test before it is reproduced.* Proofreading helps to ensure that each examinee will receive a legible copy free of typographical, spelling, and other errors. It also eliminates or reduces the time spent during the test clearing up these errors.
- *Give a practice test to reduce examinee anxiety.*

During the Test

- *Organize an efficient method for distributing and collecting the tests.* With a small group, this is seldom a concern. However, with 60 or so examinees spread out in a large room, an efficient distribution procedure is necessary so that all examinees are given approximately the same amount of time to complete the test, and an efficient collection procedure is vital to keeping the test secure.
- *Help examinees pace themselves.* This can be accomplished by quietly marking on a blackboard the time remaining as well as a rough estimate of the portion of the test on which the examinees should be working.
- *Answer individual questions carefully and privately.* To avoid disturbing others, answer an individual question at the examinee's or the proctor's desk. However, take care that your response does not give any examinee an advantage over others.
- *Control cheating.* Obviously, cheating negates the validity of a set of test scores. Of more serious concern, though, are the negative attitudes generated toward those who cheat, toward the proctor who does not control cheating, and toward testing in general.
- *Control the environment.* Any factor that prevents an examinee from doing his or her best on a written test lowers the reliability, validity, and usability of the resulting set of scores. Some of these factors—examinee motivation and reading habits—are not under the direct control of the tester, although they can be influenced. You can, however, provide adequate lighting, eliminate noise distractions, maintain a comfortable temperature, and provide adequate space in which to work.

After the Test

- *Correct the tests and report the scores as quickly as possible.* The rapidity of this operation depends, of course, on the type and length of the test administered. However, examinees generally appreciate prompt results.

TEST-TAKING SKILLS

Preparing for the Test

- Schedule your time ahead—plan when good study times are available.
- Know when, where, and how you will be tested. Ask your instructor.
- Go to your instructor when you encounter a difficult or problematic area while studying.
- Be in the best physical and mental shape possible.
- Be motivated and positive in your attitude toward the test.
- Be mechanically prepared, with pencils, space, text, tables, and notes.
- Practice, practice, practice by finding practice tests to take. Generally people who are more familiar with taking tests perform better. Practice efforts are generally better on timed tests. The less the interval between practice and testing, the better the practice effect.
- Carefully read the summaries of each chapter. Look at highlighted text, figures, and tables.
- Study with classmates.
- Avoid cramming.
- Be test-wise, even though it may not help with well-developed tests. It will help on poorly constructed tests.
- If studying for an essay exam, make up and answer practice questions in advance.
- If studying for a completion test, find out if spelling will count.
- If studying for a matching test, find out if you can use an answer in more than one place.
- If preparing for an open-book test, mark particularly important pages or sections with tags to make them easy to find during the test.
- If studying for a take-home test, find out what sources you will be permitted to use for help.
- Look over all materials related to the course. (But don't stay up all night.)
- Get a good night's rest.
- Do not use stimulants or tranquilizers.
- Do not drink a lot of fluids or eat a big meal just before the test.
- Get to the testing area early and familiarize yourself with the area.
- Avoid last-minute panic questions. Panic is contagious. Do not talk with friends immediately before the examination.
- Relax.

Getting Started and Taking the Test

- Sit where you feel comfortable—whether that means near a window, near an outlet, or where you generally sit in class. Don't sit near annoying people.
- Read and listen to instructions carefully. There is important information in the instructions, including oral directions or corrections.
- Find out how the test will be scored, whether some items are worth more than others, whether guessing is penalized, and whether neatness counts.
- Know how much time you have to complete the examination, and be aware of time remaining during the examination.
- Look quickly through the test before starting so that you can plan ahead and gauge your time.
- Check to see that you have all the pages and items before beginning.
- Pace yourself, budget your time, and don't spend too much time on any one item.
- Concentrate on the test; do not pay attention to others in the room.
- Think positively.
- Stay calm if you don't know the answer; make an educated guess.
- Ask your instructor if you do not understand something.
- If you are stuck, move on to the next question and come back to a difficult one later. Activity reduces anxiety.
- Be aware of when time is almost up so that you can review and check the test.
- Do not worry about other students (e.g., if they leave earlier than you, the questions they ask).
- If you are using a separate answer sheet, check often that you are in the correct column and row. Check that you answered all the items. If there is time, reread the questions and your answers.
- Just before you submit the answer sheet, count the number of answers you have selected. Make sure that the number of answers selected equals the number of test items.

After Taking the Test

- Write down what you remember about the test.
- If you think you did poorly, go to your instructor's office and review the test with the instructor.
- Appeal when you feel that you are correct about an answer that was marked as incorrect.

@ Go to the WSG to complete Student Activity 8.10.

• ***Report test scores anonymously.*** Let each examinee decide whether to make his or her score known to others. Use a confidential identification number system if you post scores.

• ***Avoid misusing and misinterpreting test scores.*** By following the suggestions in this section, you will be able to improve the reliability and validity of your written tests. However, remember that no test is perfectly reliable and valid. Because of this, do not base crucial decisions on the results of one written test. For example, do not interpret a 1-point variation between two examinees' scores on a written test as showing a significant difference between the examinees. (Refer back to chapter 6 for information about the standard error of measurement to appreciate the level of confidence you can have about the precision of an individual test score.) Such an interpretation is a misuse of test scores. Along with other forms of measurement, consider written test results when evaluating people, but allow these results to influence the evaluations only to the degree their accuracy permits.

ANALYZING THE TEST

To determine the amount of confidence that can be placed in the set of scores resulting from a test administration, examine the reliability and the validity of the test. This is based on how closely (validity) and consistently (reliability) the test actually measures what is intended. Evidence for the reliability and validity of a test is both global (overall test performance) and specific (quality of individual questions).

Reliability

If a test were perfectly reliable, each examinee's observed score would be an exact representation of his or her true level of achievement of whatever the test measures. Each observed score would be a true score, uncontaminated by error. In actuality, of course, an observed score consists of two parts: the true score and the error score (recall this from chapter 6). The error score may be positive or negative, increasing or decreasing the observed score. As the error portion for the observed score increases, reliability decreases. Unfortunately, there are several sources of error in written tests:

• ***Inadequate sampling.*** The questions that appear on a test represent only a sample of the infinite population of possible questions that could have been selected. Error is introduced if the sample selected does not adequately represent the desired population of possible questions. An examinee's failure to be credited with understanding, or being penalized for not comprehending, a particular notion because no question was included on the test to measure that comprehension is an example of how sampling error might reduce test reliability (and validity).

• ***An examinee's mental and physical condition.*** Illness, severe anxiety, overconfidence, or fatigue can alter one's score and thus lower the reliability of a test.

• ***Environmental conditions.*** Poor lighting, poor temperature control, excessive noise, or any other similar variable that negatively affects concentration can cause observed scores to misrepresent true scores.

• ***Guessing.*** Because each examinee has, in theory at least, the same chance for good luck (and bad luck) when blindly guessing during an objective test, it would seem that in the long run the total effect would balance out and there would be no error introduced. However, one administration of a test does not represent the long run, and test reliability might be reduced because some examinees could, on one administration of a test, be luckier guessers than their peers.

- ***Changes in the field.*** Sometimes error is introduced not by the measuring instrument but by the fact that the variable being measured is changeable. Lack of a consistent definition (e.g., disagreement by authorities on the definition of physical fitness) and fluctuations in the amount of the attribute being measured (e.g., attitude toward physical activity can change from time to time) make constructing a reliable test difficult in some areas.

Thus, many factors, some of which are at least partially under your control, can introduce error and consequently reduce the reliability of a written test. As indicated in chapter 6, there are several methods of calculating a coefficient to express the reliability of a test; each of these methods reflects one or more of the sources of error. If test questions are scored correct (1) or incorrect (0), the alpha coefficient (identical to the Kuder–Richardson formula 20, or KR_{20}) can be used to estimate test reliability. The KR_{20} is actually the average of all possible split-halves reliability coefficients and as such is a relatively conservative estimate of a test's reliability. Obtaining a satisfactory reliability coefficient when using a conservative procedure is good because using other less conservative procedures would only result in higher estimates. KR_{20} is defined as:

$$\mathrm{KR}_{20} = \frac{K}{K-1}\left[1 - \frac{\Sigma pq}{s^2_{\text{total}}}\right] \tag{8.1}$$

where K is the number of test items, s^2_{total} is the variance of the test scores, and Σpq is the sum of the difficulty (p) times q where q is defined as $(1 - p)$. You will learn more about p (Difficulty or Diff.) in the Item Analysis section that follows.

Another method of estimating a written test's reliability when it is reasonable to assume that all items on the test are equally difficult and discrimination is the KR_{21}. The formula follows here:

$$\mathrm{KR}_{21} = \frac{K}{K-1}\left[1 - \frac{M\left(1-\bar{p}\right)}{s^2_{\text{total}}}\right] \tag{8.2}$$

where K is the number of questions on the test, s^2_{total} is the variance of the test scores, M is the mean test score, and $\bar{p}$ is the average difficulty defined as M/K. Note the similarity between KR_{20}, KR_{21}, and the alpha coefficient (see equation 6.3). The alpha coefficient is actually equivalent to KR_{20}. The KR_{21} reliability estimate is relatively easy to calculate, but the assumption of equally difficult and discriminating items is rarely true. Violation of this assumption results in this formula's underestimating the test's reliability; therefore, the KR_{21} formula is a more conservative estimate of the test reliability. Thus KR_{20} will always be greater than or equal to KR_{21}. Obtaining a satisfactory reliability coefficient when using a conservative procedure is a good idea because using other less conservative procedures would only result in higher estimates.

Mastery Item 8.5

Use the KR_{21} formula to estimate the reliability of a 60-item test having a mean of 45 and a standard deviation of 6.

Dataset Application

The chapter 8 large dataset in the WSG consists of 400 responses to 10 items. Open the chapter 8 large dataset and do the following:

1. Use Analysis → Scale → Reliability to estimate KR_{20} for the 10 items.
2. Use Analysis → Descriptive Statistics → Descriptives and obtain the variances (chapter 3) for each of the items and the total score. Substitute these values into the KR_{20} formula just presented (or the alpha coefficient found in chapter 6) and confirm the results that you obtained in step 1.
3. Use the results from the SPSS output to calculate KR_{21} using formula 8.2.
4. Are you pleased with the reliability you have obtained? If not, what would you do? (For a hint, turn to the Spearman-Brown prophecy formula found in chapter 6).

Validity

If a written test does not measure what it is designed to measure (even though it may measure something consistently), the resulting test scores are of little value. As noted in chapter 6, there are various types of validity and several methods of assessing them.

For a written test, one of the most important types of validity is content validity. This is generally determined subjectively by the extent to which the individual test items represent a sufficient sample of the educational and content objectives included in a course of instruction. By examining a copy of a test, you determine the degree of content validity the test has for the particular situation. Following the proper procedures for constructing a written test, especially using a table of specifications, helps ensure that your test will have content validity. Having another content expert review your test items provides additional input on the test's content validity.

The preceding presentations on test reliability and validity refer to the overall test. But the quality of the entire test is determined by the quality of the individual items. We now turn to item analysis, which will help determine the quality of the individual items and how they might contribute to overall test reliability and validity.

ITEM ANALYSIS

Analyzing the responses to the test items is important for several reasons but especially to continually improve the items and consequently the test. The difficulty level and the discriminating power (ability of the question to separate strong from weak examinees) of each item are the keys to item improvement. **Item analysis** can also improve your instruction by identifying weaknesses in the examinees as a group, in instructional methods, or in the curriculum. It can also improve your skill in constructing written tests. Most of the illustrations and examples presented involve multiple-choice questions because there are efficient methods for analyzing them. Software programs exist that can calculate these values for you if you use scannable answer sheets. However, you can modify most of the following steps of item analysis for other types of objective items, and you can apply the principles involved to most types of questions. The procedures for completing an item analysis by hand follow.

- ***Step 1.*** Score the tests.
- ***Step 2.*** Arrange the answer sheets in order from high to low score.

- ***Step 3.*** Separate the answer sheets into three subgroups: (a) the upper group, which consists of the upper 27% (approximately) of the answer sheets; (b) the middle group, which consists of the middle 46% (approximately); and (c) the lower group, which consists of the same number of answer sheets as placed in the upper group. You will use only the answer sheets of the two extreme groups—the upper and the lower—in the item analysis. Test authorities suggest that to include as many responses as possible and maximize the differences between the types of responses, the upper and lower groups should each include 27% of the answer sheets. Generally, as long as there is an equal number in each of these groups, use the most convenient number of answer sheets between 25% and 33%. For example, if 60 answer sheets were available for analysis, the highest and lowest 15 to 20 could be used.
- ***Step 4.*** Count and record for each item the frequency of selection of each possible response by the upper group.
- ***Step 5.*** Count and record for each item the frequency of selection of each possible response by the lower group.

Steps 4 and 5 are the most time-consuming portions of the item analysis. Several procedures can reduce the tedium of this task:

- Use previously prepared scorecards for each item.
- Use a computer to speed the process of recording responses.
- Cooperate with another scorer, with one person reading and the other recording, or use an optical scanner to have a computer accomplish these steps.

An example of a possible organization of the resulting data is shown in figure 8.3. (These data were obtained from a question included on a nationally standardized test of physical fitness knowledge administered to college senior physical education majors.)

At the completion of step 5, the necessary data are available to calculate an **index of difficulty** and an **index of discrimination** for each item. The data in figure 8.3 illustrate the calculation of these two indexes and how change suggested by the response pattern can improve an item. In this example, the left side of the figure contains the initial draft of the question and the data (as described previously) resulting from its administration to

Source: *Handbook of Physical Fitness* — Topic: Physical fitness

First draft: In the opinion of most authorities, three of the following factors have contributed to a lowering of the national level of physical fitness. Which has NOT had this effect?

A. An increase in life span
B. A decrease in the physical effort required for daily living
C. An increase in the number of occupations involving sedentary activity
*D. An increase in school consolidation

Item 5 | Test: Form D trial | Date: 6/98 | *n* = 185

Responses	A	B	C	D*	E	Omit	Diff.	Net D
Upper 27% = 50	28	2	1	19		0		
Lower 27% = 50	24	8	1	17		0	36%	4%

Revision: In the opinion of most authorities, three of the following have contributed to a lowering of the national level of physical fitness. Which has NOT had this effect?

A. An increase in the number of senior citizens
B. A decrease in the physical effort required for daily living
C. An increase in the number of occupations involving sedentary activity
*D. An increase in school consolidation

Item 25 | Test: Final form A | Date: 9/00 | *n* = 1112

Responses	A	B	C	D*	E	Omit	Diff.	Net D
Upper 27% = 300	69	10	5	216		0		
Lower 27% = 300	89	52	54	104		1	53%	37%

Figure 8.3 One way to organize data for item analysis.

approximately 185 examinees. The right side contains the revised question and the data resulting from its administration to more than 1000 examinees.

- ***Step 6.*** Calculate and record the index of difficulty for each item; this is the estimated percentage of examinees who answered the item correctly. The formula is as follows:

$$\text{Diff} = \frac{U_c + L_c}{U_n + L_n} \times 100 \tag{8.3}$$

where Diff is the index of difficulty, U_c is the number of examinees in the upper group answering the question correctly, L_c is the number of examinees in the lower group answering the question correctly, U_n is the number of examinees in the upper group, and L_n is the number of examinees in the lower group (recall that $U_n = L_n$).

Inspection of this formula reveals that the index of difficulty is the percentage of examinees answering the question correctly; thus, the higher the index, the easier the question. The following examples illustrate the use of the index of difficulty formula (see figure 8.3). You can compare item difficulties to course objectives and student learning objectives to help determine what content objectives students are meeting and those that might need additional reinforcement or benefit from changes in course delivery strategies.

First-draft results: $n = 185$; therefore $U_n = L_n = 185 \times 0.27 = 50$.

Revision results: $n = 1112$; therefore $U_n = L_n = 1112 \times 0.27 = 300$.

The maximum amount of discrimination can occur only when an item has an index of difficulty of exactly 50%. If this criterion were met by every question on a test, the mean score of the test would be equal to one-half the number of items on the test. For example, the mean score of a test containing 80 items would be 40. This ideal, however, assumes that no element of chance is involved. On an 80-item multiple-choice test on which each item has four possible responses, random marking of the answer sheet should produce approximately 20 correct responses (i.e., 1/4 × 80 = 20). Considering chance, the mean score of the test just described should be 50. This value is obtained by determining what score lies halfway between the chance score and the highest possible score (80 items – 20 correct by chance = 60 items; if an examinee answers 50% of these 60 items correctly, he or she would have 30 items correct, plus the 20 by chance, resulting in a score of 50). If the index of difficulty of each of the 80 items was 62.5%, the mean score for the test would be 50 (80 × 0.625 = 50).

It is not possible, especially on the first draft, to produce an item having exactly some predetermined difficulty index. *The point is, to maximize an item's discrimination power, an attempt should be made to write each item in such a manner that half or slightly more than half of the examinees will be able to answer it correctly.* One further point should be noted. Maximum discrimination can occur only for an item of middle difficulty, but meeting this condition does not necessarily guarantee that it will occur. Figure 8.4 describes the relationship between discrimination and difficulty and shows that as difficulty increases from 0 to 0.50, the *potential* discrimination increases. However, as difficulty continues to increase from 0.50 to 1.0, *potential* discrimination decreases.

- ***Step 7.*** Calculate and record the index of discrimination for each item; this is an estimate of how well an item discriminates among examinees who have been categorized by some criterion.

$$\text{Net D} = \frac{U_c - L_c}{U_n} \times 100 \tag{8.4}$$

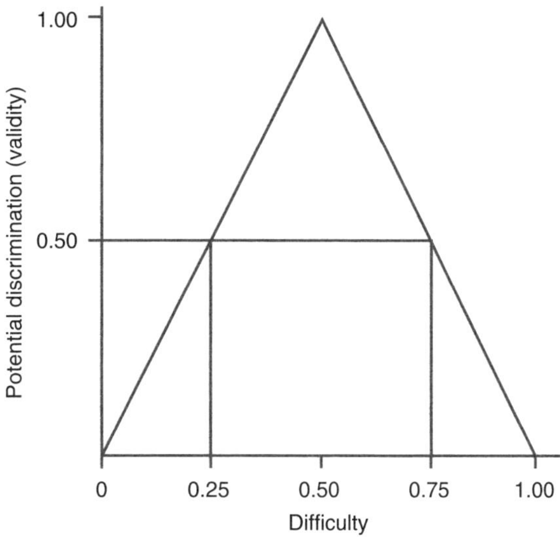

Figure 8.4 The relationship between discrimination and difficulty.

where Net D is the index of discrimination. (Note that either U_n or L_n may be used in the denominator.) The index of discrimination presented here, known as the Net D, is only one of nearly 100 discrimination indexes that have been devised. The most commonly cited discrimination indexes are correlational techniques for quantifying the relationship between the score on a particular item and a criterion score (usually the total test score). However, we use the Net D because it is relatively simple to calculate, uses the same data as required to determine the difficulty index, and is fairly simple to interpret. The following examples, again using the data presented in figure 8.3, illustrate the use of the Net D formula.

First-draft results: $n = 185$; therefore $U_n = L_n = 50$.

Revision results: $n = 1112$; therefore $U_n = L_n = 300$.

The criteria usually used to examine the discriminating power of an item are the scores on the entire test on which the item appears. In general, if the examinees who performed well on the entire test did well on the item and the examinees who did poorly on the entire test did poorly on the item, the item is considered a good discriminator. If approximately the same number of high-performing and low-performing examinees answer an item correctly, it is considered to have little or no discriminatory power. If the item is answered correctly by more of the low-performing examinees than by the high-performing examinees, it is considered a negative discriminator. *Discrimination is the most important characteristic of an item. A test cannot be reliable or valid unless the individual items discriminate among examinees.*

Note that the higher the value of Net D, the higher the discriminating power of the item, and note also that the Net D formula could produce a negative number indicating an item that discriminates negatively. In fact, the value obtained is actually the net percentage of good, or positive, discriminations achieved by an item, thus the name, Net D. Figure 8.5 illustrates this concept.

No discrimination occurred between Bill, Kelly, Pete, Alicia, Judy, and Gregg, because all answered the item correctly. Similarly, no discrimination occurred between Fred, Michelle, Dave, and Stephanie, because all answered the item incorrectly. The discrimination that occurred between Bill (or Kelly, Pete, or Alicia) and Michelle (or Dave or Stephanie) is con-

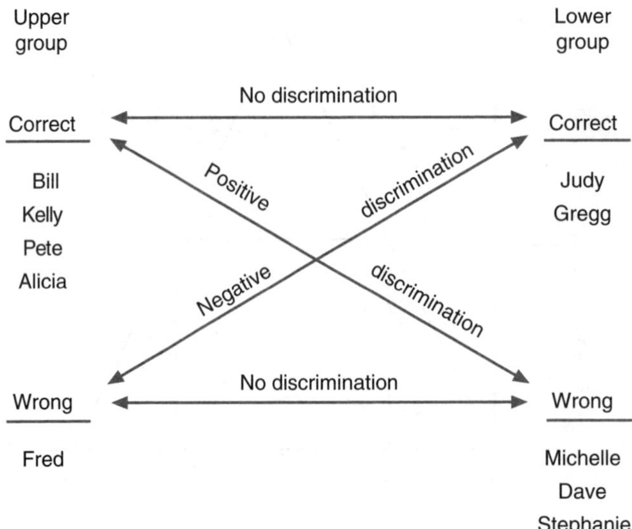

Figure 8.5 Positive and negative discrimination.

sidered a good, or positive, discrimination because of the groups in which these examinees have been placed based on their total test scores. Altogether a total of 12 (4 × 3) positive discriminations occurred. Conversely, the discrimination that occurred between Fred and Judy (or Gregg) is considered a bad, or negative, discrimination because Fred is in the upper group and Judy and Gregg are in the lower group. A total of two (2 × 1) negative discriminations occurred. The maximum number of discriminations possible with five examinees in each group is 25 (5 × 5). Of these 25, 12 were positive, 2 were negative, and 11 did not occur. Subtracting the 2 negative discriminations from the 12 positive discriminations results in a net of 10 positive discriminations. The ratio of net positive discriminations to the total possible (10/25) is 40%. Using equation 8.4 for Net D results in this same value:

$$\text{Net D} = \frac{4-2}{5} \times 100 = 40\%$$

Attempt to keep the index of discrimination of an item on an achievement test as high as possible. Most test construction authorities agree that an item with a discrimination index of 40% or higher is a good item. Discrimination indexes between 20% and 40% are acceptable but may indicate the need for revision, especially as the value approaches 20%. Items with an index of discrimination below 20%, and especially those with negative discrimination indexes, are poor and should probably be discarded from future tests.

- ***Step 8.*** Examine the pattern of responses to determine how an item might be improved.

According to the previous suggestions for retaining and discarding questions based on their discrimination index, the initial draft of the question in figure 8.3 probably should have been discarded. However, the response pattern of the examinees revealed a possible solution. Although it is often difficult to understand why certain responses are selected or ignored and even more difficult to determine possible alterations of the responses or stem that will improve an item, examining the response patterns often suggests possibilities. For example, response A to the first draft of the item displayed in figure 8.3 was chosen by more than 50% of the combined upper and lower groups of examinees, even though it is incorrect. Rewording this distractor in the revision resulted in the keyed response becoming more attractive than the first response, especially to the examinees in the upper group. The positive changes in the difficulty and discrimination indexes indicate that the alteration of this one response improved the item considerably.

Mastery Item 8.6

How many answer sheets should be used in an item analysis for a written test taken by 250 examinees?

Mastery Item 8.7

Calculate the index of difficulty and the Net D discrimination index for a multiple-choice question answered correctly by 40 of the 60 examinees in the upper group and by 10 of the 60 examinees in the lower group.

Mastery Item 8.8

To demonstrate the relationship between the difficulty and potential discrimination of an item, calculate the indexes of difficulty and discrimination for the following five items:

Item no.	Upper group $n = 10$	Lower group $n = 10$
1	2 correct	0 correct
2	5 correct	5 correct
3	10 correct	5 correct
4	10 correct	0 correct
5	5 correct	10 correct

Table 8.3 illustrates the lowest, highest, and desired values for difficulty and discrimination.

Positive discrimination values result in increased test reliability, negative discriminations result in decreased test reliability, and zero discriminations result in no change in test reliability. Figure 8.6 illustrates how the correlation between the item score (0 is incorrect and 1 is correct) and the total test score is also an estimate of item discrimination. Note that the three items' total score correlations presented in Figure 8.6 are for positive (a), negative (b), and no (c) discrimination. These result in increased reliability (a), decreased reliability (b), and no effect on reliability (c).

Table 8.3 Difficulty and Discrimination Indexes

	Difficulty index	Discrimination index
Lowest	.00 (0%)	–1.00 (–100%)
Highest	1.00 (100%)	1.00 (100%)
Desired	.50 (50%)	1.00 (100%)*

*As indicated in the text, values above .40 (40%) are considered quite good and discriminating.

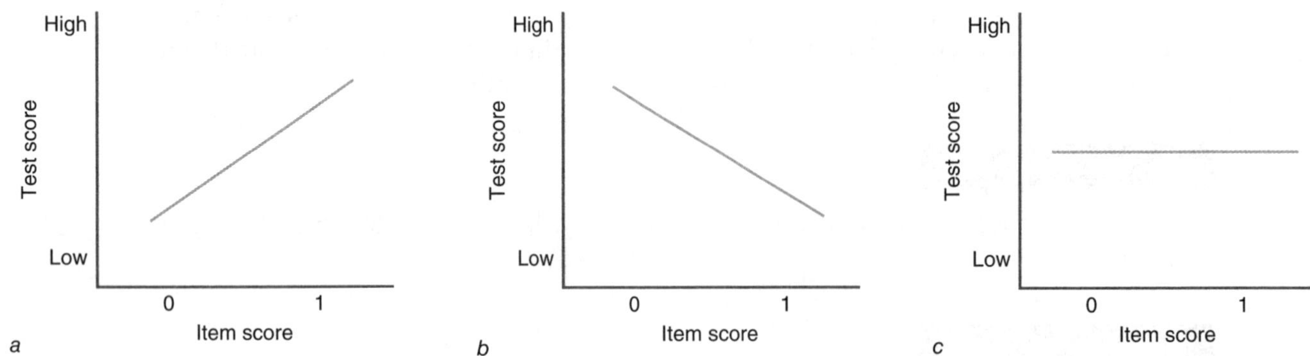

Figure 8.6 Illustration of how correlation can be used to estimate test item discrimination.

SOURCES OF WRITTEN TESTS

When a written test is given in the field of human performance, the chances are great that the examination was constructed locally. Generally, the number of sources for written tests in our discipline is relatively limited. With a few exceptions, nationally standardized written tests are not available.

When you are constructing a written test, it is often helpful to examine similar tests to obtain ideas for questions. Some possible sources for similar tests are professionally constructed tests, textbooks, periodicals, theses, and dissertations. Zhu, Safrit, and Cohen (1999) have published a test on physical fitness knowledge that was developed using many of the concepts presented in this chapter.

QUESTIONNAIRES AND SURVEYS

The questionnaire is a close relative of the written test. Both of these data collection instruments require careful construction and thoughtful analysis of the data they produce. The written test is primarily designed to assess the amount of knowledge a student has and to discriminate among students on the basis of their cognitive behavior, whereas questionnaires are typically used to measure affective domain concerns such as attitudes, opinions, and behaviors. For example, you might conduct a survey to determine how many minutes of moderate-to-vigorous physical activity (MVPA) people engage in each week to determine if they meet public health recommendations, or you might want to assess attitudes toward physical activity before and after exposure to a particular curriculum.

Questionnaire responses provide the independent and dependent variables for survey research. Cox and Cox (2008) provided an extensive presentation of questionnaire development. Thomas, Nelson, and Silverman (2015) listed eight steps for conducting survey research:

1. Determining the objectives
2. Delimiting the sample
3. Constructing the questionnaire
4. Conducting the pilot study
5. Writing the cover letter
6. Sending the questionnaire

7. Following up
8. Analyzing the results and preparing the report

Using a questionnaire to collect information has both advantages and disadvantages. On the plus side, the questionnaire can be relatively efficient in terms of money and time. Because a questionnaire can be sent to all respondents simultaneously, data collection can be completed over a period of a few weeks. As an alternative to mailing a questionnaire, it is now increasingly common to use the Internet to gather information. For both written and Internet surveys, respondents can be widely spread geographically, and they can respond at their own convenience. Each respondent is exposed to exactly the same instrument. For the best response to a written or electronic questionnaire, keep the survey short and specific.

On the negative side, the value of questionnaire data can be reduced by low response rate, inability to clarify a question that the respondent finds ambiguous, unanswered questions, and lack of assurance regarding who actually completes the questionnaire. Some of these concerns can be addressed through careful planning, but they can never be totally eliminated.

Planning the Questionnaire

There are now numerous online survey companies available on the Internet. Many of them have surveys already constructed for many common concerns such as customer satisfaction, faculty satisfaction, and educational outcomes. These companies also offer help in designing your own questionnaire and collecting and analyzing the data those questionnaires produce.

Time spent planning a questionnaire is invaluable. Before constructing a questionnaire, clarify the purpose of the instrument by studying and formulating relevant hypotheses so it will be possible to determine specifically which data the items on the questionnaire are designed to obtain. Unfortunately, this direct link between the items on the questionnaire and their exact purpose is not always carefully considered, resulting in either the collection of unnecessary information, the inability to respond to some hypotheses, or both. To prevent this, ask yourself exactly how each item on the questionnaire is to be analyzed. If you can't answer this question for a particular item, it should be omitted.

As with a question on a written test, it is difficult to know how an item on a questionnaire will function the first time it is used. This is why it is necessary to do a few pilot studies before finalizing your questionnaire. *Perhaps the best advice is to conduct pilot work with the questionnaire as you are developing it.* In the first trial run, ask some colleagues to examine potential items for ambiguities, personal idiosyncrasies, and problems in the directions to the respondents. After these are addressed, get feedback on the next draft of the instrument from a small sample of people (a focus group) who are in the potential respondent pool. Their task is not only to respond to the questionnaire but also to indicate any problems they encounter. Then address these problems and attempt to input and analyze the resulting data to determine if the correct information to address the hypotheses is being secured and if there are any data entry problems (e.g., multiple responses to items, inappropriate responses).

Go to the WSG to complete Student Activity 8.11.

Constructing the Questionnaire

One of the first decisions in constructing a questionnaire is to decide whether to use open-ended or closed-ended questions. Open-ended questions are those for which response categories are not specified. Essentially, the respondent provides an essay-type answer to the question. An example is "What benefits do children gain from participation in an organized sports program?" Closed-ended questions are those requiring the respondent to select one or more of the listed alternatives—for example, "How many days a week should elementary school children participate in physical education classes? 1 2 3 4 5." Both types of question have advantages and disadvantages.

Open-Ended Questions

The advantages of open-ended questions are that they

- allow for creative answers and allow the respondent freedom of expression,
- allow the respondent to answer in as much detail as desired,
- can be used when it is difficult to determine all possible answer categories, and
- are probably more efficient than closed-ended questions when complex issues are involved.

The disadvantages of open-ended questions are that they

- do not result in standardized information from each respondent, making analysis of the data more difficult;
- require more time of the respondent, which ultimately could reduce the return rate of the questionnaire;
- are sometimes ambiguous because they attempt to solicit a general response, which can result in the respondent not being certain what is being asked; and
- can provide data that are not relevant.

Closed-Ended Questions

The advantages of closed-ended questions are that they

- are easy to code for computer analysis;
- result in standard responses, which are easiest to analyze and to use in making comparisons among respondents;
- are usually less ambiguous for the respondent; and
- the ease of the respondent's task increases the return rate of the questionnaire.

The disadvantages of the closed-ended questions are that they

- may frustrate a respondent if an appropriate category is omitted,
- can result in a respondent selecting a category even if he or she doesn't know the answer or have an opinion,
- may require too many categories to cover all possible responses, and
- are subject to possible recording errors (e.g., the respondent checks B but meant to check C).

Deciding which type of question to use depends on several factors, such as the complexity of the issues involved, the length of the questionnaire, the sensitivity of the information sought, and the time available for construction and analysis of the questionnaire. In general, closed-ended questions work best when the responses are discrete, nominal in nature, and few in number. Descriptive information such as gender, years of education, and marital status most often are measured with closed-ended questions. Closed-ended questions may have some advantages when seeking sensitive information. For example, a respondent may be willing to provide information about annual income if it is done by checking the appropriate range (e.g., between $80,000 and $100,000) rather than by listing the salary specifically. Closed-ended questions should be used if you have more time to construct the questionnaire than to analyze it. Open-ended questions are relatively simple to construct, but the coding and interpreting necessary once the responses are returned are anything but simple.

Other Item Concerns

Questionnaire items need to be simple and avoid containing more than one element. Do not include two questions in one. For example, how would you answer yes or no to this question: "Do you think physical education or music should be kept or dropped from the curriculum?" Avoid ambiguous questions. Consider the responses to this item: "Do you think the punishment was appropriate? Yes or No." If a person answers no, you do not know if they thought the punishment was too light or too harsh. To avoid misinterpretation on the part of the respondent, avoid vague terms, slang, colloquial expressions, and lengthy questions. Be certain that the level of wording is appropriate for those who will be responding. Avoid the use of leading questions. For example, how would you likely respond to this item: "Most experts believe that regular moderate exercise produces health benefits. Do you agree?"

Go to the WSG to complete Student Activity 8.12.

Factors Affecting the Questionnaire Response

Besides the instrument itself, many ancillary materials and techniques affect the success of collecting data with a questionnaire. Computers are widely used to obtain survey data. However, these data may still be problematic in that the sample choosing to respond may not be representative of the population about which you want to generalize, and not everyone feels comfortable answering personal questions on the computer because of perceived security, identity theft, and privacy issues.

Cover Letter

Next to the questionnaire itself, the most important item sent to the respondent is the cover or introductory letter or online message. This brief document has the important task of describing the nature and objectives of the questionnaire and soliciting the cooperation of the respondent. If possible, personalize cover letters or online survey invitations (address them to the respondent rather than "Dear Sir or Madam"), use some slight form of flattery (e.g., "Because of your extensive background . . ."), and include the endorsement of someone known to the respondent. In addition, be sure that the cover letter or online message is short, neat, and attractive. These strategies enhance the probability that the questionnaire will be returned.

A practice that is becoming increasingly common with computer-disseminated questionnaires is alerting the possible respondents via e-mail a few days before sending the questionnaire and including in this e-mail the importance of their response. Such an introductory e-mail can take the place of the cover letter in a mailed questionnaire.

Ease of Return

For mailed questionnaires, provide clear instructions as to how and when the questionnaire is to be returned; enclosing a self-addressed postage-paid envelope has been shown to produce increased response rates. If the number of questionnaires to be returned is large, it may be advantageous to set up a postage-due arrangement with the post office. Under this arrangement, you pay only for the questionnaires that are returned rather than putting postage on return envelopes that may never be returned. The ease of mailing for the respondent is the same in either case. If information is being collected via the Internet, be sure to address the question of whether the respondent can stop and return to complete the questionnaire at a later time.

The respondents must be representative of the population that was sampled. Having data to provide evidence that the sample reflects the population and is not systematically biased in some way is important. For example, do the ethnic, age, or gender characteristics of your respondents look like the group to whom the original questionnaire was sent?

Neatness and Length

It is logical that if you spend time making the questionnaire and cover letter neat, easy to read, free of grammatical errors, and uncluttered, the respondent may be more inclined to spend time responding. The shorter you can make the questionnaire, the more likely it will be returned. If information is being collected via the Internet, make sure to consider adding ways to indicate how much the respondent has completed and how much yet remains.

Inducements

The inclusion of a pencil or a pen ("so you won't have to locate one"), a penny ("for your thoughts"), a dollar ("for a cup of coffee while you complete the questionnaire"), or a lottery ticket ("winner to be chosen from returned questionnaires") are examples of inducements people have used to encourage respondents to return mailed questionnaires. At the University of Colorado, a $2 bill was enclosed with a mailed student satisfaction survey done each year on a sample of students. The hope is to instill a sense of obligation. For some people, it may be difficult to put the money in their wallet and the survey in the wastebasket. As with mailed questionnaires, internet surveys often include the inducement that returning the survey will result in an entry into a lottery to win a relevant prize.

Timing and Deadlines

It is best not to have the questionnaire arrive just before a major holiday or other significant event (such as the beginning or end of the school year). The return rate of the University of Colorado student satisfaction survey mentioned previously would likely be low if it arrived to students during finals week. The inclusion of reasonable deadlines should enhance return rates. Receiving the questionnaire the day before (or even after) it is due back gives the respondent an easy excuse for not completing the questionnaire. Allowing too long a time before it is to be returned may cause the instrument to be put aside and never surface again.

Follow-Up

It is generally believed that at least one follow-up procedure can enhance the return rate of questionnaires. After one or two reminders, the effectiveness of follow-up procedures generally decreases dramatically. A typical procedure is to send the original questionnaire and cover letter, wait until the responses trickle down to a few, and then send out a reminder letter or e-mail. The reminder can potentially increase the response rate. If additional follow-up seems necessary, it is common to next send a duplicate questionnaire and cover letter; if still no response is received, a telephone reminder is the next possibility. Follow-up procedures beyond this are mostly without success.

Analyzing the Questionnaire Responses

How data obtained from returned questionnaires are analyzed will depend on the types of questions (open or closed) that are asked. Open-ended responses require a reader to carefully decipher each response and make subjective judgments about the answers. A reader may need to make a list of ideas that are expressed along with the frequency of the occurrence. Closed responses are most often tabulated and presented in some descriptive way such as by using percentages and/or graphs. For mailed questionnaires, creating tables or graphs requires transferring the data to a spreadsheet that is compatible with a computer program designed for calculating such descriptive statistics. For computer-sent questionnaires, the data are typically easily converted into relevant spreadsheet formats.

The mechanics of how the data are going to be analyzed should be quite straightforward if the questionnaire constructor follows the suggestions for planning the questionnaire presented earlier. Checking to see that the questions will lead to obtaining data that can be used to evaluate each of the objectives and hypotheses in the planning stage will result in clear directions for analyzing the responses.

Questionnaire Reliability

The procedures presented in chapters 6 and 7 are most often used to validate and estimate the reliability of questionnaire responses. To estimate the reliability of a single item, you must ask the specific item on at least two occasions. However, affective and cognitive domain subscales completed on a questionnaire can have their reliability estimated with the alpha coefficient. An important issue is the stability reliability of the responses. To estimate stability reliability, you must administer the questionnaire to the same people on two or more occasions. The typical time between administrations to determine stability reliability is 2 to 4 weeks. A period longer than this could result in actual changes in respondents' opinions. If there are changes in responses that reflect true changes of opinion, the estimated reliability will be reduced. The specific type of reliability estimate to use depends on the nature of the questions asked. For example, are the items nominally (e.g., "What is your gender?") or intervally scaled (e.g., a series of questions about attitudes)? See chapters 6 and 7 for the specific methods to estimate reliability for the obtained responses.

Dataset Application

Use the chapter 8 large dataset, MVPA example, in the WSG to determine the alpha reliability (chapter 6) of the results from a questionnaire that asked people to self-report weekly minutes of MVPA for 3 weeks. How reliable are these data? What would the estimated reliability be if you obtained data for 4 weeks instead of 3 weeks? (Recall the Spearman–Brown prophecy formula from chapter 6.) What steps might you take to estimate the validity of these self-reports?

Questionnaire Validity

The most important issue of a questionnaire, as with any measuring instrument, is the validity of the responses. It is important that the respondents truthfully respond to the items and do not respond based on what they believe the socially acceptable response would be. Developing quality questionnaire items, having items reviewed by experts, conducting pilot testing, and ensuring confidentiality or anonymity are ways to increase the validity of responses. Most questionnaires are validated with content-related procedures (presented in chapter 6). However, there are ways to cross-check the responses with additional data to determine if the respondent is answering truthfully. For example, if a respondent says that he or she votes for a certain candidate in an election, there is no way to determine the actual vote. However, you can verify through public records whether the respondent is actually registered to vote and voted in the specific election. Finally, whether the respondent sample is representative of the population about which one desires to generalize is an important validation issue to consider.

Booth, Okely, Chey, and Bauman (2002) provided an example of estimating the reliability and validity of a questionnaire in their examination of the Adolescent Physical Activity Recall Questionnaire.

MEASUREMENT AND EVALUATION CHALLENGE

To measure how well three groups of students have learned basic statistical concepts after each group had been exposed to different teaching methods, Kate has constructed a 60-item multiple-choice test based on a table of specifications she had developed before her research began. The table of specifications reflects the content and importance of the material. She initially developed a 100-item test and then conducted a pilot test followed by an item analysis to evaluate the individual items and the overall test. Using the item analysis, she was able to select 60 items that each demonstrated acceptable difficulty and discrimination indexes and, as a total test, an acceptable reliability. She also asked two experts who have taught statistical concepts for nearly 20 years to evaluate the items. These experts' suggestions helped her ensure that the items had acceptable content validity.

The test will be administered to students in each of the three teaching-method groups. The mean score on the test will be used as the dependent variable in an analysis of variance (see chapter 5) to see if the groups differed in their knowledge of statistics.

SUMMARY

When a research project or a human performance curriculum requires the assessment of objectives in the cognitive domain, the instrument of choice is usually a written test. When the assessment of aspects of the affective domain, such as attitudes or opinions, is desired, a questionnaire is typically used. The procedures for planning, constructing, scoring, administering, and analyzing the results from application of such instruments have been presented in this chapter. All of the procedures described in this chapter focus on making the written test or the questionnaire as objective, reliable, and valid as possible.

Go to the WSG for homework assignments and quizzes that will help you master this chapter's content.

CHAPTER

9

Physical Fitness and Activity Assessment in Adults

Allen W. Jackson, University of North Texas

OUTLINE

OBJECTIVES

After studying this chapter, you will be able to

- identify and define the components of health-related physical fitness;
- identify and define the risks for fitness testing;
- use reliable and valid methods of measurement of aerobic capacity, body composition, and muscular fitness;
- identify and use test items specifically developed for the older adult population; and
- understand the issues associated with reliable and valid measurement of physical activity in all populations.

 The lecture outline in the WSG will help you identify the major concepts of the chapter.

MEASUREMENT AND EVALUATION CHALLENGE

Jim is a new graduate with a major in kinesiology. He was seeking to become a certified teacher but then became interested in working in the health and fitness industry. He interviewed with a new branch of the YMCA for a position as physical fitness director. The executive director of the YMCA is interested in having high-quality physical activity and fitness programs but also wants to have a good fitness assessment program available to all members. Jim believes that his interview went well. His grades at school were good and he has worked part-time in fitness facilities for the last 2 years. The director asked him if he was certified in fitness instruction by any professional organizations. Jim had to respond that he was not currently certified but would certainly be happy to seek an appropriate certification. The director asked Jim to investigate the certification he would seek and outline a suggested adult fitness assessment program. They are going to meet again in a week, at which time Jim is to report on these two issues. The director concluded, "Jim, if the meeting goes well, I think that we will ask you to join our team." Jim is excited but also nervous. Because his career may depend on his responses to the director, Jim has to do some research and thinking.

No more important objective exists in the sport and exercise sciences than the attainment of physical fitness. Physical fitness is a multifaceted objective, with different meanings to different people—a cardiologist might define physical fitness differently than a gymnastics coach. Whatever the definition or understanding of physical fitness, its importance to you in your professional career is related to two primary factors:

1. The citizens and state and federal governments of many industrialized countries have taken the position that the general public should have sufficient levels of physical activity and fitness as this will provide health benefits and enable citizens to deal with the physical challenges that they may encounter. The U.S. government established public health objectives for improved levels of physical activity and fitness largely through its initiative called *Healthy People 2000.*

Physical Activity and Health: A Report of the Surgeon General, released in 1996 by the U.S. Department of Health and Human Services (USDHHS 1996), was a landmark scientific presentation of the health benefits of physical activity and fitness. The report summarized the physiological and psychosocial benefits that people of all ages gain from a physically active lifestyle.

Continuing the efforts of *Healthy People 2000,* the Centers for Disease Control and Prevention (CDC) and the U.S. Public Health Service included physical activity and fitness as a component of *Healthy People 2010* and *Healthy People 2020.* In October 2008 the U.S. Department of Health and Human Services (USDHSS 2008) released the *2008 Physical Activity Guidelines for Americans* to provide physical activity recommendations for all Americans that are consistent with the known research evidence relating physical activity to health. The World Health Organization's position on physical activity and health has led other countries, governments, and agencies to promote physical activity and fitness. Figure 9.1 provides the physical activity objectives from *Healthy People 2020.* Between 2010 (baselines) and 2020 (targets), a 10% improvement in moderate and vigorous physical activity and strength activity is desired for better public health in the United States.

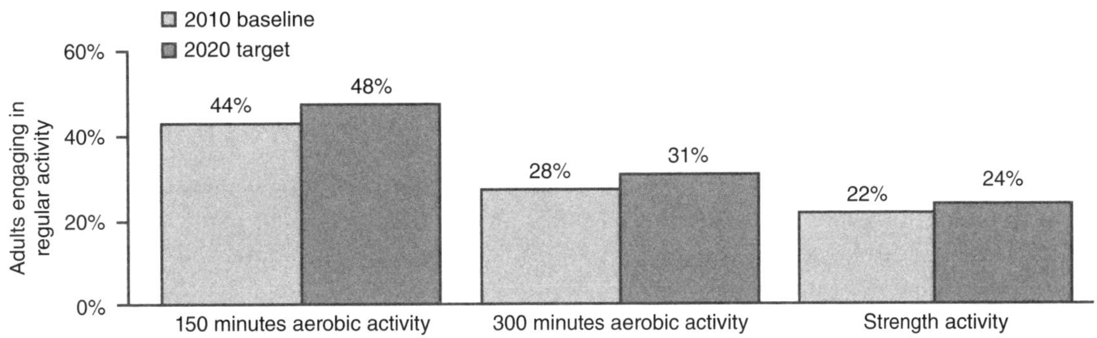

Figure 9.1 *Healthy People 2020* physical activity objectives: baselines (2010) and targets (2020) for improvement.

2. The basic justification for professions in sport and exercise sciences is the improvement and maintenance of physical activity and fitness as an important step in developing healthy lifestyle behaviors.

In 2007 the American College of Sports Medicine (ACSM) and the American Medical Association launched *Exercise is Medicine* as a national initiative to encourage health care providers to include exercise when designing treatment plans for patients. The basic assumption of the initiative is that exercise and physical activity are important in the treatment and prevention of a variety of diseases and should be assessed as part of medical care. A goal of *Exercise is Medicine* is to have health care providers discuss physical activity behaviors each time they meet with patients, regardless of the purpose of the visit.

Professionals, such as you, need to know and understand these factors and the effect they will have on your career. Many excellent sources of information on physical fitness testing are available. Indeed, entire books are devoted to physical fitness training and assessment (e.g., Golding 2000). This chapter provides examples of adult physical fitness tests and methods for estimating physical activity levels, focusing on reliability and validity.

Because physical fitness is multifaceted, an effective definition must be broad and encompassing. Two factors, the purposes of the tests and the defined population, provide a framework for defining physical fitness for any person. As you see from table 9.1, we might have many objectives (different fitness tests) for different groups of people. Different physical fitness definitions result in different levels of capacity or function. For example, one who is engaged in high-level athletic activities will need a greater level of physical

Table 9.1 Populations and Purposes of Physical Fitness Testing

Population	Health related	Motor	Diagnosis	Military preparation	Functional capacity
Youth	*	*		*	*
Adults	*				*
The aged	*				*
Special					
Persons with mental and physical disabilities	*		*		*
Athletes		*	*		
The ill or injured			*		

(performance) fitness. The purposes of fitness assessment are related to the specific population to be tested. Thus, we can define physical fitness based on who and what are to be measured. In this chapter we primarily address healthy adults. *Therefore, we define health-related physical fitness as the attainment or maintenance of physical capacities that are related to good or improved health and are necessary for performing daily activities and confronting expected or unexpected physical challenges.* This definition is consistent with the health-related physical fitness definition presented by Pate (1988) and supported by the ACSM (ACSM 2014a). In this chapter we examine tests of health-related physical fitness and basic functional capacity.

Pettee-Gabriel, Morrow, and Woolsey (2012) provide a framework for assessing physical activity and differentiate physical activity (a behavior) assessment from physical fitness (an attribute) assessment. The framework includes energy expenditure, physical fitness, physical activity, and sedentary behaviors (see figure 9.2, which was originally presented in chapter 1).

Mastery Item 9.1

Consider table 9.1. What would be some reasons for both a 30-year-old mother and a 17-year-old cross country runner to take a test to determine their cardiorespiratory endurance?

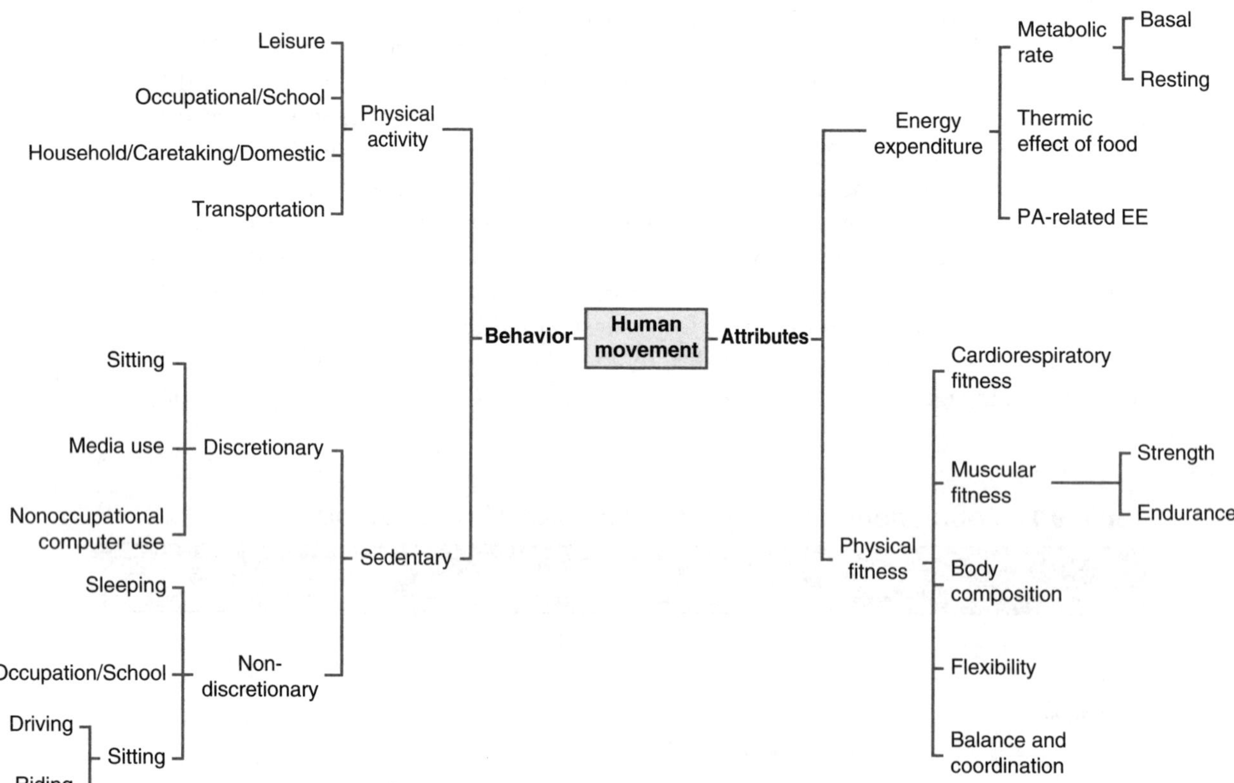

Figure 9.2 A framework for physical activity.
Adapted from Pettee-Gabriel, Morrow, and Woolsey 2012.

HEALTH-RELATED PHYSICAL FITNESS

The ACSM has identified three fitness factors that are health related; these are listed in table 9.2 and defined in subsequent sections of this chapter. The evidence to support these factors as related to health has come from the branch of medicine called **epidemiology**, which examines the incidence, prevalence, and distribution of disease. For example, a large majority of epidemiologic studies have indicated that physically active groups have lower relative risks of developing fatal cardiovascular disease (CVD) than sedentary groups (Caspersen 1989). **Relative risk** refers to the risk of mortality (death) or morbidity (disease) associated with one group compared with another. Physically active groups, logically, should have higher levels of **cardiorespiratory endurance**, which is the body's ability to extract and use oxygen in a manner that permits continuous exercise, physical work, or physical activities. Studies have shown an inverse relationship between death rates and cardiovascular endurance (Blair, Kohl, et al. 1989; Blair et al. 1996; Ekelund et al. 1988). Figure 9.3 demonstrates the findings of Ekelund and colleagues.

Table 9.2 Health-Related Fitness Factors and Benefits

Factor	Benefits
Cardiorespiratory endurance	Reduction in risk of cardiovascular disease and all-cause mortality Reduction in risk for morbidity and mortality for some types of cancer
Body composition	Reduction in risk of cardiovascular disease, type 2 diabetes, and metabolic syndrome
Muscular fitness, including muscular strength, muscular endurance, and flexibility	Reduction in risk of all-cause mortality Reduction in risk of lower-back pain and injury Reduction in the prevalence and incidence of obesity Reduction in the prevalence and incidence of type 2 diabetes and metabolic syndrome Maintenance of or increase in bone mass Improvement in posture and functional capacity Improvement in glucose tolerance Maintenance of ability to conduct daily activities Increase in free fat mass and resting metabolic rate

Data from ACSM 2014

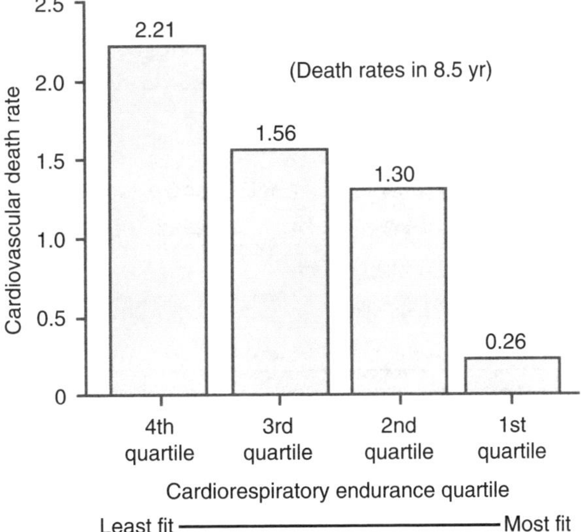

Figure 9.3 Relationship between cardiorespiratory endurance and cardiovascular death rate.

 Go to the WSG to complete Student Activity 9.1 and view Video 9.1.

The poorest cardiorespiratory endurance quartile death rate was 8.5 times higher than the most fit quartile. People who suffer from obesity have higher rates of CVD, cancer, and diabetes. Thus, body composition is included in a health-related fitness battery to determine percent body fat and the presence of obesity (ACSM 2010). Muscular fitness, including muscular strength, muscular endurance, and flexibility, is related to good health; maintaining a minimum level of muscular fitness is essential for accomplishing daily activities and being prepared to deal with expected or unexpected physical challenges (see table 9.2). Research has indicated inverse relations between muscular strength, obesity, and all-cause mortality after controlling for cardiorespiratory endurance (FitzGerald et al. 2004; Ruiz et al. 2008; Jackson et al. 2010).

ESTABLISHING THE RISK FOR FITNESS TESTING

In adult fitness assessment, one of the critical issues is establishing criteria for testing that do not require medical clearance or physician supervision. The criteria established by the ACSM set the risks associated with fitness testing (ACSM 2014a). People with low risk have at most only one of the major CVD risk factors listed in table 9.3 (ACSM 2014a). A person can conduct a self-guided screening of risk for physical activity and fitness by administering the Physical Activity Readiness Questionnaire (PAR-Q) developed by the Canadian Society for Exercise Physiology and supported by the Public Health Agency of Canada (ACSM 2014a; see table 9.4). Table 9.5 provides the recommendations for physician supervision of exercise testing in the adult population (ACSM 2014a).

In the following sections of this chapter, we examine some of the tests and protocols available for assessing health and fitness of adults. It is impossible to cover every test, but we emphasize some of the more important test protocols and measurement issues related to them. First we provide examples of laboratory (criterion) methods and then provide examples of field, or surrogate, measures that are often used because they are more feasible to administer. Field tests, by their nature, will have less validity than laboratory methods. It is important to be aware of this decline in validity and to acknowledge it when using a field test.

Table 9.3 Major Risk Factors and Classifications for Cardiovascular Disease

Family history	Father or brother: CVD <55 years of age Mother or sister: CVD <65 years of age
Age	Men ≥45 yr; women ≥55 yr
Cigarette smoking	Current or recently quit smoking
Hypertension	Systolic blood pressure ≥140 or diastolic blood pressure ≥90
Dyslipidemia	Total cholesterol ≥200 mg/dl or HDL <40 mg/dl or LDL ≥130 mg/dl
Prediabetes	Fasting blood glucose ≥100 mg/dl
Obesity	BMI ≥30 or waist girth >102 cm men; >88 cm women
Sedentary lifestyle	<3 days per week, 30 minutes per day, moderate physical activity, for 3 months
Risk classification	
Low risk	Men or women who are asymptomatic and meet no more than one risk factor threshold
Moderate risk	Men and women who are asymptomatic and meet two or more of the risk factors
High risk	Symptomatic or those with known cardiovascular, pulmonary, or metabolic disease

Note: HDL = high-density lipoprotein; BMI = body mass index; LDL = low-density lipoprotein.

Data from American College of Sports Medicine 2014, *ACSM's guidelines for exercise testing and prescription*, 8th ed. (Philadelphia: Lea & Febiger).

Table 9.4 Physical Activity Readiness Questionnaire

THESE ARE THE TYPES OF QUESTIONS THAT YOU WILL FIND ON THE PAR-Q & YOU FORM, DEVELOPED BY THE CANADIAN SOCIETY FOR EXERCISE PHYSIOLOGY AND HEALTH CANADA		
Yes	**No**	**Question**
		Do you feel pain in your chest when you do physical activity?
		In the past month, have you had chest pain when you were not doing physical activity?
		Do you lose your balance because of dizziness or do you ever lose consciousness?
		Is your doctor currently prescribing drugs (for example, water pills) for your blood pressure or heart condition?

Data from Canadian Society for Exercise Physiology and Health Canada.

Table 9.5 Conditions Requiring Exercise Stress Test, Medical Exam, and Medical Supervision

Risk classification (see table 9.3)		Low risk <2 risk factors	Moderate risk ≥2 risk factors	High risk symptomatic or known disease
Medical exam before exercise	Moderate exercise	No	No	Yes
	Vigorous exercise	No	Yes	Yes
Exercise test before exercise	Moderate exercise	No	No	Yes
	Vigorous exercise	No	No	Yes
Medical supervision of exercise test if done	Submaximal test	No	No	Yes
	Maximal test	No	No	Yes

Data from American College of Sports Medicine 2014, *ACSM's guidelines for exercise testing and prescription*, 8th ed. (Philadelphia: Lea & Febiger).

PHYSICAL ACTIVITY OR PHYSICAL FITNESS?

Healthy People 2020 and the *2008 Physical Activity Guidelines for Americans* have goals and objectives related to increases in the process or behavior of physical activity but not specifically about the outcome of physical activity, physical fitness (health-related physical fitness). In this chapter we examine the reliable and valid measurement of both physical activity and physical fitness. An illustration of the relationship of physical activity and fitness with morbidity (disease) and mortality (death) outcomes is presented in figure 9.4. Higher levels of physical activity and fitness are both related to lower risks of morbidity and mortality. However, higher levels of physical fitness produce greater reductions in the risks of morbidity and mortality.

Two possibilities may explain this difference:

- Physical fitness may provide a true greater inverse association with chronic diseases and the mortality that those diseases cause.
- The measurement of the attribute of physical fitness tends to more reliable and valid than the measurement of the behavior of physical activity. There is more error in the measurement of physical activity. Therefore, it would be more difficult in research studies to assess accurate relationships between physical activity and health outcomes than between physical fitness and the same health outcomes.

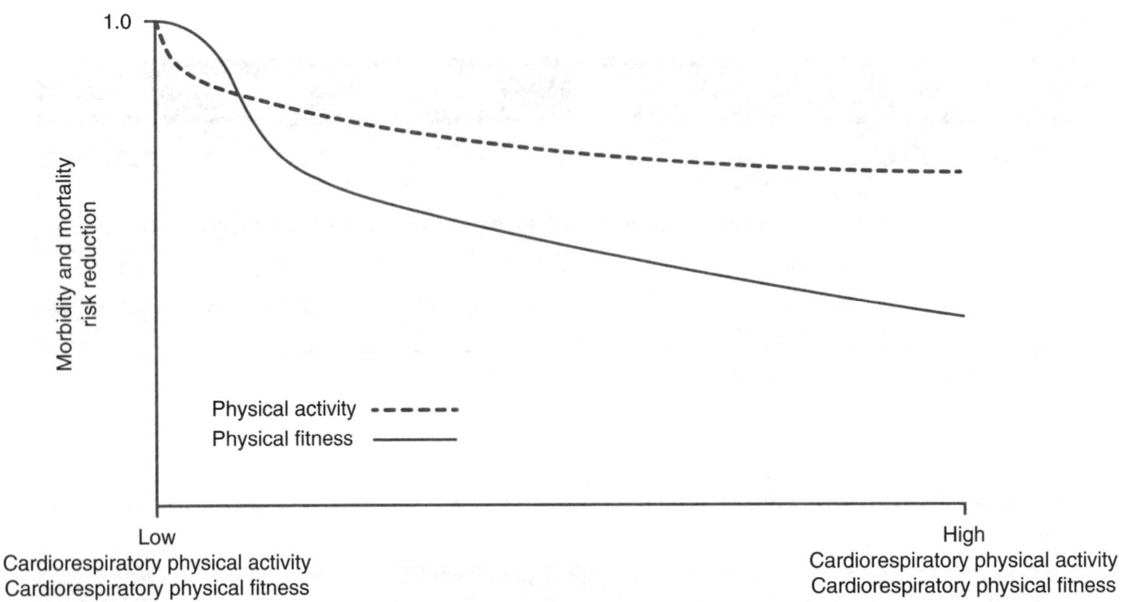

Figure 9.4 Impact of physical activity and physical fitness on morbidity and mortality risk reduction.

MEASURING AEROBIC CAPACITY

As mentioned earlier, physical activity and cardiorespiratory endurance are related to the risk of CVD. The exercise physiologist's concept of cardiorespiratory endurance is a person's aerobic capacity, or **aerobic power**, which is the ability to supply oxygen to the working muscles during physical activity.

Laboratory Methods

Fitness assessment in laboratories and clinical settings involves expensive and sophisticated equipment and exacting test protocols. From a measurement perspective, these tests are often criterion referenced.

Measuring Maximal Oxygen Consumption

The single most reliable ($r_{xx'}$ >.80) and valid measure of aerobic capacity is the **maximal oxygen consumption**, or $\dot{V}O_2$max (ACSM 2010; Safrit et al. 1988). $\dot{V}O_2$max is a measure of the maximal amount of oxygen that can be used by a person during exhaustive exercise.

In laboratory testing of $\dot{V}O_2$max, participants perform a **maximal exercise test** on an ergometer, such as a treadmill, stationary cycle, step bench, swimming flume, or arm crank device (figure 9.5). Each participant performs the exercise under a specific protocol until he or she reaches exhaustion. While the participant is exercising, expired gases are monitored with a gas analysis system. Most modern exercise physiology laboratories use an automated and computerized metabolic system.

A variety of exercise protocols are available in the literature for determining $\dot{V}O_2$max; all focus on increments in work rate until a participant reaches exhaustive levels of physical exertion (ACSM 2014a). *$\dot{V}O_2$max is achieved when the work rate is increased, but the oxygen consumption ($\dot{V}O_2$max does not increase or has reached a plateau.* Other indications of $\dot{V}O_2$max are a respiratory exchange ratio (RER) greater than 1.1 and heart rates near age-predicted maximal levels. When these physiological criteria are not clearly achieved, then

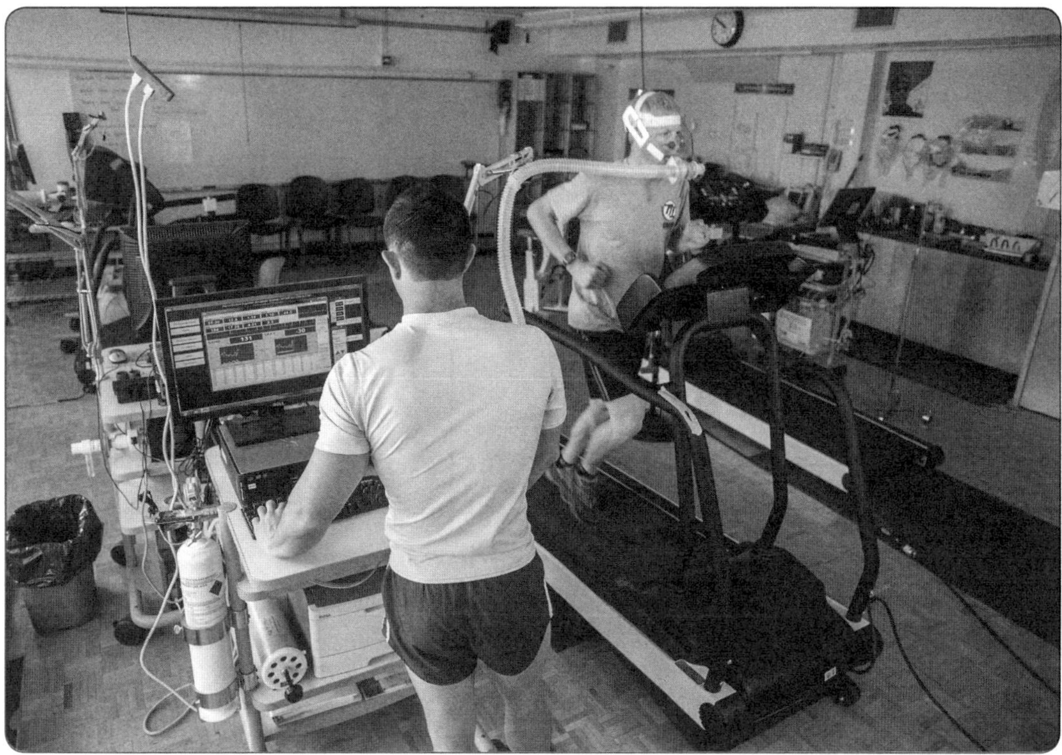

Figure 9.5 Gas exchange analysis during a maximal exercise.

the maximal oxygen consumption measured during the test is called $\dot{V}O_2$peak. $\dot{V}O_2$max and $\dot{V}O_2$peak are highly correlated and represent a valid measure of a participant's aerobic capacity. Blair, Kohl, and colleagues (1989) estimated that $\dot{V}O_2$max values of 31.5 ml · kg^{-1} · min^{-1} for females and 35 ml · kg^{-1} · min^{-1} for males represent the minimal levels of aerobic capacity associated with a reduced risk of disease and death for a wide range of adult ages. In table 9.6, minimal $\dot{V}O_2$max values associated with lower risks of cardiovascular diseases for specific age groups are provided (Sui et al. 2007). Figure 9.6 demonstrates the Balke treadmill protocol for determining $\dot{V}O_2$max. Table 9.7 provides evaluative norms for $\dot{V}O_2$max.

Table 9.6 Minimum Levels of Aerobic Capacity for Reduced Risks of Morbidity and Mortality

	AGE GROUP			
Gender	20-39 yrs	40-49 yrs	50-59 yrs	60+ yrs
Men	36.4	34.7	29.8	25.2
Women	28.7	26.6	23.5	20.3

$\dot{V}O_2$max as ml · kg^{-1} · min^{-1}.

Mastery Item 9.2

What level of $\dot{V}O_2$max would you like to have based on the norms in table 9.7?

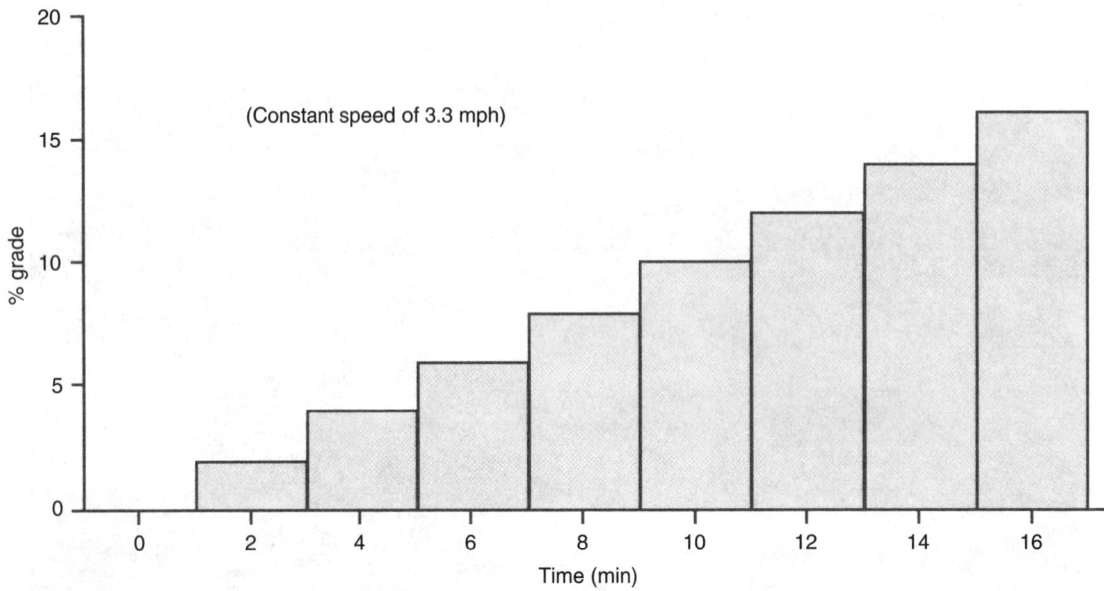

Figure 9.6 The Balke treadmill protocol.

Table 9.7 Male and Female Norms for $\dot{V}O_2max$ (ml · kg⁻¹ · min⁻¹)

	AGE (YEARS)					
Male rating	**18-25**	**26-35**	**36-45**	**46-55**	**56-65**	**66+**
Excellent	100-65	95-60	90-55	83-49	65-43	53-38
Good	60-53	55-50	49-45	45-40	40-37	34-32
Above average	50-48	48-44	43-40	39-36	35-33	31-29
Average	45-43	42-39	38-36	35-32	32-30	28-26
Below average	42-38	38-34	35-31	31-29	28-26	25-23
Poor	36-32	33-30	30-27	27-25	25-22	22-20
Very poor	30-20	27-15	24-14	24-13	21-12	18-10
Female rating						
Excellent	95-59	95-58	75-50	72-45	58-40	55-34
Good	56-50	53-48	46-42	41-36	36-33	31-29
Above average	47-44	45-43	41-37	35-32	32-30	28-26
Average	42-39	41-37	36-33	31-29	28-26	25-23
Below average	38-35	36-34	32-29	28-26	25-23	22-20
Poor	33-30	32-28	28-25	25-22	22-19	19-17
Very poor	27-15	25-14	24-12	20-11	18-10	16-10

Data based on Golding 2000.

Estimating $\dot{V}O_2max$

Although $\dot{V}O_2max$ is the criterion measure of aerobic capacity, it is a difficult measure to determine because it requires expensive metabolic equipment, exhaustive exercise performance, and a lot of time. Consequently, researchers in exercise science have developed techniques for estimating, or predicting, $\dot{V}O_2max$ reliably and validly. The estimations are calculated from measurements of maximal or submaximal exercise performance or submaximal heart rate; the same or similar exercise protocols and ergometers discussed previously are used.

Maximal Exercise Performance $\dot{V}O_2$max can be accurately estimated from the maximum exercise time of a maximal treadmill exercise test (Pollock et al. 1976). Although this procedure requires exhaustive exercise, it does not require the metabolic measurement of expired gases. Thus, the test is greatly simplified, and expensive metabolic equipment is not necessary. Published correlations (concurrent validities) between $\dot{V}O_2$max and maximal exercise time exceed .90. Baumgartner, Jackson, Mahar, and Rowe (2016) provide $\dot{V}O_2$max estimates for maximal treadmill times for several treadmill protocols.

Submaximal Exercise Testing Submaximal estimates of $\dot{V}O_2$max are based on the linear relationship among heart rate, workload, and $\dot{V}O_2$max. Such estimates are based on **submaximal exercise tests**, which require less than maximal effort. As figure 9.7 indicates, a participant with good aerobic capacity has a higher $\dot{V}O_2$max than one with poor aerobic capacity (both have a maximal heart rate of 200 beats/minute). The slopes of the lines representing the linear relationship between heart rate and $\dot{V}O_2$max are different for each participant. In figure 9.8, we see the difference in workloads each participant can achieve for a fixed submaximal heart rate, 160 beats/minute.

Several exercise test protocols are available for estimations of $\dot{V}O_2$max (ACSM 2014a). These estimations are based on the linear relationships of workload, heart rate, and oxygen consumption. One of the classic procedures is referred to as the Åstrand–Rhyming nomogram (Åstrand and Rhyming 1954). This was originally established as a cycle ergometer test that coordinated workload and heart rate responses into a prediction of $\dot{V}O_2$max. Baumgartner and colleagues (2016) converted the nomogram into an equation that can be used to produce the same predictions of $\dot{V}O_2$max from cycling or treadmill tests (Jackson et al. 1990), allowing computer calculations of predicted aerobic capacity. The ACSM (2014a, 2010) and the YMCA (Golding 2000) provide descriptions of specific treadmill and cycle test protocols for estimating $\dot{V}O_2$max.

Go to the WSG to complete Student Activity 9.2 and view Video 9.2.

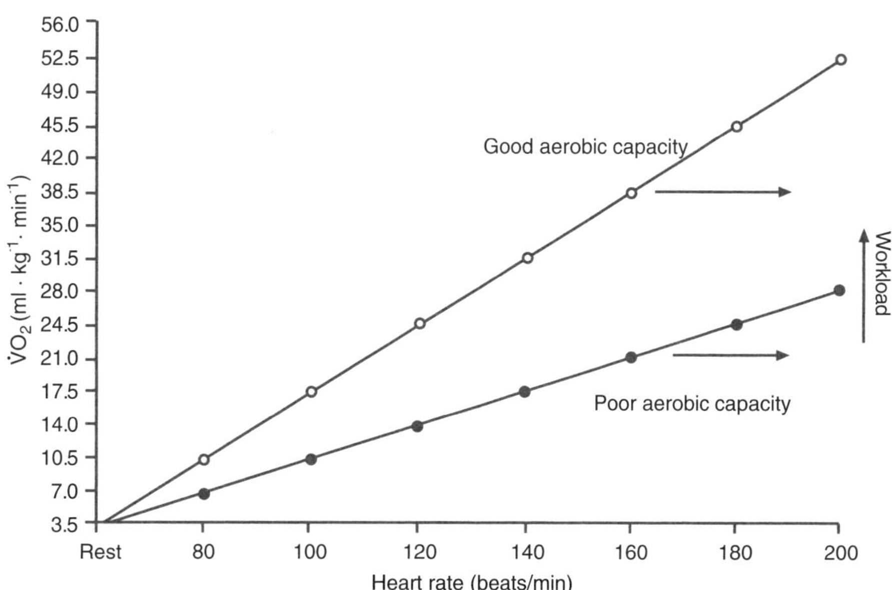

Figure 9.7 Linear relationship among oxygen consumption, heart rate, and workload.

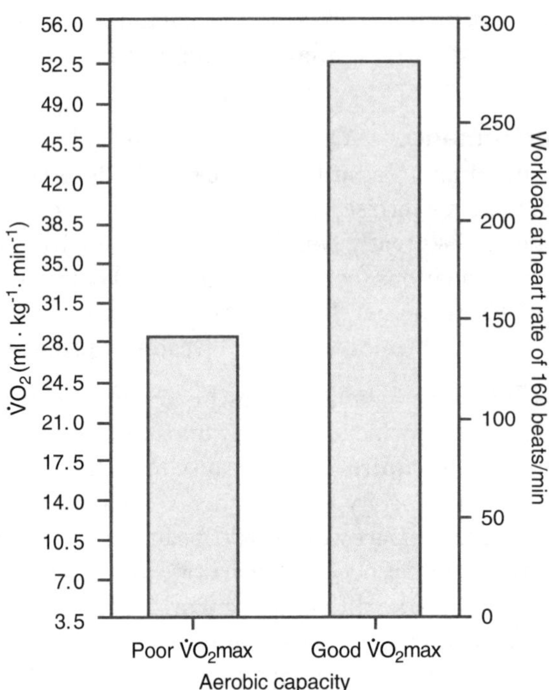

Figure 9.8 How maximal oxygen consumption affects a submaximal workload.

Mastery Item 9.3

In figure 9.7, which participant would achieve the higher heart rate and rating of perceived exertion (RPE) for any submaximal workload?

Perceptual Effort During Exercise Testing

Borg (1962) pioneered the measurement of perceptual effort, or **perceived exertion**, during exercise testing. Perceived exertion is the mental perception of the intensity of physical work. The measurement of perceived physical effort or stress has been named *rating of perceived exertion (RPE)*. Borg (1998) presented RPE scales for assessment of perceived effort during exercise testing. Several versions of the RPE scale with specific purposes and applications are available. RPE scale values correlate with exercise variables such as heart rate, ventilation, lactic acid production, percent $\dot{V}O_2$max, and workload (ACSM 2014b). The participant simply gives a verbal or visual score from the scale during the exercise test as the workload increases or as time progresses. RPE is typically monitored during exercise tests and is used in exercise prescription to control exercise intensity. The *2008 Physical Activity Guidelines for Americans* presents a simple relative intensity scale (figure 9.9) to aid a participant in setting the intensity of a physical activity or exercise period (USDHHS 2008). The relative intensity is a person's level of effort relative to his or her fitness level.

- Relatively moderate-intensity activity is a level of effort of 5 or 6 on a scale of 0 to 10, where 0 is the level of effort of sitting, and 10 is maximal effort.
- Relatively vigorous-intensity activity is a 7 or 8 on this scale.

Laboratory tests for assessment of aerobic capacity tend to be reliable and valid. However, sources of measurement error are present even in these laboratory situations as in any other testing situation. Participants, the test and protocol, or test administrators can

0	1	2	3	4	5	6	7	8	9	10
Sitting					Moderate intensity		Vigorous intensity			Maximal effort

Figure 9.9 Relative intensity scale.

Data from U.S. Department of Health and Human Services 2008.

all be sources of measurement error. The following list provides some important facts you should know concerning laboratory assessment of aerobic capacity.

- Equipment, treadmills, cycles, and gas analysis systems should be calibrated and checked regularly.
- Test administrators should be trained and qualified.
- Practice test administrations should be required for participant and administrator to become familiar with test protocols and equipment.
- Standardized test procedures should be established and followed; this creates a focused test environment.
- Treadmill $\dot{V}O_2max$ values will be greater than values from cycle ergometer tests for most participants.
- Many Americans seldom ride bicycles, so cycle exercise tests can produce artificially low values of $\dot{V}O_2max$ attributable to test cessation from localized fatigue in the legs.
- Typically, submaximal estimates of $\dot{V}O_2max$ have a standard error of estimate greater than 5.0 $ml \cdot kg^{-1} \cdot min^{-1}$

Field Methods

Field methods include ways to assess aerobic capacity and are feasible for mass testing. Generally, field methods require little equipment and are less expensive in time and costs than laboratory methods.

Distance Runs

Distance runs to achieve the fastest possible time or greatest distance covered in a fixed time are some of the more popular field tests of aerobic capacity. For adults, distances of 1 mile (1.6 km) or greater are used. Safrit and colleagues (1988) indicated that distance runs tend to be reliable ($r_{xx'}$ > .78) and have a general concurrent validity coefficient of .741 ± .14. The 12-minute run for distance developed by Cooper (1968) is an example of a distance-run test. AAHPERD has published norms for the 1-mile (1.6 km) run for college students (AAHPERD 1985; see table 9.8).

Table 9.8 Percentile Norms for the 1-Mile (1.6 km) Run for College Students (min:sec)

Percentile	Males	Females
90	5:44	7:26
75	6:12	8:15
50	6:49	9:22
25	7:32	10:41
10	8:30	12:00

Data based on American Association of Health, Physical Education, Recreation and Dance 1985.

Distance runs are useful for educational situations in which testing entire classes is required in a short amount of time. However, to ensure reliability, validity, and safety (i.e., to correct pacing and provide proper physical conditioning), participants should receive aerobic training and practice trials on the test. It is important that distances, timers, and recording procedures be used in a standardized manner. Older adults, those 65 years or older, or people with poor aerobic capacity should undergo one of the other field tests or procedures discussed next.

Step Tests

Several step-test protocols are available for estimating aerobic capacity. These tests are based on the linear relationships among workload, heart rate, and $\dot{V}O_2$max discussed previously. Generally, each participant steps up and down to an up, up, down, down cadence until a specific workload, heart rate, or time is achieved. The aerobic capacity is then estimated from the heart rate response or the recovery heart rate. Participants with higher aerobic capacity will have a faster return to lower heart rates. The YMCA 3-Minute Step Test is one of the simplest step tests to administer and is useful for initial testing of participants who may not be fit.

YMCA 3-MINUTE STEP TEST

Purpose

To assess aerobic fitness in mass testing situations with adults.

Objective

To step up and down to a set cadence for 3 minutes and take the resulting heart rate.

Equipment

12-inch (30.5 cm) high bench

Metronome set at 96 beats/minute

Watch or timer

Stethoscope (carotid pulse can be used)

Instructions

The participant listens to the metronome to become familiar with the cadence and begins when ready and the time starts. The participant steps up, up, down, down to the 96 beat/minute cadence, which allows 24 steps/minute. This continues for 3 minutes. After the final step down, the participant sits down and the heart rate is counted for 1 minute.

Scoring

The 1-minute recovery heart rate is the score for the test. Table 9.9 provides evaluative norms for test results.

Rockport 1-Mile (1.6 km) Walk Test

Kline and colleagues (1987) presented a field method for estimating $\dot{V}O_2$max that has been called the Rockport 1-Mile (1.6 km) Walk Test. The procedure involves using the time of a 1-mile walk, gender, age, body weight, and ending heart rate to estimate $\dot{V}O_2$max. The 1-mile walk requires participants to walk as fast as possible; their heart rates are taken

Table 9.9 Male and Female Norms for Recovery Heart Rate After the 3-Minute Step Test (beats/min)

	AGE (YEARS)					
Male rating	**18-25**	**26-35**	**36-45**	**46-55**	**56-65**	**66+**
Excellent	50-76	51-76	49-76	56-82	60-77	59-81
Good	79-84	79-85	80-88	87-93	86-94	87-92
Above average	88-93	88-94	92-98	95-101	97-100	94-102
Average	95-100	96-102	100-105	103-111	103-109	104-110
Below average	102-107	104-110	108-113	113-119	111-117	114-118
Poor	111-119	114-121	116-124	121-126	119-128	121-126
Very poor	124-157	126-161	130-163	131-159	131-154	130-151
Female rating						
Excellent	52-81	58-80	51-84	63-91	60-92	70-92
Good	85-93	85-92	89-96	95-101	97-103	96-101
Above average	96-102	95-101	100-104	104-110	106-111	104-111
Average	104-110	104-110	107-112	113-118	113-118	116-121
Below average	113-120	113-119	115-120	120-124	119-127	123-126
Poor	122-131	122-129	124-132	126-132	129-135	128-133
Very poor	135-169	134-171	137-169	137-171	141-174	135-155

Data based on Golding 2000.

immediately at the end of the walk. The 1-mile walk test was shown to be reliable ($r_{xx'}$ = .98; Kline et al. 1987). The prediction equation (equation 9.1) produced a concurrent validity coefficient of .88 with a standard error of estimate of 5.0 ml · kg^{-1} · min^{-1}:

$$\dot{V}O_2\text{max} = 132.853 - (.0769) \times \text{wt} - (.3877) \times \text{age} + (6.315) \times \text{gv} - (3.2469) \times 1 \text{ mi walk time} - (.1565) \times \text{heart rate} \quad (9.1)$$

with wt as weight in pounds, age in years, gv (gender values) of 0 for females and 1 for males, 1 mi (1.6 km) walk time in minutes (to the hundredths of a minute), heart rate in beats per minute at end of walk, and $\dot{V}O_2$max in ml · kg^{-1} · min^{-1}. Notice that this is a multiple regression (prediction) equation introduced in chapter 4.

The original study used a sample with an age range of 30 to 69 years. Further research supported the validity (r_{xy} = .79; s_e = 5.68 ml · kg^{-1} · min^{-1}) of the equation for adults aged 20 to 29 years (Coleman et al. 1987). As with any physical performance test, the 1-mile (1.6 km) walk can have improved reliability and validity if a practice trial is administered (Jackson, Solomon, and Stusek 1992). Evaluative norms for people aged 30 to 69 years for the 1-mile (1.6 km) walk test and percentile norms for people aged 18 to 30 years are provided in tables 9.10 and 9.11, respectively.

Predicting $\dot{V}O_2$max Without Exercise

Jackson and colleagues (1990) developed an equation for estimating $\dot{V}O_2$max without an exercise test of any kind. The equations had reasonable validity coefficients (r_{xy} > .79) and standard errors of estimation (s_e < 5.7 ml · kg^{-1} · min^{-1}); the latter is comparable to submaximal exercise test and field test standard errors of estimation. This, too, is a multiple regression equation introduced in chapter 4. This technique allows for accurate estimation

Table 9.10 Norms for the 1-Mile (1.6 km) Walk Test (Participants Aged 30 to 69 Years; min:sec)

Rating	Males (*N* = 151)	Females (*N* = 150)
Excellent	<10:12	<11:40
Good	10:13-11:42	11:41-13:08
High average	11:43-13:13	13:09-14:36
Low average	13:14-14:44	14:37-16:04
Fair	14:45-16:23	16:05-17:31
Poor	>16:24	>17:32

Based on Kline et al. 1987.

Table 9.11 Norms for the 1-Mile (1.6 km) Walk Test (Participants Aged 18 to 30 Years; min:sec)

Percentile	Males (*N* = 400)	Females (*N* = 426)
90	11:08	11:45
75	11:42	12:49
50	12:38	13:15
25	13:38	14:12
10	14:37	15:03

Adapted from Jackson and Soloman 1994.

of aerobic capacity in situations in which large numbers of participants need to be evaluated, such as in epidemiologic research. The equation (equation 9.2) is as follows:

$$\dot{V}O_2\text{max} = 50.513 + 1.589 \times \text{self-reported physical activity} - .0289 \times \text{age in years} - 0.522 \times \text{percent body fat} + 5.863 \times \text{gender (female = 0, male = 1)} \quad (9.2)$$

Wier and colleagues (2006) demonstrated that body mass index (BMI), percent body fat, or waist girth could be used interchangeably in producing estimations of $\dot{V}O_2$maxwith essentially the same level of accuracy. Jurca et al. (2005) used data from three large international datasets and confirmed that cardiorespiratory fitness level could be estimated well (validities >.75) with a nonexercise test model, including gender, age, BMI, resting heart rate, and self-reported physical activity.

MEASURING BODY COMPOSITION

Obesity is a risk factor in the development of CVD, cancer, and adult-onset diabetes. As a consequence—and because the United States has a large percentage (>35%) of adults who are obese—measuring obesity in an accurate manner is an important measurement goal.

The term *obesity* refers specifically to overfatness, not overweight. A well-muscled athlete who is extremely fit may be considered overweight on a height and weight table but may actually be quite lean. In health-related fitness, the measurement of **body composition** involves estimating a person's percent body fat, which requires that his or her body density be determined. A good method for conceptualizing body composition is to divide the body into two compartments: *lean*, which includes muscle, bone, and organs and is of high density, and *fat*, which is of low density. For a fixed body weight, a leaner person with a lower percent body fat will have a higher body density than a fatter person of the same weight. In body density estimation, lean tissue is assumed to have an average density of 1.10 g/cm^3, whereas fat tissue is assumed to have an average density of .90 g/cm^3. This assumption leads to one of the errors in body composition measurement: Lean tissue and fat tissue do not have the same density, and types of lean tissue (e.g., bone vs. muscle) have different densities. This variable source of measurement error is present in the methods discussed next.

A variety of methods of body composition measurement are available, including the following:

- Hydrostatic weighing
- Dual-energy X-ray absorptiometry (DXA)
- Air displacement plethysmography
- Computed tomography scans and magnetic resonance imaging
- Isotopic dilution
- Ultrasound
- Anthropometry (skinfolds and girths)
- Bioelectrical impedance
- Total body electrical conductivity (TBEC)
- Near-infrared interactance

ACSM (2014b) provided a summary of the advantages and disadvantages of these techniques of measurement of body composition. These methods, their reliability, and their validity are results of the development of relatively new technologies.

Laboratory Methods

As indicated earlier, there are a variety of laboratory procedures for assessing body composition. Hydrostatic weighing and DXA techniques are currently in wide use.

Hydrostatic Weighing

Hydrostatic weighing (underwater weighing), which is based on Archimedes' principle, is a popular method of laboratory assessment of body density (figure 9.10). This method has provided the criterion measurements for validating such field methods as skinfold and girth measurements. The rationale behind underwater weighing is to determine one's body volume by determining the amount of water displaced during the procedure. For two people of the same weight but different percent body fat, the leaner person, who has a higher body density, will have a higher underwater weight than the fatter person. Equation 9.3 can help you calculate body density from hydrostatic weighing:

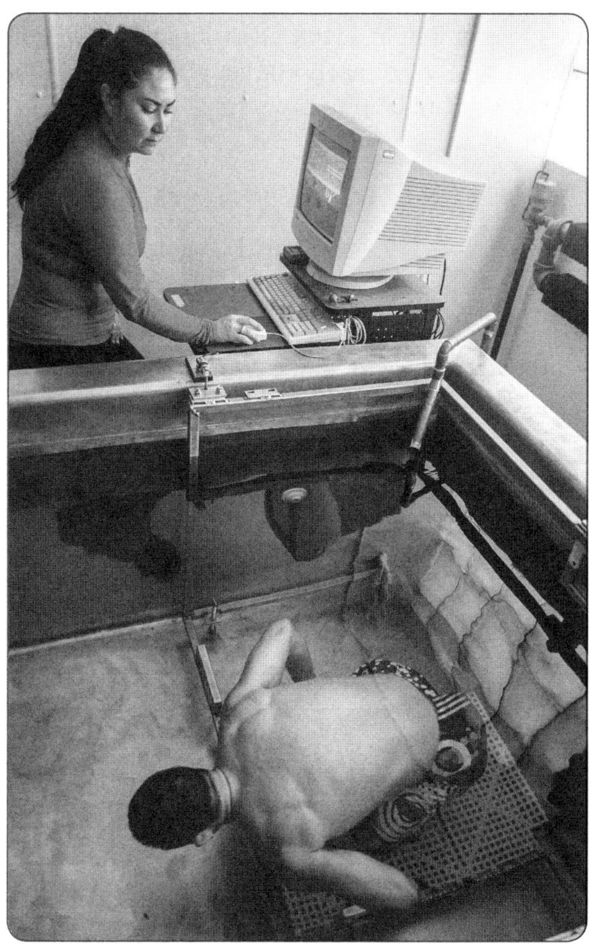

Figure 9.10 Hydrostatic weighing.

$$BD = \frac{Wt_d}{\frac{(Wt_d - Wt_w)}{D_w} - (RV + 100ml)} \tag{9.3}$$

where BD = body density in g/cm^3, Wt_d = participant's dry weight in kilograms, Wt_w = his or her weight under water in kilograms, D_w = density of water in g/cm^3 at the temperature of measurement, and RV = residual volume in liters.

To ensure reliability and validity of the body density measured by hydrostatic weighing, underwater weighing should be repeated as many as 10 times or until a consistent weight is determined, and the **residual volume**, which is the air left in the lungs after maximal forced expiration of air, should be measured, not predicted or estimated. Measuring residual volume is a complex laboratory process that requires sophisticated equipment; however, if you use estimated residual volume, your measurement error of body density will be quite large (Morrow, Jackson, Bradley, and Hartung 1986). Nieman (1995) provides an excellent step-by-step description of the hydrostatic weighing technique.

Using Body Density

Once the body density is determined, the percent body fat can be estimated from Siri's (1956) equation:

$$\%fat = (495 \div BD) - 450 \tag{9.4}$$

Estimating percent body fat allows the calculation of other useful measures of body weights: fat weight, lean weight, and **target weight**. The target weight is the weight a person should achieve to reach a target percent body fat. This establishes an easily measured goal for a weight-reduction program. The following equations are used for determining these weights:

$$\text{fat weight} = (\%fat \div 100) \times \text{body weight} \tag{9.5}$$

$$\text{lean weight} = \text{body weight} - \text{fat weight} \tag{9.6}$$

$$\text{target weight} = \text{lean weight} \div [1 - (\text{target } \%fat \div 100)] \tag{9.7}$$

Table 9.12 provides an example of these calculations.

Table 9.12 Example Calculation of Fat, Lean, and Target Weights

EXAMPLE: MALE = 150 LB (68 KG); PERCENT BODY FAT = 30%; TARGET PERCENT BODY FAT = 25%		
Component	**Calculation**	**Result**
Fat weight	(30/100) × 150	45 lb (20.4 kg)
Lean weight	150 – 45	105 lb (47.6 kg)
Target weight	105/(1 – 25/100)	140 lb (63.6 kg)

Mastery Item 9.4

A male weighs 200 pounds (90.7 kg) and has 30% body fat. His target percent body fat is 15%. What is his target weight?

Dual-Energy X-Ray Absorptiometry (DXA)

DXA (previously called DEXA) use in clinical and research applications is growing as the equipment becomes more affordable. The technique is based on a three-compartment model of body mineral stores, fat-free mass, and fat mass. DXA is an X-ray technology that passes rays at two energy levels through the body. The attenuation or changes of those rays as they pass through bone, organs, muscle, and fat provide estimates of bone mass, fat-free mass, and fat mass, which allows estimation of bone mineral density as well as percent body fat. DXA data can be used in research studies related to osteoporosis and body composition change in weight-loss or weight-control interventions. Data on specific areas of the body, such as abdominal body fat, can be gathered. Estimates of total percent body fat have a reported standard error of less than 2% (ACSM 2014b). Sources of measurement error are the lack of standardization between equipment manufacturers, variability between measurements from machines of the same manufacturer, and different software used by the same machine. Thus, for high reliability of measurements, it is strongly recommended that DXA scans be performed on the same instrument for repeated intra-subject measurements (ACSM 2014b).

Field Methods

Field methods for body composition assessment include skinfold measurements, the BMI, and the waist–hip girth ratio.

Skinfolds

Determining body composition with hydrostatic weighing or DXA is necessary in research studies, but it is not a feasible method for body composition measurement in the field. One of the most feasible, reliable, valid, and popular methods of field estimation of body composition is the skinfold technique, which is the measurement of skinfold (actually, fatfold) thicknesses at specific body sites. The measurements are performed with skinfold calipers, such as those manufactured by Lange and Harpenden (figure 9.11).

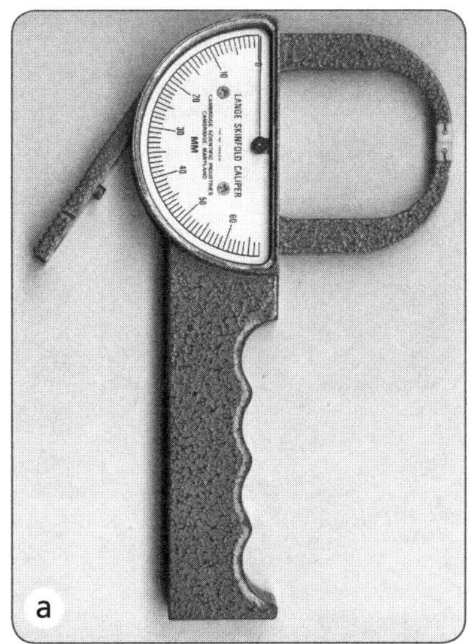

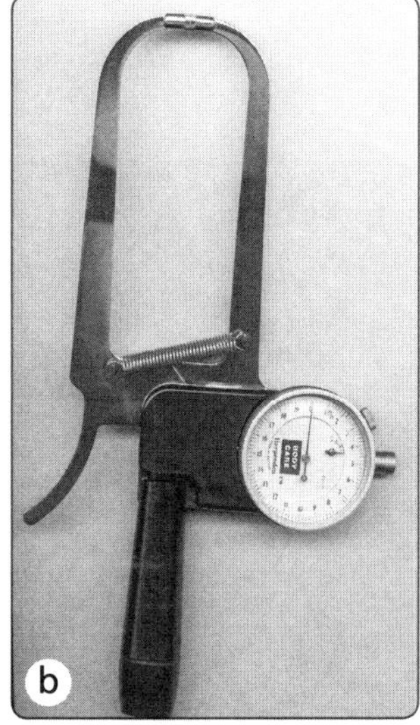

Figure 9.11 Skinfold calipers: *(a)* Lange and *(b)* Harpenden.

Two research studies (Jackson and Pollock 1978; Jackson, Pollock, and Ward 1980) developed valid generalized equations for predicting body density from skinfold measurements for males and females with an age range of 18 to 61 years. The equations were adapted for the YMCA (Golding 2000) and provided predictions of percent body fat. The seven skinfolds used were the chest, axilla, triceps, subscapular, abdomen, suprailium, and thigh. Figure 9.12 illustrates the measures used for the 3- and 4-site prediction equations. The skinfolds were each highly correlated ($r > .76$) with hydrostatically determined body density. During their analysis, the researchers found that the skinfolds had a nonlinear, quadratic relationship to body density; age was also a useful predictor. Table 9.13 provides the relevant equations. Using skinfolds, gender, and age, the concurrent validity of skinfold equations for men exceeds .90 and for women is about .85.

Skinfold measurements predict body density and percent body fat in a valid manner. *However, to ensure reliability of your skinfold measures, you should have plenty of practice.* Properly trained testers should be able to produce measurements with high reliability ($r_{xx'} > .90$). The recommended steps for taking skinfold measures are as follows:

1. Lift skinfolds two or three times before placing the skinfold caliper and taking a measurement.

2. Place the calipers below the thumb and fingers and perpendicular to the fold so that the dial can be easily read; release the caliper grip completely; and read the dial 1 to 2 seconds later.

3. Repeat the process at least three times; the measures should not vary by more than 1 millimeter (.04 in). The median value should be used as the measure. An interval of at least 15 seconds should occur between each measurement to allow the site to return to normal. If you get inconsistent values, you should go to another site and then return to the difficult one. As many as 50 to 100 practice sessions on participants may be needed to develop reliable skinfold techniques; with proper preparation, you can achieve reliable measures with a variety of skinfold calipers (Morrow, Fridye, and Monaghen 1986). Nieman (1995) provided specific rules for skinfold measurements.

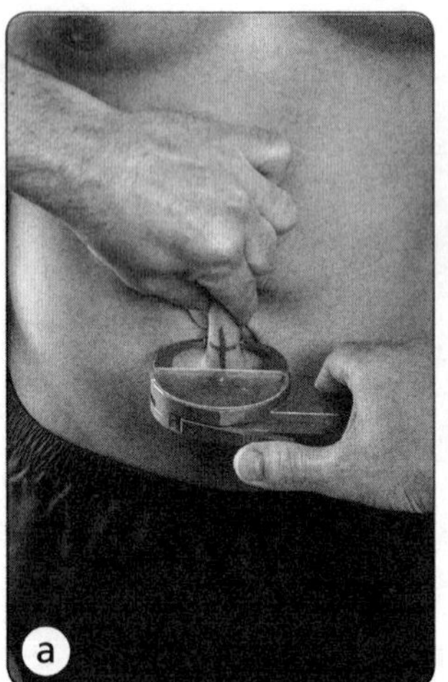

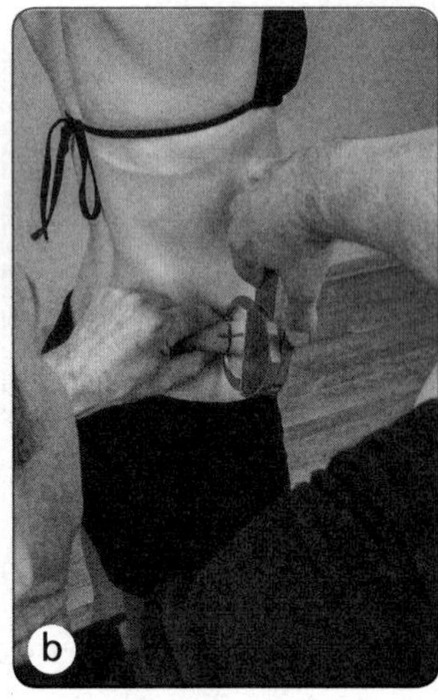

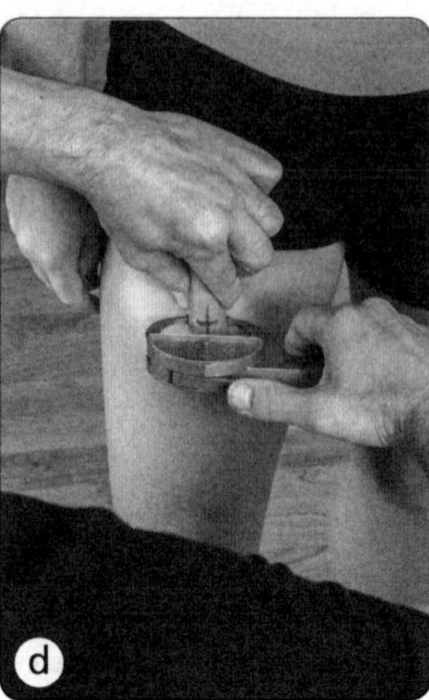

Figure 9.12 Where to measure skinfolds: *(a)* abdomen, *(b)* ilium or hip, *(c)* triceps, and *(d)* thigh.

Properly taken skinfold measures are useful field estimates of body composition. Keep in mind that, as the data in table 9.13 show, there is a standard error of estimate of up to 3.98% fat. When you report a participant's percent body fat, it is a good idea to inform the participant that it is an estimate and to state the potential error of that estimate—for example, "Your percent body fat is 15 with a potential range of 11 to 19." In dealing with very obese participants, you may not be able to take skinfolds. You may need to use another technique.

Table 9.13 YMCA Equations for Estimation of Percent Body Fat

FOUR SITES: ABDOMEN, SUPRAILIUM, TRICEPS, AND THIGH
Males
%fat = .29288 × (sum of 4) – .0005 × (sum of 4)^2 + .15845 × (age) – 5.76377 r = .901 s_e = 3.49%
Females
%fat = .29669 × (sum of 4) – .00043 × (sum of 4)^2 + .02963 × (age) + 1.4072 r = .846 s_e = 3.89%
THREE SITES: ABDOMEN, SUPRAILIUM, AND TRICEPS
Males
%fat = .39287 × (sum of 3) – .00105 × (sum of 3)^2 + .15772 × (age) – 5.18845 r = .893 s_e = 3.63%
Females
%fat = .41563 × (sum of 3) – .00112 × (sum of 3)^2 + .03661 × (age) + 4.03653 r = .825 s_e = 3.98%

Data based on Golding 2000.

YMCA SKINFOLD TEST

Purpose

To estimate a person's percent body fat.

Objective

To provide a field method of accurately estimating body composition characteristics.

Equipment

Skinfold calipers

Instructions

Take skinfolds at the abdomen, ilium, triceps, and thigh sites with the procedures described previously.

Scoring

Convert the skinfold measures to percent body fat using the equations in table 9.13. Compare the values to the recommended percent body fat levels and the evaluative norms provided in tables 9.14 and 9.15.

Table 9.14 ACSM-Recommended Levels of Percent Body Fat

	AGE (YEARS)					
Gender	Essential	Minimal	Athletic	≤34 yr	35-55 yr	≥56
Male	3-5	5	5-13	8-22	10-25	10-25
Female	8-12	10-12	12-22	20-35	23-38	25-38

Data based on American College of Sports Medicine 2014, *ACSM's guidelines for exercise testing and prescription*, 8th ed. (Philadelphia: Lea & Febiger).

Table 9.15 Norms for Percent Body Fat in Males and Females

	AGE (YEARS)					
Male rating	18-25	26-35	36-45	46-55	56-65	66+
Excellent	3-7	4-10	5-13	8-16	11-17	12-18
Good	8-10	11-13	15-17	17-19	19-21	19-20
Above Average	11-12	14-16	18-20	20-22	22-23	21-22
Average	13-15	17-19	21-22	23-24	24-25	23-24
Below Average	16-18	20-22	23-25	25-27	26-27	25-26
Poor	19-21	23-26	26-28	28-30	28-29	27-29
Very Poor	23-35	27-38	29-39	31-40	31-40	30-39
Female rating						
Excellent	9-17	7-16	9-18	12-21	12-22	11-20
Good	18-19	18-20	19-22	23-25	24-26	22-25
Above Average	20-21	21-22	23-25	26-28	27-29	26-28
Average	22-23	23-25	26-28	29-30	30-32	29-31
Below Average	24-26	26-28	29-31	31-33	33-35	32-34
Poor	27-30	29-32	32-35	34-37	36-38	35-37
Very Poor	32-43	34-46	37-47	39-50	39-49	38-45

Data based on Golding 2000.

Body Mass Index

The body mass index (BMI) is a simple measure expressing the relationship of weight to height that is correlated with fatness. It is used in epidemiological research and has a moderately high correlation (r_{xy} = .69) with body density. It is easily calculated from the following formula:

$$\text{BMI} = \frac{\text{Weight}}{\text{Height}^2} \tag{9.8}$$

where weight is measured in kilograms and height in meters.

The following ratings have been applied to the BMI by the National Heart, Lung, and Blood Institute of the National Institutes of Health (table 9.16).

In the field, the BMI can serve as an acceptable substitute for skinfold measurements on very obese participants; however, do not use it on participants who are lean or of normal weight, for whom skinfolds are more accurate. Table 9.17 provides the BMI for a given height and weight.

Go to the WSG to complete Student Activity 9.3 and view Videos 9.3 and 9.4.

Table 9.16 Disease Risk Relative to Normal Weight and Waist Circumference

	BMI (kg/m²)	Obesity class	Men 102 cm (40 in) or less Women 88 cm (35 in) or less	Men >102 cm (40 in) Women >88 cm (35 in)
Underweight	<18.5		–	–
Normal weight	18.5-24.9		–	–
Overweight	25.0-29.9		Increased	High
Obesity	30.0-34.9	I	High	Very high
	35.0-39.9	II	Very high	Very high
Extreme obesity	40.0+	III	Extremely high	Extremely high

Data based on National Heart, Lung, and Blood Institute of the National Institutes of Health.

Distribution of Body Fat

It is well known that excessive body fat is a health risk, but another factor is the distribution of this body fat. People with excessive body fat on the trunk (android obesity) as compared with the lower body (gynoid obesity) have a higher risk of coronary heart disease (CHD). CHD is a component (along with stroke) of cardiovascular disease (CVD). As shown in table 9.16, men with a waist circumference greater than 102 centimeters (40 in.) and women with a waist circumference greater than 88 centimeters (35 in.) are considered to have increased risk for type 2 diabetes, hypertension, and CVD by the National Heart, Lung, and Blood Institute. A simple measure of this body composition risk factor is the waist–hip girth ratio; the circumference of the waist is divided by the hip circumference. Ratios greater than 1.0 for males and 0.80 for females are associated with a significantly increased risk of CHD (American Heart Association 1994).

MEASURING MUSCULAR STRENGTH AND ENDURANCE

Many people, even professionals in exercise science, use the terms *strength, force, power, work, torque,* and *endurance* almost interchangeably. However, it is important for you as a measurement specialist to understand that each of the terms has a distinct meaning. Work is the result of the physical effort that is performed. It is defined by the following equation:

$$\text{work (W)} = \text{force (F)} \times \text{distance (D)} \tag{9.9}$$

For example, the equation 150 ft lb (~203.3 Nm) = 150 lb × 1 foot (0.3 m) means that a weight of 150 pounds (68 kg) was moved 1 foot (0.3 m) in distance. Power is the amount of work performed in a fixed amount of time. It is defined by the following equation:

$$\text{power (P)} = (\text{F} \times \text{D}) \div \text{time (T)} = \text{W} \div \text{T} \tag{9.10}$$

A power value of 150 ft lb/s (~203.3 Nm/s) means that a weight of 150 pounds (68 kg) was moved 1 foot (0.3 m) in distance in 1 second. Muscular strength is the force that can be generated by the musculature that is contracting. Torque is the effectiveness of a force for producing rotation about an axis. Many of the computerized dynamometers report the torque produced during muscular contractions as well as the force or strength.

Table 9.17 Body Mass Index

	HEIGHT (IN.)																
Weight (lb)	**48**	**49**	**50**	**51**	**52**	**53**	**54**	**55**	**56**	**57**	**58**	**59**	**60**	**61**	**62**	**63**	**Weight (kg)**
100	30.6	29.3	28.2	27.1	26.1	25.1	24.2	23.3	22.5	21.7	21.6	20.2	19.6	18.9	18.3	17.8	**45.5**
105	32.1	30.8	29.6	28.4	27.4	26.3	25.4	24.5	23.6	22.8	22.0	21.3	20.5	19.9	19.2	18.6	**47.7**
110	33.6	32.3	31.0	29.8	28.7	27.6	26.6	25.6	24.7	23.9	23.0	22.3	21.5	20.8	20.2	19.5	**50.0**
115	35.2	33.7	32.4	31.2	30.0	28.8	27.8	26.8	25.8	24.9	24.1	23.2	22.5	21.8	21.1	20.4	**52.3**
120	36.7	35.2	33.8	32.5	31.3	30.1	29.0	27.9	27.0	26.0	25.1	24.3	23.5	22.7	22.0	21.3	**54.5**
125	38.2	36.7	35.2	33.9	32.6	31.4	30.2	29.1	28.1	27.1	26.2	25.3	24.5	23.7	22.9	22.2	**56.8**
130	39.8	38.1	36.6	35.2	33.9	32.6	31.4	30.3	29.2	28.2	27.2	26.3	25.4	24.6	23.8	23.1	**59.1**
135	41.3	39.6	38.0	36.6	35.2	33.9	32.6	31.4	30.3	29.3	28.3	27.3	26.4	25.6	24.7	24.0	**61.4**
140	42.8	41.1	39.5	37.9	36.5	35.1	33.8	32.6	31.5	30.4	29.3	28.3	27.4	26.5	25.7	24.9	**63.6**
145	44.3	42.5	40.9	39.3	37.8	36.4	35.0	33.8	32.6	31.4	30.4	29.3	28.4	27.5	26.6	25.7	**65.9**
150	45.9	44.0	42.3	40.6	39.1	37.6	36.2	34.9	33.7	32.5	31.4	30.4	29.4	28.4	27.5	26.6	**68.2**
155	47.4	45.5	43.7	42.0	40.4	38.9	37.5	36.1	34.8	33.6	32.5	31.4	30.3	29.3	28.4	27.5	**70.5**
160	48.9	47.0	45.1	43.3	41.7	40.1	38.7	37.3	35.9	34.7	33.5	32.4	31.3	30.3	29.3	28.4	**72.7**
165	50.5	48.4	46.5	44.7	43.0	41.4	39.9	38.4	37.1	35.8	34.6	33.4	32.3	31.2	30.2	29.3	**75.0**
170	52.0	49.9	47.9	46.6	44.3	42.6	41.1	39.6	38.2	36.9	35.6	34.4	33.3	32.3	31.2	30.2	**77.3**
175	53.5	51.4	49.3	47.4	45.6	43.9	42.3	40.8	39.3	37.9	36.7	35.4	34.2	33.1	32.1	31.1	**79.5**
180	55.0	52.8	50.7	48.8	46.9	45.1	43.5	41.9	40.4	39.0	37.7	36.4	35.2	34.1	33.0	32.0	**81.8**
185	56.6	54.3	52.1	50.1	48.2	46.4	44.7	43.1	41.6	40.1	38.7	37.4	36.2	35.0	33.9	32.8	**84.1**
190	58.1	55.8	53.5	51.5	49.5	47.7	45.9	44.3	42.7	41.2	39.8	38.5	37.2	36.0	34.8	33.7	**86.4**
195	59.6	57.2	55.0	52.8	50.8	48.9	47.1	45.4	43.8	42.3	40.8	39.5	38.2	36.9	35.7	34.6	**88.6**
200	61.2	58.7	56.4	54.2	52.1	50.2	48.3	46.6	44.9	43.4	41.9	40.5	39.1	37.9	36.7	35.5	**90.9**
205	62.7	60.2	57.8	55.5	53.4	51.4	49.5	47.7	46.1	44.5	42.9	41.5	40.1	38.8	37.6	36.4	**93.2**
210	64.2	61.6	59.2	56.9	54.7	52.7	50.7	48.9	47.2	45.5	44.0	42.5	41.1	39.8	38.5	37.3	**95.5**
215	65.7	63.1	60.6	58.2	56.0	53.9	51.9	50.1	48.3	46.6	45.0	43.5	42.1	40.7	39.4	38.2	**97.7**
220	67.3	64.6	62.0	59.6	57.3	55.2	53.2	51.2	49.4	47.7	46.1	44.5	43.1	41.7	40.3	39.1	**100.0**
225	68.8	66.0	63.4	60.9	58.6	56.4	54.4	52.4	50.5	48.8	47.1	45.5	44.0	42.6	41.2	39.9	**102.3**
230	70.3	67.5	64.8	62.3	59.9	57.7	55.6	53.6	51.7	49.9	48.2	46.6	45.0	43.5	42.2	40.8	**104.5**
235	71.9	69.0	66.2	63.7	61.2	58.9	56.8	54.7	52.8	51.0	49.2	47.6	46.0	44.5	43.1	41.7	**106.8**
240	73.4	70.4	67.6	65.0	62.5	60.2	58.0	55.9	53.9	52.0	50.3	48.6	47.0	45.4	44.0	42.6	**109.1**
245	74.9	71.9	69.0	66.4	63.8	61.5	59.2	57.1	55.0	53.1	51.3	49.6	47.9	46.4	44.9	43.5	**111.4**
250	76.4	73.4	70.5	67.7	65.1	62.7	60.4	58.2	56.2	54.2	52.4	50.6	48.9	47.3	45.8	44.4	**113.6**
	1.22	1.24	1.27	1.30	1.32	1.35	1.37	1.40	1.42	1.45	1.47	1.50	1.52	1.55	1.57	1.60	
	HEIGHT (M)																

(continued)

Table 9.17 Body Mass Index *(continued)*

	HEIGHT (IN.)															
Weight (lb)	64	65	66	67	68	69	70	71	72	73	74	75	76	77	78	**Weight (kg)**
100	17.2	16.7	16.2	15.7	15.2	14.8	14.4	14.0	13.6	13.2	12.9	12.5	12.2	11.9	11.6	**45.5**
105	18.1	17.5	17.0	16.5	16.0	15.5	15.1	14.7	14.3	13.9	13.5	13.2	12.8	12.5	12.2	**47.7**
110	18.9	18.3	17.8	17.3	16.8	16.3	15.8	15.4	14.9	14.5	14.2	13.8	13.4	13.1	12.7	**50.0**
115	19.8	19.2	18.6	18.0	17.5	17.0	16.5	16.1	15.6	15.2	14.8	14.4	14.0	13.7	13.3	**52.3**
120	20.6	20.0	19.4	18.8	18.3	17.8	17.3	16.8	16.3	15.9	15.4	15.0	14.6	14.3	13.9	**54.5**
125	21.5	20.8	20.2	19.6	19.0	18.5	18.0	17.5	17.0	16.5	16.1	15.7	15.2	14.9	14.5	**56.8**
130	22.4	21.7	21.0	20.4	19.8	19.2	18.7	18.2	17.7	17.2	16.7	16.3	15.9	15.4	15.1	**59.1**
135	23.2	22.5	21.8	21.2	20.6	20.0	19.4	18.9	18.3	17.8	17.4	16.9	16.5	16.0	15.6	**61.4**
140	24.1	23.3	22.6	22.0	21.3	20.7	20.1	19.6	19.0	18.5	18.0	17.5	17.1	16.6	16.2	**63.6**
145	24.9	24.2	23.5	22.8	22.1	21.5	20.8	20.3	19.7	19.2	18.7	18.2	17.7	17.2	16.8	**65.9**
150	25.8	25.0	24.3	23.5	22.9	22.2	21.6	21.0	20.4	19.8	19.3	18.8	18.3	17.8	17.4	**68.2**
155	26.7	25.8	25.1	24.3	23.6	22.9	22.3	21.7	21.1	20.5	19.9	19.4	18.9	18.4	17.9	**70.5**
160	27.5	26.7	25.9	25.1	24.4	23.7	23.0	22.4	21.7	21.2	20.6	20.0	19.5	19.0	18.5	**72.7**
165	28.4	27.5	26.7	25.9	25.1	24.4	23.7	23.1	22.4	21.8	21.2	20.7	20.1	19.6	19.1	**75.0**
170	29.2	28.3	27.5	26.7	25.9	25.2	24.4	23.8	23.1	22.5	21.9	21.3	20.7	20.2	19.7	**77.3**
175	30.1	29.2	28.3	27.5	26.7	25.9	25.2	24.5	23.8	23.1	22.5	21.9	21.3	20.8	20.3	**79.5**
180	31.0	30.0	29.1	28.3	27.4	26.6	25.9	25.2	24.5	23.8	23.2	22.5	22.0	21.4	20.8	**81.8**
185	31.8	30.8	29.9	29.0	28.2	27.4	26.6	25.9	25.1	24.5	23.8	23.2	22.6	22.0	21.4	**84.1**
190	32.7	31.7	30.7	29.8	28.9	28.1	27.3	26.6	25.8	25.1	24.4	23.8	23.2	22.6	22.0	**86.4**
195	33.5	32.5	31.5	30.6	29.7	28.9	28.0	27.3	26.5	25.8	25.1	24.4	23.8	23.2	22.6	**88.6**
200	34.4	33.4	32.3	31.4	30.5	29.6	28.8	28.0	27.2	26.4	25.7	25.1	24.4	23.8	23.2	**90.9**
205	35.3	34.2	33.2	32.2	31.2	30.3	29.5	28.7	27.9	27.1	26.4	25.7	25.0	24.4	23.7	**93.2**
210	36.1	35.0	34.0	33.0	32.0	31.1	30.2	29.4	28.5	27.8	27.0	26.3	25.6	25.0	24.3	**95.5**
215	37.0	35.9	34.8	33.7	32.8	31.8	30.9	30.0	29.2	28.4	27.7	26.9	26.2	25.5	24.9	**97.7**
220	37.8	36.7	35.6	34.5	33.5	32.6	31.6	30.7	29.9	29.1	28.3	27.6	26.8	26.1	25.5	**100.0**
225	38.7	37.5	36.4	35.3	34.3	33.3	32.4	31.4	30.6	29.7	28.9	28.2	27.4	26.7	26.1	**102.3**
230	39.6	38.4	37.2	36.1	35.0	34.0	33.1	32.1	31.3	30.4	29.6	28.8	28.1	27.3	26.6	**104.5**
235	40.4	39.2	38.0	36.9	35.8	34.8	33.8	32.8	31.9	31.1	30.2	29.4	28.7	27.9	27.2	**106.8**
240	41.3	40.0	38.8	37.7	36.6	35.5	34.5	33.5	32.6	31.7	30.9	30.1	29.3	28.5	27.8	**109.1**
245	42.1	40.9	39.6	38.5	37.3	36.3	35.2	34.2	33.3	32.4	31.5	30.7	29.9	29.1	28.4	**111.4**
250	43.0	41.7	40.4	39.2	38.1	37.0	35.9	34.9	34.0	33.1	32.2	31.3	30.5	29.7	29.0	**113.6**
	1.63	1.65	1.68	1.7	1.73	1.75	1.78	1.80	1.83	1.85	1.88	1.91	1.93	1.96	1.98	
	HEIGHT (M)															

Muscular endurance is the physical ability to perform work. The pull-up test is often called a test of arm and shoulder strength, but in reality it is a measure of muscular endurance if you can do more than one. You count the maximum number of pull-ups performed, which is a measure of the amount of work that was completed. Muscular endurance can be categorized as relative endurance and absolute endurance. **Relative endurance** is a measurement of repetitive performance related to maximum strength. **Absolute endurance** is a measurement of repetitive performance at a fixed resistance. For example, measuring a participant's maximum strength on the bench press and then having him or her perform as many repetitions as possible at 75% of maximum strength is a relative endurance test. Performing the maximum number of repetitions at a fixed weight of 100 pounds (45.4 kg) would be a test of absolute endurance. Absolute endurance is highly correlated with maximum strength, but relative endurance has a low correlation with maximum strength.

Muscular actions are defined and categorized by specific terms, the most common of which are *concentric, eccentric, isometric, isotonic,* and *isokinetic.* The terms are defined as follows:

- ***Concentric contraction.*** The muscle generates force as it shortens.
- ***Eccentric contraction.*** The muscle generates force as it lengthens.
- ***Isometric contraction.*** The muscle generates force but remains static in length and causes no movement.
- ***Isotonic contraction.*** The muscle generates enough force to move a constant load at a variable speed through a full range of motion.
- ***Isokinetic contraction.*** The muscle generates force at a constant speed through a full range of motion.

In laboratory testing, researchers assess muscular performance generally by measuring the force, torque, work, and power generated in concentric, eccentric, isokinetic, and isometric contractions. In field situations, muscular performance is assessed with concentric, isotonic contractions. The term **repetition maximum (RM)** is used to indicate how many repetitions one can do. For example, 10RM means the maximum weight that can be lifted 10 times.

Laboratory Methods

Similar to laboratory measurements of aerobic capacity, measuring muscular strength and endurance in the laboratory requires expensive and sophisticated equipment and precise testing protocols. They are the most valid assessments but are difficult to administer.

Computerized Dynamometers

State-of-the-art muscular fitness measurement techniques involve using *computerized dynamometers,* which integrate mechanical ergometers, electronic sensors, computers, and sophisticated software. Computerized dynamometers allow for detailed measurements of force, work, torque, and power generated in terms not only of maximal values but of values throughout a range of motion. Isokinetic dynamometers allow the speed of movement to be controlled during testing. These devices are used in orthopedic clinics by physicians, physical therapists, and athletic trainers in rehabilitation programs for patients recovering from orthopedic injury or surgery; researchers in exercise science use the devices in strength and endurance studies. Strength is measured in terms of peak force and endurance in terms of fatigue rates in force production during a set of repetitions. Each of these devices has its own advantages and disadvantages. Because devices can be expensive, selection of a device requires an analysis of cost–benefit ratios.

The Biodex, a computerized isokinetic dynamometer (Biodex System 2 and Advantage Software 4.0, Biodex Medical Systems, Inc., Shirley, NY) is a device in widespread clinical use. Figure 9.13 illustrates graphic output of forces or torques that the system is able to generate.

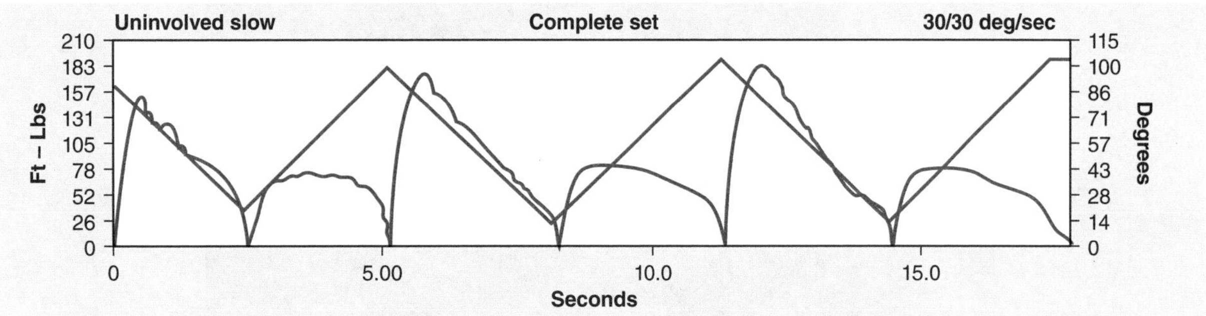

Figure 9.13 Graphic output of forces or torques produced on the Biodex dynamometer.

The Biodex can produce reliable ($r_{xx'}$ > .90) force measurements. However, as with any physical performance test, practice trials for both testers and test participants are needed to reduce measurement error. Mayhew and Rothstein (1985) provided a thorough discussion of the measurement issues associated with muscular performance assessed with dynamometers. These devices must be calibrated regularly to ensure reliable and valid measurements. Computerized dynamometry for muscular strength and endurance assessment should be used to provide criterion measures for concurrent validity research on more feasible field tests of muscular fitness. This application of computerized dynamometry would be valuable, because most field assessments in muscular fitness have only content validity to support their use.

Back Extension Strength Test

Lower-back pain is a serious, prevalent health problem in the adult population. Lack of strength and endurance in the extensor muscles of the lower back has been associated with lower-back pain (Suzuki and Endo 1983). Graves and colleagues (1990) examined measurement issues associated with isometric strength of the back extensors (figure 9.14). Their test protocol has participants begin at 72° of lumbar flexion and continue at every 12°, down to 0° of lumbar flexion. At each angle of flexion, each participant exerts a maximal isometric contraction of the extensor muscles of the back. The maximal isometric torque is measured by the dynamometer's automated system. This protocol has produced reliable ($r_{xx'}$ > .78) torque results. Average strength values for this test are provided in table 9.18.

Noncomputerized Dynamometers

Handheld dynamometers (HHDs) are also used to assess muscle strength. Although HHDs cannot provide as much information as isokinetic dynamometers, they have an advantage over isokinetic dynamometers in that they are less expensive and are portable. These devices provide an objective measurement of isometric force produced by a muscle or muscle group and can easily be used in the clinic, home, or office. HHDs come in several shapes and sizes. The tester instructs each participant to perform the desired motion and then, holding the HHD in one hand and providing stabilization to unwanted motions with the other hand, keeps the HHD steady as the participant pushes or pulls against it. The tester must hold the HHD stable as the participant presses or pulls against it with as much force as he

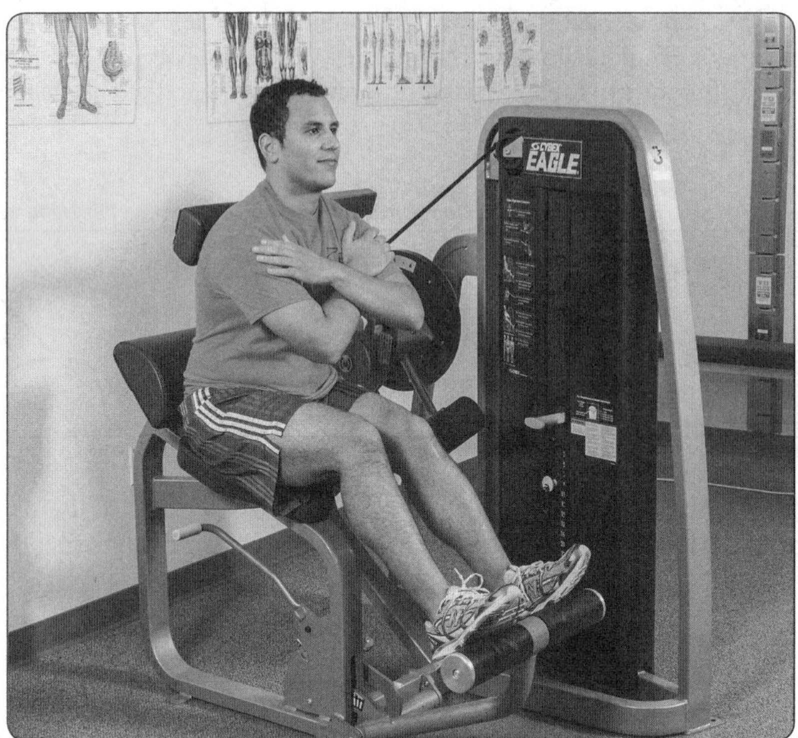

Figure 9.14 Back extension.

Table 9.18 Mean Isometric Torque Values for Lumbar Extension ($N \cdot$ m/kg body weight)

	DEGREES OF LUMBAR FLEXION						
Gender	**0**	**12**	**24**	**36**	**48**	**60**	**72**
Male	3.0	3.8	4.4	4.8	5.2	5.5	6.0
Female	2.2	2.7	3.0	3.1	3.3	3.5	3.9

Data based on Graves et al. 1990.

or she can generate. This type of test is known as a *make test*. The HHD will record and hold the digital display of the maximum force produced during the test in kilograms or pounds so that the tester can easily read the display after the test is completed. The major disadvantage of HHD use, that the tester must be physically strong enough to effectively stabilize the device during test administration, is most problematic during administration of strength tests of the larger muscles of the lower extremity. For this reason, many clinicians and scientists use straps to stabilize the HHD when testing muscles in the lower extremity whenever possible. The use of straps effectively removes the issue of tester strength and has been shown to dramatically improve reliability of testing lower-extremity muscles with use of a HHD. Research has demonstrated that HHD test results can be reliable and valid, but the force or torque results between HHDs of the same make and model and between HHDs of different makes and models can be inconsistent (Trudelle-Jackson et al. 1994). Thus for reliable and valid results, practitioners should use one HHD when comparing results across repeated measurements of any kind.

Manual Muscle Testing

A precursor to HHD testing was manual muscle testing (MMT) techniques, which are still used most often to assess muscle strength in clinical situations. MMT techniques originated in the 1940s; although the techniques have been refined over the years, they still retain the original basic testing principles. These techniques are used extensively in the clinic because they require no equipment and can be administered quickly. When using MMT, the tester asks a patient to move the joint through its available range of motion against gravity. If the person is able to move through the full range, the tester applies resistance at the end of the range. The tester begins to apply resistance, then gradually increases the resistance and notes how well the person is able to hold the muscle contraction against the increasing resistance. After applying resistance for 4 or 5 seconds, the tester ceases to resist and rates the amount of resistance that was applied. The person's effort is graded on a scale of 0 to 5:

- Grade 5: Muscle contracts normally against full resistance.
- Grade 4: Muscular strength is reduced, but muscular contraction can still move the joint against resistance.
- Grade 3: Muscular strength is further reduced such that the joint can be moved only against gravity with the tester's resistance completely removed (e.g., the elbow can be moved from full extension to full flexion starting with the arm hanging down at the side).
- Grade 2: The muscle can move only if the resistance of gravity is removed (e.g., the elbow can be fully flexed only if the arm is maintained in a horizontal plane).
- Grade 1: Only a trace or flicker of movement is seen or felt in the muscle, or fasciculations are observed in the muscle.
- Grade 0: No movement is observed.

A system of pluses and minuses is often used to further sensitize the grading system. If a patient was not able to complete the motion against gravity, no resistance is applied and instead the patient is repositioned so that gravity is minimized. In this instance, the tester assesses whether that person can complete the joint range of motion when gravity is not opposing the motion. For example, when the quadriceps muscle is being assessed, the person is in a seated position and asked to fully extend the knee (this motion is against gravity). If the person completes the motion, then resistance is applied and graded accordingly. If the person is not able to fully extend the knee against gravity, then he or she is repositioned to side-lying with the knee flexed and asked to fully extend the knee from this position. Tester experience as well as tester strength plays a role when MMT techniques are used to assess muscular strength.

Field Methods

Field assessment of muscular strength and endurance involves lifting external weights or the repetitive movement of the body. The measure of maximum strength is the 1-repetition maximum (1RM), which is the maximum amount of weight a person can lift for one repetition. Muscular endurance is assessed by performing either the maximum number of repetitions of a weightlifting exercise with a submaximal weight load or the maximum number of repetitions of a body movement exercise such as sit-ups.

Upper- and Lower-Body Strength and Endurance

The ACSM (2014a) recommends 1RM values of the bench press and the leg press strength tests as valid measures of upper- and lower-body strength. The ACSM further suggests that the 1RM values be divided by a person's body weight to present a strength measure that is equitable across weight classes.

The following steps present a method for assessing the 1RM value for any given exercise. These steps were used to produce reliable ($r_{xx'}$ > .92) and valid measures of upper- and lower-body strength (Jackson, Watkins, and Patton 1980).

1. The participant warms up with stretching and light lifting.
2. The participant performs a lift below what is an estimate of his or her maximum. A practice session is extremely useful for novice participants.
3. To prevent fatigue, the participant rests at least 2 minutes between lifts.
4. The weight is increased by a small increment, 5 or 10 pounds (2.3 or 4.5 kg), depending on the exercise and weight increments available.
5. The process is continued until the participant fails an attempt.
6. The last weight successfully completed is the 1RM weight.
7. Divide the 1RM by the participant's body weight.

If more than five repetitions are needed to determine the 1RM value, the participant is retested after a day's rest with a heavier starting weight.

Tables 9.19 and 9.20 provide norms for the bench press and leg press tests. These norms reflect measurements taken from the Universal Gym Weight Lifting Machine, a rack-mounted weight machine. These standards are not valid if your testing is done with free weights or another type of machine. One of the difficulties of strength measurement is that each strength-testing device—free weights, computerized dynamometers, or rack-mounted devices—produces different results. Use standards appropriate for your testing situation; it may be necessary for you to develop your own local standards.

Table 9.19 Bench Press Strength (1RM lb/lb [kg/kg] body weight)

	AGE (YEARS)				
Rating	20-29	30-39	40-49	50-59	60+
	MEN				
Excellent	>1.26	>1.08	>0.97	>0.86	>0.78
Good	1.17-1.25	1.01-1.07	0.91-0.96	0.81-0.85	0.74-0.77
Average	0.97-1.16	0.86-1.00	0.78-0.90	0.70-0.80	0.64-0.73
Fair	0.88-0.96	0.79-0.85	0.72-0.77	0.65-0.69	0.60-0.63
Poor	<0.87	<0.78	<0.71	<0.64	<0.59
	WOMEN				
Excellent	>0.78	>0.66	>0.61	>0.54	>0.55
Good	0.72-0.77	0.62-0.65	0.57-0.60	0.51-0.53	0.51-0.54
Average	0.59-0.71	0.53-0.61	0.48-0.56	0.43-0.50	0.41-0.50
Fair	0.53-0.58	0.49-0.52	0.44-0.47	0.40-0.42	0.37-0.40
Poor	<0.52	<0.48	<0.43	<0.39	<0.36

Data based on The Cooper Institute 2002.

Table 9.20 Leg Press Strength (1RM lb/lb [kg/kg] body weight)

	AGE (YEARS)				
Rating	**20-29**	**30-39**	**40-49**	**50-59**	**60+**
	MEN				
Excellent	>2.08	>1.88	>1.76	>1.66	>1.56
Good	2.00-2.07	1.80-1.87	1.70-1.75	1.60-1.65	1.50-1.55
Average	1.83-1.99	1.63-1.79	1.56-1.69	1.46-1.59	1.37-1.49
Fair	1.65-1.82	1.55-1.62	1.50-1.55	1.40-1.45	1.31-1.36
Poor	<1.64	<1.54	<1.49	<1.39	<1.30
	WOMEN				
Excellent	>1.63	>1.42	>1.32	>1.26	>1.15
Good	1.54-1.62	1.35-1.41	1.26-1.31	1.13-1.25	1.08-1.14
Average	1.35-1.53	1.20-1.34	1.12-1.25	0.99-1.12	0.92-1.07
Fair	1.26-1.34	1.13-1.19	1.06-1.11	0.86-0.98	0.85-0.91
Poor	<1.25	<1.12	<1.05	<0.85	<0.84

Based on The Cooper Institute 2002.

The YMCA Bench Press Test (Golding 2000) is used to assess upper-body endurance. The Canadian Standardized Test of Fitness uses a push-up test to exhaustion to measure upper-body endurance. These tests are described in the following highlighted boxes.

Johnson and Nelson (1979) indicated that you can achieve a high reliability ($r_{xx'}$ = .93) when these standardized test procedures are followed. *The results of this test will be negatively correlated with the body weight of the person tested.*

YMCA BENCH PRESS TEST

Purpose

To assess the absolute endurance of the upper body.

Objective

To perform a set of bench press repetitions to exhaustion.

Equipment

Barbells (35 and 80 pounds [15.9 and 36.4 kg])

Metronome set at 60 beats/minute

Weightlifting bench

Instructions

The weight is 35 pounds (15.9 kg) for women and 80 pounds (36.4 kg) for men. The consecutive repetitions are performed to a cadence of 60 beats/minute, with each sound indicating a movement up or down.

Scoring

The test continues until exhaustion or until the participant can no longer maintain the required cadence. Table 9.21 provides norms for this test. Because this is an absolute endurance test, it is positively correlated with maximal strength and body weight or size.

Table 9.21 Male and Female Norms for the YMCA Bench Press Test (Number Completed)

	AGE (YEARS)					
Male rating	**18-25**	**26-35**	**36-45**	**46-55**	**56-65**	**66+**
Excellent	64-44	61-41	55-36	47-28	41-24	36-20
Good	41-34	37-30	32-26	25-21	21-17	16-12
Above average	33-29	29-26	25-22	20-16	14-12	10-10
Average	28-24	24-21	21-18	14-12	11-9	8-7
Below average	22-20	20-17	17-14	11-9	8-5	6-4
Poor	17-13	16-12	12-9	8-5	4-2	3-2
Very poor	10-0	9-0	6-0	2-0	1-0	1-0
Female rating						
Excellent	66-42	62-40	57-33	50-29	42-24	30-18
Good	38-30	34-29	30-26	24-20	21-17	16-12
Above average	28-25	28-24	24-21	18-14	14-12	10-8
Average	22-20	22-18	20-16	13-10	10-8	7-5
Below average	18-16	17-14	14-12	9-7	6-5	4-3
Poor	13-9	13-9	10-6	6-2	4-2	2-0
Very poor	6-0	6-0	4-0	1-0	1-0	0-0

Data based on Golding 2000.

CANADIAN STANDARDIZED TEST OF FITNESS—PUSH-UP TEST

Purpose

To assess upper-body endurance.

Objective

To perform push-ups to exhaustion.

Equipment

Mat

Instructions

Females perform the test with the knees bent and touching the ground; males perform the test with the toes touching the ground (figure 9.15).

Scoring

The number of correct repetitions is compared with the norms provided in table 9.22.

Figure 9.15 *(a)* Traditional and *(b)* modified push-ups.

Table 9.22 Male and Female Norms for the Push-Up Test (Number Completed)

	AGE (YEARS)					
Male rating	**15-19**	**20-29**	**30-39**	**40-49**	**50-59**	**60-69**
Excellent	39+	36+	30+	22+	21+	18+
Above average	29-38	29-35	22-29	17-21	13-20	11-17
Average	23-28	22-28	17-21	13-16	10-12	8-10
Below average	18-22	17-21	12-16	10-12	7-9	5-7
Poor	17-	16-	11-	9-	6-	4-
Female rating						
Excellent	33+	30+	27+	24+	21+	17+
Above average	25-32	21-29	20-26	15-23	11-20	12-16
Average	18-24	15-20	13-19	11-14	7-10	5-11
Below average	12-17	10-14	8-12	5-10	2-6	1-4
Poor	11-	9-	7-	4-	1-	1-

Data based on *Canadian standardized test of fitness operations manual*, 3rd ed. 1986.

Trunk Endurance

The most universally used test of abdominal endurance is some type of a sit-up test. The YMCA's half sit-up test protocol is described here.

Reliability estimates for sit-up tests are typically ≥.68. Acceptable reliability requires the use of standardized procedures and practice trials. The participant's body weight is negatively correlated to the results of this test.

Robertson and Magnusdottir (1987) presented an alternative field test of abdominal endurance similar to the YMCA half sit-up test. Their curl-up test requires the participant to lift the head and upper back a distance that allows the extended arms and fingers to move 3 inches (7.6 cm) parallel to the ground rather than sit up completely. This requires less use of the hip flexors and more use of the abdominals than the standard sit-up test. The participant completes as many of these as possible in 60 seconds. The authors report excellent reliability ($r_{xx'}$ = .93) for the test.

Go to the WSG to complete Student Activity 9.4 and view Videos 9.5, 9.6, and 9.7.

YMCA HALF SIT-UP TEST

Purpose

To assess abdominal endurance.

Objective

To perform the maximum number of half sit-ups in 1 minute.

Equipment

Stopwatch

Mat

Tape

Instructions

The participant lies face up on mat or rug with knees at a right angle (90°) and feet flat on the ground. The feet are not held down. The participant places hands, palms facing down, on the mat or rug with the fingers touching the first piece of tape (3.5 in. [8.9 cm] between pieces of tape) (see figure 9.16*a*). The participant flattens the lower back to the mat or rug and then sits up halfway so that the fingers move from the first piece of tape to the second (figure 9.16*b*). The participant then returns the shoulders to the mat or rug and repeats the movement as described. The head does not have to touch the surface. The lower back is to be kept flat on the mat or rug during the movements; arching the back can cause injury. The number of half sit-ups performed in 1 minute are counted. Participants should be instructed to pace themselves so they can do half sit-ups for 1 minute.

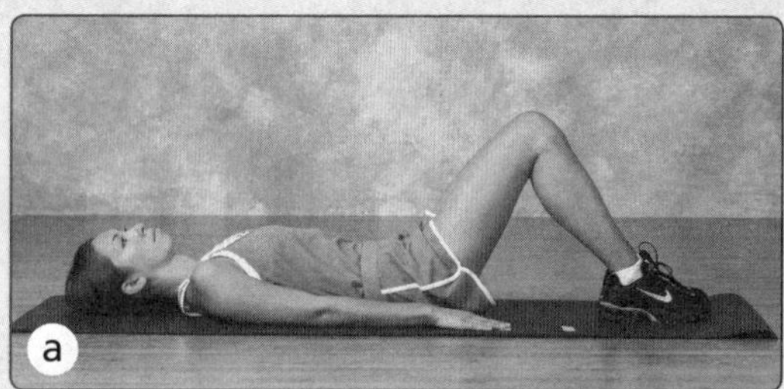

Figure 9.16 The sit-up test: *(a)* beginning position and *(b)* up position.

Scoring

See norms in table 9.23.

Table 9.23 Male and Female Norms for the YMCA Half Sit-Up Test (Number Completed)

	AGE (YEARS)					
Male	**18-25**	**26-35**	**36-45**	**46-55**	**56-65**	**66+**
Excellent	99-77	80-62	79-60	78-61	77-56	66-50
Good	72-61	58-53	57-48	57-52	53-48	44-38
Above average	57-52	52-44	45-43	51-44	46-41	25-31
Average	49-43	41-37	39-33	41-36	39-33	30-26
Below average	41-37	36-33	32-29	33-29	32-28	24-22
Poor	35-29	32-26	28-24	25-21	25-21	21-15
Very poor	27-14	21-7	21-6	16-6	20-5	12-5
Female						
Excellent	91-68	70-54	74-54	73-48	63-44	54-34
Good	64-58	50-44	48-42	44-37	42-35	33-31
Above average	57-51	41-37	38-35	36-33	32-27	29-26
Average	48-41	36-33	32-30	32-30	25-23	25-21
Below average	38-34	32-28	28-23	28-25	22-18	20-16
Poor	33-28	26-22	22-19	23-19	15-11	13-10
Very poor	25-11	20-7	16-4	13-2	8-1	9-0

Data based on YMCA of the USA.

MEASURING FLEXIBILITY

The measurement of flexibility, or range of motion of a joint or group of joints, is an important aspect of assessment of patients recovering from orthopedic surgery or orthopedic injuries. *Because flexibility is specific to a joint and its surrounding tissues, there are no valid tests of general flexibility.* For example, if you have flexible ankle joints, you may not have flexible shoulder joints.

Laboratory Methods

Miller (1985) summarized the measurement issues associated with the assessment of joint range of motion. He described clinical measurement techniques, including the following:

- Goniometry—manual, electric, and pendulum goniometers
- Visual estimation
- Radiography
- Photography
- Linear measurements
- Trigonometry

Miller asserted that radiography is the most reliable and valid method but that it has limited feasibility owing to problems with radiation. Goniometry is the most feasible method of clinical assessment of flexibility; it can be reliable and valid if proper test procedures are followed. Reliability estimates for goniometric measurement of hamstring flexibility have exceeded .90 in past research (Jackson and Baker 1986; Jackson and Langford 1989). Clinical tests of flexibility should serve as criterion measures for validity studies of field tests of flexibility for health-related fitness. Such concurrent validity studies improve the validity of measurement of flexibility in the field.

Field Methods

As reported previously, the ACSM (2014a) indicates that lower-back pain is a prominent health problem in the United States. In theory, lack of flexibility in the lower back should be associated with lower-back pain, but valid research has yet to establish a relationship between the two (ACSM 2014a). Flexibility of the lower back is measured with tests of trunk flexion and extension. Most adult fitness batteries include a test of trunk flexion that is a version of the sit-and-reach test.

Trunk Flexion

The sit-and-reach test is a universally used test of trunk flexion. Specifically, it was designed to measure the flexibility of the lower-back and hamstring muscles. The YMCA adult fitness test version of the sit-and-reach test is described next.

Keep in mind that flexibility of the lower back has not been empirically related to lower-back pain. Research has shown that the sit-and-reach test is a valid measure of hamstring flexibility and is highly reliable ($r_{xx'} > .90$) but is poorly correlated with a clinical measure of lower-back flexibility and probably is not a valid field test of lower-back flexibility (Jackson and Baker 1986; Jackson and Langford 1989). Further research (Jackson et al. 1998) documented no relationship between reported lower-back pain and performance on the sit-and-reach or sit-up tests. These observations provide little support for including sit-and-reach tests in health-related fitness assessments, but these tests are often present in both adult and youth fitness batteries.

YMCA ADULT TRUNK FLEXION (SIT-AND-REACH) TEST

Purpose

To assess trunk flexibility.

Objective

To reach as far forward as possible.

Equipment

Yardstick

Masking tape

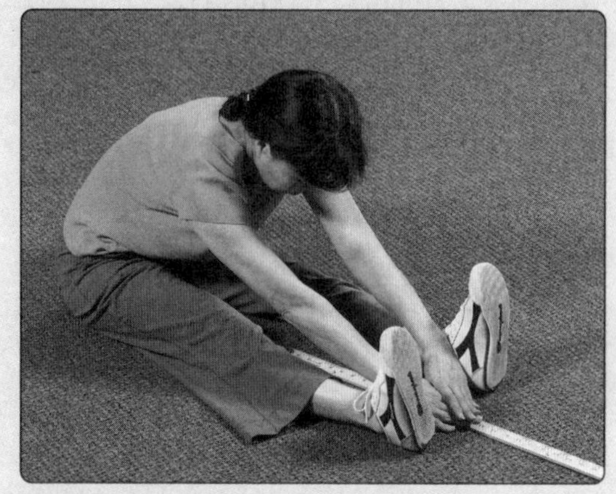

Figure 9.17 The sit-and-reach test.

Instructions

A yardstick is placed on the floor with an 18-inch (45.7 cm) piece of tape across the 15-inch (38.1 cm) mark on the yardstick. The tape should secure the yardstick to the floor. The participant sits with the 0 end of the yardstick between the legs. The participant's heels should almost touch the tape at the 15-inch (38.1 cm) mark and be about 12 inches (30.5 cm) apart. With the legs held straight, the participant bends forward slowly and reaches with parallel hands as far as possible and touches the yardstick. The participant should hold this reach long enough for the distance to be recorded (figure 9.17).

Scoring

Perform three trials. The best score, recorded to the nearest quarter inch (.6 cm), is compared with the norms supplied in table 9.24.

Table 9.24 Male and Female Norms for the YMCA Trunk Flexion Test (Inches)

	AGE (YEARS)					
Male rating	**18-25**	**26-35**	**36-45**	**46-55**	**56-65**	**66+**
Excellent	28-22	28-21	28-21	26-19	24-17	24-17
Good	21-20	19-19	19-18	18-16	16-15	16-14
Above average	19-18	17-17	17-16	15-14	13-13	13-12
Average	17-16	16-15	15-15	13-12	11-11	11-10
Below average	15-14	14-13	13-13	11-10	9-9	9-8
Poor	13-12	12-11	11-9	9-8	8-6	7-6
Very poor	11-2	9-2	7-1	6-1	5-1	4-0
Female rating						
Excellent	29-24	28-23	28-22	27-21	26-20	26-20
Good	22-22	22-21	21-20	20-19	19-18	19-18
Above average	21-20	20-20	19-18	18-17	17-16	17-17
Average	19-19	19-18	17-17	16-16	15-15	16-15
Below average	18-17	17-16	16-15	14-14	14-13	14-13
Poor	16-16	15-14	14-13	13-12	12-10	12-10
Very poor	14-7	13-5	12-4	10-3	9-2	9-1

Note: For metric measurements, refer to a site such as this: http://www.worldwidemetric.com/measurements.html.

Data based on Golding 2000.

Trunk Extension

Measuring trunk flexion with the sit-and-reach test is prevalent in health-related fitness testing; however, trunk extension assessments are not generally included in health-related fitness testing, and little research has been conducted on developing a valid field test of this ability. The Fitnessgram (see chapter 10) includes measurement of trunk extension. Jensen and Hirst (1980) described a field test in which the participant lies prone with the hands clasped near the small of the back. The participant then raises the upper body off the floor as far as possible while an aide holds the legs down. The score is the distance from the suprasternal notch to the floor multiplied by 100 and divided by the trunk length. Safrit (1986) suggested determining the trunk length while the participant is seated; trunk length is then the distance from the suprasternal notch to the seat. There is no reliability and validity information on this test. Furthermore, body weight and back extensor strength would be related to the scores on this test, which would limit it as a valid measure of flexibility. Developing a reliable and valid measure of lower-back extension requires additional research.

HEALTH-RELATED FITNESS BATTERIES

Several organizations have grouped health-related fitness test items together into test batteries. These test batteries and their documentation provide you with the ability to administer reliable and valid fitness tests, interpret the results, and convey fitness information to your program participants.

YMCA Physical Fitness Test Battery

Throughout this chapter we have highlighted the fitness test battery used by the YMCA (Golding 2000). The test battery is used in the physical fitness assessments that the YMCA performs on its members. The test is available and easily adaptable to many adult physical fitness testing situations. The entire test battery includes these measurements:

- Height
- Weight
- Resting heart rate
- Resting blood pressure
- Body composition
- Cardiovascular evaluation
- Flexibility
- Muscular strength and endurance

We have described and provided norms for the 3-minute step test (cardiorespiratory endurance), the skinfold estimation of body composition, the sit-and-reach test (flexibility), the bench press test (muscular strength), and the half sit-up test (muscular endurance).

Canadian Standardized Test of Fitness

The Canadian Standardized Test of Fitness, a physical fitness test battery, was the result of the Canadian Fitness Survey conducted in 1981 on thousands of participants to develop an understanding of the fitness level of the Canadian population. The test battery included the following components:

- Resting heart rate
- Resting blood pressure
- Body composition (skinfolds)
- Cardiorespiratory endurance (A variety of test results can be used, including treadmill time, cycle ergometer results, and distance-run times.)
- Flexibility (sit-and-reach test)
- Abdominal endurance (with the 1-minute sit-up test)
- Upper-body strength and endurance (push-up test)

The President's Challenge: Adult Fitness Test

In May 2008, the President's Council on Fitness, Sports & Nutrition launched an online (www.adultfitnesstest.org) adult fitness self-test. The test battery was designed to be self-administered. However, a partner is needed to complete the entire test battery. The test is designed to measure health-related physical fitness, including aerobic fitness, muscular strength and endurance, flexibility, and body composition. The battery includes the following tests:

- Aerobic fitness
    - 1-mile (1.6 km) walk test
    - 1.5-mile (2.4 km) run
- Body composition
    - Body mass index
    - Waist girth
- Muscular strength and endurance
    - Standard or modified push-up
    - Half sit-up test
- Flexibility
    - Sit-and-reach test

 Go to the WSG to view Video 9.8.

PHYSICAL FITNESS ASSESSMENT IN OLDER ADULTS

An older adult is defined as someone aged 65 years and over. In 2011 there were approximately 41.4 million older adults in the United States, about 13% of the entire population, and this number is projected to grow to 80 million older adults by the year 2040. The increase in the number of older adults in the United States is similar to that in other industrialized countries.

Older adults have the highest rates of chronic diseases such as cardiovascular disease, cancer, diabetes, osteoporosis, and arthritis. Health care costs for older adults are contributing to health financing problems. Medicare, the U.S. national health care program for older adults, is facing an uncertain financial future that may have grave implications for future generations of Americans. As professionals in human performance, we serve an important role in helping improve the physical activity and fitness levels of young, middle-aged, and older adults. Improved health of our population derived from increased physical activity and fitness would help alleviate the health care financing shortages facing the United States. Aging is related to

- decreased sensations of taste, smell, vision, and hearing;
- decreased mental abilities (e.g., memory, judgment, speech);
- decreased organ function of the digestive system, urinary tract, liver, and kidneys;
- decreased bone mineral content and muscle mass resulting in less lean body weight; and
- decreased physical fitness (e.g., cardiorespiratory endurance, strength, flexibility, muscular endurance, reaction and movement times, balance).

This last factor is of primary concern to anyone entering a health- and fitness-related profession. Studies show that older adults respond to appropriate endurance- and strength-training programs in a manner similar to younger adults. Hagberg and colleagues (1989) demonstrated a 22% increase in $\dot{V}O_2max$ as the result of endurance training in 70- to 79-year-old males and females. Fiatarone and colleagues (1994) conducted a high-intensity program of strength training in older males and females with an average age of 87 years. The participants were described as frail nursing home residents. Their study demonstrated strength gains of 113% and lean body mass gains of 3%. More important, the participants dramatically improved their walking speed and stair-climbing power, which are clinical measures of functional capacity. **Functional capacity**, the ability to perform the normal activities of daily living, is of prime importance for older adults if they are to maintain independent living status and a high quality of life (USDHHS 1996). The relationships among physical activity, physical fitness, and functional capacity produce the need for reliable and valid measurement of physical activity and physical fitness in older adults.

We have defined health-related physical fitness in adults as including cardiorespiratory endurance, body composition, and musculoskeletal fitness. Are the parameters of health-related physical fitness the same in older adults? Although the same factors are still important, an older person's health and fitness should also include motor fitness factors such as balance, reaction time, and movement time in order to focus on maintaining functional capacity, activities of daily living, and overall quality of life.

For example, falls are a major health problem in older adults. Many older adults suffer fractures from falls, and the risk for mortality increases after falls occur. Indeed, many older adults become fearful of falling again, and they restrict their movements and lifestyle because of that fear. Strength and balance training in older adults improves fitness parameters that increase lean body and bone mass and lower the risk of falling and of suffering a fracture if one does fall (USDHHS 1996).

In concert with this broader definition of health-related physical fitness in older adults, Rikli and Jones (1999a, 1999b), at the Ruby Gerontology Center at the University of California at Fullerton, have developed a fitness battery for older adults. The test battery incorporates measures of strength, flexibility, cardiorespiratory endurance, motor fitness, and body composition (table 9.25). Local and national advisory panels of health and fitness experts have assisted in selecting and validating the test items. The theme of item selection was the important role that physical fitness has in delaying frailty and maintaining mobility in older adults, both of which are key factors in a healthy older population with a high quality of life. Test items were selected based on whether they met the following criteria: Does the item

- represent major functional fitness components (i.e., key physiologic parameters associated with the functions required for independent living);
- have acceptable test–retest reliability (>.80);

- have acceptable validity, with support for at least two to three types of validation: content-related, criterion-related, or construct-related;
- reflect usual age-related changes in physical performance; and
- require minimum equipment and space?

Table 9.25 Fitness Parameters and Items of the Older Adult Fitness Test

Physical fitness parameter	Test item
Lower-body strength	30-s chair stand
Upper-body strength	Arm curl
Lower-body flexibility	Chair sit-and-reach
Upper-body flexibility	Back scratch
Cardiorespiratory endurance	6-min walk or 2-min step-in-place test
Motor fitness Composite measure of power, speed, agility, and balance	8-foot (2.4 m) up-and-go
Body composition	Body mass index (BMI)

In addition, is the test item

- able to detect physical changes attributable to training or exercise;
- able to be assessed on a continuous scale across wide ranges of functional ability (frail to highly fit);
- easy to administer and score;
- capable of being self- or partner-administered in the home setting;
- safe to perform without medical release for the majority of community-residing older adults;
- socially acceptable and meaningful; and
- reasonably quick to administer, with individual testing time of no more than 30 to 45 minutes?

Reliability and validity for the test battery are acceptable. Test–retest reliability estimates exceeded .80 for all tests for older males and females. Criterion-related validity coefficients exceeded .70 for five of the seven performance tests for both males and females, and all seven performance tests demonstrated construct validity. Thus, the battery has been established with content validity and feasibility as guidelines. Further research has established sufficient reliability and validity for the battery.

 Go to the WSG to view Video 9.9.

OLDER ADULT FITNESS BATTERY

By definition, older people would not be determined apparently healthy. But Rikli and Jones (1999a) found that ambulatory, community-residing older adults with "no medical conditions that would be contraindicated for submaximal testing" (ACSM 2014a) who had not been advised to refrain from exercise could be safely tested. Rikli and Jones (1999b) stated that persons who should not take the tests without physician approval are those who

- have been advised by their doctors not to exercise because of a medical condition;
- have experienced chest pain, dizziness, or exertional angina (chest tightness, pressure, pain, heaviness) during exercise;

- have experienced congestive heart failure; or
- have uncontrolled high blood pressure (greater than 160/100).

Thus, with careful screening and caution, this test battery should be a valid and feasible tool for fitness assessment in older persons. The test items are presented in the pages that follow. Tables 9.26 and 9.27 provide percentile norms for females and males, respectively, on the eight items on the Older Adult Fitness Test Battery (Rikli and Jones 1999b).

Rikli and Jones (2013a) extended their work with senior adults by developing criterion-referenced fitness standards (recall chapter 7; see table 9.28). For each of their tests, they identified a cut score that was associated with functioning independently for those between the ages of 60 and 94 years of age. Reliabilities and validities for the criterion-referenced standards were between .79 and .97, indicating that the standards reflected good consistency (i.e., reliability) and predictability (i.e., validity). See Rikli and Jones (2013a; 2013b) for a complete list of these standards by fitness test and age groups.

30-SECOND CHAIR STAND

Purpose

To assess lower-body strength.

Objective

To complete as many stands from a sitting position as possible in 30 seconds.

Equipment

Stopwatch

Straight-back or folding chair (without arms) with seat height approximately 17 inches (43.2 cm)

Instructions

For safety purposes, the chair is placed against a wall or stabilized in some other way to prevent it from moving during the test. The participant begins the test seated in the middle of the chair, back straight, and feet flat on the floor, arms crossed at the wrists and held against the chest (figure 9.18). On the go signal, the participant rises to a full stand and then returns to a seated position. The participant is encouraged to complete as many full stands as possible within 30 seconds. After a demonstration by the tester, the participant should do a practice trial of one or two repetitions as a check for proper form, followed by one 30-second test trial.

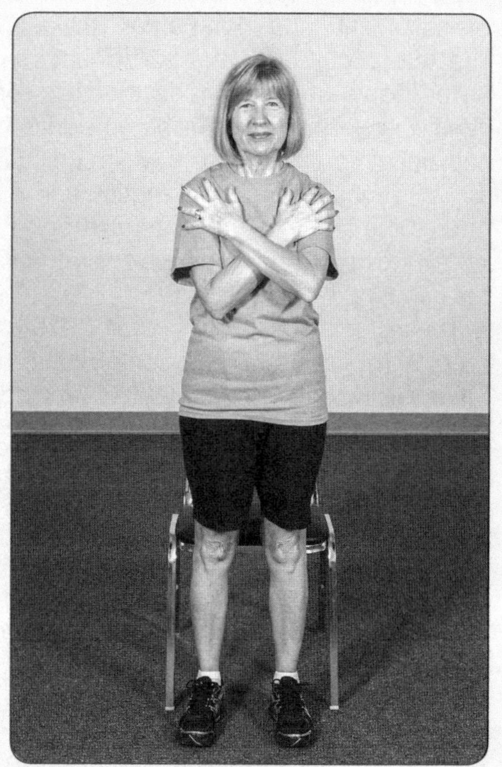

Figure 9.18 30-second chair stand test.

Scoring

The score is the total number of stands executed correctly within 30 seconds. If the participant is more than halfway up at the end of 30 seconds, it counts as a full stand.

ARM CURL

Purpose

To assess upper-body strength.

Objective

To perform as many correctly executed arm curls as possible in 30 seconds.

Equipment

Wristwatch with a second hand

Straight-back or folding chair

Hand weight (dumbbell): 5 pounds (2.3 kg) for women; 8 pounds (3.6 kg) for men

Instructions

The participant sits on the chair, back straight, feet flat on the floor, and with the dominant side of the body close to the edge of the chair. The participant holds the weight at the side in the dominant hand, using a handshake grip. The test begins with the participant's arm in the down position beside the chair, perpendicular to the floor. At the go signal, the participant turns the palm up while curling the arm through a full range of motion and then returns the arm to a fully extended position. At the down position, the weight returns to the handshake grip position.

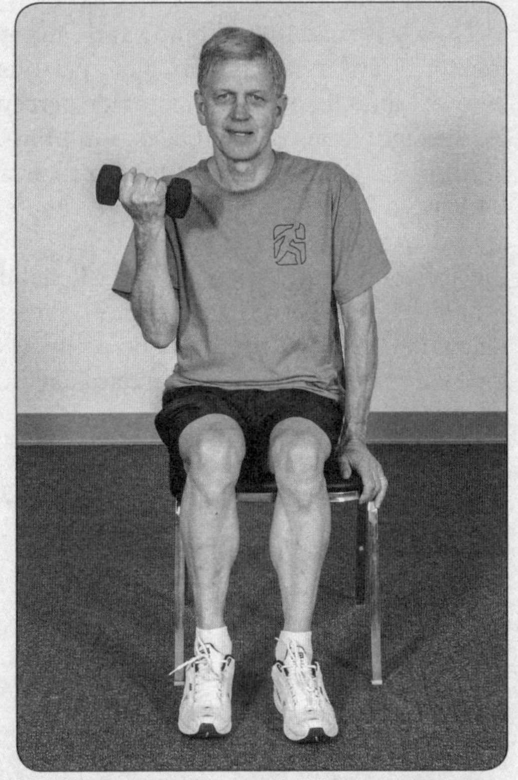

Figure 9.19 Arm curl test.

The tester kneels (or sits in chair) next to the participant on the dominant arm side, placing his or her fingers on the person's midbiceps to stabilize the upper arm and to ensure that a full curl is made (participant's forearm should squeeze examiner's fingers; figure 9.19). The participant's upper arm must remain still throughout the test.

The tester may also need to position his or her other hand behind the participant's elbow to help gauge when full extension has been reached and to prevent a backward swinging motion of the arm.

The participant is encouraged to execute as many curls as possible within the 30-second time limit. After a demonstration, give a practice trial of one or two repetitions to check for proper form, followed by one 30-second trial.

Scoring

The score is the total number of curls made correctly within 30 seconds. If the arm is more than halfway curled at the end of the 30 seconds, it counts as a curl.

Adapted, by permission, from R.E. Rikli and C.J. Jones, 1999, "Development and validation of a functional fitness test for community-residing older adults," *Journal of Aging and Physical Activity* 7(2): 129-161.

6-MINUTE WALK TEST

Purpose

To assess aerobic endurance.

Objective

To assess the maximum distance a participant can walk in 6 minutes along a 50-yard (45.7 m) course.

Equipment

- Stopwatch
- Long measuring tape (over 20 yards [18.3 m])
- 4 cones
- 20 to 25 Popsicle (craft) sticks for each participant
- Adhesive name tags (for each participant)
- Pen
- Chalk or masking tape to mark the course

Set-Up

A 50-yard (45.7 m) course is set up and marked in 5-yard (4.6 m) segments with chalk or tape (figure 9.20). The walking area needs to be well lit and have a nonslippery, level surface. For safety purposes, chairs are positioned at several points along the outside of the walkway.

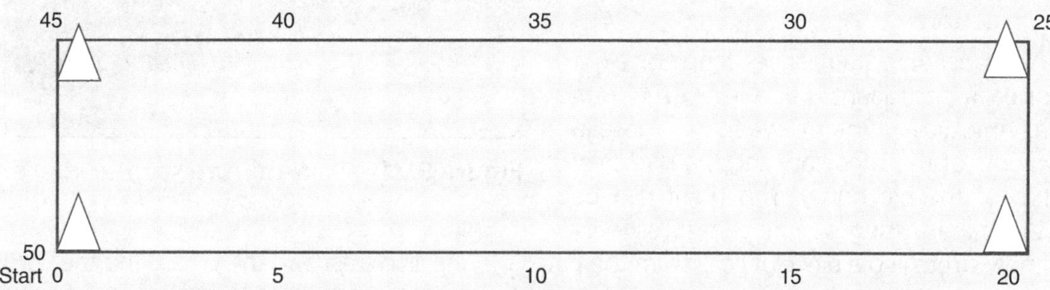

Figure 9.20 The 6-minute walk test uses a 50-yard (45.7 m) course measured into 5-yard (4.6 m) segments.

Instructions

On the go signal, participants walk as quickly as possible (without running) around the course as many times as they can within the time limit. The tester gives a Popsicle stick (or other similar object) to the participant each time he or she completes one lap, or someone tallies laps. If two or more participants are tested at once, starting times are staggered 10 seconds apart so participants do not walk in clusters or pairs. Each participant is assigned a number to indicate the order of starting and stopping (self-adhesive name tags to number each participant can be used).

During the test, participants may stop and rest (sitting on the provided chairs) if necessary and then resume walking. The tester should move to the center of the course after all participants have started and should call out elapsed time when participants are approximately half done, when 2 minutes are left, and when 1 minute is left. At the end of each participant's respective 6 minutes, participants are instructed to stop and move to the right where an assistant will record the score. A practice test is conducted before the test day to assist with proper pacing and to improve scoring accuracy.

The test is discontinued if at any time a participant shows signs of dizziness, pain, nausea, or undue fatigue. At the end of the test, participants should slowly walk around for about a minute to cool down.

Scoring

The score is the total number of yards (m) walked (to the nearest 5 yards [4.6 m]) in 6 minutes.

Adapted, by permission, from R.E. Rikli and C.J. Jones, 1999, "Development and validation of a functional fitness test for community-residing older adults," *Journal of Aging and Physical Activity* 7(2): 129-161.

2-MINUTE STEP-IN-PLACE

Purpose

An alternative test to assess aerobic endurance.

Objective

To assess the maximum number of steps in place a participant can complete in 2 minutes.

Equipment

Stopwatch

Tape measure or length of 30-inch (76.2 cm) cord

Masking tape

Mechanical counter (if possible) to ensure accurate counting of steps

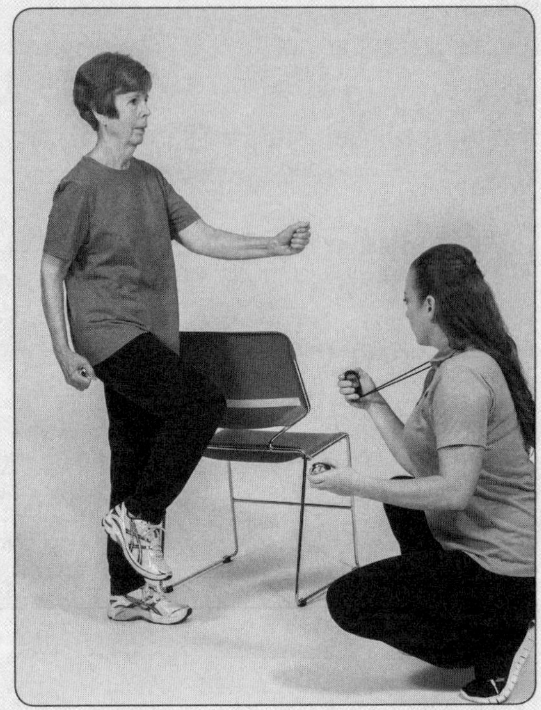

Figure 9.21 Two-minute step-in-place.

Instructions

The proper (minimum) knee-stepping height for each participant is at a level even with the midway point between the patella (middle of the knee cap) and the iliac crest (top of the hip bone). This point can be determined using a tape measure or by simply stretching a piece of cord from the patella to the iliac crest and then doubling the cord over to determine the midway point. To monitor correct knee height when stepping, books can be stacked on an adjacent table, or a ruler can be attached to a chair or wall with masking tape to mark proper knee height.

On the go signal, the participant begins stepping (not running) in place, completing as many steps as possible within 2 minutes (figure 9.21). The tester counts the number of steps completed, serves as a spotter in case of loss of balance, and ensures that the participant maintains proper knee height. As soon as proper knee height cannot be maintained, the participant is asked to stop or to stop and rest until proper form can be regained. Stepping may be resumed if the 2-minute period has not elapsed. If necessary, the participant can place one hand on a table or chair to assist in maintaining balance. To assist with pacing, participants should be told when 1 minute has passed and when there are 30 seconds to go. At the end of the test, each participant should slowly walk around for about a minute to cool down.

A practice test before the test day can assist with proper pacing and improve scoring accuracy. On test day, the examiner should demonstrate the procedure and allow the participants to practice briefly.

Scoring

The score is the total number of steps taken within 2 minutes. Count only full steps (i.e., each time the right knee reaches the minimum height).

Adapted, by permission, from R.E. Rikli and C.J. Jones, 1999, "Development and validation of a functional fitness test for community-residing older adults," *Journal of Aging and Physical Activity* 7(2): 129-161.

CHAIR SIT-AND-REACH TEST

Purpose

To assess lower-body (primarily hamstring) flexibility.

Objective

To sit in a chair and attempt to touch the toes with the fingers.

Equipment

Straight-back or folding chair (seat height approximately 17 inches [43.2 cm])

Instructions

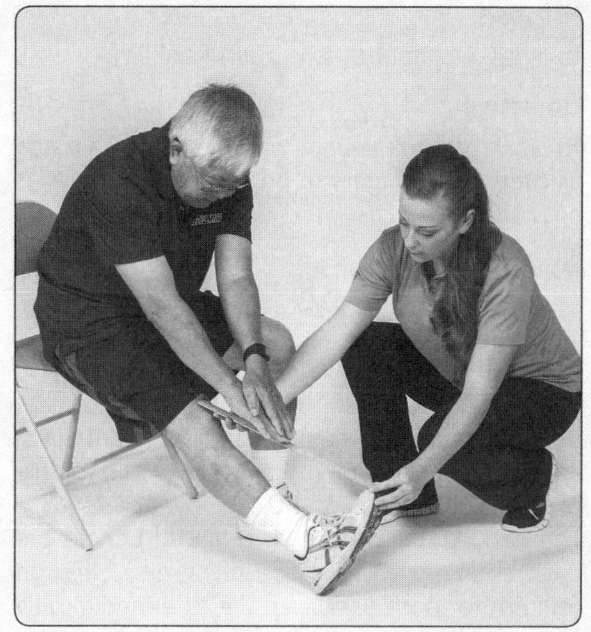

Figure 9.22 Chair sit-and-reach test.

For safety, a chair is placed against a wall and checked to see that it remains stable when a person sits on the front edge. The participant sits in the chair and moves forward until sitting on the front edge. The crease between the top of the leg and the buttocks should be even with the edge of the chair seat. Keeping one leg bent and the foot flat on the floor, the participant extends the other leg (the preferred leg) straight in front of the hip, with the heel flat on the floor and the foot flexed at approximately 90° (figure 9.22).

With the leg as straight as possible (but not hyperextended), the participant slowly bends forward at the hip joint, keeping the spine as straight as possible and the head in line with the spine, not tucked. The participant attempts to touch the toes by sliding the hands, one on top of the other with the tips of the middle fingers even, down the extended leg. The reach must be held for 2 seconds. If the extended knee starts to bend before scoring, the participant is asked to slowly sit back until the knee is straight. Participants are reminded to exhale as they bend forward, to avoid bouncing or rapid, forceful movements, and to never stretch to the point of pain.

After the tester provides a demonstration, participants are asked to determine the preferred leg—the leg that yields the better score. Then each participant is given two practice (stretching) trials on that leg, followed by two test trials.

Scoring

Using an 18-inch (45.7 cm) ruler, the tester records the number of inches a person is short of reaching the toe (minus score) or reaches beyond the toe (plus score). The middle of the toe at the end of the shoe represents a zero score. Both test scores are recorded to the nearest 0.5 inch (1.3 cm) and circle the best score. Minus or plus are also indicated on the scorecard.

BACK SCRATCH

Purpose

To assess upper-body (shoulder) flexibility.

Objective

To reach behind the back with the hands to touch or overlap the fingers of both hands as much as possible.

Equipment

18-inch (45.7 cm) ruler (half of a yardstick)

Instructions

In a standing position, the participant places the preferred hand over the same shoulder, palm down and fingers extended, reaching down the middle of the back as far as possible (elbow pointed up). The hand of the other arm is placed behind the back, palm up, reaching up as far as possible in an attempt to touch or overlap the extended middle fingers of both hands.

Without moving the participant's hands, the tester helps to see that the middle fingers of each hand are directed toward each other (figure 9.23). Participants are not allowed to grab their fingers together and pull.

After a demonstration, the participant determines the preferred hand and is given two stretching trials followed by two test trials.

Figure 9.23 Back scratch test.

Scoring

The tester measures the distance of overlap, or distance between the tips of the middle fingers to the nearest 0.5 inch (1.3 cm). Minus scores represent the distance short of touching middle fingers; plus scores represent the degree of overlap of middle fingers. The tester records both test scores and circles the best score. The best score is used to evaluate performance. Minus or plus is also indicated on the scorecard. (It is important to work on flexibility on both sides of the body, but only the better side has been used in developing norms.)

Adapted, by permission, from R.E. Rikli and C.J. Jones, 1999, "Development and validation of a functional fitness test for community-residing older adults," *Journal of Aging and Physical Activity* 7(2): 129-161.

8-FOOT (2.4 M) UP-AND-GO

Purpose

To assess physical mobility (involves power, speed, agility, and dynamic balance).

Objective

To stand, walk 16 feet (4.9 m), and sit back down in the fastest possible time.

Equipment

Stopwatch

Tape measure

Cone (or similar marker)

Straight-back or folding chair (seat height approximately 17 inches [43.2 cm])

Instructions

The chair is positioned against a wall or in some other way to secure it during testing. The chair needs to be in a clear unobstructed area, facing a cone marker exactly 8 feet (2.4 m) away (measured from a point on the floor even with the front edge of the chair to the back of the marker). There should be at least 4 feet (1.2 m) of clearance beyond the cone to allow ample turning room for the participant.

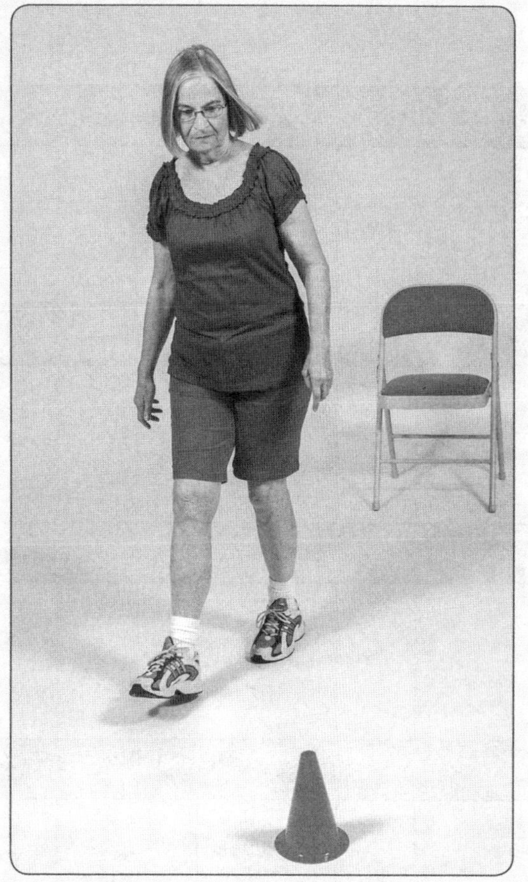

Figure 9.24 8-foot (2.4 m) up-and-go test.

The tester reminds the participant that this is a timed test and that the objective is to walk as quickly as possible (without running) around the cone and back to the chair. The participant starts in a seated position in the chair with erect posture, hands on thighs, and feet flat on the floor with one foot slightly in front of the other. On the go signal, the participant gets up from the chair (may push off thighs or chair), walks as quickly as possible around the cone, and returns to the chair and sits down (figure 9.24). The tester should serve as a spotter, standing midway between the chair and the cone, ready to assist the participant in case of loss of balance. For reliable scoring, the tester must start the stopwatch on go, whether or not the person has started to move, and stop the stopwatch at the exact instant the person sits in the chair.

After a demonstration, the participant should walk through the test one time to practice and then perform two test trials. Participants are reminded that the stopwatch will not be stopped until a participant is fully seated in the chair.

Scoring

The score is the time elapsed from the go signal until the participant returns to a seated position on the chair. Both test scores are recorded to the nearest tenth of a second. The best score (lowest time), which will be the score used to evaluate performance, is circled.

Adapted, by permission, from R.E. Rikli and C.J. Jones, 1999, "Development and validation of a functional fitness test for community-residing older adults," *Journal of Aging and Physical Activity* 7(2): 129-161.

Table 9.26 Age-Group Percentile—Female

	Chair stand		60-64 (*n* = 595)	65-69 (*n* = 1027)	70-74 (*n* = 1240)	75-79 (*n* = 937)	80-84 (*n* = 502)	85-89 (*n* = 305)	90-94 (*n* = 141)
percentile	10th	no.	9	9	8	7	6	5	2
	25th		12	11	10	10	9	8	4
	50th		15	14	13	12	11	10	8
	75th		17	16	15	15	14	13	11
	90th		20	18	18	17	16	15	14
	Arm curl		**60-64 (*n* = 598)**	**65-69 (*n* = 1034)**	**70-74 (*n* = 1258)**	**75-79 (*n* = 953)**	**80-84 (*n* = 519)**	**85-89 (*n* = 329)**	**90-94 (*n* = 146)**
percentile	10th	no.	10	10	9	8	8	7	6
	25th		13	12	12	11	10	10	8
	50th		16	15	15	14	13	12	11
	75th		19	18	17	17	16	15	13
	90th		22	21	20	20	18	17	16
	6-min walk		**60-64 (*n* = 356)**	**65-69 (*n* = 617)**	**70-74 (*n*= 728)**	**75-79 (*n* = 513)**	**80-84 (*n* = 276)**	**85-89 (*n* = 152)**	**90-94 (*n* = 79)**
percentile	10th	yd	495	440	420	365	310	260	195
	25th		545	500	480	430	385	340	275
	50th		605	570	550	510	460	425	350
	75th		660	635	615	585	540	510	440
	90th		710	695	675	655	610	595	520
	2-min step		**60-64 (*n* = 264)**	**65-69 (*n* = 491)**	**70-74 (*n* = 597)**	**75-79 (*n* = 489)**	**80-84 (*n* = 279)**	**85-89 (*n* = 167)**	**90-94 (*n* = 61)**
percentile	10th	no.	60	57	53	52	46	42	31
	25th		75	73	68	68	60	55	44
	50th		91	90	84	84	75	70	58
	75th		107	107	101	100	91	85	72
	90th		122	123	116	115	104	98	85
	Chair sit-and-reach		**60-64 (*n* = 591)**	**65-69 (*n* = 1037)**	**70-74 (*n* = 1250)**	**75-79 (*n* = 954)**	**80-84 (*n* = 514)**	**85-89 (*n* = 332)**	**90-94 (*n* = 151)**
percentile	10th	in.	–3.0	–3.0	–3.5	–4.0	–4.5	–4.5	–7.0
	25th		–0.5	–0.5	–1.0	–1.5	–2.0	–2.5	–4.5
	50th		2.0	2.0	1.5	1.0	0.5	–0.5	–2.0
	75th		5.0	4.5	4.0	3.5	3.0	2.5	1.0
	90th		7.0	6.5	6.0	5.5	5.0	4.5	3.5
	Back scratch		**60-64 (*n* = 592)**	**65-69 (*n* = 1030)**	**70-74 (*n* = 1246)**	**75-79 (*n* = 946)**	**80-84 (*n* = 517)**	**85-89 (*n* = 323)**	**90-94 (*n* = 148)**
percentile	10th	in.	–5.5	–6.0	–6.5	–7.5	–8.0	–10.0	–11.5
	25th		–3.0	–3.5	–4.0	–5.0	–5.5	–7.0	–8.0
	50th		–0.5	–1.0	–1.5	–2.0	–2.5	–4.0	–4.5
	75th		1.5	1.5	1.0	0.5	0.0	–1.0	–1.0
	90th		4.0	3.5	3.0	3.0	2.5	2.0	2.0
	8-ft (2.4 m) up-and-go		**60-64 (*n* = 594)**	**65-69 (*n* = 1033)**	**70-74 (*n* = 1244)**	**75-79 (*n* = 938)**	**80-84 (*n* = 497)**	**85-89 (*n* = 306)**	**90-94 (*n* = 142)**
percentile	10th	s	6.7	7.1	8.0	8.3	10.0	11.1	13.5
	25th		6.0	6.4	7.1	7.4	8.7	9.6	11.5
	50th		5.2	5.6	6.0	6.3	7.2	7.9	9.4
	75th		4.4	4.8	4.9	5.2	5.7	6.2	7.3
	90th		3.7	4.1	4.0	4.3	4.4	5.1	5.3
	Body mass index		**60-64 (*n* = 572)**	**65-69 (*n* = 1016)**	**70-74 (*n* = 1213)**	**75-79 (*n* = 916)**	**80-84 (*n* = 504)**	**85-89 (*n* = 337)**	**90-94 (*n* = 149)**
percentile	10th	kg/m2	19.6	19.8	20.3	19.8	19.6	19.5	18.3
	25th		22.8	23.0	23.1	22.5	22.0	21.8	21.1
	50th		26.3	26.5	26.1	25.4	24.7	24.3	24.1
	75th		29.8	30.0	29.1	28.3	27.4	26.8	27.1
	90th		33.0	33.2	31.9	31.0	30.0	29.0	29.5

Note: For metric measurements, refer to a site such as this: http://www.worldwidemetric.com/measurements.html.

Adapted, by permission, from R.E. Rikli and C.J. Jones, 1999a, "Functional fitness normative scores for community-residing older adults, ages 60-94," *Journal of Aging and Physical Activity* 7(2): 162-181.

Table 9.27 Age-Group Percentile—Male

	Chair stand		60-64 (*n* = 230)	65-69 (*n* = 460)	70-74 (*n* = 498)	75-79 (*n* = 434)	80-84 (*n* = 226)	85-89 (*n* = 108)	90-94 (*n* = 71)
percentile	10th		11	9	9	8	7	6	5
	25th		14	12	12	11	10	8	7
	50th	no.	16	15	15	14	12	11	10
	75th		19	18	17	17	15	14	12
	90th		22	21	20	19	18	17	15
	Arm curl		**60-64 (*n* = 229)**	**65-69 (*n* = 458)**	**70-74 (*n* = 498)**	**75-79 (*n* = 440)**	**80-84 (*n* = 232)**	**85-89 (*n* = 113)**	**90-94 (*n* = 71)**
percentile	10th		13	12	11	10	10	8	7
	25th		16	15	14	13	13	11	10
	50th	no.	19	18	17	16	16	14	12
	75th		22	21	21	19	19	17	14
	90th		25	25	24	22	21	19	17
	6 min walk		**60-64 (*n* = 144)**	**65-69 (*n* = 281)**	**70-74 (*n*= 294)**	**75-79 (*n* = 230)**	**80-84 (*n* = 130)**	**85-89 (*n* = 60)**	**90-94 (*n* = 48)**
percentile	10th		555	500	480	395	370	295	215
	25th		610	560	545	470	445	380	305
	50th	yd	675	630	610	555	525	475	405
	75th		735	700	680	640	605	570	500
	90th		790	765	745	715	680	660	590
	2 min step		**60-64 (*n* = 92)**	**65-69 (*n* = 211)**	**70-74 (*n* = 225)**	**75-79 (*n* = 226)**	**80-84 (*n* = 119)**	**85-89 (*n* = 50)**	**90-94 (*n* = 38)**
percentile	10th		74	72	66	56	56	44	36
	25th		87	88	80	73	71	59	52
	50th	no.	101	101	95	91	87	75	69
	75th		115	116	110	109	103	91	86
	90th		128	130	125	125	118	106	102
	Chair sit-and-reach		**60-64 (*n* = 228)**	**65-69 (*n* = 461)**	**70-74 (*n* = 494)**	**75-79 (*n* = 434)**	**80-84 (*n* = 231)**	**85-89 (*n* = 113)**	**90-94 (*n* = 74)**
percentile	10th		−6.0	−6.0	−6.5	−7.0	−6.0	−8.0	−9.0
	25th		−2.5	−3.0	−3.5	−4.0	−5.5	−5.5	−6.5
	50th	in.	0.5	0.0	−0.5	−1.0	−2.0	−2.5	−3.5
	75th		4.0	3.0	2.5	2.0	1.5	0.5	0.5
	90th		6.5	6.0	5.5	5.0	4.5	3.0	2.0
	Back scratch		**60-64 (*n* = 228)**	**65-69 (*n* = 457)**	**70-74 (*n* = 489)**	**75-79 (*n* = 430)**	**80-84 (*n* = 226)**	**85-89 (*n* = 113)**	**90-94 (*n* = 73)**
percentile	10th		−10.0	−10.5	−11.0	−12.0	−12.5	−12.5	−13.5
	25th		−6.5	−7.5	−8.0	−9.0	−9.5	−10.0	−10.5
	50th	in.	−3.5	−4.0	−4.5	−5.5	−5.5	−6.0	−7.0
	75th		0.0	−1.0	−1.0	−2.0	−2.0	−3.0	−4.0
	90th		2.5	2.0	2.0	1.0	1.0	0.0	−1.0
	8-ft (2.4 m) up-and-go		**60-64 (*n* = 229)**	**65-69 (*n* = 461)**	**70-74 (*n* = 492)**	**75-79 (*n* = 436)**	**80-84 (*n* = 227)**	**85-89 (*n* = 106)**	**90-94 (*n* = 72)**
percentile	10th		6.4	6.5	6.8	8.3	8.7	10.5	11.8
	25th		5.6	5.7	6.0	7.2	7.6	8.9	10.0
	50th	s	4.7	5.1	5.3	5.9	6.4	7.2	8.1
	75th		3.8	4.3	4.2	4.6	5.2	5.3	6.2
	90th		3.0	3.8	3.6	3.5	4.1	3.9	4.4
	Body mass index		**60-64 (*n* = 228)**	**65-69 (*n* = 460)**	**70-74 (*n* = 491)**	**75-79 (*n* = 429)**	**80-84 (*n* = 230)**	**85-89 (*n* = 114)**	**90-94 (*n* = 69)**
percentile	10th		22.0	22.1	21.6	21.4	21.7	21.8	20.2
	25th		24.6	24.7	24.0	23.8	23.8	23.3	22.4
	50th	kg/m2	27.4	27.5	26.6	26.4	26.1	24.9	24.9
	75th		30.2	30.3	29.2	29.0	28.4	26.5	27.4
	90th		32.8	32.9	31.6	31.4	30.5	28.0	29.6

Note: for metric measurements, refer to a site such as this: http://www.worldwidemetric.com/measurements.html.

Adapted, by permission, from R.E. Rikli and C.J. Jones, 1999a, "Functional fitness normative scores for community-residing older adults, ages 60-94," *Journal of Aging and Physical Activity* 7(2): 162-181.

Table 9.28 Criterion-Referenced Fitness Standards for Maintaining Physical Independence in Older Adults

	AGE GROUPS							% of decline reflected over 30 years
	60-64	65-69	70-74	75-79	80-84	85-89	90-94	
Lower-body strength (number of chair stands in 30 s)								
Women	15	15	14	13	12	11	9	40.0
Men	17	16	15	14	13	11	9	47.1
Upper-body strength (number of arm curls in 30 s)								
Women	17	17	16	15	14	13	11	35.3
Men	19	18	17	16	15	13	11	42.1
Aerobic endurance (yards [m] walked in 6 min)								
Women	625	605	580	550	510	460	400	36.0
Men	680	650	620	580	530	470	400	41.2
Alternate aerobic endurance (number of steps in 2 min)								
Women	97	93	89	84	78	70	60	38.1
Men	106	101	95	88	80	71	60	43.4
Agility/dynamic balance (8-ft [2.4 m] up-and-go, s)								
Women	5.0	5.3	5.6	6.0	6.5	7.1	8.0	37.5
Men	4.8	5.1	5.5	5.9	6.4	7.1	8.0	40.0

Note: The proposed fitness standards were developed for use with the Senior Fitness Test (SFT) battery (Rikli & Jones, 2013a, 2013b). The standards are based on actual SFT scores obtained by moderate-functioning older adults in previously published cross-sectional database (Rikli & Jones, 1991b) with scores adjusted as appropriate to reflect other relevant information in the literature, including an increased rate of decline over the years when performance is tracked longitudinally versus cross-sectionally.

Importantly, the concepts that have been presented throughout your text are important to assessing physical fitness and physical activity in older adults. In fact, Vingren et al. (2014) indicate that the fundamental concepts important to assessing physical fitness and physical activity in older adults are reliability, objectivity, relevance, validity, generalizability, error, norm-referenced, criterion-referenced, and formative and summative evaluation. Many of these terms are central to measurement and evaluation of human performance and used repeatedly in your text.

SPECIAL POPULATIONS

Special populations include people with physical or mental disabilities or both. Lack of knowledge and understanding has clouded the history of society's care and attention to people with disabilities. Laws and changes in attitudes have provided a more enlightened approach to interactions with people with disabilities. However, reliable and valid fitness assessment of adults with disabilities is not a well-researched or well-understood topic compared with work with other adults. Shephard (1990), in his book *Fitness in Special Populations*, provided a comprehensive and detailed presentation of exercise and fitness issues for people with disabilities. He stated that the assessment of fitness should include the following:

- Anaerobic capacity and power
- Aerobic capacity
- Electrocardiographic response to exercise
- Muscular fitness, including strength, endurance, and flexibility
- Body composition

Appropriate fitness assessment requires accurate classification of the person's disability and proper test and protocol selection. For example, if you wish to test the aerobic capacity of a person with paraplegia, you can select a wheelchair ergometer test or an arm ergometer test. Test protocols for those who are deaf would not be applicable to those who are blind. Specific training and education are required for fitness measurement in special populations. Shephard (1990) provided fitness information for people who are disabled and in wheelchairs (e.g., those with paraplegia and amputations), blind, or deaf, or those with mental retardation or who have autism, cerebral palsy, muscular dystrophy, or multiple sclerosis.

Meeting the fitness needs of persons with disabilities is a challenge for the future and an area of professional opportunity. From a measurement and evaluation perspective, research on reliable and valid fitness tests in populations with specific disabilities is needed. Research is needed to provide a better understanding of fitness levels and the necessary level of fitness for improved health in groups with disabilities.

MEASURING PHYSICAL ACTIVITY

The measurement and evaluation of physical fitness have long been an area of research and practice in exercise science. Given the epidemiological evidence supporting the role of physical activity—the behavior of bodily movement that requires the contraction of muscles and the expenditure of energy—in health promotion and chronic disease prevention, the reliable and valid measurement of physical activity is essential in determining

- the amount of physical activity people get,
- the amount of sedentary behavior,
- the role of physical activity in health status,
- factors that relate to physical activity behavior, and
- the effect of interventions to promote physical activity.

Of course, physical fitness and physical activity are related in that genetic factors and physical activity behaviors are what determine one's physical fitness. Education and health promotion interventions can be developed to directly improve the quality and quantity of physical activity and as a result improve physical fitness. The Surgeon General's Report on Physical Activity and Health concluded that many Americans do not get enough physical activity to promote health and lower the risk of a variety of chronic diseases (USDHHS 1996). Is it the attribute of physical fitness or the behavior of physical activity that is health promoting? This question may never be answered. However, we must be able to measure both physical fitness and physical activity behaviors in reliable and valid manners.

The physical activity pyramid shown in figure 9.25 presents guidelines for physical activity. The basic concept is for people to develop physically active lifestyles that will produce sufficient levels of health-related fitness and will promote physical and mental health and limit sedentary behaviors. The ACSM suggests that a person must expend 150 kilocalories (kcal) per day (or 1000 kcal per week) to achieve a health benefit from physical activity. In *2008 Physical Activity Guidelines for Americans* (USDHHS 2008) the following is a key

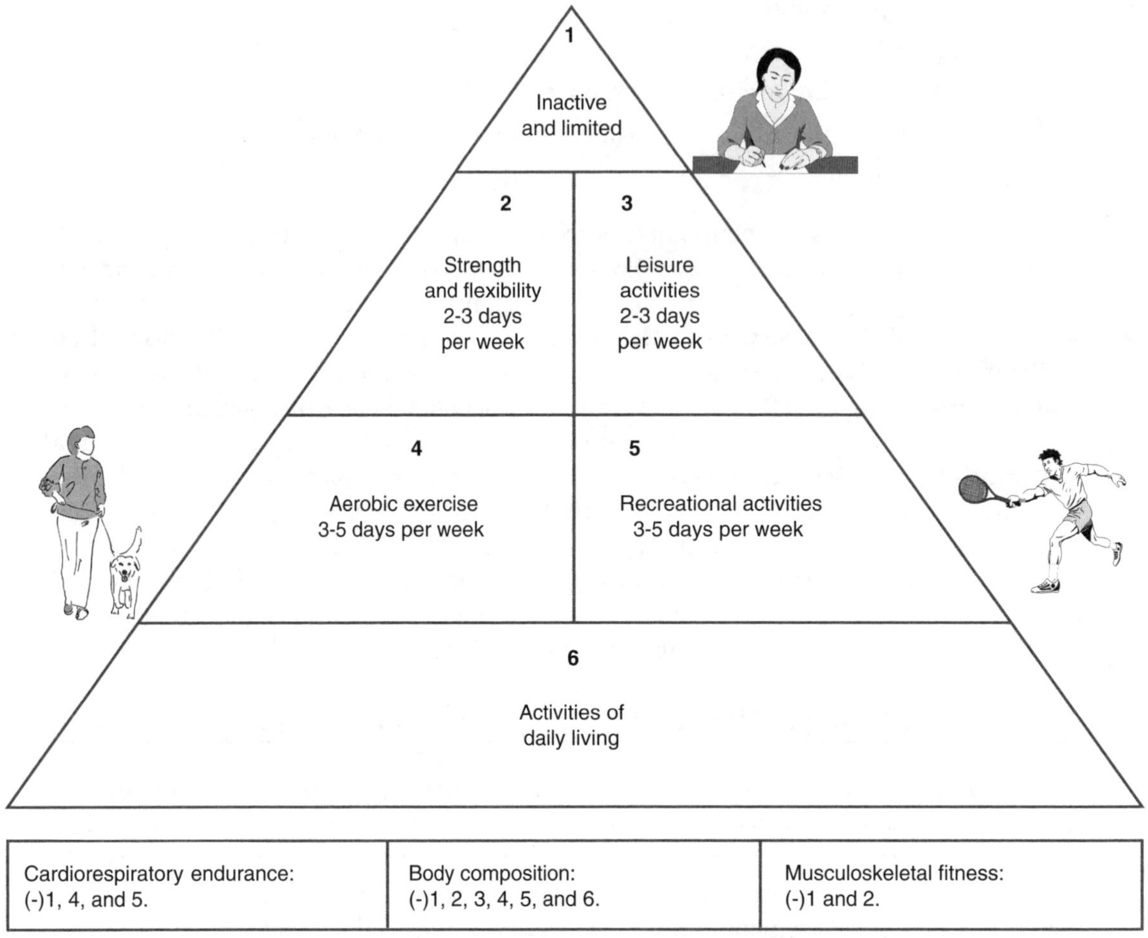

Figure 9.25 The relationship of health-related fitness to the physical activity pyramid.

guideline: For substantial health benefits, adults should do at least 150 minutes (2 hours and 30 minutes) a week of moderate-intensity, or 75 minutes (1 hour and 15 minutes) a week of vigorous-intensity aerobic physical activity, or an equivalent combination of moderate- and vigorous-intensity aerobic activity. Aerobic activity should be performed in episodes of at least 10 minutes, and preferably, it should be spread throughout the week. A variety of strategies are available for achieving a lifestyle that includes healthful amounts of physical activity. From a health standpoint, the most important message of the pyramid may be to limit the time a person is inactive because, as the figure indicates, physical inactivity is negatively related to all components of health-related physical fitness.

Physical activity, like other health behaviors, is a difficult behavior to assess with high reliability and validity. It is typically measured through either self-reports (e.g., diaries, logs, recall surveys, retrospective quantitative histories, global self-reports) or direct monitoring (e.g., behavioral observation by a trained observer, electronic monitoring of heart rate or body motion, physiological monitoring using direct calorimetry in a metabolic chamber, indirect calorimetry using a portable gas analysis system; doubly labeled water; USDHHS 1996). Only limited success and research have occurred in establishing the reliability and validity of physical activity measurements, whereas, the surgeon general's report states, measures of physical fitness have good to excellent accuracy and reliability (USDHHS 1996). See Montoye and colleagues (1996) for a review of physical activity assessment tools.

One of the principal difficulties in establishing the validity of a physical activity measure is the lack of a suitable 'gold-standard' criterion measure for comparison (USDHHS 1996, p. 35). Because of this problem, many validation studies have used the physical fitness attribute of cardiorespiratory endurance as the criterion for concurrent validity estimation. The median correlation or concurrent validity coefficient between measures of cardiorespiratory fitness and self-report survey measures of physical activity was .41 in 12 research studies that used two popular physical activity survey instruments (USDHHS 1996, p. 35). Although the magnitude of this validity coefficient may seem low, remember that genetics accounts for at least 30% of cardiorespiratory endurance and that physical activity is still the most important predictor of cardiorespiratory endurance (Blair, Kannel, Kohl, and Goodyear 1989; Perusse, Tremblay, Leblanc, and Bouchard 1989). Not all measures of physical activity are appropriate for all populations. Table 9.29 provides a list of self-report and direct-monitoring assessment procedures and the age groups for which they are suited. One needs to be cautious when using self-report measures of physical activity. Troiano et al. (2008) showed that approximately 50% of adults self-report meeting national physical activity guidelines. However, when objective measures are obtained with accelerometry, less than 5% actually achieve the minimum moderate-to-vigorous physical activity (MVPA) guidelines.

Reliability of measures of physical activity is obviously an important issue. In a comprehensive study, Jacobs, Ainsworth, Hartman, and Leon (1993) examined the reliability and validity of 10 commonly used physical activity questionnaires and estimated the reliability of self-report measures of total physical activity and subsets of physical activity (i.e., light, moderate, heavy, leisure, and work). Test–retest reliability estimates across a 12-month interval for all questionnaires and a 1-month interval for two of the questionnaires were reported. As expected, the reliability estimates for the 12-month interval were extremely variable, ranging from a low of .12 to a high of .93. The estimates for the month interval were much higher, ranging from .63 to .95.

An accurate assessment of physical activity includes measuring not only frequency and duration of the activity but also the intensity and context of the activity. Accurate determination of energy expenditure requires measures of all three components. Ainsworth and colleagues (1993, 2000) developed the Compendium of Physical Activities to aid in the

Table 9.29 Physical Activity Assessment in Age Groups of the Population

Type	Instrument	Children	Adults	Older persons
Self-report survey	Task-specific diary	No	Yes	Yes
	Recall questionnaire	No	Yes	Yes
	Quantitative history	No	Yes	Yes
	Global self-report	No	Yes	Yes
Direct monitoring	Behavioral observation	No	Yes	Yes
	Job classification	No	Yes	No
	Heart rate monitor and motion sensor (ONE Tool)	Yes	Yes	Yes
	Heart rate monitor	Yes	Yes	Yes
	Electronic motion sensor	No	Yes	Yes
	Pedometer	No	Yes	Yes
	Gait assessment	Yes	Yes	Yes
	Accelerometers	Yes	Yes	Yes
	Horizontal time monitor	Yes	Yes	Yes
	Stabilometers	No	No	No
	Direct calorimetry	Yes	Yes	Yes
	Indirect calorimetry	No	Yes	Yes
	Doubly labeled water	Yes	Yes	Yes

Data based on Laporte, Montoye, and Caspersen 1985.

assessment of energy expenditure attributable to physical activity. The compendium is a coding scheme for classifying a wide variety of physical activities into energy expenditure levels. The coding scheme uses five digits to classify physical activity by purpose, specific type, and intensity. The compendium standardizes measurements of physical activity and allows a better comparison and evaluation of research findings.

A simple self-report method of assessing physical activity has demonstrated excellent reliability and construct validity, comparable to those of more complicated assessment techniques (Jackson et al. 2007). The method uses a single-response scale to assess physical activity behaviors. Figure 9.26 depicts the five-level version of this single-response approach. The scale is based on the stages of change in physical activity behavior. Test–retest reliability estimates have exceeded .90 for the scale, and the correlation (r = .57) between the scale responses and treadmill assessments of aerobic capacity was consistent with correlations for other self-report techniques.

There is great interest in using motion sensors (e.g., pedometers, accelerometers) to measure physical activity. Pedometers are inexpensive, unobtrusive devices that are worn on the belt to measure steps taken. A criticism of pedometer measurement of physical activity is that pedometers cannot measure intensity of the steps taken. Thus, a slowly moving step counts the same as a running step. Accelerometers are devices that can count steps and more accurately differentiate between slow, moderate, and quickly moving physical activity. Like the accelerometer on your car, they differentiate between speeds of movement. Accelerometers can provide time of moderate or vigorous physical activity and produce estimates of caloric expenditure. Unfortunately, accelerometers are rather expensive and not generally used in nonresearch settings.

Hybrid pedometers, which combine some capacities of traditional pedometers and accelerometers, are available that can be programmed to differentiate between walking and moderate or vigorous steps by adjusting a steps per minute function. Thus, steps taken above a user-defined count of steps per minute are recorded as MVPA. These crosses between pedometers and accelerometers are relatively affordable and record steps, minutes of physical activity, and minutes of MVPA. Tudor-Locke and Bassett (2004) suggest the following pedometer steps per day as indices of public health for adults:

<5000	Sedentary lifestyle index
5000-7499	Low active
7500-9999	Somewhat active
≥10,000	Active
≥12,000	Very active

I don't exercise or walk regularly now, and I do not plan to start in the near future.
I don't exercise or walk regularly now, but I have been thinking of starting.
I am doing moderate physical activities fewer than 5 times a week or vigorous ones fewer than 3 times a week.
I have been doing moderate physical activities 5 or more times a week or vigorous ones at least 3 times a week for the last 1 to 6 months.
I have been doing moderate physical activities 5 or more times a week or vigorous exercise at least 3 times a week for 7 months or longer.

Figure 9.26 The single-response physical activity scale (five levels).

Reliable and valid measurement of physical activity is a difficult but not impossible task. Selecting appropriate instruments, using standardized procedures, and performing pilot studies are necessary steps for achieving reliable and valid measurements. A simple construct-related validity concept holds true for assessments of physical activity: People classified as physically active with accepted measures in the epidemiological literature have lower rates of morbidity and mortality attributable to chronic disease than those who are inactive (USDHHS 1996). This consistent and persistent finding in the research literature supports the reachable goal of reliable and valid assessment of physical activity. Welk (2002) compiled detailed presentations of the variety of measurement techniques used in the assessment of physical activity. This is an excellent source for developing a comprehensive picture of this challenging measurement area.

Because of the increase in sedentary behaviors, many researchers are interested in the mechanisms for assessing discretionary (e.g., sitting, media use) and nondiscretionary (e.g., occupational, school) sedentary behaviors. See the Pettee-Gabriel et al. (2012) framework mentioned at the beginning of this chapter (figure 9.2) for how sedentary behaviors fit into the movement framework. Clark et al. (2013) investigated the reliability and validity of past-day recall of sedentary behavior in adults. They report intraclass reliability (see chapter 6 for a discussion of this concept) of .50 and concurrent validity of .57 with accelerometer-measured sit–lie time. Hales et al. (2013) reported similar reliabilities but lower validities using the Home Self-Administered Tool for Environmental Assessment of Activity and Diet (HomeSTEAD) when parents completed the form for children ages 3 to 12 on three separate occasions separated by 12 to 18 days.

Dataset Application

Use the chapter 9 large dataset in the WSG. Assume that the data represent the number of steps taken per day for each of 2 days and that you are interested in whether people take at least 7500 steps per day each day (a value that defines them as somewhat active as explained earlier). Calculate the descriptive statistics for each day (chapter 3). Do males and females differ in the number of steps they take (chapter 5)? Correlate the results for the 2 days (chapter 4). This is an interclass reliability illustration (chapter 6). Calculate the alpha coefficient for the 2 days (chapter 6). Variables have been created to indicate if people have met the criterion for each day. Use the procedures in chapter 7 to determine the percent agreement, phi coefficient, chi-square, and Kappa for meeting the criterion across these 2 days. Comment on the criterion-referenced reliability obtained.

CERTIFICATION PROGRAMS

The ACSM is a U.S. organization that leads in the research and promotion of all areas of exercise science. Any professional who is serious about a career in adult fitness programs that include reliable and valid fitness assessment should be an active member in this organization.

The ACSM offers a certification program that includes the following:

- Health Fitness Certifications
    - ACSM Certified Group Exercise Instructor
    - ACSM Certified Personal Trainer
    - ACSM Certified Exercise Physiologist

- Clinical Certifications
    - ACSM Certified Clinical Exercise Specialist
    - ACSM Registered Clinical Exercise Physiologist
- Specialty Certifications
    - Exercise Is Medicine Credential
    - ACSM/ACS Certified Cancer Exercise Trainer
    - ACSM/NCHPAD Certified Inclusive Fitness Trainer
    - ACSM/NPAS Physical Activity in Public Health Specialist

MEASURING MUSCLE-STRENGTHENING PHYSICAL ACTIVITY

Healthy People 2020 and the *2008 Physical Activity Guidelines for Americans* have objectives for muscle-strengthening activities. The health benefits of muscular strength and endurance are listed in Table 9.2. The specific objective for muscle-strengthening activities from the 2008 Guidelines (DHHS, 2008) is as follows:

> *Adults should also do muscle-strengthening activities that are moderate or high intensity and involve all major muscle groups on 2 or more days a week, as these activities provide additional health benefits.*

Typically, muscle-strengthening activities are assessed by self-report questions. An example is the question asked by the Centers for Disease Control and Prevention's Behavioral Risk Factor Surveillance System telephone survey:

> *"During the past month, how many times per week or per month did you do physical activities or exercises to STRENGTHEN your muscles? Do NOT count aerobic activities like walking, running, or bicycling. Count activities using your own body weight like yoga, sit-ups, or push-ups and those using weight machines, free weights, or elastic bands*
>
> *Times per week* _________
>
> *Times per month* _________
>
> *Never* _________
>
> *Don't know / Not sure*
>
> *Refused* _________"

These self-report questions provide little information about the respondent's actual muscle-strengthening behaviors. What type of activities? What and how many muscle groups are involved? How many sets and repetitions are performed? Researchers at the University of North Texas and the Cooper Institute developed Tracking Resistance Exercise and Strength Training (TREST), an online system with questions and visual displays to gather more detailed and specific information about people's participation in muscle-strengthening activities. Initial studies have demonstrated the system to be reliable and valid.

 Go to the WSG to view Videos 9.10 and 9.11.

The National Strength and Conditioning Association (NSCA) offers recognized certifications programs, including these:

- Certified Strength and Conditioning Specialist (CSCS)
- Certified Special Population Specialist (CSPS)
- NSCA-Certified Personal Trainer (NSCA-CPT)
- Tactical Strength and Conditioning-Facilitators (TSAC-F)

The Cooper Institute (Dallas, Texas) offers on-line certification for personal trainers (CI-CPT).

By directing your education and preparation toward achieving these certifications, you will have verified, from a measurement and evaluation perspective, your ability to conduct reliable and valid fitness assessments. Howley and Franks (2007) and the ACSM (2010) provide excellent and comprehensive resources for the professional in adult fitness and evaluation programs.

MEASUREMENT AND EVALUATION CHALLENGE

Jim had two major tasks before meeting again with the YMCA executive director: to outline a suggested adult fitness assessment program and to investigate the certification he would seek. One of these tasks is easily solved, given the resources the Y already has available: He can suggest implementing the Y's fitness testing program. The other, deciding the certification he should seek, is also fairly easily accomplished. Because he will be working with apparently healthy people who are not athletes, he probably should seek the ACSM's Certified Health Fitness Specialist certification. The fitness staff he may employ should have or seek certifications such as those for the ACSM's, the NSCA's, or the Cooper Institute's certified personal trainer. Additionally, in his research Jim has learned that the specific test he might administer depends on the purposes of the test and the age and gender of the participants. He is on his way to having an effective fitness promotion and evaluation program staffed by qualified personnel.

SUMMARY

The material presented in this chapter should provide you with a sound basis for understanding the various factors involved in assessing adult fitness and physical activity reliably and validly. Mastering the material does not make you qualified to administer fitness tests in all adult situations but is an important step to achieving those qualifications. If adult fitness programs are to be an important part of your professional career, seek appropriate education and training to obtain ACSM or other professional certifications. Adult fitness testing, as with any other testing situation, requires appropriate test selection, preparation, practice trials, and attention to detail for sound measurement and evaluation of performance.

 Go to the WSG for homework assignments and quizzes that will help you master this chapter's content.

CHAPTER 10

Physical Fitness and Activity Assessment in Youth

Allen W. Jackson, University of North Texas

OUTLINE

OBJECTIVES

After studying this chapter, you will be able to

- discuss the status of youth fitness in the United States;
- differentiate athletic, or motor fitness, and health-related fitness in youth;
- describe similarities and differences in youth fitness batteries;
- administer fitness tests for youth in a reliable and valid manner;
- address issues in physical fitness assessment in special populations; and
- assess physical activity in youth.

 The lecture outline in the WSG will help you identify the major concepts of the chapter.

MEASUREMENT AND EVALUATION CHALLENGE

Jo is an elementary school physical education teacher. For years, her school district has been using a locally developed fitness test to recognize children with a physical fitness award. The district's test included 600-yard (548.6 m) walk–run, shuttle run, 50-yard (45.7 m) dash, standing long jump, and sit-ups. The new physical education coordinator for the district has instituted a policy that requires the use of the Fitnessgram across the district because it is a health-related fitness test and is the official test of the Presidential Youth Fitness Program established by the President's Council on Fitness, Sports & Nutrition. Jo is unfamiliar with the Fitnessgram—its philosophy, components, and administration. Additionally, she has heard that the Fitnessgram assesses physical activity levels as well as physical fitness. The achievement system for the Fitnessgram is reportedly different from the one with which she is familiar. Previously, student awards were based on percentile performance, but the Fitnessgram uses health-related criterion-referenced standards. Jo sees that she has much to do to prepare for the new school year.

The levels of physical activity and fitness of youth in the United States are a controversial issue in the field of human performance. Various political leaders, physical educators, and fitness experts proclaim that the nation's youth are dangerously inactive and physically unfit. Historically, awareness of and concern for poor physical fitness levels in youth can be tracked to World Wars I and II, when many young men were deemed not physically fit enough to serve in the armed services during times of national emergency. In the early 1950s, physical fitness testing indicated that European children had higher levels of physical fitness than American children. This led President Eisenhower to establish what has become the President's Council on Fitness, Sports & Nutrition (PCFSN). The council, along with the American Alliance of Health, Physical Education and Recreation (AAHPER), established a national youth fitness testing program, which included the Presidential Physical Fitness Award. This program also developed the AAHPER Youth Fitness Test. This test is an example of a **youth fitness battery**, which combines several fitness test items to provide an overall assessment of physical fitness. In terms of measurement and evaluation, there has been a strong shift toward an increased emphasis on the promotion and assessment of physical activity (the behavior) instead of on physical fitness assessment (the outcome). See Mood et al. (2007) for a summary of physical fitness and physical activity changes and Morrow and colleagues (2009) for a review of 50+ years of youth fitness testing.

Mastery Item 10.1

Were you given physical fitness tests during your youth? What tests did you take? What items did you complete? Did the test results provide a reliable and valid estimate of your physical fitness?

The National School Population Fitness Survey, conducted and published by the PCFSN (Reiff et al. 1985), concluded in part that

> *the physical performance of children and youth in 1985 was not much different from that of youth in 1975. Extrapolated to the entire population, the study data suggested there is*

still a low level of performance in important components of physical fitness by millions of our youth. (p. 45)

The same survey indicated that large numbers of boys and girls had a low level of performance on cardiorespiratory endurance tests. The Office of Disease Prevention and Health Promotion, an agency of the U.S. Public Health Service, conducted the National Children and Youth Fitness Study (NCYFS; Pate, Ross, Dotson, and Gilbert 1985) to evaluate the fitness levels of children. The skinfold measures in 1985 were significantly greater than skinfold measurements of children from the 1960s (Pate, Ross, Dotson, and Gilbert 1985). Thus, those children born later appeared to have higher amounts of body fat. This information indicated that many youth in the United States were unfit and that fitness levels might be decreasing. The state of Texas passed legislation in 2007 requiring the assessment of health-related physical fitness of all Texas children enrolled in public schools. The Fitnessgram test battery was selected, and six fitness tests were administered in the 2007-2008 academic year to approximately 2.6 million students. The percentage of students reportedly achieving a healthy level of fitness on all six tests ranged from a high of 32% in third grade to a low of 8% in twelfth grade. California and New York City also test hundreds of thousands of school-age children annually. See the September 2010 supplement of the *Research Quarterly for Exercise and Sport* for a summary of the Texas physical fitness testing on nearly 3 million children and youth. Also see Morrow and Ede (2009) for issues related to large-scale physical fitness testing in schools.

However, conflicting information indicates that youth fitness levels are not necessarily poor and may not be decreasing. Looney and Plowman (1990) conducted a study in which they applied the criterion-referenced standards of the Fitnessgram youth fitness battery to the test results of the NCYFS I (1985) and NCYFS II (1987). They found that 80% to 90% of the children could pass the body composition standards, depending on the age of the child. Boys and girls demonstrated high passing rates for the 1-mile (1.6 km) run test in some age groups (figure 10.1). Corbin and Pangrazi (1992) also concluded that the majority of American children and youth were able to meet criterion-referenced standards on individual Fitnessgram test items. Morrow (2005) provided commentary on the status of fitness levels in the United States. Powell et al. (2009) reported that substantial numbers

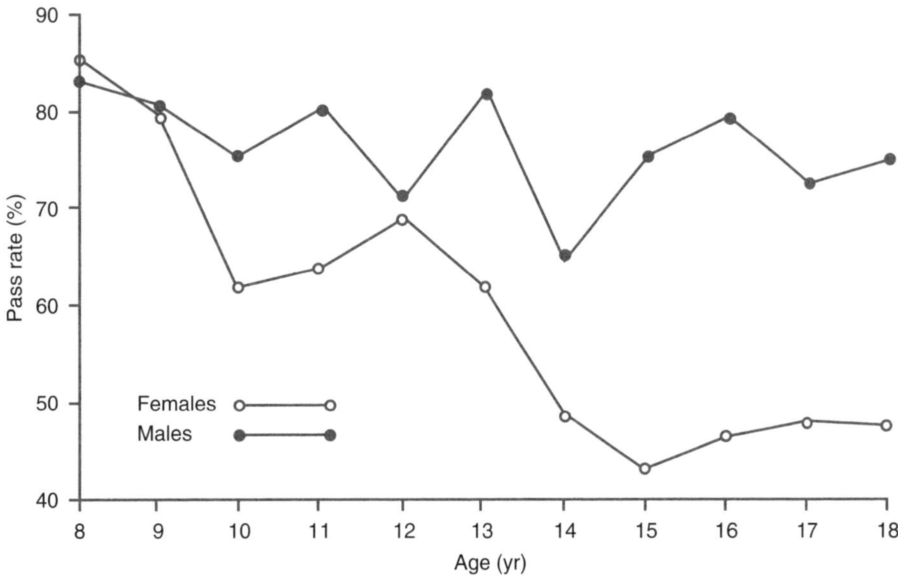

Figure 10.1 Passing rates for the Fitnessgram 1-mile (1.6 km) run.

of fifth- and seventh-grade students in Georgia exhibited unhealthy levels of physical fitness based on Fitnessgram testing.

Rowland (1990), a respected researcher in exercise science with children, summarized this issue by stating that there was conflicting information. The conclusion that youth fitness levels were low and declining is not clearly supported in the literature. In 1992 a forum on youth fitness was published in *Research Quarterly for Exercise and Sport* that included seven papers by recognized experts. These papers did not present a consensus on the levels of youth fitness and the trend of those fitness levels. Blair (1992) estimated that 20% of American youth, or between 8 and 9 million children, had potentially unhealthy fitness levels. The National Center for Health Statistics (NCHS) has presented an alarming trend in overweight children and adolescents in the United States (NCHS 2013). This trend indicates a growth in the percentage of overweight children and adolescents from 4% to 6% in the 1970s to more than 18% from 2007 to 2010 (NCHS 2013). In relation to physical activity, some of the data have been provided by the Centers for Disease Control and Prevention (CDC) through the Youth Risk Behavior Surveillance System (YRBSS). This surveillance system monitors health behaviors of children in grades 9 through 12 through school-based surveys conducted across the United States. Figure 10.2 shows that large percentages of students in 2011 were not meeting recommended physical activity levels, not attending daily physical education classes, and spending a lot of time in sedentary behaviors (http://www.cdc.gov/HealthyYouth/yrbs/index.htm). In the *2008 Physical Activity Guidelines for Americans,* key guidelines for children and adolescents were as follows:

- Children and adolescents should do 60 minutes or more of physical activity daily.
  - *Aerobic:* Most of the 60 or more minutes a day should be either moderate- or vigorous-intensity aerobic physical activity, and should include vigorous-intensity physical activity at least 3 days a week.
  - *Muscle strengthening:* As part of their 60 or more minutes of daily physical activity, children and adolescents should include muscle-strengthening physical activity on at least 3 days of the week.
  - *Bone strengthening:* As part of their 60 or more minutes of daily physical activity, children and adolescents should include bone-strengthening physical activity on at least 3 days of the week.
- It is important to encourage young people to participate in physical activities that are appropriate for their age, that are enjoyable, and that offer variety (USDHHS, 2008).

As a professional in human performance, you should know that there are many children with low levels of fitness and poor physical activity levels. National and international leaders and public health professionals are calling for improved physical activity and fitness for those children who are inactive and unfit. *Healthy People 2020* has established goals for increases in physical activity and physical education experiences for U.S. children (see figure 10.3).

Mastery Item 10.2

How would you describe the status of physical fitness of American youth?

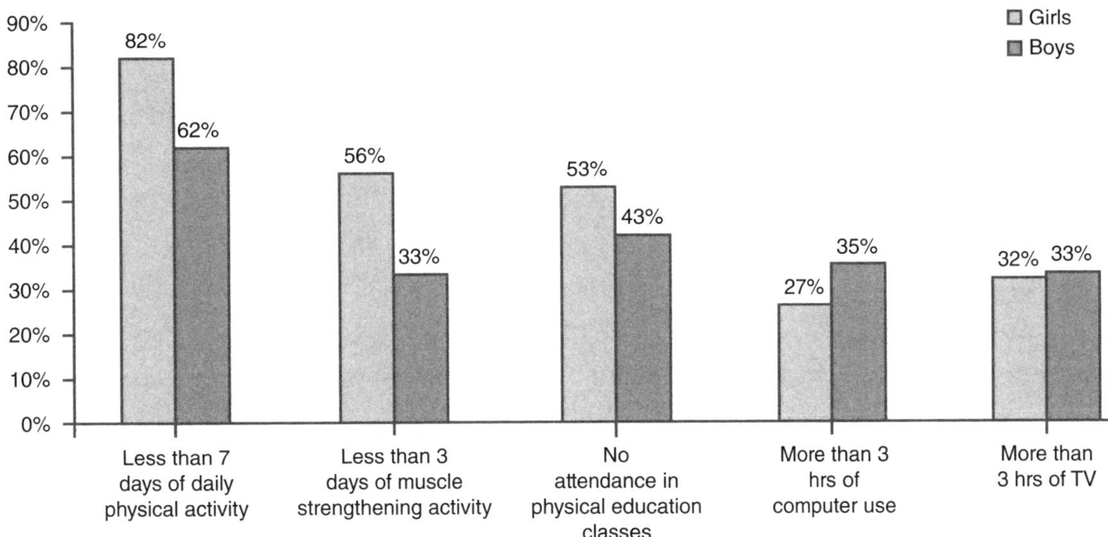

Figure 10.2 Youth Risk Factor Behavior Survey results 2011.

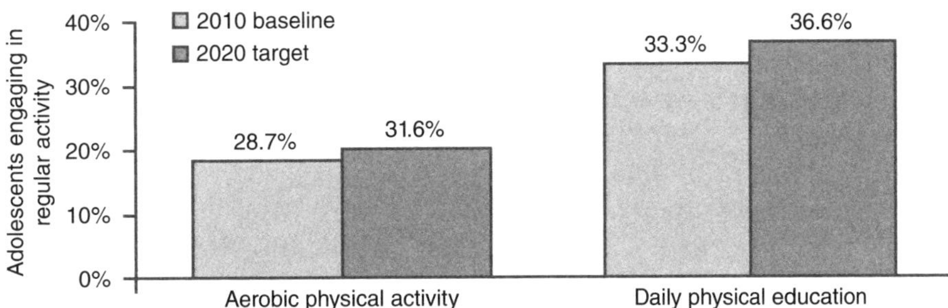

Figure 10.3 ***Healthy People 2020*** **physical activity objectives: baselines (2010) and targets (2020) for improvement in adolescents.**

Healthy People 2020. Available: www.healthypeople.gov/2020/topics-objectives/topic/physical-activity/objectives.

HEALTH-RELATED FITNESS AND MOTOR FITNESS

The original AAHPER Youth Fitness Test developed in the 1950s and revised in the 1960s and 1970s included not only tests of health-related fitness described in chapter 9 but also tests of **athletic fitness**, or **motor fitness**—physical fitness related to sport performance. The test included the following items:

- Pull-ups for boys and flexed arm hang for girls
- 1-minute sit-up test
- 600-yard (548.6 m) run
- Shuttle run
- Standing long jump
- 50-yd (45.7 m) dash

As you can see, only the first three items might be considered health-related based on the ACSM's (2014a) definition given in chapter 9. However, even the 600-yard (548.6 m) run is not a good measure of aerobic capacity. The other items relate to the ability to participate in sports and athletics.

In the 1970s another youth fitness test was developed for the American Alliance of Health, Physical Education, Recreation and Dance (AAHPERD). This test was called the AAHPERD Health-Related Physical Fitness Test (AAHPERD 1980) and included measures of the following components:

- Cardiovascular endurance (distance runs)
- Body composition (skinfolds)
- Musculoskeletal function

The test items in this battery are components consistent with the present definition of health-related fitness. In the future, youth fitness evaluation will focus on administering health-related fitness batteries. The principal programs in the United States are the Fitnessgram (Cooper Institute for Aerobics Research 1987, 1992, 1999, 2004) and the President's Challenge (PCPFS 1999). We will discuss each of these testing programs later in the chapter.

NORM-REFERENCED VERSUS CRITERION-REFERENCED STANDARDS

Another important change in youth fitness testing has been the change from **norm-referenced fitness standards**—levels of performance relative to a specifically defined group—to **criterion-referenced fitness standards**—specific predetermined levels of performance—for determining fitness achievement. The AAHPERD Health-Related Physical Fitness Test was published with percentile norms (AAHPERD 1980). Children who scored below the 50th percentile on a test or tests were encouraged to train to be able to achieve that level. The Fitnessgram was the first nationally recognized fitness test battery that used health-related criterion-referenced fitness standards (Cooper Institute for Aerobics Research 1987, 1992, 1999, 2007). In the past, the PCFSN (PCFSN 1999) used criterion-referenced standards for evaluation of youth fitness scores in the President's Challenge program. Safrit and Looney (1992) and others have taken the position that health-related criterion-referenced standards are useful and appropriate but that normative data on youth fitness scores are also useful for (a) evaluating a program, (b) identifying excellence in achievement, and (c) identifying the current status of people either locally or nationally.

NORMATIVE DATA

Results of the NCYFS I and II, mentioned earlier, were published in the *Journal of Physical Education, Recreation and Dance* in two segments (NCYFS I: Ross, Dotson, Gilbert, and Katz 1985; NCYFS II: Ross et al. 1987). Supported by the U.S. Public Health Service, the NCYFS strove to develop the best normative data on youth fitness. The children in the study represented a national probability sample in order to provide a sample representative of children in the United States. Previous databases had been considered questionable because they represented large convenience samples that may have biased estimates or had not fully represented the population of children in the United States. NCYFS I resulted in norms for youth aged 10 to 18; NCYFS II produced norms for children aged 6 to 9.

Pate and colleagues (2006) used test results from the National Health and Nutrition Examination Survey (NHANES) 1999-2002 to provide estimates of cardiorespiratory fitness levels of American youth aged 12 to 19 years. NHANES uses a mobile examination center that measures health-related variables throughout the United States. The protocol included a submaximal treadmill exercise test, which was used to provide estimates of $\dot{V}O_2$max. Boys' mean $\dot{V}O_2$max (46.4 ml · kg^{-1} · min^{-1}) was significantly higher than girls'

(38.7 ml · kg^{-1}· min^{-1}) but did not differ across racial or ethnic groups. For boys, the $\dot{V}O_2$max increased significantly ($p < .001$) from 44.6 (12-13 yr) to 47.6 (18-19 yr). For girls, the $\dot{V}O_2$max decreased significantly ($p < .02$) from 39.7 (12-13 yr) to 37.5 (18-19). Approximately two-thirds of males and females met recommended health-related fitness standards based on the Fitnessgram Healthy Fitness Zones.

Although youth fitness evaluation has moved away from the use of norm-referenced standards to criterion-referenced standards, we agree with Safrit and Looney (1992) that the availability of normative data has value in youth fitness assessment. Tables 10.1 through 10.5 provide you with percentile norms for the various tests and measurements gathered in NCYFS I and II. Taken together, these tables give you a good summary of the best normative data available on youth in the United States. Unfortunately, nationwide youth physical fitness testing has not been conducted in over a quarter of a century. See Morrow (2005) for commentary on nationwide youth fitness testing.

Table 10.1 Percentile Norms for Males and Females on Distance Runs (min:sec)

	AGE (YEARS)												
Male percentile	6	7	8	9	10	11	12	13	14	15	16	17	18
90	4:27	4:11	8:46	8:10	8:13	7:25	7:13	6:48	6:27	6:23	6:13	6:08	6:10
75	4:52	4:33	9:29	9:00	8:48	8:02	7:53	7:14	7:08	6:52	6:39	6:40	6:42
50	5:23	5:00	10:39	10:10	9:52	9:03	8:48	8:04	7:51	7:30	7:27	7:31	7:35
25	5:58	5:35	12:14	11:44	11:00	10:32	10:13	9:06	9:10	8:30	8:18	8:37	8:34
10	6:40	6:20	14:05	13:37	12:27	12:07	11:48	10:38	10:34	10:13	9:36	10:43	10:50
Female percentile													
90	4:46	4:32	9:39	9:08	9:09	8:45	8:34	8:27	8:11	8:23	8:28	8:20	8:22
75	5:13	4:54	10:23	9:50	10:09	9:56	9:52	9:30	9:16	9:28	9:25	9:26	9:31
50	5:44	5:25	11:32	11:13	11:14	11:15	10:58	10:52	10:32	10:46	10:34	10:34	10:51
25	6:14	6:01	12:59	12:45	12:52	12:54	12:33	12:17	11:49	12:18	12:10	12:03	12:14
10	6:51	6:38	14:48	14:31	14:20	14:35	14:07	13:45	13:13	14:07	13:42	13:46	15:18

Note: Participants aged 6 to 7 run 1/2 mile (0.8 km) ; others run 1 mile (1.6 km).

Data from Ross et al.1985; NCYFS II; Ross et al.1987.

Table 10.2 Percentile Norms for Males and Females for Sum of Skinfolds (mm)

	AGE (YEARS)												
Male percentile	6	7	8	9	10	11	12	13	14	15	16	17	18
90	12	12	12	12	12	12	12	11	12	12	12	13	13
75	14	14	14	15	14	14	14	13	13	14	14	14	15
50	16	17	18	21	17	18	17	17	17	17	17	17	18
25	20	22	24	29	24	25	24	23	22	22	22	22	24
10	27	32	37	40	35	36	38	34	33	32	30	30	30
Female percentile													
90	15	15	15	16	13	14	15	15	17	19	19	20	19
75	18	18	19	20	16	17	18	19	20	23	22	23	22
50	21	22	24	26	20	21	22	24	26	28	26	28	27
25	27	28	33	35	27	30	29	31	33	34	33	36	34
10	33	37	43	45	36	40	40	43	40	43	42	42	42

Note: For participants aged 6 to 9, values represent the sum of triceps and calf skinfolds; for other participants, values are the sum of triceps and subscapular skinfolds.

Data from Ross et al.1985; NCYFS II; Ross et al.1987.

Table 10.3 Percentile Norms for Males and Females on the 1-Minute Bent-Knee Sit-Up Test (number completed)

Male percentile	6	7	8	9	10	11	12	13	14	15	16	17	18
90	28	32	35	39	47	48	50	52	52	53	55	56	54
75	24	28	30	33	40	41	44	46	47	48	49	50	50
50	19	23	26	28	34	36	38	40	41	42	43	43	43
25	14	18	20	23	28	30	32	32	35	36	38	37	36
10	9	12	15	16	22	22	25	28	30	31	32	31	31
Female percentile													
90	28	33	34	36	43	42	46	46	47	45	49	47	47
75	23	27	29	31	37	37	40	40	41	40	40	40	40
50	18	21	25	26	31	32	33	33	35	35	35	36	35
25	14	16	19	21	25	26	28	27	29	30	30	30	30
10	6	11	13	15	20	20	21	21	23	24	23	24	24

Data from Ross et al.1985; NCYFS II; Ross et al.1987.

Table 10.4 Percentile Norms for Males and Females on the Pull-Up Test (Number Completed)

	AGE (YEARS)												
Male percentile	**6**	**7**	**8**	**9**	**10**	**11**	**12**	**13**	**14**	**15**	**16**	**17**	**18**
90	15	19	20	20	8	8	8	10	12	14	14	15	16
75	10	13	14	15	4	5	5	7	8	10	12	12	13
50	6	8	10	10	1	2	3	4	5	7	9	9	10
25	3	4	6	6	0	0	0	1	2	4	6	5	6
10	1	1	3	3	0	0	0	0	0	1	2	2	3
Female percentile													
90	2	13	16	17	17	3	3	2	2	2	2	2	2
75	1	9	11	11	12	1	1	1	1	1	1	1	1
50	0	6	7	8	9	0	0	0	0	0	0	0	0
25	0	3	4	4	4	0	0	0	0	0	0	0	0
10	0	0	1	1	1	0	0	0	0	0	0	0	0

Data from Ross et al.1985; NCYFS II; Ross et al.1987.

Table 10.5 Percentile Norms for Males and Females on the Sit-and-Reach Test (in.)

	AGE (YEARS)												
Male percentile	**6**	**7**	**8**	**9**	**10**	**11**	**12**	**13**	**14**	**15**	**16**	**17**	**18**
90	16.0	16.0	16.0	15.5	16.0	16.5	16.0	16.5	17.5	18.0	19.0	19.5	19.5
75	15.0	15.0	14.5	14.5	14.5	15.0	15.0	15.0	15.5	16.5	17.0	17.5	17.5
50	13.5	13.5	13.5	13.0	13.5	13.0	13.0	13.0	13.5	14.0	15.0	15.5	15.0
25	12.0	11.5	11.5	11.0	11.5	11.5	11.0	11.0	11.0	12.0	13.0	13.0	13.0
10	10.5	10.0	9.5	9.5	10.0	9.5	8.5	9.0	9.0	9.5	10.0	10.5	10.0
Female percentile													
90	16.5	17.0	17.0	17.0	17.5	18.0	19.0	20.0	19.5	20.0	20.5	20.5	20.5
75	15.5	16.0	16.0	16.0	16.5	16.5	17.0	18.0	18.5	19.0	19.0	19.0	19.0
50	14.0	14.5	14.0	14.0	14.5	15.0	15.5	16.0	17.0	17.0	17.5	18.0	17.5
25	12.5	13.0	12.5	12.5	13.0	13.0	14.0	14.0	15.0	15.5	16.0	15.5	15.5
10	11.5	11.5	11.0	11.0	10.5	11.5	12.0	12.0	12.5	13.5	14.0	13.5	13.0

Note: Participants aged 6 to 9 performed modified pull-ups. For metric measurements, refer to a site such as this: http://www.worldwidemetric.com/measurements.html.

Data from Ross et al.1985; NCYFS II; Ross et al.1987.

YOUTH FITNESS TEST BATTERIES

As we have discussed, assessing physical fitness in youth has changed from a motor-fitness emphasis to a health-related emphasis in the United States and abroad. Youth fitness test batteries are used to measure and evaluate physical fitness. Historically, several nationally used test batteries were available for youth fitness assessment. Table 10.6 lists two youth fitness batteries, the items contained on each, and the organization where you can obtain the battery. Three elements are present in the test batteries:

1. Health-related fitness items
2. Criterion-referenced standards for each test
3. Motivational awards

Youth fitness test batteries use criterion-referenced standards for passing or failing a test and achieving the fitness awards. Safrit and Pemberton (1995) provided a complete guide to youth fitness testing.

In the United States a major change in youth fitness programming and testing occurred in 2013. The PCFSN began the Presidential Youth Fitness Program (PYFP), which is a comprehensive school-based program that promotes health and regular physical activity for America's youth. The PYFP partners are the U.S. Centers for Disease Control and Prevention (CDC), the Cooper Institute, The National Foundation on Fitness, Sports and Nutrition, the PCFSN, and the Society of Health and Physical Educators (SHAPE America). The program includes these essential elements:

- Adoption of the Cooper Institute's fitness assessment, Fitnessgram
- Professional development, such as monthly webinars, led by SHAPE America, a leading organization of professionals involved in physical education
- The expertise of the CDC in leading the development of a plan to track and evaluate the PYFP

Table 10.6 Youth Fitness Test Batteries

	Fitnessgram	Eurofit
Aerobic capacity	PACER (r) Mile run Walk test	Endurance shuttle run Bike ergometer test
Body composition	Skinfolds (r) Body mass index	Height Weight Skinfolds
Abdominal strength and endurance	Curl-ups (r)	Sit-ups
Upper-body strength and endurance	Push-ups (r) Modified pull-ups Pull-ups Flexed arm hang	Hand grip Bent-arm hang
Trunk extensor strength and endurance	Trunk lift (r)	
Flexibility	Back saver sit-and-reach Shoulder stretch	Sit-and-reach
Running speed and agility		Shuttle run
Speed of limb movement		Plate tapping
Power		Standing long jump
Balance		Flamingo balance

Note: r = recommended; PACER = progressive aerobic cardiovascular endurance run. Test battery contact information is in the Youth Fitness Test Contact Information highlight box.

The key feature was the adoption of the Fitnessgram for the assessment of youth fitness in the United States. Thus, the United States now has one nationally recognized youth fitness testing battery.

YOUTH FITNESS RESOURCES

Fitnessgram
Human Kinetics
1607 N. Market St
Champaign, IL 61820
www.fitnessgram.net
www.usgames.com

Eurofit
Vrije Universiteit en Universiteit van Amsterdam
Meibergdreef 15-1105
AZ Amsterdam
www.topendsports.com/testing/eurofit.htm

The Eurofit physical fitness test battery, which is a combination of health-related and motor fitness components, is also presented in table 10.6. The Eurofit is used on the European continent. Eurofit combines health-related and performance-related items with additional motor ability items (e.g., plate tapping, power, balance).

Jo, our teacher from the measurement challenge, will need to talk with her colleagues and discuss the implications for testing and evaluation that will result from using the Fitnessgram.

Go to the WSG to complete Student Activity 10.1 and view Videos 10.1 and 10.2.

FITNESSGRAM

We highlight the Fitnessgram physical fitness battery because it has been validated, used by millions of children, and is now the national youth fitness test battery as recognized by the PYFP. The Fitnessgram is a physical fitness battery that includes health-related criterion-referenced standards. These criterion-referenced standards identify up to three fitness zones—Healthy Fitness Zone, Needs Improvement Zone, and Needs Improvement Health Risk Zone—depending on the selected test. The program includes optional tests, computer software, and support. Following are five items from the battery. The criterion-referenced standards for each test are provided in tables 10.7 and 10.8.

Table 10.7 Fitnessgram Health-Related Criterion-Referenced Standards: Boys

BOYS											
Age	Aerobic capacity $\dot{V}O_2max$ (ml • kg^{-1} • min^{-1})			Percent body fat				Body mass index			
	PACER, 1-mile (1.6 km) run, and walk test										
	NI-Health Risk	NI	HFZ	Very lean	HFZ	NI	NI-Health Risk	Very lean	HFZ	NI	NI-Health Risk
5	Completion of test. Lap count or time standards not recommended.			≤8.8	8.9-18.8	18.9	≥27.0	≤13.8	13.9-16.8	16.9	≥18.1
6				≤8.4	8.5-18.8	18.9	≥27.0	≤13.7	13.8-17.1	17.2	≥18.8
7				≤8.2	8.3-18.8	18.9	≥27.0	≤13.7	13.8-17.6	17.7	≥19.6
8				≤8.3	8.4-18.8	18.9	≥27.0	≤13.9	14.0-18.2	18.3	≥20.6
9				≤8.6	8.7-20.6	20.7	≥30.1	≤14.1	14.2-18.9	19.0	≥21.6
10	≤37.3	37.4-40.1	≥40.2	≤8.8	8.9-22.4	22.5	≥33.2	≤14.4	14.5-19.7	19.8	≥22.7
11	≤37.3	37.4-40.1	≥40.2	≤8.7	8.8-23.6	23.7	≥35.4	≤14.8	14.9-20.5	20.6	≥23.7
12	≤37.6	37.4-40.2	≥40.3	≤8.3	8.4-23.6	23.7	≥35.9	≤15.2	15.3-21.3	21.4	≥24.7
13	≤38.6	38.7-41.0	≥41.1	≤7.7	7.8-22.8	22.9	≥35.0	≤15.7	15.8-22.2	22.3	≥25.6
14	≤39.6	39.7-42.4	≥42.5	≤7.0	7.1-21.3	21.4	≥33.2	≤16.3	16.4-23.0	23.1	≥26.5
15	≤40.6	40.7-43.5	≥43.6	≤6.5	6.6-20.1	20.2	≥31.5	≤16.8	16.9-23.7	23.8	≥27.2
16	≤41.0	41.1-44.0	≥44.1	≤6.4	6.5-20.1	20.2	≥31.6	≤17.4	17.5-24.5	24.4	≥27.9
17	≤41.2	41.3-44.1	≥44.2	≤6.6	6.7-20.9	21.0	≥33.0	≤18.0	18.1-24.9	25.0	≥28.6
>17	≤41.2	41.3-44.2	≥44.3	≤6.9	7.0-22.2	22.3	≥35.1	≤18.5	18.6-24.9	25.0	≥29.3

Age	Curl-up (no. completed)	Trunk lift (in.)	90° push-up (no. completed)	Modified pull-up (no. completed)	Flexed arm hang (s)	Back saver sit-and-reach (in.)	Shoulder stretch
5	≥2	6-12	≥3	≥2	≥2	8	Healthy Fitness Zone = touching fingertips together behind back on both the right and left sides.
6	≥2	6-12	≥3	≥2	≥2	8	
7	≥4	6-12	≥4	≥3	≥3	8	
8	≥6	6-12	≥5	≥4	≥3	8	
9	≥9	6-12	≥6	≥5	≥4	8	
10	≥12	9-12	≥7	≥5	≥4	8	
11	≥15	9-12	≥8	≥6	≥6	8	
12	≥18	9-12	≥10	≥7	≥10	8	
13	≥21	9-12	≥12	≥8	≥12	8	
14	≥24	9-12	≥14	≥9	≥15	8	
15	≥24	9-12	≥16	≥10	≥15	8	
16	≥24	9-12	≥18	≥12	≥15	8	
17	≥24	9-12	≥18	≥14	≥15	8	
>17	≥24	9-12	≥18	≥14	≥15	8	

Note: For metric measurements, refer to a site such as this: http://www.worldwidemetric.com/measurements.html.

Adapted, by permission, from The Cooper Institute, 2010, *FITNESSGRAM/ACTIVITYGRAM test administration manual,* updated 4th ed. (Champaign, IL: Human Kinetics), 65.

Table 10.8 Fitnessgram Health-Related Criterion-Referenced Standards: Girls

GIRLS												
Age	Aerobic capacity $\dot{V}O_2max$ (ml • kg^{-1} • min^{-1})			Percent body fat				Body mass index				
	PACER, 1-mile (1.6 km) run, and walk test											
	NI-Health Risk	NI	HFZ	Very lean	HFZ	NI	NI-Health Risk	Very lean	HFZ	NI	NI-Health Risk
5	Completion of test. Lap count or time standards not recommended.			≤9.7	9.8-20.8	20.9	≥28.4	≤13.5	13.6-16.8	16.9	≥18.5
6				≤9.8	9.9-20.8	20.9	≥28.4	≤13.4	13.5-17.2	17.3	≥19.2
7				≤10.0	10.1-20.8	20.9	≥28.4	≤13.5	13.6-17.9	18.0	≥20.2
8				≤10.4	10.5-20.8	20.9	≥28.4	≤13.6	13.7-18.6	18.7	≥21.2
9				≤10.9	11.0-22.6	22.7	≥30.8	≤13.9	14.0-19.4	19.5	≥22.4
10	≤37.3	37.4-40.1	≥40.2	≤115.5	11.6-24.3	24.4	≥33.0	≤14.2	14.3-20.3	20.4	≥23.6
11	≤37.3	37.4-40.1	≥40.2	≤12.1	12.2-25.7	25.8	≥34.5	≤14.6	14.7-21.2	21.3	≥24.7
12	≤37.0	37.1-40.0	≥40.1	≤12.6	12.7-26.7	26.8	≥335.5	≤15.1	15.2-22.1	22.2	≥25.8
13	≤36.6	36.7-39.6	≥39.7	≤13.3	13.4-27.7	27.8	≥36.3	≤15.6	15.7-22.9	23.0	≥26.8
14	≤36.3	36.4-39.3	≥39.4	≤13.9	14.0-28.5	28.6	≥36.8	≤16.1	16.2-23.6	23.7	≥27.7
15	≤36.0	36.1-39.0	≥39.1	≤14.5	14.6-29.1	29.2	≥37.1	≤16.6	16.7-24.3	24.4	≥28.5
16	≤35.8	35.9-38.8	≥38.9	≤15.2	15.3-29.7	29.8	≥37.4	≤17.0	17.1-24.8	24.9	≥29.3
17	≤35.7	35.8-38.7	≥38.8	≤15.8	15.9-30.4	30.5	≥37.9	≤17.4	17.5-24.9	25.0	≥30.0
>17	≤35.3	35.4-38.5	≥38.6	≤16.4	16.5-31.3	31.4	≥38.6	≤17.7	17.8-24.9	25.0	≥30.0

Age	Curl-up (no. completed)	Trunk lift (in)	90° push-up (no. completed)	Modified pull-up (no. completed)	Flexed arm hang (s)	Back saver sit-and-reach (in.)	Shoulder stretch
5	≥2	6-12	≥3	≥2	≥2	9	Healthy Fitness Zone = touching fingertips together behind back on both the right and left sides.
6	≥2	6-12	≥3	≥2	≥2	9	
7	≥4	6-12	≥4	≥3	≥3	9	
8	≥6	6-12	≥5	≥4	≥3	9	
9	≥9	6-12	≥6	≥4	≥4	9	
10	≥12	9-12	≥7	≥4	≥4	9	
11	≥15	9-12	≥7	≥4	≥6	10	
12	≥18	9-12	≥7	≥4	≥7	10	
13	≥18	9-12	≥7	≥4	≥8	10	
14	≥18	9-12	≥7	≥4	≥8	10	
15	≥18	9-12	≥7	≥4	≥8	10	
16	≥18	9-12	≥7	≥4	≥8	10	
17	≥18	9-12	≥7	≥4	≥8	10	
>17	≥18	9-12	≥7	≥4	≥8	10	

Note: For metric measurements, refer to a site such as this: http://www.worldwidemetric.com/measurements.html.

Adapted, by permission, from The Cooper Institute, 2010, *FITNESSGRAM/ACTIVITYGRAM test administration manual,* updated 4th ed. (Champaign, IL: Human Kinetics), 66.

PACER

Purpose

To measure aerobic capacity.

Objective

To run as long as possible back and forth across a 20-meter (21.9 yd) space at a specified pace that gets faster each minute.

Equipment and Facilities

A flat, nonslippery surface at least 20 meters (21.9 yd) long that allows a width of 1.02 to 1.52 meters (40-60 in) for each student.

CD or cassette player with adequate volume

PACER CD or audiocassette

Measuring tape

8 or more marker cones

Pencil

Copies of score sheet A or B

Note: Students should wear shoes with nonslip soles.

Instructions

Mark the 20-meter (21.9 yd) course with marker cones to divide lanes and a tape or chalk line at each end. If using the audiotape, calibrate it by timing the 1-minute test interval at the beginning of the tape. If the tape has stretched and the timing is off by more than 0.5 seconds, obtain another copy of the tape. Make copies of score sheet A or B for each group of students to be tested.

Before the test day, allow students to listen to several minutes of the tape so that they know what to expect. Students should then be allowed at least two practice sessions.

Allow students to select partners. Have students who are being tested line up behind the start line.

The PACER CD has a music version and one with only the beeps. The PACER tape has two music versions and one beep-only version. Each version of the test will give a 5-second countdown and tell students when to start.

Students should run across the 20-m (21.9 yd) distance and touch the line with the foot by the time the beep sounds. At the sound of the beep, they turn around and run back to the other end. If some students get to the line before the beep, they must wait for the beep before running the other direction. Students continue in this manner until they fail to reach the line before the beep for a second time.

A single beep will sound at the end of the time for each lap. A triple beep sounds at the end of each minute. The triple beep serves the same function as the single beep and also alerts the runners that the pace will get faster.

The PACER also has a 15-meter (16.4 yd) option for those with restricted areas not permitting a 20-meter (21.9 yd) test. The CD is adjusted for the reduced test length.

When to Stop

The first time a student does not reach the line by the beep, he or she reverses direction immediately. Allow a student to attempt to catch up with the pace. The test is completed for a student when he or she fails to reach the line by the beep for the second time. Students just completing the test should continue to walk and stretch in the cool-down area. Figure 10.4 provides diagrams of testing procedures.

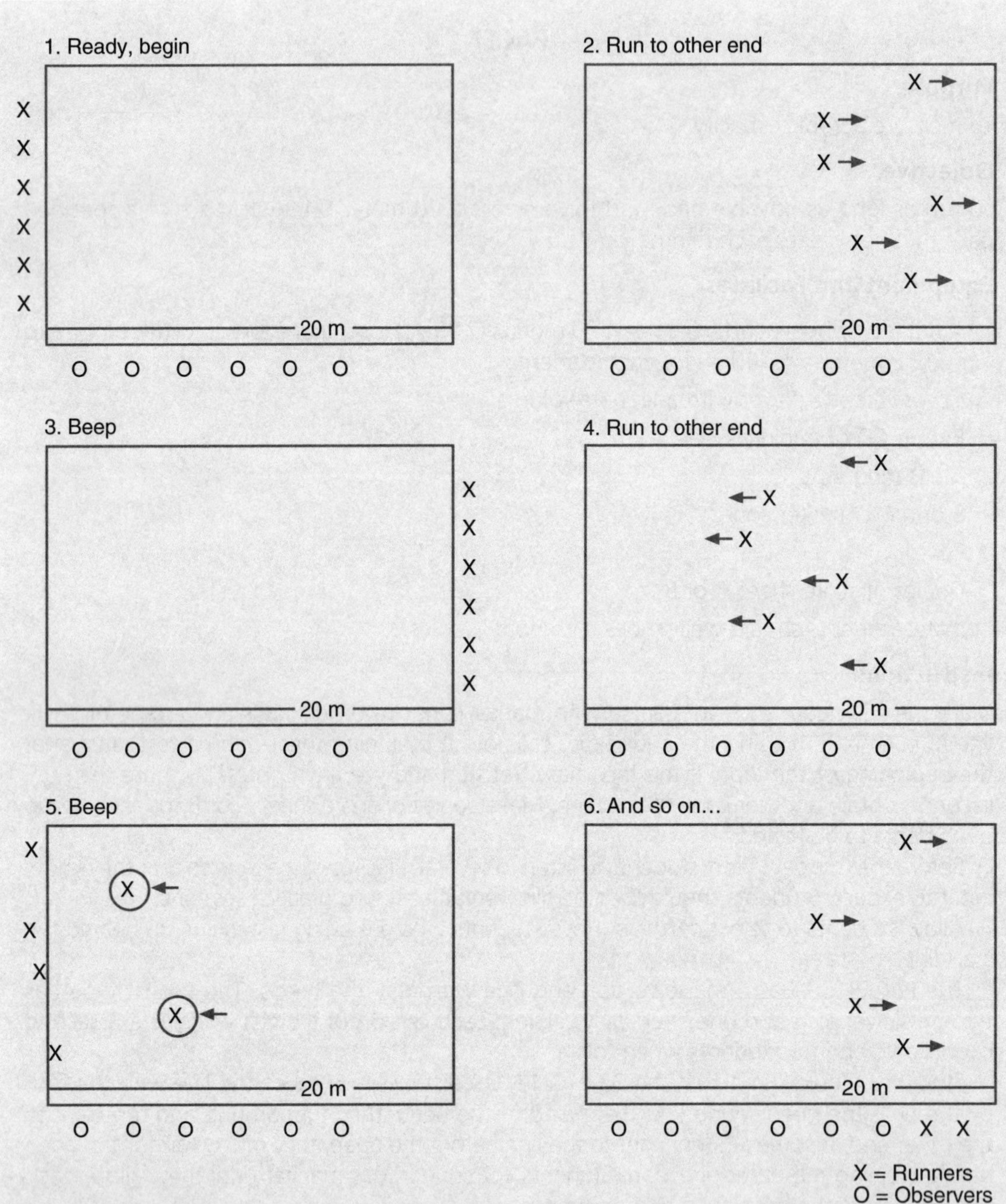

Figure 10.4 Diagram of PACER.

Reprinted, by permission, from The Cooper Institute, 2010, *FITNESSGRAM/ACTIVITYGRAM test administration manual*, updated 4th ed. (Champaign, IL: Human Kinetics), 31.

Scoring

In the PACER test, a lap is one 20-meter (21.9 yd) distance (from one end to the other). Have one student record the lap number (cross off each lap number on a PACER score sheet). The recorded score is the total number of laps completed by a student. For ease in administration, count the first lap that a student does not reach the line by the beep. Be consistent with all students and classes.

SKINFOLD MEASUREMENTS

Purpose

To measure percent body fat.

Objective

The thickness of the skinfolds at the triceps and medial calf sites is measured and used to estimate percent body fat.

Equipment

Skinfold calipers (both expensive and inexpensive calipers have been shown to provide reliable and valid measures).

Instructions

The triceps skinfold is taken on the right arm over the triceps muscle. The skinfold is vertical and midway between the acromion process of the scapula and the elbow (see figure 9.12). The medial calf skinfold is a vertical skinfold taken at the level of maximal girth of the calf on the right leg (figure 10.5). The foot rests on a stool or other device so that the knee is at a 90° angle.

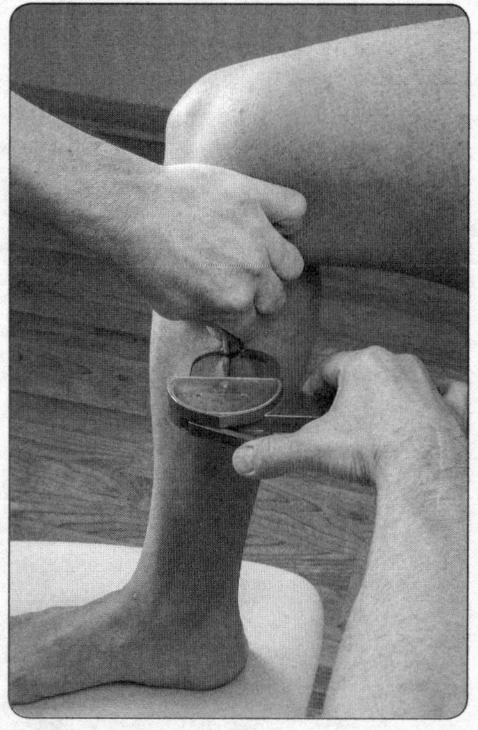

Figure 10.5 Fitnessgram calf skinfold measurement.

Scoring

Measure each skinfold three times and record the median value. Record the skinfold thickness to the nearest 0.5 mm.

Mastery Item 10.3

Review the material provided in chapter 9 on obtaining reliable and valid skinfold measurements.

CURL-UP

Purpose

To measure abdominal strength and endurance.

Objective

Students perform as many curl-ups as possible to a maximum of 75.

Equipment

Gym mats

Cardboard strips (a narrow strip [30 in. (76.2 cm) × 3 in. (7.6 cm)] for kindergarten to grade 4 and a wider strip [30 in. (76.2 cm) × 4.5 in. (11.4 cm)] for older children)

Tape player and audiotape (for controlling cadence)

Instructions

Students perform this test in groups of three. One student performs the curl-ups, the second supports the head of the performer, and the third secures the strip so that it does not move. Each student being tested lies in a supine position on the mat with knees bent at an approximate 140° angle, legs apart, and arms straight and parallel to the trunk with the palms resting on the mat. The fingers are extended and the head is in the partner's hand resting on the mat. The cardboard strip is placed under the knees with the fingers touching the nearer edge. The third student stands on the strip so that it does not move during the test. The student being tested curls up so that the fingers slide to the other side of the strip (figure 10.6). That student performs as many curl-ups as possible while maintaining a cadence of 1 curl-up every 3 seconds but stops at a maximum number of 75.

Figure 10.6 Fitnessgram curl-up test.

Scoring

Record the number of curl-ups to a maximum of 75.

TRUNK LIFT

Purpose

To measure trunk extensor strength and flexibility.

Objective

The student being tested lifts the upper body from the ground using the muscles of the back and holds the position to allow an accurate measurement from chin to the ground.

Equipment

Gym mats and yardstick or ruler

Instructions

The student being tested lies face down on a mat. The toes are pointed and the hands placed under the thighs. The student lifts the upper body, in a slow and controlled pace, to a maximum of 12 inches (30.5 cm). The student holds the position until a ruler is placed in front of him or her and the measurement is taken from the floor to the chin (figure 10.7).

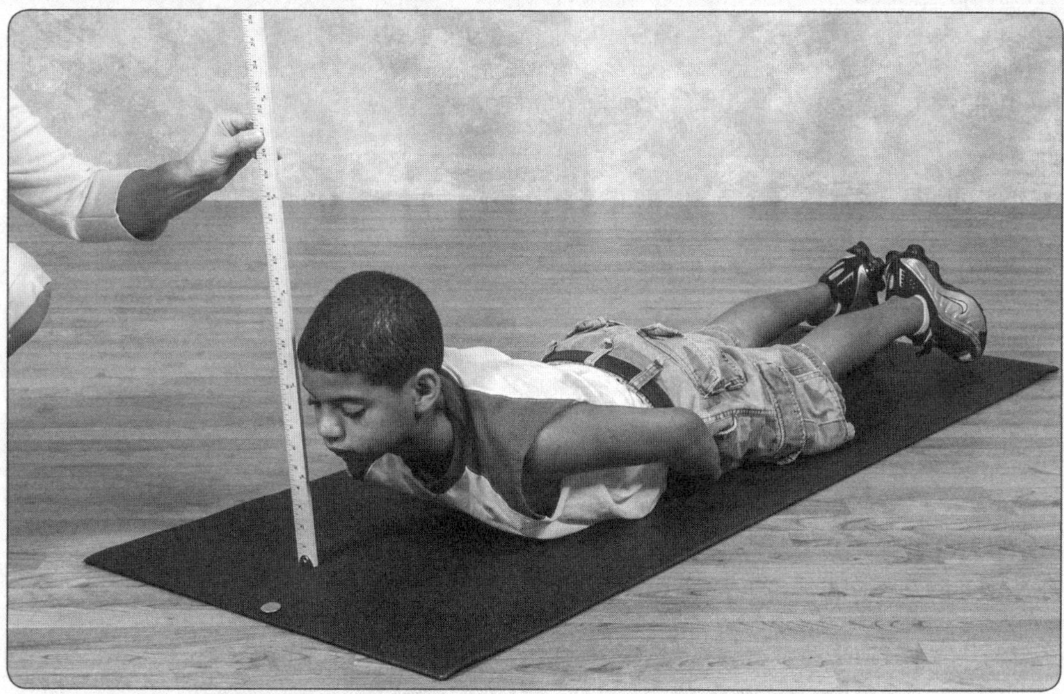

Figure 10.7 Fitnessgram trunk lift test.

Scoring

Perform two trials and record the highest score to the nearest inch (cm), up to a maximum of 12 inches (30.5 cm).

90° PUSH-UPS

Purpose

To measure upper-body strength and endurance.

Objective

Students complete as many 90° push-ups as possible.

Equipment

Tape player

Audiotape to control the cadence of one push-up every 3 seconds.

Instructions

Students work in pairs, with one counting while the other is tested. The student being tested lies face down on the ground with the hands under the shoulders; fingers stretched; legs straight, parallel, and slightly apart; and the toes tucked under the feet. At a start command, the student pushes up with the arms until they are straight (figure 10.8). The legs and back should be kept straight throughout the test. The student lowers the body using the arms until the arms bend to a 90° angle and the upper arms are parallel to the floor. This action is repeated as many times as possible following the cadence of one repetition every 3 seconds. The test is continued until the student cannot maintain the pace or demonstrates poor form.

Figure 10.8 Fitnessgram push-up test.

Scoring

Record the number of push-ups performed.

Go to the WSG to view Video 10.3.

ENHANCING RELIABLE AND VALID FITNESS TEST RESULTS WITH CHILDREN

The tests presented in the fitness batteries in table 10.6 have different levels of evidence to support their validity. Distance runs and skinfolds have demonstrated criterion-related validity in research (Lohman 1989; Safrit et al. 1988). The tests of muscular fitness, pull-ups, push-ups, modified pull-ups, flexed arm hang, sit-ups, curl-ups, and the sit-and-reach test have limited criterion-related validity support but are generally accepted on the basis of content or logical validity. Research has questioned the assumption of content validity for some tests: sit-and-reach, pull-ups, and sit-ups (Engelman and Morrow 1991; Hall, Hetzler, Perrin, and Weltman 1992; Jackson and Baker 1986). However, you can assume sufficient validity of your tests provided that you are using one of the fitness batteries presented in this chapter.

The practical issue is for you to take the appropriate steps during fitness test administration that will produce reliable test results. This is an important issue in youth fitness testing because research has shown that field tests of fitness in children can demonstrate poor reliability, especially in younger children (Rikli, Petray, and Baumgartner 1992). To enhance reliability ($r_{xx'}$ > .80), you can take a variety of practical steps to minimize measurement error, including the following:

1. Attain adequate knowledge of test descriptions.
2. Give proper demonstrations and instructions.
3. Develop effective student and teacher preparation through adequate practice trials.
4. Conduct reliability studies.

Of these steps, providing adequate practice trials is the most important. Children, especially younger ones, need several trials of a fitness test to learn how to take the test and to provide consistent results. For example, children need to practice the PACER several times to learn the appropriate pace and to provide a consistent time that is a reliable and valid representation of their cardiorespiratory endurance. If you are going to take skinfolds, you need to practice the measurement techniques discussed in chapter 9.

SPECIAL CHILDREN

One of the biggest measurement challenges that you may confront as a professional in human performance is the assessment of physical fitness in children with physical or mental disabilities. Keep in mind that the fitness test batteries discussed in this chapter would exclude or be biased against many children with **physical disabilities** (i.e., those having physical or organic limitations, such as cerebral palsy) and children with **mental disabilities** (i.e., those with mental or psychological limitations, such as autism) because of their specific disabilities. *Before you can administer or evaluate fitness test results, you must consider the participants' physical and organic limitations; neural and emotional capacity; interfering reflexes; and acquisition of prerequisite functions, responses, and abilities* (Seaman and DePauw 1989). You can develop the basic knowledge and competence that you need for assessing the fitness and activity of children with disabilities from your instruction and learning experiences in **adapted physical education**, that is, physical education adjusted to accommodate children with physical or mental limitations. The physical fitness tests selected should be appropriate for an individual student based on his or her disability as well as on the fitness capacity to be measured. Seaman and DePauw (1989) and Winnick and Short (1999) are excellent sources for detailed information on fitness assessment of special children.

The Brockport Physical Fitness Test (Winnick and Short 1999), a health-related physical fitness test for youths aged 10 through 17 with various disabilities, was developed through a research study, Project Target, funded by the U.S. Department of Education. The test battery includes criterion-referenced standards for 25 tests. The test manual helps professionals consider each student's disability and select the most appropriate test and test protocol. The test comes with Fitness Challenge software to help professionals administer the test and develop a database. Table 10.9 provides potential items for fitness assessment, the appropriate population with disabilities for the test, and reliability and validity comments for each test. The complete test kit includes a manual, software, a demonstration video, and a fitness training guide.

Table 10.9 Brockport Physical Fitness Test Items

Test item	Mental retardation	Blind with assistance	Cerebral palsy	Spinal cord injury	Congenital anomalies/ amputation	Reliability	Validity
PACER	R	R			O	Acceptable	Content Concurrent
1-mi (1.6 km) run or walk		O			R	Acceptable	Concurrent
Target aerobic movement test	R		R	R	R	Acceptable	Content
Skinfolds	R	R	R	R	R	Acceptable	Concurrent
BMI	O	O	O			Acceptable	Concurrent
Reverse curl				R		Not reported	Content
Seated push-up			R	R	R	Not reported	Content
40-m (43.7 yd) push or walk			R			Not reported	Content
Wheelchair ramp test			R/O			Not reported	Content
Bench press	O			O	R	Acceptable	Content
Dumbbell press			R/O	O	R/O	Acceptable	Content Concurrent
Extended arm hang	R					Acceptable	Content
Dominant grip strength	O		O	R	O	Acceptable	Construct
Isometric push-up	O					Acceptable	Content
Push-up		O				Acceptable	Content
Pull-up		O				Acceptable	
Modified pull-up		O				Acceptable	Content
Curl-up		R			R	Acceptable	Content
Modified curl-up		R				Acceptable	Content
Trunk lift	R	R			R	Acceptable	Content
Modified Apley test			R	R	R	Not reported	Content
Shoulder stretch	O	O				Acceptable	Content
Modified Thomas stretch			R	R		Not reported	Content
Back saver sit-and-reach test	R	R			R	Acceptable	Content
Target stretch test			R/O	R	R	Acceptable	Content Concurrent

Note: O = Option; R = Recommended.

INSTITUTE OF MEDICINE RECOMMENDATIONS FOR YOUTH FITNESS ASSESSMENT

In September 2012 the Institute of Medicine, at the request of the Robert Wood Johnson Foundation, created the Committee on Fitness Measures and Health Outcomes in Youth to assess the relationship between youth fitness items and health outcomes and recommend the best fitness test items for (1) use in national youth fitness surveys and (2) fitness testing in schools (http://www.iom.edu/Reports/2012/Fitness-Measures-and-Health-Outcomes-in-Youth.aspx). The highlight box summarizes the committee's recommendations for fitness measures.

FITNESS MEASURES

For National Youth Fitness Surveys

- To measure body composition, national surveys should include the following:
  - body mass index (BMI) as an estimate of body weight in relation to height,
  - skinfold thickness at the triceps and below the shoulder blade as indicators of underlying fat, and
  - waist circumference as an indicator of abdominal fat.
- To measure cardiorespiratory endurance, national surveys should include a progressive shuttle run, such as the 20-meter shuttle run. If physical space is limited, cycle ergometer or treadmill tests are valid and reliable alternatives.
- To measure musculoskeletal fitness, national surveys should include handgrip strength and standing long jump tests.

For Fitness Testing in Schools

- To measure body composition, schools should use BMI measures.
- To measure cardiorespiratory endurance, schools should include progressive shuttle runs.
- To measure musculoskeletal fitness, schools should include handgrip strength and standing long jump tests.

For schools, additional test items that have not yet been shown to be related to health but that are valid, reliable, and feasible may be considered as supplemental educational tools (e.g., distance or timed runs for cardiorespiratory endurance, modified pull-ups and push-ups for measuring upper-body musculoskeletal strength, curl-ups for measuring core strength). A measure of flexibility, such as the sit-and-reach test, may also be included.

From a measurement and evaluation perspective there are some important things to note:

1. The recommendations are consistent with the importance of measuring health-related physical fitness.
   a. Cardiorespiratory fitness
   b. Body composition
   c. Muscular-skeletal fitness

2. Differences exist between the assessment of body composition for national youth fitness surveys and fitness testing in schools.

a. National surveys should include BMI, skinfold measures of the triceps and shoulder blade, and a waist circumference measure.

b. Fitness testing in schools should be limited to BMI.

3. The recommendation for administering tests of handgrip strength and the standing long jump tests is not consistent with the Fitnessgram test battery, which is the fitness testing program of the PCFSN. These tests are currently not included in the Fitnessgram test battery. The Eurofit test battery does include handgrip strength and the standing long jump.

Although the Fitnessgram does not include the handgrip test, it does include measures of upper-body strength and endurance, push-ups (recommended), modified pull-ups, and flexed arm hang. However, the standing long jump test is not included.

As mentioned earlier in this chapter, the original AAHPER Youth Fitness test developed in the 1950s and revised in the 1960s and 1970s was the test battery with associated percentile norms used by the PCFSN for assessment of youth fitness. The original AAHPER Youth Fitness test battery included the standing long jump test. The test battery was considered an assessment of athletic or motor fitness, not health-related fitness. Thus, the standing long jump was removed from national youth fitness testing. However, following the recommendation of the Institute of Medicine, it seems reasonable that the standing long jump, which assesses leg power, should now be added to youth fitness assessment protocols. The test protocol for the AAHPER Youth Fitness Test is provided in Figure 10.9, and Table 10.10 presents comparative percentiles test values for boys and girls from age 6 through 17 years (recommended by the authors of this text). The 75th and 25th percentiles are normative data collected by SHAPE America (formerly AAHPERD) and the PCFSN. The 80th and 45th percentiles come from the Chrysler Fund–AU Fitness Testing Program. Those percentiles were associated with the descriptors of fitness attainment (45th) and outstanding (80th). The recommended values are expert judgment values from the textbook authors for a minimum level of performance consistent with a healthy level of leg power.

MEASURING PHYSICAL ACTIVITY IN YOUTH

The concerns about youth fitness discussed earlier in this chapter are matched by concerns about youth physical activity levels. Public health officials and physical activity researchers need reliable and valid measures of physical activity in children and youth for effective research studies that will increase youth activity and fitness levels (Sallis et al. 1993). Issues regarding the physical activity levels of youth include the following:

- Physical activity improves the overall health of children.
- Physically inactive children tend to become inactive adults with higher risks for chronic diseases.
- Conversely, physically active children tend to become active adults with lower risks for chronic diseases.

For example, Dennison, Straus, Mellits, and Charney (1988) found that children with the poorest performances on distance-run tests had the highest risk of being physically inactive as young adults. Telama, Yang, Laasko, and Viikari (1997) also reported that children and adolescents who are active tend to be more physically active as young adults. These kinds of longitudinal studies, examining the relationship between youth physical activity and

Equipment: Mat, tape measure

Description: Pupil stands with feet several inches apart and the toes behind the takeoff line. Prior to jumping, the pupil swings the arms backward and bends the knees. The jump is accomplished by simultaneously extending the knees and swinging forward the arms.

Rules:
1. Allow three trials.
2. Measure from the takeoff line to the heel or other part of the body that touches the floor nearest the takeoff line.
3. The scorer stands to the side.

Scoring: Record the best of the three trials in feet and inches to the nearest inch.

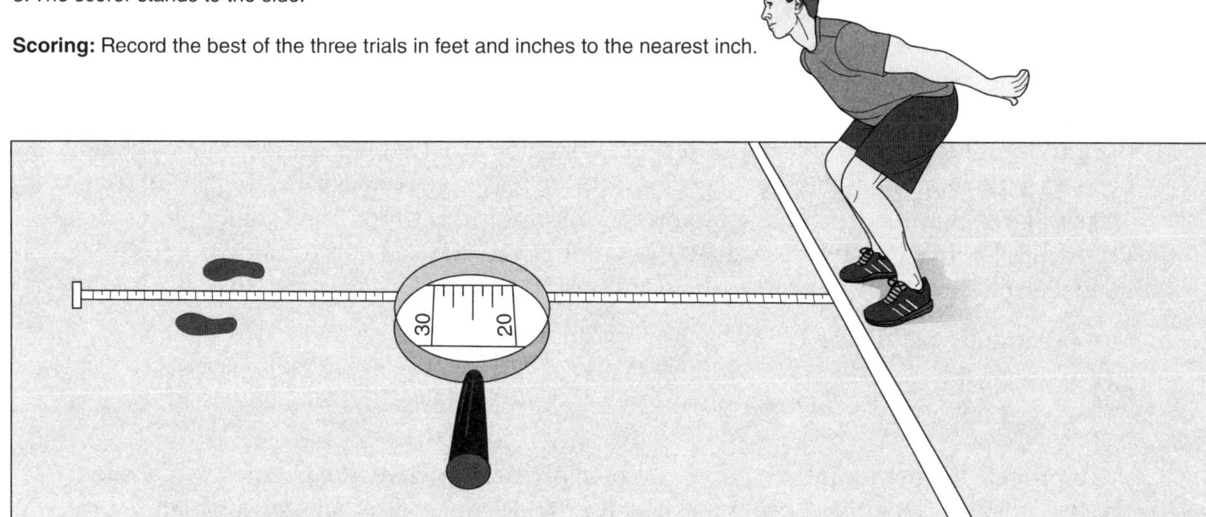

Figure 10.9 AAHPERD Youth Fitness Test (1976): standing long jump.

AAHPERD Youth Fitness Test Manual, Revised by Paul Alfred Hunsicker & guy G. Feiff. © 1976. American Alliance for Health, Physical Education, Recreation and Dance. Used with permission.

Table 10.10 Standing Long Jump Performance from Various Tests and Recommended Health Standard

	AGE IN YEARS (VALUES ARE IN)											
	6	7	8	9	10	11	12	13	14	15	16	17+
Boys												
AAHPERD 75th				64	64	67	71	75	80	86	90	93
PCFSN 75th	49	53	57	61	66	69	72	78	84	88	91	94
Chrysler Fund–AAU Fitness Test 80th	50	54	58	63	66	69	73	79	87	89	92	96
AAHPERD 25th				54	54	56	60	62	66	73	78	78
PCFSN 25th	39	42	47	50	53	57	60	64	69	73	78	82
Chrysler Fund-AAU Fitness Test 45th	43	48	51	55	59	61	64	70	76	79	83	87
Recommendation	42	46	50	53	56	60	63	68	73	77	81	85
Girls												
AAHPERD 75th				62	62	64	66	69	71	70	69	72
PCFSN 75th	45	48	52	55	60	64	67	69	70	71	71	72
Chrysler Fund–AAU Fitness Test 80th	48	51	54	59	61	64	69	71	72	73	73	74
AAHPERD 25th				49	49	52	54	57	58	59	57	59
PCFSN 25th	36	38	42	44	49	52	55	56	57	57	58	58
Chrysler Fund-AAU Fitness Test 45th	40	44	47	50	53	57	61	63	65	64	65	66
Recommendation	39	42	45	48	52	55	59	60	62	61	62	63

Note: For metric measurements, refer to a site such as this: http://www.worldwidemetric.com/measurements.html.

adult physical activity, are impossible without accurate assessments of physical activity while the study participants are children.

As discussed in chapter 9 regarding adults, assessing physical activity levels of children is most reliably and validly done via direct monitoring (e.g., accelerometers, pedometers, direct observation). But lack of feasibility limits applying such procedures in large-scale studies. In Sallis et al. (1993), three self-report instruments—a 7-day recall interview, a self-administered survey, and a simple activity rating—were examined for test–retest reliability and validity. Reliability ranged from .77 to .89 on the three self-reports over all participants. However, as would be expected, reliability was higher for older children than for younger children. The 7-day recall had moderate concurrent validity (r = .44–.53) when related to a criterion of monitored heart rate. Again, validity improved as the age of the children increased. Sallis and colleagues concluded that self-report techniques could be used with youth of high school age, but caution was necessary with younger children.

The PCFSN has recognized the importance of physical activity. The Council offers the Presidential Active Lifestyle Award (PALA+). The award is based on the following requirements for physical activities recorded:

- 60 minutes per day or 12,000 steps
- 5 days per week
- 6 weeks out of 8 weeks

To provide further incentives for youth to be physically active, the Council established the Presidential Champions program, which provides an awards scheme for higher and longer amounts of physical activity:

- Bronze Award—40,000 physical activity points
- Silver Award—90,000 physical activity points
- Gold Award—160,000 physical activity points

The Council has also established a website where people of all ages can track their physical activity behaviors. You can use the site to track all of your physical activities, including the use of pedometers. People can receive the PALA+ or the Presidential Champions awards based on the physical activity levels presented previously.

How many children complete the 12,000 steps per day that are necessary for achieving the PALA+ from the PCFSN? Figure 10.10 shows the percentage of children and adolescents completing 12,000 pedometer steps per day from a large scale Canadian study (Craig, Cameron, and Tudor-Locke 2013). The percentage of boys is always above the percentage of girls at any age. The percentage of boys and girls drops considerably from age 11 to age 16.

Go to the WSG to complete Student Activity 10.2 and view Video 10.4.

Dataset Application

In preparation for entering data into the President's Challenge website, a group of elementary school children wore pedometers for a week to measure their physical activity levels. The daily number of steps was calculated for each student. Use the data provided in the chapter 10 large dataset in the WSG. What is the alpha coefficient for the 7 days (review chapter 6 if necessary)? Determine if there is a significant difference in the mean number of steps taken by boys and girls for the entire week (use an independent t test from chapter 5). Also determine the correlation (from chapter 4) between the number of steps taken and the students' weights. Finally, determine if there is a relationship (from chapter 4) between number of steps taken per week and the number of Fitnessgram Healthy Fitness Zones achieved.

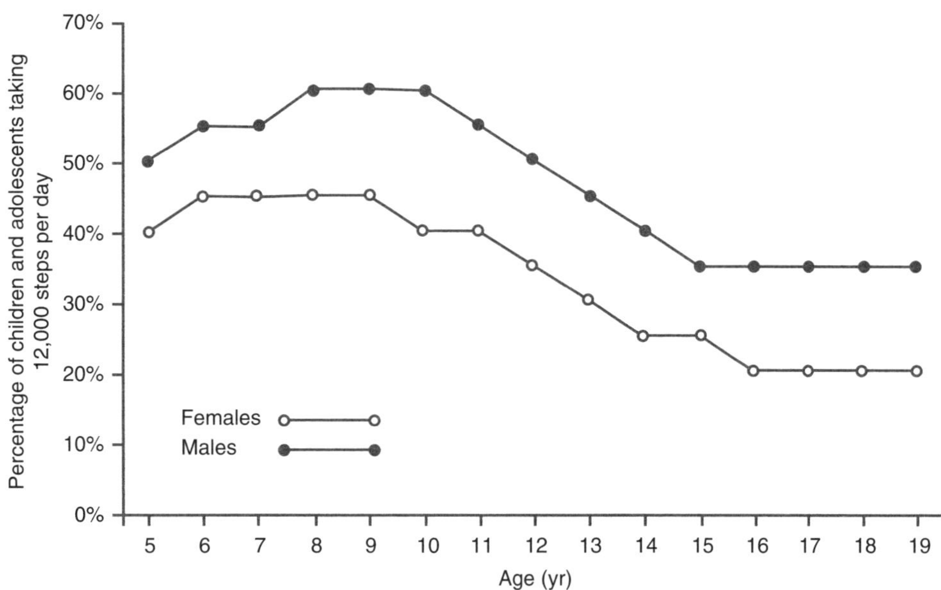

Figure 10.10 The approximate percentage of Canadian children and adolescents taking 12,000 steps per day.

The Cooper Institute has developed an assessment of physical activity to accompany the Fitnessgram physical fitness testing program. The Activitygram assessment has two approaches. The first, the Fitnessgram Physical Activity Questionnaire, is based on the 2008 Physical Activity Guidelines for Children and Youth.

1. On how many of the past 7 days did you participate in any **aerobic physical activity** for a total of 60 minutes or more over the course of a day? This includes moderate activities as well as vigorous activities. Running, hopping, skipping, jumping rope, swimming, dancing, and bicycling are all examples of aerobic activites. (0, 1, 2, 3, 4, 5, 6, 7 days)

2. On how many of the past 7 days did you do **muscle-strengthening physical activity** to strengthen or tone your muscles? Examples include activities such as playing on playground equipment, climbing, lifting weights, or working with resistance bands. (0, 1, 2, 3, 4, 5, 6, 7 days)

3. On how many of the past 7 days did you do **bone-strengthening physical activity** to strengthen your bones? Examples include sports and activities that involve jumping, running, hopping, or skipping. (0, 1, 2, 3, 4, 5, 6, 7 days)

Using students' responses to the questions, Fitnessgram and Activitygram software provides individualized feedback to each student on the standard Fitnessgram and Activitygram assessment shown in figure 10.11. Morrow et al (2013) have shown that increased number of physically active days is related to achieving the Fitnessgram Healthy Fitness Zone.

The second Activitygram approach to physical activity assessment is based on the Previous Day Physical Activity Recall (Weston, Petosa, and Pate 1997). This segmented-day approach requires each child or youth to report activities in 30-minute blocks from 7:00 a.m. to 11:00 p.m. Students report the frequency, intensity, time, and type of physical activity, which provides content validity for the activity recall in that the relevant factors of exercise prescription and participation are assessed. The logging chart that each student completes is illustrated in figure 10.12. The data from the chart are entered into the Fitnessgram computer software, and an Activitygram report is generated and given to each student. An example of the report is presented in figure 10.13.

Joe Jogger
Grade: 6 Age: 12
Central School
Instructor(s): Joanna Watson

	Date	**Height**	**Weight**
Current:	7/22/2010	5'4"	130 lbs

AEROBIC CAPACITY

Needs Improvement (High Risk | Some Risk) | Healthy Fitness Zone

Aerobic Capacity (VO2 Max)
Current: 41.3

Your score for Aerobic Capacity is based on the number of PACER laps and your BMI. It shows your ability to do activities such as running, cycling, and sports at a high level.

	PACER Laps	**BMI**
Current:	20	22.3

MUSCLE STRENGTH, ENDURANCE, & FLEXIBILITY

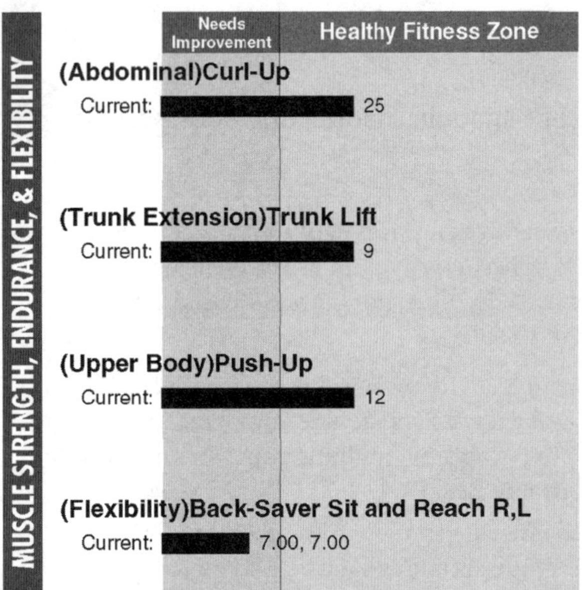

BODY COMPOSITION

Percent Body Fat

Healthy Fitness Zone | Needs Improvement (Some Risk | High Risk)

Current: 16.0

Being too lean or too heavy may be a sign of (or lead to) health problems.

ACTIVITY

	Number of Days
On how many of the past 7 days did you participate in physical activity for a total of 30-60 minutes, or more, over the course of the day?	3
On how many of the past 7 days did you do exercises to strengthen or tone your muscles?	3
On how many of the past 7 days did you do exercises to loosen up or relax your muscles?	1

MESSAGES

Although your aerobic capacity score is in the Healthy Fitness Zone now, you are not doing enough physical activity. Try to do more moderate or vigorous activity (at least 60 minutes, each day) to feel good and remain healthy.

Your abdominal, trunk, and upper-body strength are in the Healthy Fitness Zone. To maintain your fitness, be sure that your strength-training activities include resistance exercises for all of these areas. Abdominal and trunk exercises should be done 3 to 5 days each week. Strength activities for other areas should be done 3 days.

Improve your flexibility by stretching slowly 3 or 4 days each week, holding the stretch 20-30 seconds.

Joe, Good News. Your body composition score is in the Healthy Fitness Zone but you are not getting enough physical activity. To maintain this healthy level, do the following:
-Try to get more activity (at least 60 minutes every day).
-Limit time spent watching TV or playing video games.
-Eat a healthy diet including fresh fruits and vegetables.
-Limit foods with solid fats and added sugars.

Healthy Fitness Zone for 12 year-old boys
Aerobic Capacity: >= 40.3 ml/kg/min
Curl-Up: >= 18 repetitions
Trunk Lift: 9-12 inches
Push-Up: >= 10 repetitions
Back-Saver Sit and Reach: At least 8 inches on R & L
Percent Body Fat: 8.4% - 23.6%

To be healthy and fit it is important to do some physical activity almost every day. Aerobic exercise is good for your heart and body composition. Strength and flexibility exercises are good for your muscles and joints.

Good job! You are doing some aerobic activity and strength exercises. Add some flexibility exercises to improve your overall fitness.

Figure 10.11 Fitnessgram report.

Reprinted with permission from The Cooper Institute, 2010, *FITNESSGRAM/ACTIVITYGRAM test administration manual*, updated 4th Ed. (Champaign, IL: Human Kinetics), 106.

Name__________________________**Teacher**___________________________**Grade**______**Date**____________

Record the main activity that you did during each 30-minute time period by writing the activity type and activity number in the appropriate box (types and numbers can be found in the box located at the bottom of the page). You may have done many things in each 30-minute time period, but try to pick the activity you did for most of the time. Then, check the box that describes how it felt (light/easy [**L**], moderate/medium [**M**], vigorous/hard [**V**]). Note: for all rest activities, use the **Rest** box and you can leave the **L**, **M**, or **V** columns blank. In the Time column, write the amount of time that the activity felt this hard or easy: **S** (some), **M** (most), or **A** (all).

Time	Type	Number	Rest	L	M	V	Time	Time	Type	Number	Rest	L	M	V	Time
7:00								3:00							
7:30								3:30							
8:00								4:00							
8:30								4:30							
9:00								5:00							
9:30								5:30							
10:00								6:00							
10:30								6:30							
11:00								7:00							
11:30								7:30							
12:00								8:00							
12:30								8:30							
1:00								9:00							
1:30								9:30							
2:00								10:00							
2:30								10:30							

Activity types and numbers

Lifestyle activity (LA)	Aerobic sports (AE)	Flexibility activity (FA)
1. Walk, bike, skate 2. Housework/yardwork 3. Active games/play 4. Active job 5. Other lifestyle activity	11. Field sports 12. Court sports 13. Racket sports 14. Aerobic sports–PE 15. Other aerobic sports	21. Martial arts 22. Stretching 23. Yoga 24. Ballet dance 25. Other flexibility
Aerobic activity (AA)	**Muscular activity (MA)**	**Resting (R)**
6. Aerobic class/dancing 7. Aerobic gym 8. Aerobic activity 9. Aerobic activity in PE 10. Other aerobic activity	16. Gymnastics 17. Muscular sports 18. Weightlifting 19. Wrestling 20. Other muscular	26. Schoolwork 27. Computer/TV 28. Eating/resting 29. Sleeping 30. Other rest

Figure 10.12 Activitygram logging chart.

Reprinted with permission from The Cooper Institute, 2010, *FITNESSGRAM/ACTIVITYGRAM test administration manual*, updated 4th Ed. (Champaign, IL: Human Kinetics), 107.

Madison, Riley

IN PARTNERSHIP WITH NFL Play60 THE NFL MOVEMENT FOR AN ACTIVE GENERATION

March Assignment: 3/9/2010
Westside Middle School
District 2 Test

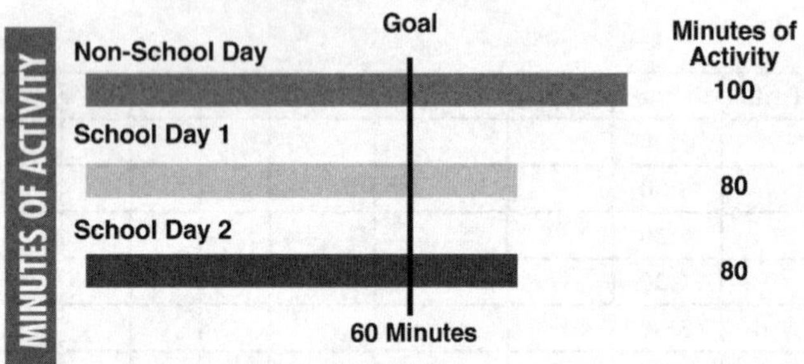

MESSAGES • MESSAGES • MESSAGES

The chart shows the number of minutes that you reported doing moderate (medium) or vigorous (hard) activity on each day. Congratulations, your log indicates that you are doing at least 60 minutes of activity on most every day. This will help to promote good fitness and wellness. For fun and variety, try some new activities that you have never done before.

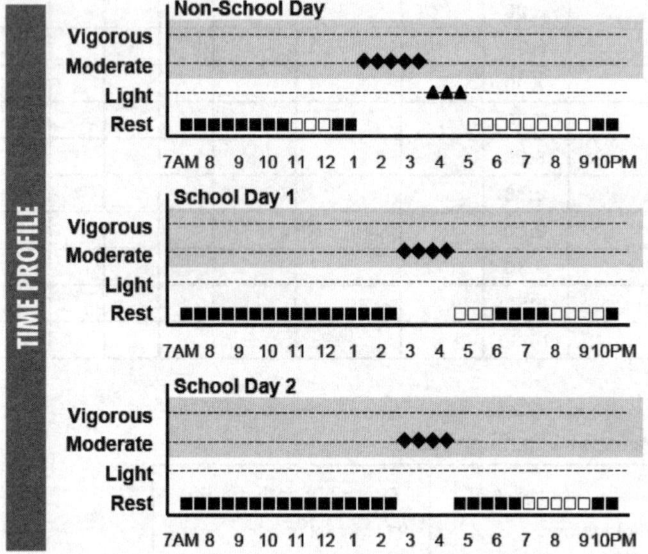

LEGEND:
◆ Most of the time (20 minutes) ■ All of the time (30 minutes)
▲ Some of the time (10 minutes) □ TV/Computer Time

The time profile shows the activity level you reported for each 30 minute period of the day. Your results show that you were not active during school but that you were active after school and on weekends. If it is not possible to be active during school in PE or recess then try to be more active after school. Keep up the good work.

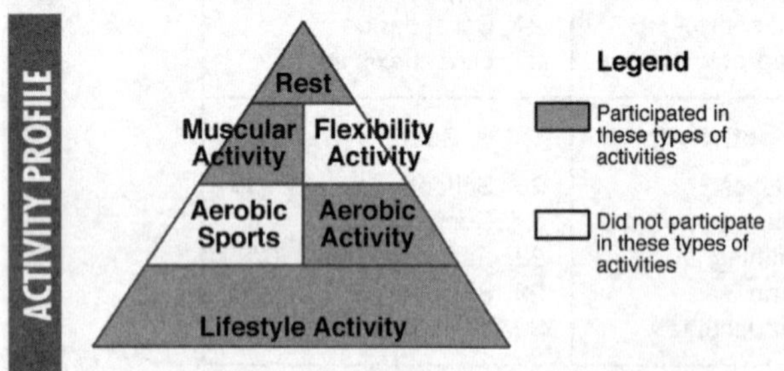

The activity pyramid reveals the different types of activity that you reported doing over a few days. Your results indicate that you participated in regular lifestyle activity as well as some activity from the other levels. This is great! Try to stay active with a variety of different activities on a regular basis.

Your results indicate that you spend an average of 4 hours per day watching TV or working on the computer. While some time on these activities is okay, you should try to limit the total time to less than 2 hours.

ACTIVITYGRAM provides information about your normal levels of physical activity. The ACTIVITYGRAM report shows what types of activity you do and how often you do them. It includes the information that you previously entered for two or three days during one week.

Figure 10.13 Activitygram report.

Reprinted with permission from The Cooper Institute, 2010, *FITNESSGRAM/ACTIVITYGRAM test administration manual*, updated 4th Ed. (Champaign, IL: Human Kinetics), 82.

 Go to the WSG to complete Student Activity 10.3 and view Videos 10.5 and 10.6.

Figure 10.14 provides the physical activity questions used in the national 2013 YRBSS conducted by the CDC as part of the YRBSS. The five questions include three about physical activity behaviors and two about sedentary behaviors. Research about physical activity and physical inactivity and health has become increasingly directed toward the amount of inactive time a person regularly spends on a daily basis. Researchers and public health professionals want to know the activity and inactivity patterns of children and youth. Television watching and computer use are sedentary behaviors that may be contributing to the increased overweight condition observed in children and adolescents.

During the past 7 days, on how many days were you physically active for a total of at least 60 minutes per day? (Add up all the time you spent in any kind of physical activity that increased your heart rate and made you breathe hard some of the time.)

A. 0 days
B. 1 day
C. 2 days
D. 3 days
E. 4 days
F. 5 days
G. 6 days
H. 7 days

On an average school day, how many hours do you watch TV?

A. I do not watch TV on an average school day
B. Less than 1 hour per day
C. 1 hour per day
D. 2 hours per day
E. 3 hours per day
F. 4 hours per day
G. 5 or more hours per day

On an average school day, how many hours do you play video or computer games or use a computer for something that is not school work? (Include activities such as Nintendo, Game Boy, PlayStation, Xbox, computer games, and the Internet.)

A. I do not play video or computer games or use a computer for something that is not school work
B. Less than 1 hour per day
C. 1 hour per day
D. 2 hours per day
E. 3 hours per day
F. 4 hours per day
G. 5 or more hours per day

In an average week when you are in school, on how many days do you go to physical education (PE) classes?

A. 0 days
B. 1 day
C. 2 days
D. 3 days
E. 4 days
F. 5 days

During the past 12 months, on how many sports teams did you play? (Include any teams run by your school or community groups.)

A. 0 teams
B. 1 team
C. 2 teams
D. 3 or more teams

Figure 10.14 Youth Risk Behavior Surveillance System questions for 2013.
Data from CDC 2015.

Self-report measures of physical activity behaviors in children and youth can be misleading. LeBlanc and Janssen (2010) report that 46% and 30% of boys and girls aged 12 to 19, respectively, self-report meeting the 60-minute per day physical activity guideline. However, the percentages drop to 4% (boys) and 1% (girls) when objective measures are obtained with accelerometers. Direct observation of physical activity behaviors can be expensive in terms of time and labor. However, new technologies including portable computers and measurement systems such as BEACHES, SOFIT, and SOPLAY have allowed the reliable, valid, and feasible measurements using direct observation of physical activity. BEACHES (McKenzie, Sallis, Patterson, et al. 1991) was designed to record physical activity, eating behaviors, and related environmental factors. SOFIT (McKenzie, Sallis, and Nader 1991) was developed to record physical activity, lesson context, and teacher behavior during physical education classes. SOPLAY (McKenzie, Marshall, Sallis, and Conway 2000) was designed to measure the physical activity of groups of people, as was the case for BEACHES and SOFIT. SOPLAY allows recording information about the environment that may play a role in the observed physical activity behavior of the group being studied.

Table 10.11 summarizes SOFIT, BEACHES, SOPLAY, and SOPARC, direct observation instruments for assessing physical activity behaviors in school, home, park, and recreation settings. These observation instruments were created by physical activity researchers at San Diego State University. Note the reliability and validation support based on information presented in chapters 6 and 7. See references listed for McKenzie, Cohen, et al. (2006) and Ridgers, Stratton, and McKenzie (2010) at the back of your textbook.

Table 10.11 Instruments Used for Observation of Physical Activity Behaviors

Instrument	Purpose	Reliability	Validity
SOFIT (System for Observing Fitness Instruction Time)	To obtain data on student activity levels, lesson content, and teacher interactions related to physical activity.	Interobserver agreement ranges from 82% to 99%.	Correlations with various external criteria range from .42 to .99.
BEACHES (Behaviors of Eating and Activity for Children's Health: Evaluation System)	Allows an integrated assessment of eating and physical activity, including a wide range of potentially modifiable environmental and social factors.	Interobserver agreement ranges from 90% to 99%. Observers continue training until agreement exceeds 80%-85%.	BEACHES items correlate with external criteria (e.g., heart rate increased with each activity code increment).
SOPLAY (System for Observing Play and Leisure Activity in Youth)	To evaluate and measure the relationship between children's physical activity and their play environment (e.g., accessible, usable, organized, supervised, equipped).	Interobserver agreement ranges from .90 to .99. Intraclass correlations range from .75 to .98.	Content validation based on the SOFIT and BEACHES systems.
SOPARC (System for Observing Play and Active Recreation in Communities)	To obtain data on the characteristics of participants and their physical activity levels in park and recreational settings, also park characteristics.	Interobserver reliabilities range from 88% to 99.8%.	Content validation based on the SOFIT and BEACHES systems.

MEASUREMENT AND EVALUATION CHALLENGE

Jo has learned a great deal about youth fitness and physical activity assessment. She is now familiar with the differences between various factors. She recognizes the difference between health-related fitness and motor fitness. She understands the benefits of criterion-referenced standards compared with the norm-referenced standards the district previously used. She is also now aware of the importance of measuring physical activity in addition to measuring physical fitness. She sees that the Fitnessgram and Activitygram provide many opportunities to assess fitness and activity that were unavailable to her with the district's previous fitness assessment.

SUMMARY

The reliable and valid assessment of youth physical activity and fitness is a major objective of the assessment of human performance. This is especially true for public school or private school physical educators. Although there is not universal agreement about the levels of fitness in youth, it is clear that there are many children and youth whose fitness and physical activity levels are not consistent with good health (Blair 1992; Craig, Cameron, and Tudor-Locke 2013). Reliable and valid fitness testing of youth involves selecting tests related to program objectives, receiving appropriate training, and administering sufficient practice trials to ensure stable test results. A good resource with detailed information about fitness and activity assessment for the professional is *Measurement in Pediatric Exercise Science* (Docherty 1996). Testing children with special needs requires particular attention to test selection and administration. Professional physical educators need appropriate education for training and testing children with special needs.

Although the validity of a test and its passing standard cannot be associated with a reduced risk of a specific disease, the tests that are administered communicate an important message. When we test for cardiovascular endurance, we are telling children and youth that this is an important attribute. This is true for the other health-related fitness tests as well. In general, it has been scientifically documented that higher levels of physical activity and fitness are associated with reduced risks of certain diseases. Effectively measuring and evaluating the physical fitness and activity levels of American youth will aid in developing a healthy and physically educated population.

Go to the WSG for homework assignments and quizzes that will help you master this chapter's content.

CHAPTER 11

Assessment of Sport Skills and Motor Abilities

OUTLINE

OBJECTIVES

After studying this chapter, you will be able to

- differentiate between skills and abilities;
- apply sound testing procedures in ability and skill assessment;
- develop psychomotor tests with sufficient reliability and validity;
- differentiate among and be able to use various types of sport skills tests;
- define and delineate the basic motor abilities; and
- develop tests and test batteries to select, classify, and diagnose athletes based on psychomotor tests.

The lecture outline in the WSG will help you identify the major concepts of the chapter.

MEASUREMENT AND EVALUATION CHALLENGE

Scott is a graduate assistant for the men's volleyball team. He is currently enrolled in a graduate measurement class. He has been asked to complete a project for the class, and he's interested in connecting the project to his work with the volleyball team. The team has some good athletes who are not especially tall, as well as some members that are quite tall but not especially quick. Team members have been tested on a series of motor performance tests related to volleyball playing potential. Scott would like to use this test battery to design individualized physical training programs for the players. He is not sure how to go about this. He has decided to talk with his measurement and evaluation instructor.

The measurement of sport skills and motor abilities is one of the fundamental aspects of the measurement of human performance. Fleishman (1964) provided the modern foundation for work in this area, including the delineation between skills and abilities. According to Fleishman, a **skill** is a learned trait based on the abilities that a person has; **abilities** are more innate than skills; skills are more sport specific, whereas abilities are more general. Battinelli (1984) summarized this relationship in an article tracing the history of the debate over generality versus specificity of motor ability, stating the following:

> *The evident trend that emerges from the literature . . . over the years seems to demonstrate that the acquirement of motor abilities and motor skills through motor learning processes is dependent upon general as well as specific factors. The general components of motor ability (muscle strength, muscle endurance, power, speed, balance, flexibility, agility and cardiovascular endurance) have become the practical physical supports to motor learning, while skill specificities have been shown to be representative of the neural physiological processes exemplified in such learning. (p. 111)*

The definition of the trait to be measured is important in determining the manner in which it will be measured. In chapters 9 and 10, we noted that not only has the definition of physical fitness changed, but also has the manner in which it is measured. In sport skills and motor performance testing, the distinction emphasized by Fleishman and the advent of computers and advanced statistical techniques have also brought about changes. These changes, as well as the accepted practices associated with the measurement of sport skills and motor abilities, are presented in this chapter.

GUIDELINES FOR SPORT SKILLS AND MOTOR PERFORMANCE TESTS

When selecting or devising tests to measure sport skills or motor ability, you should follow certain accepted testing procedures. Whether you use standardized tests or your own tests depends upon your expertise and the specific use of the test. If you will be making comparisons with other groups, use some type of standardized test. If the information is simply for you, then you can modify a standard test or develop a new one to suit your purposes.

The American Alliance for Health, Physical Education, Recreation and Dance (AAHPERD), now SHAPE America, has provided guidelines for skills test development (Hensley 1989); these are the basis for the development of the AAHPERD skills test series and also apply to the measurement of motor abilities. The guidelines state that skills tests should

- have at least minimally acceptable reliability and validity,
- be simple to administer and to take,
- have instructions that are easy to understand,
- require neither expensive nor extensive equipment,
- be reasonable in terms of preparation and administration time,
- encourage correct form and be gamelike but involve only one performer,
- be of suitable difficulty (neither so difficult that they cause discouragement nor so simple that they are not challenging),
- be interesting and meaningful to performers,
- exclude extraneous variables as much as possible,
- provide for accurate scoring by using the most precise and meaningful measure,
- follow specific guidelines if a target is the basis for scoring,
- require a sufficient number of trials to obtain a reasonable measure of performance (tests that have accuracy as a principal component require more trials than tests measuring other characteristics), and
- yield scores that provide for diagnostic interpretation whenever possible.

The guidelines also indicate that if a target is the basis for scoring, it should have the capacity to encompass 90% of the attempts; near misses may need to be awarded points. Determining target placement should be based on two principal factors: (a) the developmental level of each student (e.g., the height of a given target may be appropriate for a 17-year-old but not for a 10-year-old) and (b) the allocation of points for various strategic aspects of the performance (e.g., the placement of a badminton serve to an opponent's backhand should score higher than an equally accurate placement to the opponent's forehand).

Because most tests of sport skills or motor abilities can be objectively measured, you can expect reliability and validity coefficients higher than those associated with written instruments. The AAHPERD guidelines suggest that reliability and validity coefficients exceed .70. However, although many reliability coefficients exceed this lower-bound value, it is difficult to develop tests whose validity coefficients reach it. A validity coefficient (r) of .70 would mean that nearly 50% of the variance of the test would be associated with the criterion ($r^2 = .49$; recall the coefficient of determination from chapter 4). When examining the validity of a test, consider not only its statistical relationship with the criterion, but also its practical relevance. Also consider test feasibility as you select instruments to use.

Go to the WSG to view Video 11.1.

EFFECTIVE TESTING PROCEDURES

The procedures for psychomotor testing are the same as for written testing. These procedures may be classified as pretest, testing, and posttest duties. Thinking through all aspects of the testing procedures in detail is imperative so that you can gather consistent, accurate results. When testing in an academic setting, tests are normally administered as a pretest or at the end of a unit for summative evaluation. When testing athletes, it is important to remember that "effective testing can be done at any time during the training competition program depending on what you are looking for" (Goldsmith 2005, p. 15).

Pretest Duties

Pretest planning is the first element of preparing to administer tests. The tester must be completely familiar with the test, the items to be administered within the test, the facilities available, and the necessary equipment and markings. For a physical performance test, those being tested need to be exposed to the exact test items and allowed time to practice so that they can learn how to pace themselves and how to perform the test in the most efficient manner. This way, when the test is finally administered, the measures are accurate estimates of the actual amount of learning that has taken place rather than of the students' abilities to perform in an unfamiliar situation.

With batteries of performance tests, there are other elements to consider. The first is the order of administration. If there are several physically taxing items in a test battery, spread them out across several days so that the test participants are not unduly fatigued when they participate or perform the test. Also, pair time-consuming items with items that do not take as long.

Consider whether you will need assistants to help record the data. This is important when you determine the type of score sheets to use. Two types of recording sheets are used when multiple testers and multiple stations are involved. With one form, all the data for a test participant are recorded on the participant's sheet (figure 11.1), and these sheets are carried from station to station by participants. One problem with this method is that the sheets can be misplaced or damaged as they go from station to station. This problem can be minimized if all forms for one group are placed on a single clipboard and the clipboard rotates from station to station. The second method is to provide each tester with a master list of all those taking the test; as the participants rotate through the stations, the testers simply record the data on the master list (figure 11.2). The problem with the second approach is the lengthy transcription process in the posttest recording stage. If only one tester is available, we recommend using the master list method.

You will need to develop and write standardized instructions so that each tester knows exactly what to do. Testers themselves should be familiar with the test. The instructions should be read to the test participants so that each is given exactly the same information for the test. Consider giving each student to be tested a written copy of the instructions before testing. The students should be familiar with the test procedures and should have even practiced the exact test that they will take before taking the actual test.

Testing Duties

The second phase is the actual testing. Prepare the test location as early as possible. Make sure that the surfaces are clean so that students will be able to do their best. Also, address any safety concerns at this time, such as objects close to the testing area or unsafe surfaces or equipment. Give students the opportunity to warm up for physical performance tests, to allow them to reach their potential. Present the instructions to the students, being sure that any hints or motivational comments are consistent across all groups and all those being tested.

Posttest Duties

The final phase of the testing procedure involves transcribing the test results and analyzing the scores. The method of transcription depends upon how the data were actually gathered, and the data need to be verified any time they are transferred from one medium to another. Each time the data are transcribed, they need to be proofed by another person. This is best done by involving another tester, or test participant helpers can call out scores.

Soccer skills test score sheet

Name: ____________________

Class: __________

Speed dribble (# of touch violations in parentheses)

T1 __________ (   )

T2 __________ (   )

Control dribble

T1 __________

T2 __________

Passing		Left foot	Right foot	Shooting		
10 yd	1	_____	_____	Left target	1	_____
	2	_____	_____		2	_____
15 yd	1	_____	_____		3	_____
	2	_____	_____	Right target	1	_____
					2	_____
					3	_____

Figure 11.1 Sample individual score sheet.

Fielding, throwing, and running

Grade ______

	Name	Gender	Fielding trials					Throwing trials			Running trials		
			1	2	3	4	Total	1	2	Total	1	2	Total
1.													
2.													
3.													
4.													
5.													
6.													
7.													
etc.													

Figure 11.2 Example of a tester's master sheet.

The analysis you use depends on the purpose of the testing. If individual performance within the group is important, use norm-referenced analytical procedures. On the other hand, if the purpose of the test is to compare a person's performance with a standard, use criterion-referenced analysis.

Keep all test participants' scores as confidential as possible. If test participants help record the data, position them directly next to the tester. This can alleviate the problem of having to call out scores.

Go to the WSG to view Video 11.2.

DEVELOPING PSYCHOMOTOR TESTS

Often, teachers, coaches, and researchers are interested in administering test batteries for a certain sport or motor performance area. In the analysis of athletic performance, tests of sport skills and motor performance are often used in conjunction with each other to give a teacher or coach additional information about prospective athletes. Strand and Wilson (1993) presented a 10-step flowchart for the construction of test batteries of this nature; we present a modified version of their flowchart in figure 11.3. Although the flowchart was developed primarily for sport skills, it is also applicable to motor ability testing. The steps are as follows:

- *Step 1.* Review the criteria for a good test. Basically, these criteria concern the statistical aspects of reliability, validity, and objectivity. The tester needs to be familiar with the equipment, personnel, space, and time available for testing. A test must be appropriate for the age and gender of the students and must be closely associated with the skill in question. Consider safety aspects also at this time.
- *Step 2.* Analyze the sport to determine the skills or abilities to be measured. If you are trying to evaluate current performance, then skills tests are most appropriate, but if you are trying to determine student potential for playing, then motor ability tests might be more useful. Combination batteries can also be useful, depending on the specific purpose of the administration of the battery.
- *Step 3.* Review the literature. Once you have analyzed the sport for its salient areas, review previous test batteries and literature associated with the specific skills or motor performance areas. Consult experts (such as colleagues, textbooks, professionals, teachers, researchers) at this point.
- *Step 4.* Select or construct test items. Ensure that the test items are (a) representative of the performances to be analyzed, (b) administered with relative ease, (c) as closely associated with the actual performance as possible, and (d) practically important. Each test or item should measure an independent area. It is not an efficient use of time to have several tests that measure the same basic skill or ability.
- *Step 5.* Determine the exact testing procedures. This includes selecting the number of trials necessary for each test item, the trials that will be used to establish the criterion score, and the order of the test items.
- *Step 6.* Include peer review. Have experts such as other teachers or coaches who are familiar with the activity examine the test battery.
- *Step 7.* Conduct pilot testing. If the selected test items are based on experts' opinions and a review of the literature, the pilot study analysis will help you determine the appropriateness of the test items. Pilot testing is an important step before the test item is finalized; it helps determine the total time of administration and the clarity of instructions and can uncover possible flaws in the test.

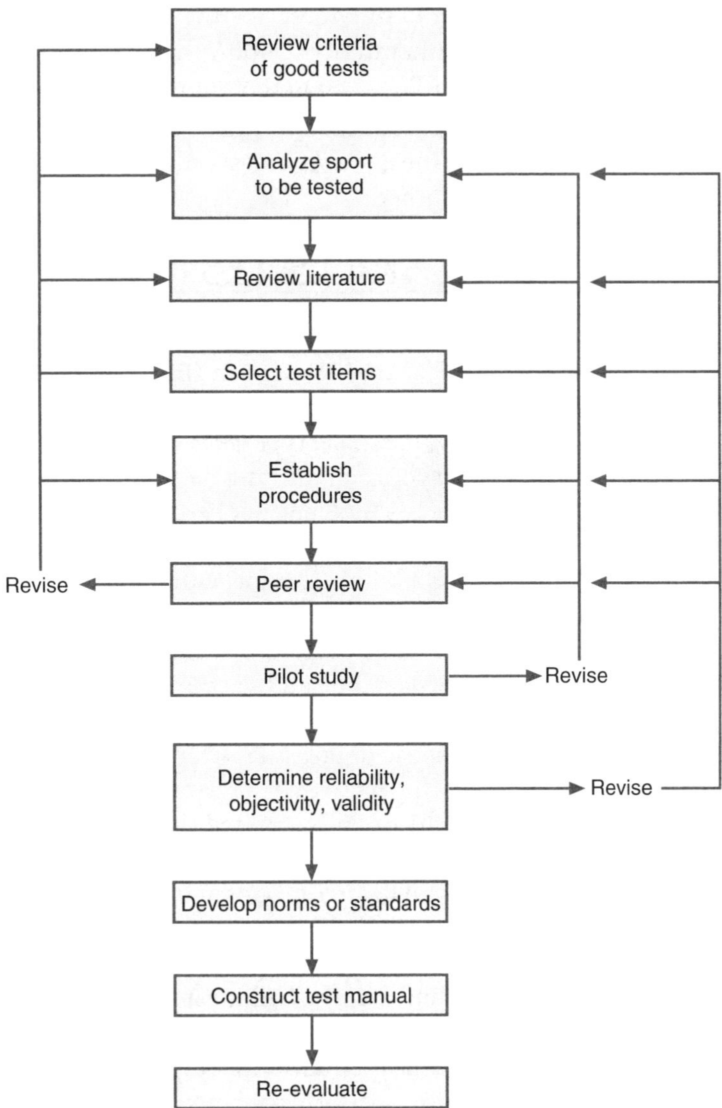

Figure 11.3 Flowchart for constructing motor performance tests.

• *Step 8*. Determine the statistical qualities of each test item—its reliability, validity, and objectivity. The reliability coefficients are estimates and are accurate only for the groups tested. They can be group specific, especially with adolescents, so it is important that both genders and all ages and skill levels included in the normative sample are tested. The methods for validating skill or motor ability tests are presented in chapters 6 and 7. Content validity is important before the statistical evaluation of the tests. You can determine concurrent validity or construct validity at this time by performing the appropriate procedures. Remember that validity coefficients are estimates and are only appropriate for groups comparable to those tested. The stability of the motor performance tests is normally established with reliability coefficients that involve repeated measures. This is the one phase of motor performance testing that separates it from the preparation phases of many psychological or educational tests. Eliminate or modify potential items at this point if they have poor reliability or validity.

• *Step 9*. Establish norms for norm-referenced tests or determine standards for criterion-referenced tests.

• *Step 10.* Construct the test manual to fully describe the test, the scoring procedures, and its statistical qualities (reliability and validity). Follow the guidelines suggested by the American Psychological Association (APA 1999) in developing a test manual.

• *Step 11.* Reevaluate the instrument from time to time. As time passes, your students may have different preparation levels; thus, norms and standards that were appropriate at one time may not be appropriate at another.

ISSUES IN SKILLS TESTING

The usual measurement issues of reliability and validity are important in skills testing. However, there are two other issues that we highlight in this chapter. One is feasibility. Skills tests typically take time to administer. You have to ask yourself *Is it more important for me to be teaching the skills and having the students perform the skills or for me to spend the time evaluating the skills?* The second issue is determining the best way to evaluate the skill. Will you need to have highly objective tests that may not be as gamelike, or is it possible to use subjective tests that are more gamelike but may have lower reliabilities? If you prefer the latter approach, see chapter 14 for additional information about performance-based assessment.

With skills testing, there often has to be a compromise between selecting a test that is extremely objective and reliable but not gamelike and selecting one that is highly valid but less objective and more time consuming. For example, consider Scott's challenge at the beginning of this chapter. He might choose to measure several skills primarily associated with volleyball performance, along with conducting some motor performance tests. One such skill is the forearm pass. Assume he initially selected three tests to measure forearm passing: the self-pass, the wall pass, and the court pass. The self-pass and the wall pass are both simple tests that can be administered to large groups and require a minimal amount of space. They both produce consistently high degrees of reliability. The problem with these tests is that they are not gamelike. For the self-pass, performers pass the ball repeatedly to themselves, the minimum criterion being that the ball must be passed at least 10 feet (3.0 m) off the ground on every repetition. For the wall pass, a performer passes the ball above a target line, again 10 feet (3.0 m) above the ground, while staying behind a restraining line 6 feet (1.8 m) from the wall. In both cases, performers can practice the test by themselves and can administer a self-test. The tests are designed for group administration and partner scoring; they are feasible, involving minimal court markings and class time. However, most volleyball experts would agree that the tests may not readily transfer to game skills—the ability to pass a serve or a hard-driven spike. The court pass test, on the other hand, places performers in a position in the backcourt, where they are most likely to receive the serve. They are asked to pass a ball that has been served or tossed by a coach or a tester to a target area. A point system is used to determine the accuracy and skill involved in the passing of 10 balls. This test has good reliability and is valid in terms of content validity. However, it requires lengthy administration time; it takes much longer to administer this gamelike court pass test than it does to administer the self-pass or wall pass tests. Also, the testers must produce reasonably consistent serves or tosses. Scott chose to go with the court pass to measure his team because of the small number of participants and their skill level. Had Scott desired to test beginners, either the self-pass or wall pass would test as effectively as the court pass.

Go to the WSG to complete Student Activities 11.1 and 11.2.

Over the past 30 years, the AAHPERD basketball test (Hopkins, Shick, and Plack 1984) and softball test (Rikli 1991) have been revised, and a tennis test (Hensley 1989) has been developed (table 11.1). AAHPERD's comprehensive testing program provides physical education instructors and coaches with a battery of reliable and valid tests that can be administered in a minimal amount of time with a minimal amount of court markings. The AAHPERD test selection guidelines indicate that skills tests should include the primary skills involved in performing a sport, usually not to exceed four. The tests of a battery should have acceptable reliability and validity and not have high intercorrelations. The battery in general should be able to validly discriminate among performance levels, thus demonstrating construct-related validity. The AAHPERD tests are designed primarily for the beginning-level performer.

Table 11.1 AAHPERD Skills Test Batteries

Basketball	Tennis	Softball
Speed spot shooting	Ground stroke	Batting
Passing	Forehand and backhand	Fielding ground balls
Control dribble	Serve	Overhand throwing
Defensive movement	Volley	Base running

Both the softball and tennis skills test manuals contain rubrics for the skills that can be substituted for or used along with the objective tests.

Go to the WSG to complete Student Activity 11.3 and view Video 11.3.

SKILLS TEST CLASSIFICATION

When you construct skills tests, consider whether to use objective or subjective procedures. Be sure that you are measuring only one trait at a time. The AAHPERD skills test batteries include only objective tests; however, subjective ratings are included as alternate forms for the tennis and softball batteries. You must consider a number of factors, including time, facilities, number of testers, and number and type of tests to be used when deciding on a specific approach to skills testing.

Objective Tests

There are four primary classifications for objective skills tests:

- Accuracy-based skills tests
- Repetitive-performance tests
- Total body movement tests
- Distance or power performance tests

Some tests may be combinations of two of these classifications. Each classification involves specific measurement issues.

Accuracy-Based Tests

Accuracy-based skills tests usually involve the skill of throwing or hitting an object, such as a volleyball, tennis ball, or badminton shuttlecock. They may also involve some other test of accuracy: throwing in football or baseball, free throws or other shots in basketball, or kicking goals in soccer. *The primary measurement issue associated with accuracy tests is the development of a scoring system that will provide reliable yet valid results.* Consider a soccer goal-scoring test. The test is set up so that a performer must kick the ball into the goal from 12 yards (11.0 m) away. For a performer to get the maximum amount of points, the ball must go between the goal posts and a restraining rope 3 feet (0.9 m) inside them. A performer is allowed six kicks: three kicks to the left side of the goal and three kicks to the right side. Two points are awarded if the ball goes into the goal in the target area, 1 point if it simply goes into the goal. The problem with this test is that good shooters often try to put the ball in the target area but end up being just slightly off, for a score of 0, whereas less competent performers may shoot every shot for the middle of the goal to ensure that they receive at least 1 point. Good shooters may thus obtain lower scores than shooters not quite as skilled. This reduces the reliability and the validity of the test.

Mastery Item 11.1

Can you detect another weakness with this soccer goal-scoring test that might affect its validity?

In volleyball tests, the court is often marked to delineate where the most difficult serves should be aimed. The problem with the soccer goal-scoring test just mentioned is inherent in this test as well, because serves that fall outside the court are given a score of 0. To rectify this problem, point values slightly lower than those assigned to the target area can be assigned to attempts falling within some area slightly outside the target. This can, however, create a problem of feasibility—that is, of marking the court for this area.

Another consideration with accuracy-based tests is the number of repetitions necessary to produce reliable scores. When the AAHPERD Tennis Skills Test was first developed, the serving component did not attain the reliability values necessary for inclusion in the battery. The Spearman–Brown prophecy formula (see chapter 6), when applied to the tennis-serving data, indicated that to improve the reliability of the test significantly, the number of trials would need to be more than doubled. This would affect the feasibility of the test. Therefore, AAHPERD developed a modified scoring system that enabled the reliability to increase to an acceptable level. When using the multiple-trial approach in serving, testers must maximize reliability while minimizing the number of trials.

An example of an accuracy-based skills test is the North Carolina State University (NCSU) Volleyball Service Test (Bartlett, Smith, Davis, and Peel 1991). Content validity was claimed because serving is a basic volleyball skill. An intraclass reliability coefficient of .65 for college students was reported. Procedures for the administration of this test are presented next.

NCSU VOLLEYBALL SERVICE TEST

Purpose

To evaluate serving in volleyball.

Equipment

- Volleyballs
- Marking tape
- Rope
- Scorecards or recording sheets
- Pencils

Instructions

Regulation-sized volleyball courts are used for the serve test; the courts are prepared as shown in figure 11.4. The tester marks point values on the floor. The test participant stands in the serving area and serves 10 times either underhand or overhand.

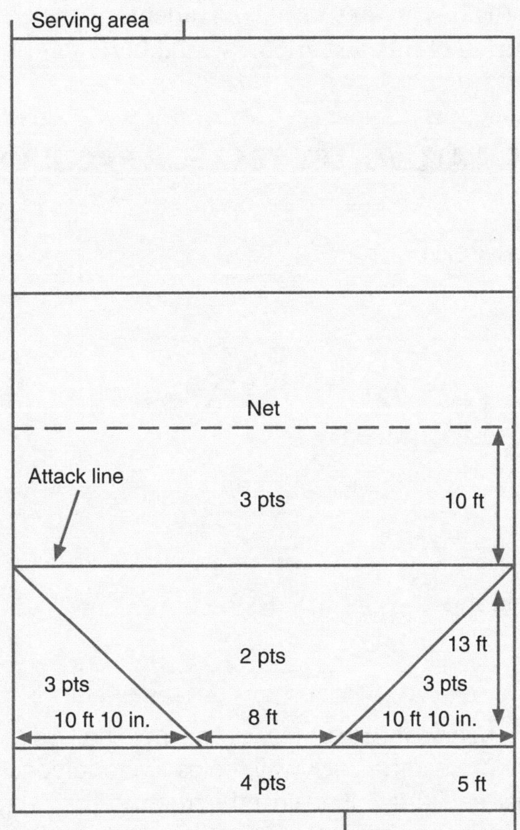

Figure 11.4 Floor plan for the NCSU Volleyball Service Test.

Scoring

The scorer gives scores of 0 to balls that contact the antennas or land out of bounds and the higher point value to balls that land on a line.

Reprinted, by permission, from J. Bartlett, L. Smith, K. Davis and J. Peel, 1991, "Development of a valid volleyball skills test battery," *Journal of Physical Education, Recreation and Dance* 62(2): 19-21.

Repetitive-Performance Tests

Repetitive-performance tests are tests that involve continuous performance of an activity (e.g., volleying) for a specified time. They are commonly called wall volleys or self-volleys and can be used to measure the strokes of racquet sports, such as the forehand or backhand stroke in tennis and volleying and passing in volleyball. Repetitive-performance tests usually have a high degree of reliability, but unless they are constructed carefully, they may not approximate the same type of action that is used in the game, in which case validity is reduced. Furthermore, because they are not gamelike, they may not transfer into actual game performance as well as some other court tests. Therefore, it is extremely important when using repetitive-performance tests that the testers make sure that the test participants use the correct form.

An example of a repetitive-performance test is the Short Wall Volley Test for racquetball (Hensley, East, and Stillwell 1979). The test was originally administered to college students but is considered appropriate for junior and senior high school students as well. Test–retest reliability coefficients are .86 for women and .76 for men. A validity coefficient of .86 has been obtained by using the instructor's rating of students as the criterion measure. Procedures for the administration of this test are presented here.

SHORT WALL VOLLEY TEST FOR RACQUETBALL

Purpose

To evaluate short wall volley skill.

Equipment

Rackets
Eye protection
4 racquetballs at each testing station
Measuring tape
Marking tape
Stopwatches
Scorecards or recording sheets
Pencils

Instructions

The test participant stands behind the short service line, holding two racquetballs. An assistant, located within the court but near the back wall, holds two additional racquetballs. To begin, the participant drops a ball and volleys it against the front wall as many times as possible in 30 seconds. All legal hits must be made from behind the short line. The ball may be hit in the air after rebounding from the front wall or after bouncing on the floor. The ball may bounce as many times as the hitter wishes before it is volleyed back to the front wall. The person being tested may step into the front court to retrieve balls that fail to return past the short line but must return behind the short line for the next stroke. If a ball is missed, a second ball may be put into play in the same manner as the first. (The missed ball may be put back into play, or a new ball can be obtained from the assistant.) Each time a volley is interrupted, a new ball must be put into play by being bounced behind the short line. Any stroke may be used to keep the ball in play. The tester may be located either inside the court or in an adjacent viewing area.

Scoring

The 30-second count should begin the instant the student being tested drops the first ball. Trial 2 should begin immediately after trial 1. The score recorded is the sum of the two trials for balls legally hitting the front wall.

Mastery Item 11.2

What else could be done to increase the reliability of the Short Wall Volley Test?

Total Body Movement Tests

Total body movement tests are often called speed tests, because they assess the speed at which a performer completes a task that involves movement of the whole body in a restricted area. A dribbling test in basketball or soccer measures this skill; base running tests in baseball and softball are also tests of this type. *These tests usually have a high degree of reliability because a large amount of interpersonal variability is associated with timed performances.* These tests can be administered quickly, but they have two inherent problems. First, a test must approximate the game performance, and in many cases flat-out speed of movement is not always required in the game. For example, in basketball, even on the fast break there must be some degree of controlled speed to allow for control of the ball. Therefore, to measure basketball dribbling skill, AAHPERD has selected a control-dribble test involving dribbling around a course of cones (Hopkins, Shick, and Plack 1984; see figure 11.5). When this type of test is used, the course must be set up to validly approximate the skill used in a game. Second, when such tests are used for evaluation, performance time is the valid

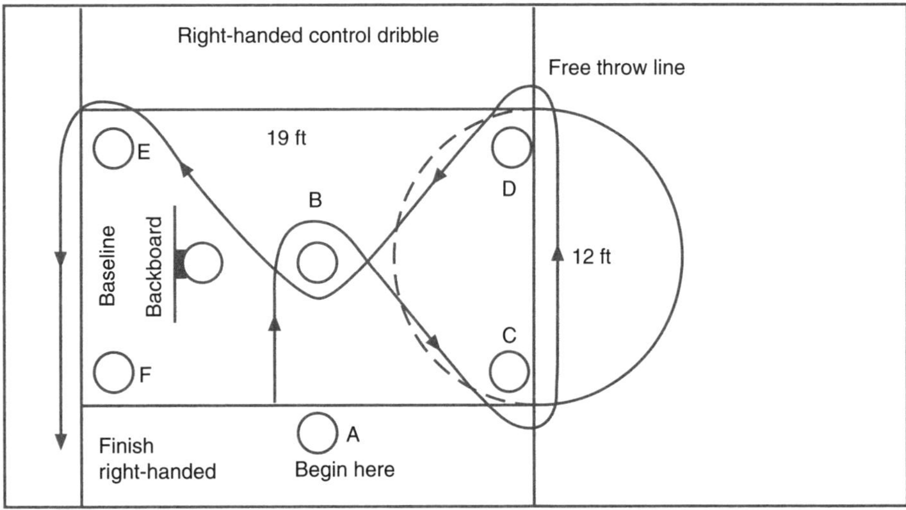

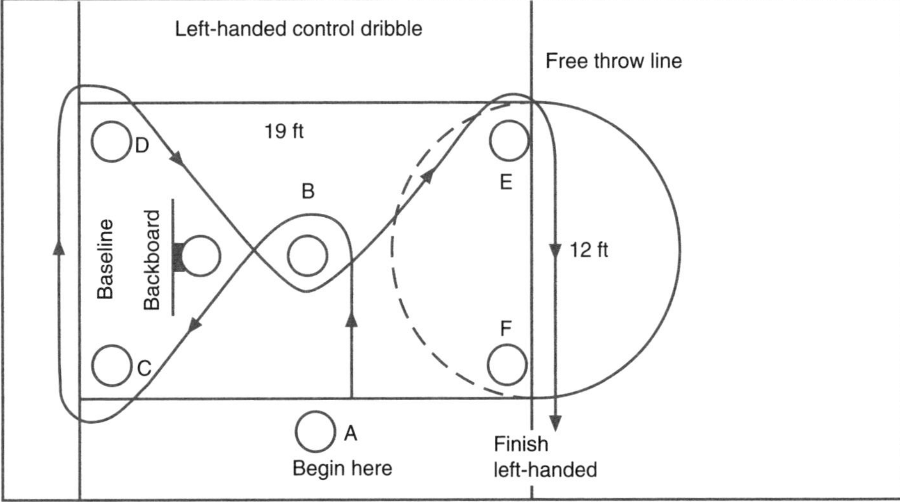

Figure 11.5 AAHPERD control-dribble test.

Reprinted, by permission, from D.R. Hopkins, J. Schick, and J.J. Plack, 1984, *Basketball for boys and girls: Skills test manual* (Reston, VA: AAHPERD).

criterion, but if you are interested in efficiency of performance, these tests would be highly related to a performer's speed. Obviously, a faster performer can complete the test in less time than a slower one, even if the slower performer has more skill in actual ball handling.

A way to eliminate the speed problem inherent in total body movement tests is to create performance ratios. You create a performance ratio by dividing a performance time by a movement time for the same participant. For example, you could compare a student's dribbling time to his or her movement time over the same course. That is, dribbling efficiency (a performance ratio) is equal to dribble time divided by movement time. Performance ratios can be extremely effective motivational tools for both highly skilled and less skilled performers because both are performing against themselves, trying to reduce the ratio to as close to 1 as possible. Theoretically, a ratio of 1 would be the minimum value on this type of test, representing the best performance. Conversely, because movement time is measured by this type of ratio, faster performers could be unduly penalized. Also, if those being tested are aware of the manner in which these ratios are created, they might not give their maximum effort on the movement time test. Doing so would allow them to achieve a better ratio score than their performance warrants. These ratios are not appropriate for use with a sports team because absolute performance is the primary measure. They are, however, an effective approach for measuring skill proficiency in a class setting.

Another example of a total body movement test is the defensive movement test from the AAHPERD Basketball Skills Test (Hopkins, Shick, and Plack 1984). Intraclass reliability coefficients above .90 were reported, and concurrent validity established for the full test battery ranged from .65 to .95. The procedures for administering this test are presented next.

Go to the WSG to complete Student Activities 11.4 and 11.5.

DEFENSIVE MOVEMENT TEST FOR BASKETBALL

Purpose

To measure performance of basic defensive movement.

Equipment

Stopwatch

Standard basketball lane

Tape for marking change-of-direction points

Instructions

The tester marks the test perimeters by the free throw boundary line behind the basket, and also marks the rebound lane lines into sections by a square and two lines (figure 11.6). Only the middle line—rebound lane marker (C in figure 11.6)—is a target point for this test. The tester marks the additional spots outside the four corners of the area at points A, B, D, and E with tape. There are three trials to the test. The first is a practice trial, and the last two are scored for the record. Each performer starts at A and faces away from the basket. On the signal, "Ready, Go!" the performer slides to the left without crossing the feet and continues to point B, touches the floor outside the lane with the left hand, executes a drop step, slides to point C, and touches the floor outside the lane with the right hand. The performer continues the course as diagrammed. The performer completes the course when both feet have crossed the finish line. Violations include foot faults (crossing of the feet during slides or turns and runs), failure of the hand to touch the floor outside the lane, and executing the drop step before the hand has touched the floor. If a performer violates the instructions, the trial stops and timing must begin again from the start.

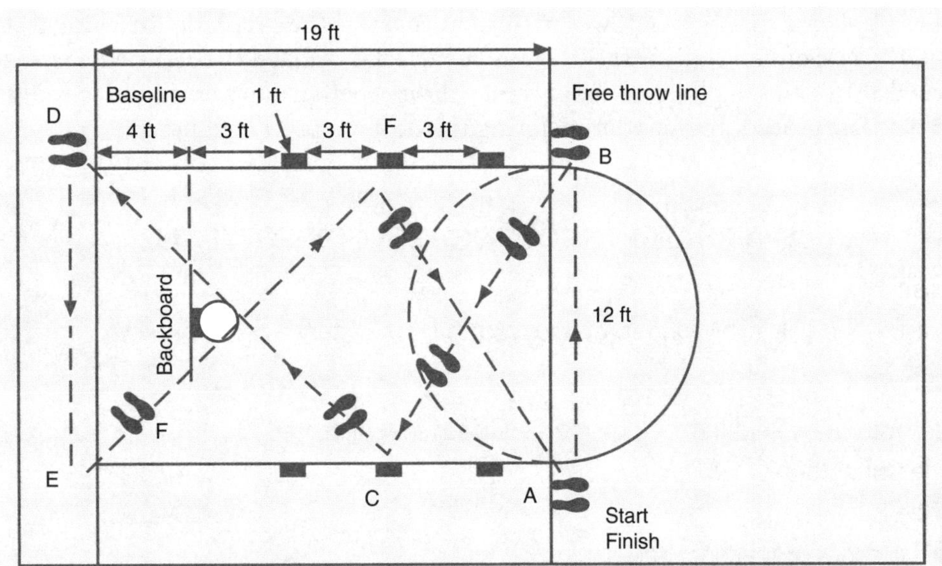

Figure 11.6 Perimeters and setup for the AAHPERD defensive movement test for basketball.

Scoring

The score for each trial is the elapsed time required to legally complete the course. Record scores to the nearest tenth of a second for each trial; the criterion score is the sum of the two latter trials.

Reprinted, by permission, from D.R. Hopkins, J. Schick, and J.J. Plack, 1984, *Basketball for boys and girls: Skills test manual* (Reston, VA: AAHPERD).

Distance or Power Tests

The final classification for objective skills tests is distance or power performance tests, which assess one's ability to project an object for maximum displacement or force. The badminton drive test for distance and the racquetball power serve test are examples of this type (Strand and Wilson 1993), as are distance throws in softball and baseball and punt, pass, and kick competitions. One problem with these tests is ensuring that they are performed in a gamelike manner. Another question with tests of distance is whether to take accuracy into account. For example, in the punt, pass, and kick contest, the distance off the line of projection is subtracted from the distance of projection. Because of this, a performer might hold back from using maximum force for fear of losing accuracy. In contrast, in track events such as the discus and shot put, these corrections are not made as long as the object is projected within a given area. Consequently, before using a test it is important to consider whether the test requires or accounts for any correction for accuracy. In the volleyball arm power test using the basketball throw (Disch 1978), accuracy is not considered an important component because arm power associated with the ability to spike a volleyball is being measured; therefore, absolute distance is the criterion. However, in other sport skills tests, such as throwing a football or baseball, the accuracy aspect may be important, so some correction needs to be used. The simplest correction is subtracting the distance off the line of projection from the total distance.

An example of a distance or power performance test is the overhand throwing test from the AAHPERD Softball Skills Test (Rikli 1991). Intraclass reliability coefficients were found to exceed .90 for all samples, and concurrent validity coefficients were found to range from .64 to .94. The procedures for administering this test are presented here.

OVERHAND THROWING TEST FOR SOFTBALL

Purpose

To measure the skill involved in throwing a softball for distance and accuracy.

Equipment

A smooth grass field that can be marked off in feet or meters

2 measuring tapes

Softballs

2 small cones or marking stakes

Instructions

The tester marks off a throwing line (or positions a measuring tape) in feet or meters down the center of a large, open field area, with a restraining line marked at one end perpendicular to the throwing line. A back boundary line is marked off 10 feet (3.0 m) behind the restraining line. The person being tested stands between the restraining line and the back boundary line, back far enough to take one or more steps before throwing. The participant, after 3 to 4 minutes of short-throw warm-ups, has two trials to throw the softball as far and as straight as possible down the throwing line, without stepping over the restraining line. Assistants or other waiting test participants are positioned in the field to indicate, using a cone or marking stake, the spot where each ball first touches the ground. If a participant steps on or over the restraining line before releasing the ball, the participant must repeat the trial.

Scoring

The net throwing score equals the throwing distance, measured at a point on the throwing line perpendicular to the spot where the ball landed, minus the error distance—the number of feet the ball landed away from the throwing line. The player's score is the better of the two throws. The tester measures both the distance and the error score to the nearest foot or meter.

Mastery Item 11.3

What are some factors that could reduce the reliability of the Overhand Throwing Test for Softball?

Subjective Ratings

Objective tests are attractive because they usually have a high degree of reliability, can be developed to produce good validity, and measure specific components of skill performance. However, **subjective ratings**, the value a rater places on a skill or performance based on personal observation, offer attractive alternatives for physical education teachers, coaches, and others interested in human performance analysis. Subjective ratings can be developed for individual skills that are often process oriented, making the ratings attractive from a teaching standpoint. A process-oriented skill is one in which the form of the skill is evaluated (e.g., in diving and gymnastics). Performers are evaluated on their preliminary position, force production phase, and follow through. Specific cues can be given to performers about where they may be losing efficiency in performance. Ellenbrand (1973) presented a classic example of a general rating scale or scoring rubric. She developed a general rating scale for the performance of gymnastics stunts (table 11.2). This scale could easily be adapted to evaluate other process-oriented performances. The performance-based evaluation procedures presented in chapter 14 are closely aligned with subjective ratings. The same reliability, validity, fairness, and feasibility issues described there are important here.

Table 11.2 Ellenbrand Gymnastics Rating Scale

The score for each item (event) is the product of a difficulty value and the executing rating. The sum of all test items in each event is the score for that event. The final test score is the sum of the scores for all events or all test items.

3 points	Correct performance. Proper mechanics. Executed in good form. Performer shows balance, control, and amplitude in movements.
2 points	Average performance. Errors evident in either mechanics or form. May show some lack of balance, control, or amplitude in movements.
1 point	Poor performance. Errors in both mechanics and form. Performer shows little balance, control, or amplitude in movement.
0 points	Improper or no performance. Incorrect mechanics or complete lack of form. No display of balance, control, or amplitude in movement.

There is no deduction for falls or repeated skills. However, a stunt that is executed with assistance receives a rating of 0.

Data from Ellenbrand 1973.

Another application of subjective ratings is to observe participants in the activity and give them a **global rating** based on their overall performance in a competitive situation. This allows the tester to evaluate several performers concurrently and possibly evaluate some intangible aspects related to game performance that are not identified by performance of individual skills in a nongame setting. The problem with this approach lies in the number of observations possible. In one game, a student may have the opportunity to contact the ball several times, but in another game that student may make few or even no contacts. Other problems with subjective ratings are defining criteria and ensuring consistency among raters. In most physical education situations, a teacher is the only available person in class to do the rating, so the evaluation may be biased by preconceived notions about the performance or the specific performer. This is also true in many coaching situations. If several ratings could be obtained, reliability would be increased. However, this would also decrease the feasibility of the testing by either requiring more raters or increasing the number of viewing sessions necessary.

Go to the WSG to complete Student Activity 11.6.

Types of Rating Scales

Verducci (1980) delineated two basic types of scales: relative scales and absolute scales. **Relative scales** are scored by comparing performance to that of others in the same group. This normative approach has the virtue of distinguishing ability well within the group but creates problems if performers are to be compared with those in other groups. Relative scales are classified as follows: rank order, equal-appearing intervals, and paired comparisons.

A widely used approach is the rank-order method. In this approach, all performers within a group are ranked on a given skill. If you are evaluating more than one skill, evaluate all the performers on one skill before moving on to the next. Rank-ordering forces differentiation among all performers, but it does not account for the degree of difference between them. Recall that the rankings result in ordinal numbers (see chapter 3).

The equal-appearing intervals method is often used when ranking groups of 20 or more participants. Using this technique, several categories are assumed to be equally spaced from one another, and the rater places participants with similar performances into the same categories. For example, categories such as best, good, average, poor, or worst might be used. A rater may decide to divide the group into the five categories. Typically, a larger percentage of participants fall into the middle categories than the extreme categories.

In the paired-comparison method, the rater compares each participant to every other participant and determines which of each pair is better than the other on the trait being assessed. When this is done for all possible pairs of participants, the results can be used to establish a relative ranking of all the participants in the group. This technique works well with groups of fewer than 10 participants.

With **absolute ratings**, a performer is evaluated on a fixed scale; performance is compared with a predetermined standard. This approach is not affected by the group in which a person is tested, and several people may end up with the same rating. Absolute scales can be classified into four types: numerical scales, descriptive scales, graphic scales, and checklists—the most widely used being numerical scales and checklists, which we will discuss in some detail. For discussions of less popular types of relative and absolute scales, see Verducci (1980, chapter 13).

Numerical scales are common. The numerical scales in tables 11.2 and 11.3 describe the levels of performance needed to earn a certain number of points. The Ellenbrand scale (table 11.2) ranges from 0 to 3, whereas the Hensley scale (table 11.3) ranges from 1 to 5. In general, numerical scales range from 1 to 9 points; it is usually difficult to accurately discriminate more than nine performance levels. Numerical scales are most useful when performers can be classified into a limited number of ordered categories, and there is consistent agreement about the characteristics of each category.

Checklists are useful if both the process and the outcome are being evaluated. **Checklists** usually mark the absence or presence of a trait. Table 11.4 presents the Level 2 Parent and Child Aquatics checklist taken from the *American Red Cross Water Safety Instructor's Manual* (American Red Cross 2009).

Table 11.3 Tennis Rating Scale: Forehand and Backhand

5 = excellent	Proper grip, good balance, footwork, and near-perfect form. Demonstrates consistent stroke mechanics. Anticipates opponent's shots. Placement appropriate for opponent's weaknesses or position.
4 = good	Proper grip, good balance, adequate footwork, and acceptable but not perfect form. Demonstrates above-average consistency of stroke mechanics. Anticipates opponent's shots. Consistent placement within court area.
3 = average	Proper grip and acceptable balance, but footwork is poor. Form is somewhat erratic and inefficient, resulting in inconsistency in shot placement. Style of play may be defensive. Little anticipation of opponent's shots.
2 = fair	Uses improper grip at times, poor footwork, and basically incorrect form. Inconsistent stroke mechanics. Defensive style of play, merely trying to get ball over net. Little anticipation of opponent's shots. Unable to sustain a rally.
1 = poor	Incorrect grip, off balance, with poor footwork. Form is very poor and erratic. Inaccurate shot placement. No anticipation of opponent's shots. Experiences difficulty in getting ball over net.

Reprinted, by permission, from L.D. Hensley (ed.), 1989, *Tennis for boys and girls: Skills test manual* (Reston, VA: AAHPERD).

Table 11.4 Parent and Child Aquatics Level 2

Skills	Completion goals
HOLDING AND SUPPORT TECHNIQUES	
Face-to-face positions	
• Hip support on front	Demonstrate
Back-to-chest position	
• Hip support on back	Demonstrate
• Back support	Demonstrate
• Arm stroke	Demonstrate
Side-to-side position	Demonstrate
Shoulder support	Demonstrate
WATER ADJUSTMENT, ENTRY, AND EXIT	
Water entry	
• Seated position	Demonstrate, with assistance
• Seated position—rolling over and sliding in	Demonstrate, with assistance
• Stepping or jumping in	Demonstrate, with assistance
• Using a ladder	Demonstrate
• Using stairs	Demonstrate
Exploring the pool	Explore, independently, in shallow water
Water exit	
• Using side of pool	Demonstrate, with assistance
• Using a ladder	Demonstrate
BREATH CONTROL	
Underwater exploration	
• Opening eyes and retrieving objects below the surface	Explore, with support, in shallow water
• Opening eyes and retrieving submerged objects	Explore, with assistance, in shallow water
Bobbing	Demonstrate
BUOYANCY ON FRONT	
Front float	Demonstrate, with assistance
Front glide	Demonstrate, with support or assistance
Front glide to the wall	Demonstrate, with assistance
BUOYANCY ON BACK	
Back float	Demonstrate, with support or assistance
Back glide	Demonstrate, with support or assistance
CHANGING DIRECTION	
Roll from front to back	Demonstrate, with assistance
Roll from back to front	Demonstrate, with assistance
SWIM ON FRONT	
Passing between adults	Demonstrate, with assistance
Drafting with breathing	Demonstrate, with assistance
Leg action—alternating or simultaneous movements	Demonstrate, with assistance
Arm action—alternating or simultaneous movements	Demonstrate, with assistance
Combined arm and leg actions on front with breathing	Explore, with assistance
SWIM ON BACK	
Leg action—alternating or simultaneous movements	Demonstrate, with assistance
Arm action—alternating or simultaneous movements	Demonstrate, with support or assistance
Combined arm and leg actions on back	Explore, with support or assistance
WATER SAFETY	
Wearing a life jacket in the water	Discuss (parent) and demonstrate (child)
Reaching assists	Discuss or demonstrate (parent)
Basic water safety rules review	Discuss (parent)
Safety at the beach and at the water park	Discuss (parent)
Water toys and their limitations	Discuss (parent)

Data from American Red Cross 2009.

Each skill is checked when the task is demonstrated or explored. This criterion-referenced approach provides a concrete evaluation of given performance levels and gives specific feedback to each performer. It is easy for an instructor to determine which competencies are lacking and to provide extra practice in those areas.

Common Errors in Rating Scales

Several common errors occur with rating scales. The most common of these is termed the **halo effect**, which is the tendency of a rater to elevate a person's score because of bias. This can occur in two ways. First, a rater may have a strong attitude about a performer that biases the evaluation in the direction of the attitude. Second, the rater may believe that the performance being evaluated is not indicative of a performer's normal level and rate that performer based on previous performances. The halo effect may also work in reverse; the rater may reduce a person's score because of negative bias.

Another common error, termed **standard error**, occurs when a rater has a standard different from that of other raters. Consider the judges' ratings in table 11.5. Inspection of the rating indicates that all three judges ordered the performance similarly; however, the numerical ratings from judge C were substantially below those from judges A and B, indicating that judge C was functioning with a standard vastly different from that used by judges A and B. This could be a major problem if all participants were not rated by all judges.

Table 11.5 Standard Error Exemplified by Three Judges' Ratings

	JUDGE		
Performer	A	B	C
1	9	8	4
2	8	9	4
3	7	7	3
4	5	6	1
5	5	5	1

A third error, called **central-tendency error**, reflects the hesitancy of raters to assign extreme ratings, either high or low. Assume that you are rating people on a scale of 1 to 5. There is a common tendency not to use the extreme categories, which reduces the effective scale to three categories (i.e., 2, 3, and 4). This not only causes scores to bunch around the mean, but also reduces the variability in the data, which can reduce reliability.

Suggestions for Improving Rating Scales

You can take several steps to alleviate many of the problems associated with rating scales.

1. Develop well-constructed scales. Here are several suggestions for doing this:
   a. State objectives in terms of observable behavior.
   b. Select traits that determine success.
   c. Define selected traits with observable behaviors.
   d. Determine the value of each trait in relation to success.
   e. Select and develop the appropriate scale for the rating instrument.
   f. Select the degrees of success or attainment for each trait and define them with observable behaviors.

g. Try out and revise the rating scale.

h. Use the rating scale in an actual test situation.

2. Thoroughly train the raters. They should have a clear understanding of the traits measured and be able to differentiate thoroughly the differences in performance levels.

3. Explain common rating errors to the raters. If they are aware of these pitfalls, they may avoid them.

4. Allow raters ample time to observe behaviors. This will increase the sampling unit of the performances.

5. Use multiple raters whenever possible. If this is not possible, then test several raters on a common group to check their objectivity. Finally, raters should rate one trait at a time, then move on to the next trait. Following this approach improves consistency.

Other Tests

In addition to objective tests and subjective ratings, there are other classifications of tests for the measurement of human performance. Performance-based testing and trials-to-criterion testing are described briefly in the following paragraphs. Performance-based testing involves actual performance of the activity that is being assessed. Trials-to-criterion testing offers instructors a way to reduce the amount of time spent testing.

Performance-Based Testing

Performance-based testing results in a classification of skill based on the actual performance that produces the score. In this situation, a concrete criterion exists. This occurs in sports such as archery, bowling, golf, and swimming. In archery, the score for a round of arrows indicates how well a performer did for that performance. Scores can be evaluated

Performance-based testing can be used in archery, where you can tell a performer's ability by looking at the total score.

to examine the stability of performance at a given distance, or performance at different distances can be evaluated to determine the concurrent validity of shooting across distance.

Bowling and golf provide a unique situation: Although you can use the total score for a game of bowling or a round of golf for assessment purposes, doing so does not evaluate specific elements of the game. Picking up a specific pin for a spare in bowling, or combining your drive, your short game, and your putting in golf, are elements that can be evaluated separately. These performances often focus on the process rather than the product, and rating scales often have a good degree of validity in these areas. The virtue of developing rating scales in these areas is that you have a concrete criterion against which to validate them. On the other hand, correlating a rating scale against an objective skills test may or may not represent validity. You may be simply getting a congruency of the two performances, neither of which may be valid.

Trials-to-Criterion Testing

An alternative approach to the measurement of skills involves the use of **trials-to-criterion testing** (Shifflett and Shuman 1982), in which a student performs a skill until he or she reaches a certain criterion performance. For example, consider a conventional free throw test involving 20 attempts. A class of 30 students would have 600 free throws to complete. Using a trials-to-criterion approach, students would be given some number of successes to complete.

For example, instead of shooting 20 free throws and counting the number made in 20 attempts, students might be instructed to shoot until they make 8 free throws. As soon as they make 8, they are to report the number of attempts it took to make the 8; the best score would be 8 attempts to make 8 shots. (At some point the test would have to be discontinued for those students who simply cannot make the 8 free throws, but for a given class, most of the shooters can probably make 8 out of 20 attempts.) If the average number of trials needed to make 8 free throws is 12 in a class of 30 students, only 360 free throws would need to be attempted (not 600), thus saving a great amount of time. If the correlation between the trials-to-criterion test scores and the scores on the conventional free throw test (involving 20 attempts) is high, you would have an attractive way of reducing the time of testing. Furthermore, in a teaching situation, this would allow a teacher to spend additional time with the weaker shooters, while the stronger shooters move on to other skills.

 Go to the WSG to complete Student Activity 11.7.

TESTING MOTOR ABILITIES

Prediction in human performance and sport has long been a popular topic of debate. Is there such a thing as a natural athlete? What physical attributes are most important for high levels of athletic performance? Is it possible to measure athletic potential and predict future athletic success?

With the advent of computers and the application of multivariate statistical techniques to the analysis of human performance, researchers are now able to explore these questions in a manner that was not possible 40 years ago. Although the statistics involved are relatively complex, the theory is basic: Attempts are made to develop test batteries that discriminate among performance levels.

Historical Overview

Early researchers operated on the theory that just as there were tests for assessing the innate ability of intelligence in the cognitive domain, there must also be a way to measure innate motor ability in the psychomotor domain. These early researchers—Rogers, Brace, Cozens, McCloy, Scott, and others—concentrated from the early 1920s through the early 1940s on determining the physical components that are basic to and necessary for successful human performance.

One of the initial attempts was the development of classification indexes for categorizing students according to ability. This was to allow physical education classes to be formed homogeneously so that they could be taught with increased efficiency. The earliest classification indexes focused on predicting ability by age, height, and weight information. McCloy (1932) developed three classification indexes in his early work that could adequately classify students:

$$\text{Elementary: } (10 \times \text{age}) + \text{weight}$$

$$\text{Junior/senior high school: } (20 \times \text{age}) + (6 \times \text{height}) + \text{weight}$$

$$\text{College: } (6 \times \text{height}) + \text{weight}$$

where age is in years, height is in inches, and weight is in pounds. By inspecting these formulas, McCloy found that at the college level, age was no longer an important factor in classification and that at the elementary level height contributed little. This was one of the first attempts to predict performance, but at this point no motor performance tests were actually used.

Next, Neilson and Cozens (1934) developed a classification index based on the same principle; however, they used powers (exponents) of height in inches, age in years, and weight in pounds. The McCloy and the Neilson and Cozens classification systems were so similar that the correlation between them was .98.

At about the same time, researchers began classifying by motor ability testing. The term **general motor ability (GMA)**—referring to overall proficiency in performing a wide range of sport-related tasks—was introduced. Rogers and McCloy developed strength indexes that included some power tests, the standing long jump, and the vertical jump, which had been found to correlate moderately with ability in a variety of activities (Clarke and Bonesteel 1935). To increase the accuracy of the prediction, test batteries were designed on the premise that certain motor abilities, such as agility, balance, coordination, endurance, power, speed, and strength, were the bases of physical performance.

An example of an early test of general motor ability is the Barrow Motor Ability Test (Barrow 1954). Although the test was initially designed for college men, norms were developed later for both junior and senior high school boys. Originally, the test included 29 test items measuring eight theoretical factors: agility, arm–shoulder coordination, balance, flexibility, hand–eye coordination, power, speed, and strength. Barrow constructed an eight-item test battery to examine the validity of the factors and obtained a multiple correlation coefficient of .92 between the composite score for the eight-item battery and a composite score based on all 29 items. Test–retest reliability coefficients ranged from .79 to .89. From a measurement standpoint, the problem associated with this validation procedure is the composite criterion: Because both the predictor tests (a sub battery) and the criterion test (the full battery) included the same tests, Barrow was simply predicting the whole from a part, giving a spuriously high correlation of .92. It is statistically invalid to use this approach. The importance of this early motor ability test battery was that Barrow started to examine the theoretical structure of the various components of the athletic ability necessary to perform a variety of sports.

@ Go to the WSG to complete Student Activity 11.8.

Larson (1941) provided another measurement approach to the examination of motor ability. He analyzed the factors underlying 27 test items and six batteries and gathered additional statistical evidence on other test batteries related to measuring motor ability. This represented an attempt to examine the construct validity of motor ability tests through the use of a statistical technique called **factor analysis**. However, he also developed these batteries using the composite criterion approach, which reduced the validity of his findings. He did find reliability coefficients in excess of .86.

The next step in the use of motor ability tests was the development of tests to measure **motor educability**—the ability to learn a variety of motor skills. These tests included track and field items as well as a number of novel stunt-like tests. One of the stunts involved having students grab one foot with the opposite hand and then jump to try to bring the other leg through the opening formed by the leg and the arm. Another item had the students jump up, spin around, and try to land facing exactly the same direction as they had faced at the start. Brace (1927) developed the Iowa Brace Test, involving 20 stunts, each scored on a pass/fail basis. McCloy later used the Iowa Brace Test as a starting point in developing a test for motor educability. McCloy's revision investigated 40 stunts and selected 21 of them. Six combinations of 10 stunts were ultimately selected to measure motor educability in the following categories: upper elementary school boys, upper elementary school girls, junior high boys, junior high girls, senior high boys, and senior high girls. A problem with tests of motor educability was the reliability of the measures. Because they were all pass/fail items, it was an either–or evaluation, which tended to reduce reliability. Furthermore, the tests tended to correlate poorly with most sport performance measures, thus casting doubt on their validity in any realistic situation.

 Go to the WSG to complete Student Activity 11.9.

The Sargent Jump was one of the first motor performance tests used to examine a person's potential for athletic performance (Sargent 1921). Named after Dudley Sargent, the test was a vertical jump purported to be a measure of leg power. This test was widely used and did measure a trait important to many power-based sports. Whereas this is a reliable test that is validly related to certain aspects of performance in a large number of sports, this single measure obviously gives an incomplete picture of overall athletic ability.

In the years that followed, the concept of **specificity**—motor abilities unique to individual psychomotor tasks—arose through the works of Henry (1956, 1958). Using correlational analysis, Henry stated that traits that have more than 50% shared variance ($r^2 > .50$; recall again the coefficient of determination from chapter 4), or, in other words, that have correlations above .70, were said to be general in nature. Any two tests that had correlations of .70 or less were said to be specific. Because of the magnitude of the correlations that Henry chose, most motor ability tests were found to be specific in nature. The cause-and-effect implication from this was that traits should be trained (and measured) specifically. Also, anyone proficient athletically would have a high degree of many of these specific traits.

From the 1940s to the 1970s, other researchers, such as Seashore, Fleishman, Cumbee, Meyer, Peterson, and Guilford, developed the notion that motor ability is specific rather than general in nature. Factors most often cited by these investigators included muscular strength and endurance, speed, power, cardiovascular endurance, flexibility, agility, and balance. In general, their theories were based on the correlations between physical factors. High correlation suggests that items have much in common, but low correlation suggests that items measure different traits. Thus, specificity of tasks can be viewed as a concurrent validity approach.

During this period, Fleishman (1964) was developing what he termed the **theory of basic abilities**. This theory serves as the basis for most of the scientific research that has been conducted subsequently in this area. Fleishman distinguished between skill and ability in the following manner: *Skills are learned traits based on the abilities that the person has, whereas abilities are more general and innate in nature than skills.* For example, the tennis serve, badminton serve, and volleyball spike are all specific skills that involve similar overarm patterns. The overarm pattern is considered the ability. Fleishman was also one of the first to examine his theory using factor analysis. His work is classic in the field.

Measurement Aspects of the Domain of Human Performance

A multitude of motor performance factors affect a person's ability to perform specific sport skills. These factors are the underlying bases for maximized human performance. You have just read about the historical development of the research in this area. Fleishman and Quaintance (1984) further examined these factors (domains) from a construct-related validity perspective (recall this term from chapter 6). They improved on earlier works by expanding the taxonomies of human performance.

Using Fleishman's work as their basis, a number of other researchers began examining the construct-related validity of various areas of human performance. The broad area of human performance is referred to as the *domain of human performance.* The structure of the areas within the broad domain—the subdomains—has been the topic of research for a number of scholars in the field. Examining the subdomains of human performance allows you to understand the qualities necessary to perform various tasks. It is important to examine these subdomains from a variety of different standpoints, because factors such as age and ability level could alter the structure of these domains. The primary subdomains of human performance are as follows:

- Muscular strength
- Speed
- Agility
- Anaerobic power
- Flexibility
- Balance
- Kinesthetic perception

Researchers have examined the construct-related validity of most of these subdomains.

Muscular strength can be classified in relation to either the body segment isolated or the method of measurement. Jackson and Frankiewicz (1975) examined the factor structure of muscular strength and concluded that there were two general dimensions: upper body and leg. To measure strength comprehensively, at least one test from each dimension should be selected. The manner in which strength is measured also affects the measurement situation. Clarke spent years examining the various aspects of isometric (static) strength (Clarke and Munroe 1970). Isotonic (now often referred to as dynamic) strength was the subject of investigation by a number of researchers, most notably Berger (1966). With the implementation of machine-based strength, the examination of isokinetic strength has become an area of research interest. Brown (2000) edited a book detailing the relationships between isokinetic strength and a variety of athletic performances and human movements. All these types of strength present different criteria for how strength is best measured. It is important when measuring strength that you select the method of measurement and the body region tested based directly on the task being evaluated.

The subdomain of *speed* is important to a number of athletic competitions and is most commonly measured by a short sprint. Disch (1979) found the subdomain of speed to consist of four dimensions: sprinting speed, controlled speed (commonly called agility), arm speed, and leg speed. Because these are distinct factors, it is necessary to measure them all if a comprehensive measure of speed is required.

 Go to the WSG to complete Student Activity 11.10.

From a mechanical standpoint, *power* is physical work divided by time, where work is defined as weight times the distance the weight is moved. Using this definition, Barlow (1970) and Considine (1970) examined the subdomain of anaerobic power, finding it to be unique to certain body regions, specifically the arms and legs. Selected tests to measure this domain are as follows:

Arm power

- One-hand shot put
- Two-hand shot put over head
- Medicine ball pitch
- Basketball throw

Leg power

- Margaria–Kalamen Leg Power Test
- Incline run

Flexibility has been found to be body-area specific. Harris (1969) examined the underlying factor structure of flexibility and found that two types of flexibility exist: movements of a limb that require only one joint action, and movements that require more than one joint action. She found 13 distinct factors and concluded that there are a number of types of flexibility within the body.

Fleishman (1964) identified a factor that he called *dynamic flexibility,* which involves a person's ability to change direction with the body parts rapidly and efficiently. This can be thought of as agility that does not involve running. This basic ability is also specific to the sport with which it is associated.

Balance is a multidimensional subdomain. Bass (1939) worked extensively with balance. Balance can be classified as either static or dynamic. Static balance is the ability to maintain total body equilibrium while standing in one spot, whereas dynamic balance involves the ability to maintain equilibrium while moving from one point to another. The types of static balance can be influenced by the restrictions of the balancing task as well as by whether the balance is performed with the eyes open or closed. Dynamic balance can be divided into simple or complex tasks based on the planes of balance involved. An example of a simple balance task involves balance on a platform stabilometer. A person has to balance in only one plane of movement. If a ball-and-socket stabilometer is used, complex balance is required in more than one plane.

Kinesthetic perception is the ability to perceive the body's position in space (Singer 1968). Although tests of it are the least objective of those that assess the subdomains of human performance, kinesthetic perception is well accepted as an area that must be considered. It is by far the most difficult to measure in terms of reliability and validity.

Perhaps the most important issue to keep in mind when you are testing in the human performance subdomains is specificity of task. This specificity relates directly to the reliability and validity of particular measurements. Remember, a test is reliable and valid only under particular circumstances (e.g., for a given gender, age, or test environment) and is not generally reliable or valid.

PURPOSES OF HUMAN PERFORMANCE ANALYSIS

Human performance analysis can be applied to several research questions. *The primary purposes of human performance analysis are selection, classification, diagnosis, and prediction.* Note that some of these concepts were initially identified in chapter 1. Analysis questions apply not only to athletic assessment, but also to job performance. Selection refers to the ability of a test to discriminate among levels of ability and thus allow someone to make choices. From an athletic standpoint, this could involve making cuts from or assigning players to teams. In job performance, selection is used in hiring. Firefighter and police recruits are often asked to complete physical performance tests that are related to the tasks they will perform in the line of duty. Classification involves clustering participants into groups for which they are best suited. In athletic situations, players are assigned to positions or events; job classification involves assignment to a task. Diagnosis concerns determining a person's shortcomings based on tests that are validly related to performance in a given area. Diagnosis is used in sport to design individualized training programs to help improve performance. In the working world, diagnostic tests could be used to examine job performance. Prediction is a fourth area. Although this area overlaps somewhat with selection and possibly classification, it does provide a slightly different approach by examining the future potential of performers. Disch and Disch (2005) reported, summarized, and provided examples of selection, classification, diagnosis, and prediction.

MacDougall and Wenger (1991) discussed benefits that an athlete can derive from motor performance testing. In their work for the Canadian Association of Sport Science, they noted the following beneficial functions of performance testing:

- Indicates athletes' strengths and weaknesses within the sport and provides baseline data for individualized training
- Provides feedback for the athlete and coach about effectiveness of the training
- Provides information about the athlete's current performance status
- Is an educational process to help the athlete and coach better monitor performance

They further stated that for testing to be effective, evaluators should follow these procedures:

- Include variables relevant to the sport.
- Select reliable and valid tests.
- Develop sport-specific test protocols.
- Control test administration rigidly.
- Maintain the athlete's right to respect.
- Repeat the testing periodically.
- Interpret the results to both coach and athlete directly.

There are two major statistical approaches to the analysis of the subdomains of human performance. The first is the correlational approach to examine the various relationships across groups. You can use multiple regression for this approach if only two groups are involved. Correlation and regression are presented in chapter 4. The second approach is the divergent group approach using discriminant analysis. The divergent group approach examines groups (can be more than two) known to differ by using variables that are logically related to performance of a skill. Tests found to discriminate between performance levels in the divergent groups are then said to be predictors from a construct-related validity standpoint. The inferential statistical procedures presented in chapter 5 can be used for this validation process also.

The following section presents examples of analysis of human performance for the purposes of selection, classification, diagnosis, and prediction.

Selection Example

An interesting study of the use of predictive validity for selection was conducted by Grove (2001). He studied 74 male baseball players who were competing at the college level: junior college (JUCO) or Division I (D1). A third group of players was formed by 16 who were drafted by the professional league within 24 months of the testing (Pro). The players were measured on run times (30 and 60 yards [27.4 and 54.9 m]), throwing speed, vertical jump, and medicine ball throw. The data were analyzed by adjusting for age differences. Significant differences ($p < .001$) were found for vertical jump, medicine ball, and throwing speed. Additional analyses indicated that the JUCO group was lower than the D1 and Pro groups on the vertical jump and medicine ball. The Pro group performed better on the throwing speed test than the other two groups. The means and standard deviations for all of the tests are presented in table 11.6. The battery was simple to administer, and with the exception of the vertical jump and medicine ball, the tests are widely used by baseball coaches and scouts at all levels. Grove concluded that the test battery used has merit as a cost-effective screening test for talent identification in baseball. He further stated that more pronounced differences may exist when players are grouped according to position as well as playing level.

Table 11.6 Descriptive Statistics and Subgroup Comparisons for the Performance Tests by Grove

	JUCO PLAYERS		DIVISION 1 PLAYERS		PRO	
Measure	Mean	SD	Mean	SD	Mean	SD
30 yd [27.4 m] time (s)	4.04	.025	4.02	0.17	3.97	0.16
60 yd [54.9 m] time (s)	7.39	.045	7.34	0.28	7.27	0.24
Radar gun (mph)**	78.14	3.45	80.61	4.63	84.40	3.49
Vertical jump (cm)*	54.96	7.53	56.04	8.32	57.33	10.70
Medicine ball (kgm)*	14.50	1.61	16.47	1.37	16.18	1.80

Note: JUCO = junior college. Group sizes for the JUCO, Div 1, and Pro players were 32, 26, and 16, respectively. For measurement conversions, refer to a site such as this: http://www.worldwidemetric.com/measurements.html.

**$p < .0005$; *$p < .001$.

Data from Grove 2001.

Another example of selection is a predictive validity study of prospective National Hockey League (NHL) players by Tarter, Kirisci, Tartar, Weatherbee, Jamnick, McGuire, and Gledhill (2009). They used aggregate physical fitness indicators to predict success in the National Hockey League. Their sample was 345 players invited to the annual combine conducted before the NHL draft. The tests selected measured upper-body strength, lower-body power, aerobic fitness, anaerobic fitness, and body composition.

Tarter and colleagues used exploratory factor analysis and receiver operating curve analysis to derive an overall composite index. Their purpose was to determine the accuracy of this index for identifying players capable of transitioning into the NHL. Tarter and colleagues also hoped to establish the construct-related validity of the test battery used. The authors' criterion for success was those players who played a minimum of five games in the NHL within a four-year period.

Tarter and colleagues found that defensemen scoring in the 80th percentile on the composite physical fitness index (CPFI) had close to a 70% probability of meeting the criterion measure. Forwards scoring in the 80th percentile had a 50% chance. When success was examined at the 90th percentile, defensemen were found to have a 72% probability and forwards a 61% probability.

Based on these results the authors developed a simplified assessment protocol that consisted of right- and left-hand grip strength, push-ups, long jump, body composition, and body weight. This simplified battery, although not as accurate as the full battery, was found to be acceptable for field testing situations. In their discussion, the authors further pointed out that physical fitness is just one ingredient required for success at the NHL level. Additional information coming from such sources as scouting reports, game statistics, and mental efficiency measures should lead to increased predictive validity. The use of this additional information will be explained in more detail in a subsequent section on sports analytics in this chapter.

Tarter and colleagues also provided a model that integrates all of the aspects they believe are important for predicting success. They concluded that using historical performance, skill observation, physical fitness, character, and neurocognitive measures together increased the accuracy of predicting success.

Go to the WSG to complete Student Activity 11.11.

Classification Example

Leone, Lariviere, and Comtois (2002) provided an excellent example of classifying athletes based on anthropometric and biomotor variables. The participants in their study were elite adolescent female athletes with a mean age of 14.3 (standard deviation of 1.3 years). The athletes competed in tennis (n = 15), figure skating (n = 46), swimming (n = 23), and volleyball (n = 16). The descriptive values for the anthropometric and biomotor variables are presented in table 11.7.

Discriminant analysis of the tests revealed three significant functions ($p < .05$). Functions are essentially the underlying dimensions that are being measured by the tests. The maximum number of significant functions possible is $K - 1$, where K is the number of groups; therefore, this test battery maximally discriminated across the four groups. Inspection of the functions revealed that the first function (i.e., dimension) discriminated between the figure skaters and all of the other sports grouped together. The variables that accounted for this discrimination were body mass, height, push-ups, and biceps girth. The second function reflected differences between volleyball players and swimmers. The variables accounting for this discrimination were body mass, biceps girth, calf girth, and height. The third function differentiated the swimmers and tennis players. The variables responsible for this discrimination were body mass, biceps girth, calf girth, sum of skinfolds, and height. The significant discriminant functions were able to classify 88% of the players into their correct sports. The classification summary table is presented in table 11.8. The results of this study indicated that elite adolescent female athletes could be properly classified into their respective sports based on the selected test battery. The anthropometric variables were primarily responsible for most significant classifications. Of course, physical and anthropometric tests alone will not perfectly classify participants. Other factors, such as motivation and desire, could affect ultimate sport performance. These factors are discussed in chapter 12.

Table 11.7 Physical Characteristics of the Athletes by Sport (Mean ± SD)

	Tennis (n = 15)	Skating (n = 46)	Swimming (n = 23)	Volleyball (n = 16)
Age (yr)	13.9 ± 1.3	14.7 ± 1.5	14.3 ± 1.3	13.8 ± 1.3
Body mass (kg)	50.6 ± 8.3	46.6 ± 8.0	54.3 ± 6.9	57.7 ± 8.3
Height (m)	1.61 ± 0.06	1.54 ± 0.07	1.62 ± 0.06	1.63 ± 0.05
Elbow (cm)	6.12 ± 0.30	5.87 ± 0.35	6.29 ± 0.26	6.40 ± 0.33
Knee (cm)	8.81 ± 0.43	8.63 ± 0.76	8.77 ± 0.34	9.31 ± 0.50
Biceps (cm)	25.5 ± 2.8	24.4 ± 2.3	27.8 ± 1.8	26.6 ± 2.2
Calf (cm)	34.0 ± 2.8	33.0 ± 2.7	34.4 ± 1.6	34.4 ± 2.2
Skinfolds (mm)	57.4 ± 17.8	47.7 ± 12.3	56.0 ± 15.0	63.1 ± 15.5
Push-ups (n)	57.8 ± 14.4	36.7 ± 13.5	62.1 ± 16.0	50.2 ± 13.5
Burpees (n)	46.1 ± 23.8	64.6 ± 33.2	52.5 ± 32.7	56.0 ± 28.4
Flexibility (cm)	37.3 ± 5.0	42.6 ± 5.1	41.0 ± 6.0	39.1 ± 6.9
$\dot{V}O_2$max (ml · kg^{-1} · min^{-1})	49.5 ± 4.4	48.3 ± 4.0	47.6 ± 3.1	48.9 ± 3.6

Data from Leone, Lariviere, and Comtois 2002.

Table 11.8 Classification for All Significant Discriminant Functions After Validation

	PREDICTED GROUP MEMBERSHIP, N (%)				
Groups	N	Tennis	Skating	Swimming	Volleyball
Tennis	15	11 (73.3)	0 (0.0)	3 (20.0)	1 (6.7)
Skating	46	0 (0.0)	46 (100)	0 (0.0)	0 (0.0)
Swimming	23	3 (13.0)	0 (0.0)	18 (78.3)	2 (8.7)
Volleyball	16	1 (6.3)	0 (0.0)	2 (12.6)	13 (81.3)

Diagnostic Example

A study by Doyle and Parfitt (1996), based on the principles of personal construct theory (Kelly 1955), examined the possibility of performance profiling athletes. This study is interesting because it presented a unique quantitative performance profiling technique that involved not only motor performance factors, but also psychological parameters. The profiling technique is presented in figure 11.7. Participants in the study were 39 track and field athletes, 22 males and 17 females, with a mean age of 20.9 years (standard deviation = 2.26). The unique aspect of this study is that the profiling technique used was to examine how each athlete currently felt about his or her preparation for competition. Instead of actually completing performance tests, the athletes were asked to respond to questions rating themselves on the various parameters displayed in the profile. The athletes were asked to respond on a scale of 1 to 10, with 1 being not important at all and 10 being of crucial importance. Their reported scores were then correlated with their performance in three upcoming competitions. To establish a criterion for success, a person's performance was recorded as a percentage of his or her personal best time divided by his or her performance time. This allowed all the athletes to be compared across various events. Multiple correlations were calculated between the performance and the event competition scores. The results of the analysis indicated that the profiling technique could validly predict competition scores. Progressively stronger relationships were found between profile scores and performance measures from competition one to competition three. It was concluded that there may be a learning process involved in the ability to rate current state more precisely.

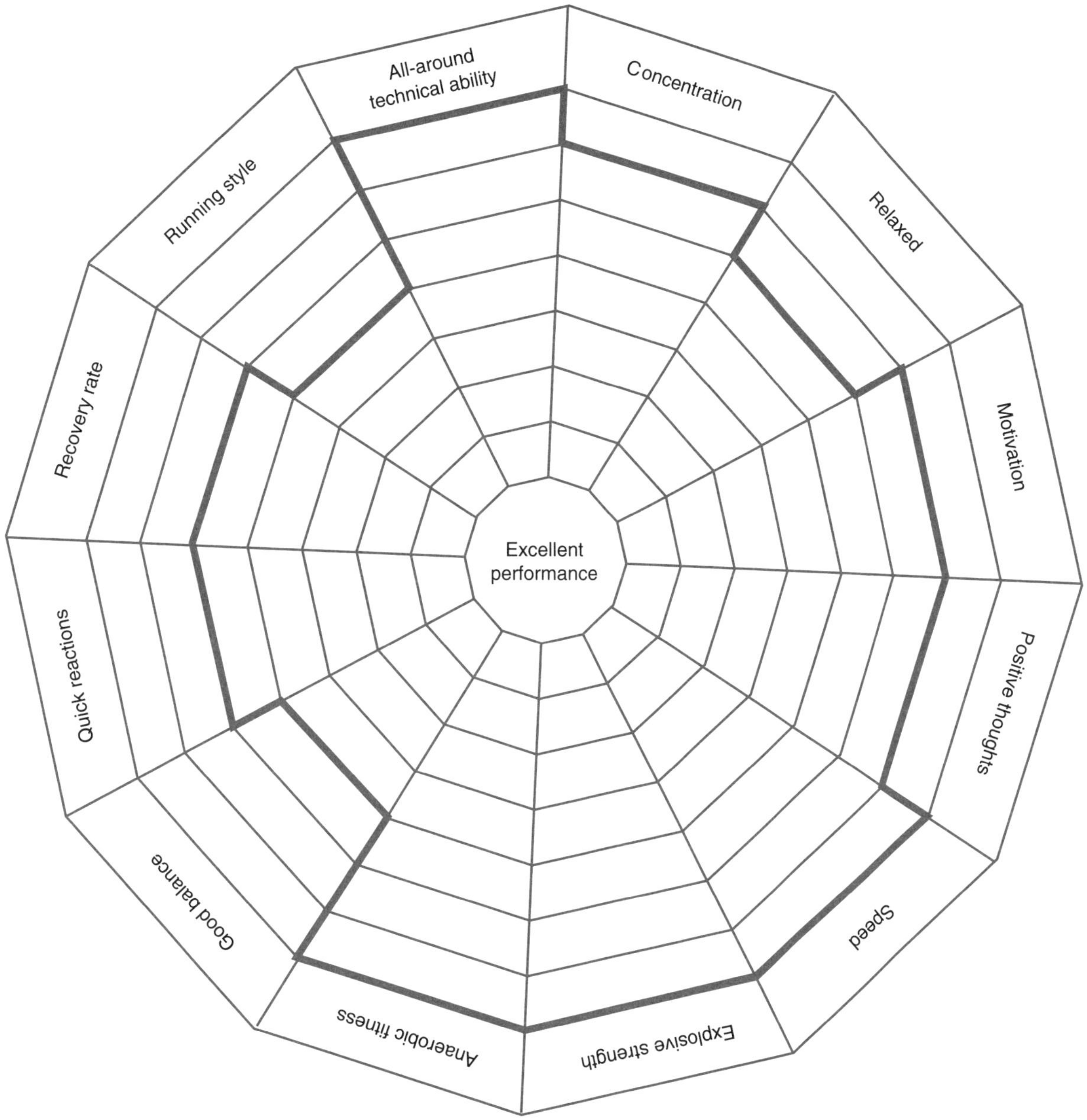

Figure 11.7 Sample performance profile.

 Go to the WSG to complete Student Activity 11.12 and view Video 11.4.

Prediction Example

Although prediction is normally accomplished through regression analysis, an interesting example of construct validity is presented by Sierer, Battaglini, Mihalik, Shields, and Tomasini (2008). They studied college football players who participated in the 2004 and 2005 National Football League (NFL) combines. The purpose of their study was to determine if selected tests used at the combine could discriminate between the drafted and nondrafted players. For the purpose of their study they characterized players into three groups: skilled, big skilled, and lineman. Skilled players were offensive and defensive backs and wide receivers. The big skilled players were linebackers, fullbacks, and tight ends. The linemen included both offensive and defensive lineman. Quarterbacks, punters, and place kickers were not included in the study. The variables included in the analysis were height, weight, 40-yard (36.6 m) dash times, bench press repetitions at 225 pounds (102 kg), vertical jump, standing long jump, pro-agility run, and the results from the three cone drill. To be included in the study, players had to have performed all of the tests. A total of 321 players met this criterion.

The data were analyzed by groups using an independent t-test. The alpha level was chosen to be .05 but was divided by three to help control for the possibility of type I errors. If the drafted players outperformed the nondrafted players, the tests were considered to possess construct-related validity. For the skilled group, the 40-yard (36.6. m) dash time, vertical jump height, pro-agility shuttle, and the results from the three cone drill were all found to significantly distinguish between the groups. For the big skilled group, the 40-yard (36.6. m) dash and the three cone drill were the only two tests that were found to be significant. For the linemen group, results from the 40-yard (36.6 m) dash, the 225-pound (102 kg) bench press, and the three cone drill were all statistically significant.

Although a number of tests were not found to significantly differentiate between the drafted and nondrafted players, all of the tests that were found to be significant were in the direction of the drafted players. The fact that different tests were found to be significant for different groups indicates the specificity of physical abilities needed to perform tasks characteristic of the various groups. It would have been interesting to use a more sophisticated statistical technique to analyze the groups and the drafted and nondrafted players at the same time. A 3 × 2 ANOVA would have allowed the authors to examine the interaction between the two independent variables.

Conclusion

Motor performance testing of athletes can be beneficial, but there are many factors that must be considered. Goldsmith (2005) listed 10 golden rules for testing competitive athletes. Those rules that someone conducting tests of this nature should remember are as follows:

1. Test for things that make sense.
2. Test because you believe it will make a difference.
3. Test with a performance-focused goal.
4. Use a battery of tests and report the results in a profile. Don't combine into a single score unless you have an extremely good reason.
5. Try to provide test results as quickly as possible. This is both for your benefit and the athlete's.
6. Although tests should measure a single trait, be aware of possible effects on other aspects of performance (e.g., does fatigue as measured by one test have an effect on technique?).
7. Don't necessarily rely on previously developed tests; develop your own!
8. Keep records. These may be useful over time. Sometimes your memory may not perceive things as they actually were.
9. Remember the Max–Min–Con Principle: Use tests that maximize the experimental variance, minimize the error variance, and control for as many other factors as possible.
10. Educate athletes about testing. Impress upon them the importance of maximal effort for optimal results.

SPORTS ANALYTICS

Sports analytics is a new term for research primarily in the area of professional sports, but with applications to sport at all levels. It came about as a result of the book *Moneyball: The Art of Winning an Unfair Game* by Michael Lewis (2003). Lewis recounted the work of Major League Baseball's Oakland A's general manager Billy Beane, who hired a statistical consultant, Paul De Podesta, as his assistant general manager to examine players' performance. Although Bill James had been working in this area for a number of years and is widely accepted as the father of Sabermetrics, Oakland was the first team to seriously apply statistical analysis to the drafting and acquisition of players.

As you've read in earlier sections of this text, most of the statistical analyses applied to testing athletes had been associated with physical or motor performance measures. The field of sports analytics examined various measures of in-game performance and also derived additional measures to help evaluate players. Beane was originally able to identify players who had been undervalued based on traditional baseball statistics: batting average, home runs, and runs batted in (RBIs). He and De Podesta began to use lesser-known measures they believed would identify undervalued players. Specifically, by using on-base percentage and slugging percentage the Oakland A's were able to discover players they could afford who could perform at the major-league level.

As a result of the Oakland A's using statistical analyses to assist in drafting players, sports analytics has become prominent in all areas of professional sport in the United States and is starting to be used by other sports and other organizations throughout the world. The use of traditional scouting departments, which was almost totally discounted in Lewis' *Moneyball,* has now become a more integral part of a comprehensive model to provide a thorough picture of prospective players.

However, there are a number of problems that occur for teams that want to use the sports analytics approach, the first of which is finding people who have both the computer and statistical background to acquire and manipulate the data. Alamar (2013) stated that the goals of sports analytics are twofold. The first goal is to save the team's decision makers time by providing all the relevant information necessary for evaluating players in an efficient manner. The second goal is to provide the same decision makers with insight into novel ways to determine who is the best player suited for that team's system. Producing this result involves team personnel at all levels buying into the use of both quantitative and qualitative measures for decision making.

According to Alamar the three essential components of a sports analytics program are (1) data management, (2) predictive models, and (3) information systems. The efficient use of these components will provide teams with the competitive advantage. Alamar states that some of the professional teams taking early advantage of analytics include the Portland Trail Blazers and the Boston Celtics of the National Basketball Association (NBA), the Philadelphia Eagles and New England Patriots of the NFL, and the St. Louis Cardinals and San Diego Padres, along with the Oakland A's, in Major League Baseball. Alamar points to the Celtics' pick of Rajon Rondo in the 2006 NBA draft as an example of using analytics to identify a player who was undervalued. The Celtics identified Rondo's ability to rebound as being an important contribution to the team.

The key to using analytics efficiently is to develop a system that works for a given team. Because of the massive amounts of data that may be gleaned from game statistics, game videos, performance testing, psychological evaluations, medical testing, and the like, it is imperative that a comprehensive plan be developed to acquire and analyze this information. Both quantitative and qualitative sources must be blended into a report that provides decision makers with usable information. Analytics simply provides information; it does not make decisions.

A number of research studies have been conducted in the area of sports analytics, but much of the information is proprietary to the team that conducted the research. The criterion for a successful model is often dichotomous, which leads to a number of statistical approaches producing successful results. Most statistical analyses in the area of sports analytics involve the use of some type of multiple regression model. The criterion may be wins and losses when examining overall team performance, or it might be the probability of success when drafting or trading for players. The key to success in developing a valid analytical model is based on the approach developed by the specific team. Alamar (2013) states that there are five basic questions that must be answered before any analysis is attempted:

1. What was the thought process that led to the analysis?
2. What is the context of the result?
3. How much uncertainty is in the analysis?
4. How does the result inform the decision making process.
5. How can we further reduce the uncertainty? (p. 55)

The Houston Rockets and the Houston Astros, professional men's basketball and baseball teams, respectively, have become deeply committed to the use of analytics on the sport side and on the business side. Daryl Morey, general manager of the Houston Rockets, was not necessarily selected because of his basketball background, but rather for his thorough knowledge of content important to sport analysts. With a degree in computer science and an emphasis in statistics from Northwestern and an MBA in management from MIT, Morey was able to bring all the quantitative skills necessary to sports analytics into his position with the Houston Rockets. Jeff Luhnow, general manager of the Houston Astros, has a similar quantitative background, with an undergraduate degree from the University of Pennsylvania in economics and engineering and an MBA from Northwestern. One of Luhnow's first hires for the Astros was former NASA engineer Sig Mejdal, who had worked for the St. Louis Cardinals for a number of years. As the Astros' director of decision sciences, Mejdal and his staff of four are charged with acquiring, manipulating, and analyzing vast amounts of information that will be useful to the Astros in developing a championship-level baseball team.

Another sports analytics text is Winston's *Mathletics* (2009), which presents a number of analytical studies in baseball, football, basketball, and other sports. It gives examples of how a variety of questions can be approached from a statistical standpoint—using variables that are commonly available—for creating measures that provide novel insights into the analysis of success in sport. Topics that can be examined through the use of analytics include predicting team success from one year to the next, comparing various players to determine their actual worth for trade purposes or salary negotiations, and analyzing various lineups to determine the probability of success in varying situations.

We are just beginning to see the widespread use of analytics by virtually all professional sports teams. Although analytics may be used at any level, the need to acquire, manipulate, manage, and analyze extremely large data sets requires extensive computer and statistical skills. The days of using only qualitative recommendations and quantitative data acquired through performance testing for sport decision making are over!

MEASUREMENT AND EVALUATION CHALLENGE

Scott found that tests should be selected that measured basic abilities pertinent to the sport of volleyball. The information he gathered from administering these tests could be used as a diagnostic technique to help him develop individualized training programs for his athletes. He calculated percentiles for the data and generated performance profiles for all of the players. An example of this type of profile is presented in table 11.9 (Disch and Disch 1979). By examining the profiles, he could see which motor performance areas needed to be developed for each player.

In table 11.9, three players are examined. Player 1 had high percentile scores on all the motor performance tests and the anthropometric measures. This player was selected as an All-American setter at the U.S. Volleyball Association Collegiate Championships. The profile of player 2 includes high scores in the motor performance characteristics but lower scores on the anthropometric characteristics. This player was an excellent setter but did not quite reach the level of performance of player 1. Player 3 was found to have favorable anthropometric characteristics, but his motor performance profile was much below that of the first two players. The data indicate that player 3 needs to concentrate on improving his motor performance characteristics, which should improve his performance on the volleyball court.

Table 11.9 Men's Volleyball Performance Profile

Percentile	Weight (lb)	Height (in.)	Reach (in.)	Percent body fat	Vertical jump (in.)	Triple hop (in.)	Agility run (s)	20-yd dash (s)
99	200	78		5.70		344	7.7	2.5
95	189	77	100	5.94	29	341		2.7
90	188		99	6.15	27	333	7.8	
85	185	76	94.5	6.58	26	330		2.8
80	183	75.5	97	6.86	25	319	7.9	
75	182		96	6.99		313		
70	181	75		7.30	24	303	8.1	
65	180		95.5	7.41		302		2.9
60	179	74	95	7.55		297		
55	174		94.5	7.6	23	296	8.3	
50	172		94	7.74		295	8.5	
45	169	73		8.09		292		
40	162		93.5	8.21		287		
35	161		93	8.47		285		3.0
30	158	72	92.5	9.68		279	8.8	
25	157	71		9.88	22	276		
20	156	70.5		10.15				
15	154		90	10.88	21	272	8.9	3.1
10	151	70	89	11.63		266	9.4	3.3
5	136	69	88	11.63	20	254	9.5	3.4

Note: Player 1: solid line; Player 2: dotted line; Player 3: dashed line.

Dataset Application

The chapter 11 large dataset in the WSG provides sample data on high school football players. There are position, anthropometric, strength, endurance, and speed variables. Use SPSS to conduct the following:

1. Calculate descriptive statistics on the variables (chapter 3).
2. Calculate the correlations among the various variables (chapter 4).
3. Determine if various position players differ on measured variables (chapter 5).
4. Use SPSS and develop histograms graphing height and weight by player position (chapter 3).
5. Use the descriptive statistics generated in #1 above and frequencies (chapter 3) to develop performance profiles for the treadmill, strength, and speed variables. Note that a lower time is better for the 30-yard (27.4 m) dash.

SUMMARY

Reliable and valid measurement of sport skills and basic physical abilities has a prominent place in human performance testing. Assessment of psychomotor ability is an essential task that may confront you as a physical therapist, personal trainer, physiologist, physical education instructor, athletic coach, or other human performance professional. Sport skills testing will be important to you as a physical education instructor, athletic coach, physical therapist, exercise scientist, or other human performance professional. A reliable and valid testing program will help you become a respected professional in human performance.

Skills testing comprises a variety of test methods, including objective procedures, subjective ratings, and direct performance assessment. An extensive presentation of a wide range of current sport skills tests is beyond the scope of this text. Those of you who are interested can find a thorough compilation of sport skills tests in Strand and Wilson (1993) and Collins and Hodges (2001). Motor ability testing, as you have read, has a long history in human performance and will take on increased importance in athletics and employment testing. *The most important consideration is selecting valid tests that meet your test objectives and are feasible in terms of time and effort.* The work of Kirby (1991) is a classic resource for descriptions and critiques of motor performance tests. Sports analytics, with its myriad variables and capabilities of analyzing large datasets quickly, has the potential to influence team and player selection, training, and performance.

Go to the WSG for homework assignments and quizzes that will help you master this chapter's content.

CHAPTER 12

Psychological Measurements in Sports and Exercise

Robert S. Weinberg, Miami University; Joseph F. Kerns, Miami University

OUTLINE

OBJECTIVES

After studying this chapter, you will be able to

- define and identify the scope of the field of sport psychology;
- illustrate the differences between the performance enhancement aspect and mental health aspect of sport psychology;
- differentiate between psychological states and traits;
- explain the differences between general and sport-specific psychological tests;
- discuss ethics and cautions in the use of psychological testing with athletes;
- describe the qualifications necessary for using and interpreting psychological tests;
- illustrate the feedback process involved in psychological testing of athletes;
- discuss the use and abuse of psychological tests for team selection;
- identify factors related to the reliability and validity of general and sport-specific psychological inventories typically used in sport and exercise settings; and
- differentiate between the research and application perspectives of psychological inventories used in sport and exercise.

@ The lecture outline in the WSG will help you identify the major concepts of the chapter.

MEASUREMENT AND EVALUATION CHALLENGE

Bill has just been hired to coach a team in the National Football League (NFL). Bill was a football player in college, but because he didn't have as much natural ability as many of his opponents or teammates, he always believed that his mental skills and competitiveness really helped him achieve the high level he had attained. For example, Bill believed that he was a self-motivated athlete, was usually able to control his emotions, was able to keep focused throughout the game, was generally confident in his abilities, and didn't let a series of bad losses get him down. Therefore, as he prepares to take over his first head coaching assignment, he believes that it's important not only to assess his players' physical abilities but also to evaluate their mental skills. However, Coach Bill has little background in assessing players' personality and mental skills and thus he has many questions. For example, what psychological inventories should he use to evaluate his players? Should he be administering and interpreting these tests, or should he hire a qualified sport psychologist to do so? When during the season should these psychological inventories be administered? Should he use interviews to find out about his players' mental skills? These are difficult questions, but if he can get the answers, Coach Bill believes that the information derived from these psychological assessments will be invaluable to him and his players in understanding and improving their mental skills.

The purpose of this chapter is to provide you with an introduction to the evolving field of sport and exercise psychology and to highlight the measurement techniques and instruments typically used to assess psychological attitudes, states, and traits. In addition, we discuss issues related to the measurement and interpretation of psychological tests along with ethical considerations when using psychological tests with athletes and sport participants.

SPORT PSYCHOLOGY: PERFORMANCE ENHANCEMENT AND MENTAL HEALTH

The field of sport psychology has developed so rapidly that many people do not have a clear understanding of what the field encompasses. Most definitions of sport psychology clearly underscore two main areas: *performance enhancement* and *mental health.* Many people see the field of sport psychology as narrow, when in fact it has a wide scope and applications to many areas of our lives. This greater breadth is represented in the field of sport and exercise psychology.

The performance enhancement aspect of sport and exercise psychology refers to the effects of psychological factors on sport performance. These psychological factors include anxiety, concentration, confidence, motivation, mental preparation, and personality (the totality of a person's distinctive psychological traits). The performance enhancement aspect of sport and exercise psychology is not restricted to elite athletes. Rather, it spans a continuum from young athletes participating in youth sports all the way to older adults playing in recreational or competitive leagues. *The key point is that the mind affects the body; therefore, the way we think and feel has a strong impact on how we physically perform.* In competitive sports, participants' physical skills are often comparable, and the difference between winning and losing is often in their mental skills.

AP Photo/Marc Serota

Sport psychology can help athletes manage frustrations and anxiety.

The other main focus of sport and exercise psychology is enhancing mental health and well-being through participation in sport, exercise, and physical activity—that is, enhancing the psychological effects of participation in sport, exercise, and physical activity. Our minds can have an important effect on our bodies; conversely, our bodies (i.e., the way we feel physically) can have an important effect on our minds. For example, research indicates that vigorous physical activity is related to a reduction in anxiety (distress and tension caused by apprehension) and **depression** (a mental condition of gloom, sadness, or dejection). Similarly, sport participation has been related to increases in self-esteem and self-confidence. *In essence, sport, exercise, and physical activity have been shown to have the capacity to increase our feelings of psychological well-being and thus exert a positive influence on our mental health.*

Participating in competitive sports can also sometimes be frustrating and upsetting when we lose or don't perform up to expectations; this can lead to increases in anxiety, depression, and aggressiveness. Sport and exercise psychologists attempt to accentuate the positive aspects of sport and exercise participation so that participants will receive positive psychological benefits.

 Go to the WSG to complete Student Activity 12.1 and view Video 12.1.

Researchers studying the performance enhancement aspect and those studying the mental health aspect of sport and exercise psychology have different objectives, so it is not surprising that their measurement objectives and the types of psychological tests they use differ considerably (although there is some overlap). For example, in studying performance enhancement, researchers are typically interested in measuring psychological factors that affect performance. Their tests might measure attentional focus, confidence, precompetitive anxiety, self-motivation, and imagery. For researchers interested in studying mental health, tests that measure anxiety, depression, self-esteem, self-concept, mood, and anger are appropriate.

Let's discuss some of the issues that sport and exercise psychologists face when measuring psychological factors that affect performance and the psychological results of participation in sport and exercise.

TRAIT VERSUS STATE MEASURES

When assessing personality and psychological variables in sport psychology, it is important to distinguish between trait and state measures. Trait psychology was the first scientific approach that evolved for the study of personality. Trait psychology is based on the assumption that personality **traits**—the fundamental aspects of personality—are relatively stable, consistent attributes that shape the way people behave. *In essence, the trait approach considers the general source of variability to reside within the person; it minimizes the role of situational or environmental factors.* This means that a person who assesses at a high level for the trait of aggressiveness would tend to act more aggressively in most situations than someone who assesses at a lower level for the same trait.

Traits, or predispositions, may be acquired through learning or be constitutionally (genetically) inherent. A well-known trait typology is extroversion–introversion. These traits relate to a person's general tendency to respond in either an outgoing or a shy manner, without regard for the situation. For example, an extrovert placed in a new situation in which he or she doesn't know anybody will likely try to meet people and be outgoing. In sport psychology, the traits that have been studied include anxiety, aggressiveness, self-motivation, confidence, and achievement motivation.

An alternative to the trait approach is to focus on the situation. In the situational, or **state** approach, behavior is expected to change from one situation to the next, and traits are given a subsidiary role in the explanation and prediction of behavior. *Psychological states are viewed as a function of the particular situation or environment in which a person is placed; thus, when the situation changes, so does the psychological state.* In essence, psychological states are transitory, potentially changing rapidly in a short time as the situation changes.

For example, assume you are a reserve member of a basketball team and usually just sit on the bench. Your team is playing in a championship game, but because you are not expected to play much, your state anxiety right before the game is low. However, your coach walks up to you and tells you that one of the starters is ill and won't be able to play and that you're going to be starting. In a few seconds, your state anxiety has greatly elevated, because you are now anxious about how well you will perform in this important game. The situation of starting on the team has caused a dramatic shift in your state anxiety level; this result has little to do with your trait anxiety.

Go to the WSG to complete Student Activity 12.2.

Although sport psychologists make the distinction between traits and states, you need to consider both when attempting to understand and predict behavior in sport and exercise settings. The idea that traits and states are codeterminants of behavior is known as the interactionalist approach; it is the approach most widely endorsed today. This approach to personality and behavior suggests that one's personality, needs, interests, and goals (i.e., traits), as well as the particular constraints of the situation (e.g., won–loss record, attractiveness of the facility, crowd support), interact to determine behavior. Thus, from a measurement perspective, it is imperative that both traits and states be considered when we are attempting to understand and predict behavior in sport and exercise settings.

An understanding of both trait and state measures can help a coach and athlete better prepare for the demands of competition.

A study conducted by Sorrentino and Sheppard (1978) demonstrated the usefulness of considering both trait and state variables in an interactionalist approach. Their study tested male and female swimmers by having them perform a 200-yd (182.8 m) freestyle time trial, once swimming individually and once as part of a relay team. The situational factor assessed was whether each swimmer had a faster individual split time when he or she swam alone or as part of a relay team. In addition, the personality characteristic of affiliation motivation was assessed. This personality trait represents whether the swimmers were more approval oriented—viewing competing with others as positive—or were more rejection threatened—feeling threatened because they might let their teammates down in a relay situation. As the researchers predicted, the approval-oriented swimmers demonstrated faster times in relays than in individual events. In contrast, the rejection-threatened swimmers swam faster in individual events than in relays. Thus, the swimmers' race times involved an interaction between their personalities (affiliation motivation) and the situation (individual versus relay).

However, it should be noted that the relationship between states and traits may not always be distinct. In fact, Brose, Lindenberger, and Schmiedek (2013) found that current affective states influenced scores on trait responses. For example, participants who felt better at the time of measurement were more likely to report above-average appraisals of their trait affect and believe that current states were more relevant than general traits. How a person feels in a particular moment can influence his or her evaluation of their general affect (i.e., how they generally feel). These findings expose a prevalent methodological (validity) concern regarding single measures of traits, which may be corrected by taking multiple measurements.

GENERAL VERSUS SPORT-SPECIFIC MEASURES

For many years, almost all the trait and state measures of personality and other psychological attributes used in sport psychology came from general psychological inventories.

These inventories measured general or global personality traits and states, with no specific reference to sport or physical activity. Examples of such inventories are the State–Trait Anxiety Inventory (Spielberger, Gorsuch, and Lushene 1970), the Test of Attentional and Interpersonal Style (Nideffer 1976), the Profile of Mood States (McNair, Lorr, and Droppleman 1971; also see Morgan 1980), the Self-Motivation Inventory (Dishman and Ickes 1981), the Eysenck Personality Inventory (Eysenck and Eysenck 1968), and the Locus of Control Scale (Rotter 1966).

Psychologists have found that situation-specific measures provide a more accurate and reliable predictor of behavior in specific situations. For example, Sarason (1975) observed that some students were doing poorly on tests, not because they weren't smart or hadn't studied, but simply because they became too anxious and froze up. These students were not particularly anxious in other situations, but taking exams made them extremely anxious. Sarason labeled such people test anxious and devised a situation-specific test called Test Anxiety that measures how anxious a person feels before taking exams. This situation-specific test was a better predictor of immediate pretest anxiety than a general test of trait anxiety. Clearly, we can better predict behavior when we have more knowledge of the specific situation and how people tend to respond to it.

Along these lines, sport psychologists have begun to develop sport-specific tests to provide more reliable and valid measures of personality traits and states in sport, exercise, and physical activity contexts. For example, your coach might not be concerned about how anxious you are before giving a speech or before taking a test, but he or she would surely be interested in how anxious you are before competition (especially if excess anxiety is detrimental to your performance). A sport-specific test of anxiety would provide a more reliable and valid assessment of an athlete's precompetitive anxiety than a general anxiety test. Some examples of psychological inventories developed specifically for use in sport and physical activity settings include the Sport Anxiety Scale (Smith, Smoll, and Schutz 1990), the Task and Ego Orientation Sport Questionnaire (Duda 1989), the Sport Motivation Scale (Briere, Vallerand, Blais, and Pelletier 1995), the Physical Self-Perception Profile (Fox and Corbin 1989), the Sport Imagery Questionnaire (Hall, Mack, Paivio, and Hausenblas 1998), the Competitive State Anxiety Inventory-2 (Martens, Vealey, and Burton 1990), the Group Environment Questionnaire (Widmeyer, Brawley, and Carron 1985), and the Trait Sport Confidence Inventory (Vealey 1986). Some sport psychologists have even gone a step further and developed tests for a specific sport rather than for sports in general, such as the Tennis Test of Attentional and Interpersonal Style (Van Schoyck and Grasha 1981), the Anxiety Assessment for Wrestlers (Gould, Horn, and Spreeman 1984), the Group Cohesion for Basketball (Yukelson, Weinberg, and Jackson 1984), and the Australian Football Mental Toughness Questionnaire (Gucciardi, Gordon, and Dimmock 2008).

To obtain better predictions of behavior, a number of sport-specific multidimensional inventories have been developed that measure a variety of psychological skills believed important for performance success. The first of these scales to receive attention was the Psychological Skills Inventory for Sports scale (Mahoney, Gabriel, and Perkins 1987). Although this scale was used a great deal after its development, it was later demonstrated that the psychometric properties were not up to standard. Thus, other measures were developed, such as the Test of Performance Strategies (TOPS; Thomas, Murphy, and Hardy 1999), which measures eight mental skills in competition and in practice (see table 12.1); the Ottawa Mental Skills Assessment Tool-3 (Durand-Bush, Salmela, & Green-Demers, 2001) assessing 12 mental skills fitting under the three general categories of cognitive skills (e.g., imagery, competition planning, refocusing), foundation skills (e.g., goal setting, self-confidence, commitment), and psychosomatic skills (e.g., relaxation, activation, fear control); and the Athletic Coping Skills Inventory-28 (ACSI-28; Smith, Schutz, Smoll, and Ptacek 1995), which has seven subscales (e.g., concentration, peaking under pressure, freedom from worry). It has been argued that the ACSI-28 measures coping skills as opposed to psychological skills, but for the purposes of this chapter it measures psychological aspects related to performance. These scales are becoming more popular because researchers often attempt to determine the effectiveness of a mental skills training program in part by showing that it can enhance a number of psychological skills. These types of scales have also been used in investigations of psychological characteristics of elite performers. For example, Taylor, Gould, and Roio (2008) used the TOPS to help discriminate between more and less successful U.S. Olympians at the 2000 Olympic Games in Sydney.

Table 12.1 Sample Items from the Test of Performance Strategies

COMPETITION STRATEGY
I talk positively to get the most out of competitions.
I perform without consciously thinking about it.
I visualize competition going exactly the way I want it to.
When the pressure is on I know how to relax.
I evaluate whether I achieve my competition goals.
PRACTICE STRATEGY
I use practice time to work on relaxation techniques.
My attention wanders while training.
I have difficulty increasing my energy level during workouts.
I talk positively to get the most out of practice.
I set goals to help me use practice time effectively.

Items are scored on the following scale: 1 = never; 2 = rarely; 3 = sometimes; 4 = often; 5 = always.

Data from Thomas, Murphy, and Hardy 1999.

Over time, the construct of perfectionism has evolved from a general to a sport-specific measure. Specifically, with the increased interest in perfectionism in sporting contexts, Dunn, Causgrove Dunn, and Syrotuik (2002) created the Sport Multidimensional Perfectionism Scale (Sport-MPS) to address the necessity to evaluate perfectionism specifically in a sport-domain context. A second version, the Sport Multidimensional Perfectionism Scale 2 (Sport-MPS-2; Gotwals and Dunn 2009), updated and expanded the original scale to include additional subscales. Subsequent research has supported the psychometric properties of the instrument and demonstrated the scale's ability to differentiate global from sport perfectionism, as well as healthy from unhealthy perfectionist tendencies (Gotwals, Dunn, Causgrove Dunn, and Gamache 2010). Multidimensional and sport-specific measures may help researchers and practitioners better understand and measure states and traits in sport (e.g., perfectionism, anxiety) and thus allow for more accurate prediction of behavior.

Go to the WSG to complete Student Activity 12.3 and view Video 12.2.

CAUTIONS IN USING PSYCHOLOGICAL TESTS

Psychological inventories are crucial to sport psychologists from theoretical and applied perspectives. Such tests help to evaluate the accuracy of different psychological theories and provide practitioners with a tool for applying theory to their practice. We focus on the use of psychological tests in applied settings because it is here that abuses of test results and misconceptions in analysis are more likely. Who is qualified to administer psychological tests to athletes? The American Psychological Association (APA) recommends that test givers have the following knowledge and offers some cautions when giving psychological tests:

1. An understanding of testing principles and the concept of measurement error. A test giver needs a clear understanding of statistical concepts such as correlation, measures of central tendency (mean, median, and mode), variance, and standard deviation. No test is perfectly reliable or valid. Tests work only in specific settings.

2. The ability and knowledge to evaluate a test's validity for the purposes (decisions) for which it is employed. A qualified test administrator will recognize that test results are not absolute or irrefutable and that there are potential sources of measurement error. He or she will do everything possible to eliminate or minimize such errors. For example, testers must be aware of the potential influences of situational factors as well as interpersonal factors that may alter the way test scores are interpreted. In addition, cultural, social, educational, and ethnic factors can all have a large impact on an athlete's test results. Finally, an athlete may answer in a socially desirable light (i.e., faking good) such as saying on a test that he or she is calm, cool, and collected, when he or she really feels nervous and tight in critical situations. As noted later, this distortion of information can render a test virtually useless.

A test not only has to be reliable and valid but also needs to be validated for the particular sample and situation in which it is being conducted. For example, you might choose a test that was developed using adults and administer it to athletes aged 13 to 15 years. However, the wording of the test might be such that the younger athletes do not fully understand the questions; thus, the results are not relevant. Similarly, a test might have been developed on a predominantly white population, and your athletes happen to be mostly African American and Hispanic. Cultural differences might lead to problems in interpreting the results with different populations.

3. Self-awareness of one's own qualifications and limitations. Unfortunately, in sport psychology there have been cases in which people were not aware of their own limitations and were thus using tests and interpreting results in a manner that was unethical and in fact potentially damaging to the athletes. For example, many psychological inventories are designed to measure psychopathology or abnormality. To interpret the test results, a test giver needs special training in psychological assessment and possibly in clinical psychology. Without this background, it would be unethical for someone to use such tests with athletes. Thus, test givers should make sure they have the appropriate training to administer and interpret the tests that they employ.

4. Some psychological tests are used inappropriately to determine whether an athlete should be drafted onto a team or to determine if an athlete has the right psychological profile for a certain position (such as a middle linebacker in U.S. football). This practice was particularly rampant in the 1960s and 1970s but seems to have abated. These unethical uses of psychological tests can cause an athlete to be hastily eliminated from a team or not drafted simply because he or she does not appear to be mentally tough. In truth, however, it is difficult to predict athletic success from the results of these tests. An athlete's or a team's performance (often measured in terms of winning and losing) is a complicated issue affected by such factors as physical ability, experience, coach–player compatibility, ability of opponents, and teammate interactions. It would be naive to think that simply knowing something about an athlete's personality provides enough information to predict whether he or she will be successful.

5. What types of psychological tests should be given to athletes, and what conditions should be established for test administration and feedback? Psychological tests have been abused in sport settings both during their administration and when feedback is provided to the athletes. In several cases, athletes have been given psychological tests with no explana-

tion as to why they were taking the test; furthermore, these athletes received no feedback about the results and interpretation of the tests. This, again, is unethical and violates the rights of those taking the tests. *Before they actually take the tests, athletes should be told the purpose of the tests, what the tests measure, and how the tests are going to be used.* In most cases, the tests should be used to help coaches and athletes better understand the athletes' psychological strengths and weaknesses so that they can focus on building up the athletes' strengths and improving their weaknesses. In addition, athletes should be provided with specific feedback on the results of the testing process. *Feedback should be provided in a way that athletes can gain more insight into themselves and understand what the test indicates.* The results and feedback can then serve as a springboard to stimulate positive changes.

6. If athletes are not told the reason for the testing, they will typically become suspicious regarding what the test is going to be used for. In such cases it is not unusual for athletes to start thinking that a coach will use the test to select the starters or weed out the undesirables. Given these circumstances, athletes will be more likely to attempt to exaggerate their strengths and minimize their weaknesses. This response style of faking good can distort the true results of the test and make its interpretation virtually useless. Thus, it is important that athletes be assured of confidentiality in whatever tests they take, because this increases the likelihood that they will answer truthfully. *Coaches should typically stay away from giving and interpreting psychological tests unless they have specific training in this area.* A sport psychology consultant who has formal training in psychological assessment and measurement is the best person to administer and interpret psychological tests.

It is often a mistake to compare an athlete's psychological test results with the norms, as explained earlier, when norms were used to draft players in American football years ago. Rather the more critical point is how the athlete feels in relation to him- or herself, which represents an intrapersonal approach. In essence, the information gleaned by the use of psychological tests should be used to help athletes improve their skills relative to their own standards rather than compare themselves against others.

Go to the WSG to complete Student Activities 12.4 and 12.5.

QUANTITATIVE VERSUS QUALITATIVE MEASUREMENT

Now that you know something about the field of sport and exercise psychology, including some measurement issues as well as cautions in using psychological tests, we discuss the two approaches that sport psychologists use to gain a better understanding of the psychological factors involved in sport, exercise participation, and physical activity. These two general approaches are quantitative and qualitative methodologies.

Quantitative research, numerical in nature and the more traditional of the two approaches, involves experimental and correlational designs that typically use precise measurements, rigid control of variables (often in a laboratory setting), and statistical analyses. There are usually objectively measured independent and dependent variables, and psychological states and traits are assessed by reliable and valid psychological inventories. In quantitative research, the researcher tries to stay out of the data-gathering process by using laboratory measurements, questionnaires, and other objective instruments. The quantitative data are statistically analyzed with computations performed by computers. The obtained scores are often interval or ratio in scale (see chapter 2).

Although qualitative research, which is textual in nature, is often depicted as the antithesis of the more traditional quantitative methods, it should be seen as a complementary method. Qualitative research methods generally include field observations, ethnography, and interviews seeking to understand the meaning of the experience to the participants in a particular setting and how the components mesh together as a whole. To this end, qualitative research focuses on the essence of the phenomenon and relies heavily on people's perceptions of their world. Hence, the objectives are primarily description, understanding, and meaning. Qualitative data are rich, because they provide depth and detail; they allow people to be understood in their own terms and in their natural settings (Patton 1990). The researcher does not manipulate variables through experimental treatment but rather is interested more in the process than in the product. Through observations and interviews, relationships and theories are allowed to emerge from the data rather than being imposed on them; thus, induction is emphasized. This approach is in contrast to quantitative research, in which deduction is primary. Finally, in qualitative research, the researcher interacts with the subjects, and the sensitivity and perception of the researcher play crucial roles in procuring and processing the observations and responses.

A study by Holt and Sparkes (2001) provided an example of the use of ethnography to study group cohesion in a team over the course of a season. The researcher (one of the authors) spent a season with a soccer team as a player and coach and collected data via participant observation, formal and informal interviews, a field diary, and a reflective journal. Having the researcher embedded in a group for an entire season allowed a richness of data and understanding that could not occur through the use of quantitative data collection techniques alone.

Additionally, data gathered from qualitative methods often serve as a foundation for the development or refinement of quantitative instruments and operational definitions. For example, the concept of mental toughness has suffered from conceptual and definitional ambiguity and has consequently been the subject of much debate. To address this issue, Jones, Hanton, and Connaughton (2002) interviewed international performers to gather data regarding the athletes' perceptions of what attributes comprise mental toughness, and how it should be defined. Not only did this study provide practical information for the evolving debate regarding mental toughness, but it also exploited the often overlooked insights of the athletes themselves. Furthermore, the study inspired and provided procedural guidance for subsequent sport-specific research directed toward an understanding of mental toughness in soccer (Thelwell, Weston, and Creenlees 2005), Australian football (Gucciardi, Gordon, and Dimmock 2008), and cricket (Bull, Shambrook, James, and Brookes 2005).

A mixed-methods approach (both qualitative and quantitative) may allow precise measurement and statistical analyses through quantitative investigation, while simultaneously including complimentary qualitative data to help explain and expand upon the numbers in a more holistic and comprehensive manner. Such an approach can allow an investigator to consider the idiosyncrasies of each person's responses, which might otherwise go unexplored. An example of a mixed-methods design can be seen in a study by Gould, Dieffenbach, and Moffett (2002) in which Olympic champions were interviewed and completed a battery of psychological inventories. The researchers were able to evaluate a diverse range of psychological characteristics based on quantitative and qualitative data. Not surprisingly, the results provided an extensive amount of information that has helped sport psychologists better understand what underlying factors influence these elite athletes. For instance, the quantitative methods (questionnaires) found that the athletes scored high in confidence. Through interviews, it was determined that a sport program or organization that fostered, nurtured, and instilled positive attitudes and skills in athletes helped build their confidence. Similarly, athlete confidence was also found to be influenced by having an optimistic, achievement-oriented environment while growing up as well as by having coaches express confidence in an athlete's development.

It should be noted that qualitative studies have become popular (e.g., Bloom, Stevens, and Wickwire 2003; Culver, Gilbert, and Trudel 2003; Sparkes 1998; Strean 1998; Stuart 2003). For a review of the role of qualitative research in sport and exercise psychology, see Brewer, Vose, Raalte, and Petitpas (2011). Most of the assessment of psychological traits and states in sport psychology has taken place through using inventories and questionnaires that have been carefully developed to provide high reliability and validity—a quantitative approach. We will take a close look at a number of these scales, including their psychometric development and their use in research and applied settings.

Quantitative Methods

As noted earlier, most psychological assessments in the field of sport and exercise psychology use the traditional quantitative methodology of the questionnaire (see Tenenbaum, Eklund, and Kamata 2012 for a summary of advances in sport and exercise psychology measurement). There are a number of quantitative questionnaires to choose from; two of the most popular employ Likert scales and semantic differential scales. Each of these scales can help define the multidimensional nature of the construct being assessed.

Likert Scales

A Likert scale is a 5-, 7-, or 9-point scale (1-5 points, 1-7 points, or 1-9 points) that assumes equal intervals between responses. In the example provided here, the difference between *strongly agree* and *agree* is considered equivalent to the difference between *disagree* and *strongly disagree*. This type of scale is used to assess the degree of agreement or disagreement with statements and is widely used in attitude inventories. An example of an item using a Likert scale is as follows:

All high school students should be required to take 2 years of physical education classes.

Strongly agree	Agree	Undecided	Disagree	Strongly disagree
5	4	3	2	1

A principal advantage of scaled responses is that they permit a wider choice of expression than categorical responses, which are typically dichotomous—that is, offering such choices as *yes* and *no* or *true* and *false*. The five, seven, or more intervals also increase the reliability of the instrument. In addition, different response words can be used in scaled responses (figure 12.1): *excellent, good, fair, poor,* and *very poor,* or *very important, important, somewhat important, not very important,* and *of no importance.*

Semantic Differential Scales

Another popular measuring technique involves using a semantic differential scale, which asks participants to respond to bipolar adjectives—pairs of adjectives with opposite meanings, such as *weak–strong, relaxed–tense, fast–slow,* and *good–bad*—with scales anchored at the extremes. Using bipolar adjectives, respondents are asked to choose a point on the continuum that best describes how they feel about a specific concept. Refer to table 12.2 on attitudes toward physical activity; notice that you can choose any of seven points that best reflects your feelings about the concept.

The process of developing semantic differential scales involves defining the concept you want to evaluate and then selecting specific bipolar adjective pairs that best describe respondents' feelings and attitudes about the concept. Research has indicated that the semantic differential

1	2	3	4	5
Never	Sometimes	Often	Frequently	Always

1	2	3	4	5
Strongly agree	Agree	No opinion	Disagree	Strongly disagree

1	2	3	4	5
Not important at all				Extremely important

1	2	3	4	5	6	7
Always						Never

1	3	4	5	6	7	8	9	10
Agree								Disagree

Figure 12.1 Examples of scaled responses.

technique measures three major factors. By far, the most frequently used factor is evaluation—the degree of goodness you attribute to the concept or object being measured. Potency involves the strength of the concept being rated, and the activity factor uses adjectives that describe action. The following list shows some examples of bipolar adjectives that measure the evaluation components; this is an example of a differential scale for measuring attitudes toward competitive sports for children.

Evaluation

- Pleasant–unpleasant
- Fair–unfair
- Honest–dishonest
- Good–bad
- Successful–unsuccessful
- Useful–useless

Table 12.2 Semantic Differential Scales for Measuring Attitudes Toward Physical Activity

PHYSICAL ACTIVITY								
Good	____	____	____	____	____	____	____	Bad
Pleasant	____	____	____	____	____	____	____	Unpleasant
Relaxed	____	____	____	____	____	____	____	Tense
Hot	____	____	____	____	____	____	____	Cold
Healthy	____	____	____	____	____	____	____	Unhealthy
Nice	____	____	____	____	____	____	____	Awful
Delicate	____	____	____	____	____	____	____	Rugged
Active	____	____	____	____	____	____	____	Passive

Potency

- Strong–weak
- Hard–soft
- Heavy–light
- Dominant–submissive
- Rugged–delicate
- Dirty–clean

Activity

- Steady–nervous
- Happy–sad
- Active–passive
- Dynamic–static
- Stationary–moving
- Fast–slow

 Go to the WSG to complete Student Activity 12.6.

Qualitative Methods

Qualitative methods are becoming popular in sport psychology research because they provide a richness of information that is usually untapped when traditional questionnaires are used.

Interviews

The interview is undoubtedly the most common source of data in qualitative research. Interviews range from a highly structured style, in which questions are determined before the interview actually starts, to open-ended interviews, which allow for free responses. The most popular mode of interview used in sport psychology research is the semistructured interview. Each participant responds to a general set of questions, but a test administrator uses different probes and follow-up questions depending on the nature of a participant's response. A good interviewer must first establish rapport with participants to allow them to open up and describe their true thoughts and feelings. It is also important that an interviewer remain nonjudgmental regardless of the content of a participant's responses. Above all, an interviewer has to be a good listener.

Using a tape recorder is probably the most common method of recording interviews, because it preserves the entire interview for subsequent data analysis. Although a small percentage of participants are initially uncomfortable being recorded, this uneasiness typically disappears quickly. Taking notes during an interview is another frequently used method to record data; sometimes note-taking is used in conjunction with recording when an interviewer wants to highlight certain important points. One drawback with taking notes without recording is that it keeps the interviewer busy, thus interfering with his or her thoughts and observations of the participant.

A good example of the use of interviews to collect qualitative data in sport psychology is the work of Gould, Dieffenbach, and Moffett (2002) noted earlier. To better understand the psychological characteristics of Olympic athletes and their development, the researchers incorporated qualitative methods (along with some quantitative psychological inventories) and an inductive analytic approach to research. They studied 10 former Olympic champions (winners of 32 medals); 10 Olympic coaches; and 10 parents, guardians, or significant others (one for each athlete). The interviews focused on the mental skills needed to become an elite athlete as well as how these mental skills were developed during the early, middle, and later years of the athletes' careers. The interviews, transcribed verbatim from recordings, were analyzed by content analysis, a method that organizes the interview into increasingly more complex themes and categories representing psychological characteristics and their development. Among the numerous findings, it was revealed that these Olympic athletes were characterized by the ability to focus and block out distractions, maintain their competitiveness, the ability to set and achieve goals, mental toughness, the ability to control anxiety, and confidence. In addition, coaches and parents were particularly important (although others were also involved to a lesser or greater extent) in assisting the development of these athletes. Specifically, coaches, parents, and others provided support and encouragement, created an achievement environment, modeled appropriate behavior, emphasized high expectations, provided motivation, taught physical and psychological skills, and enhanced confidence through positive feedback. Such depth could not have been accomplished through the strict use of questionnaires and other psychological inventories.

Observation

Observation is second only to the interview as a means of qualitative data collection. Although most early studies relied on direct observation with note-taking and coding of certain categories of behavior, the current trend is to videotape. Just as using a tape recorder to conduct an interview allows a researcher to review everything a participant said, videotaping likewise captures a participant's behavior for future analysis. During field observations, it is important that the observer be unobtrusive to the participants under study. Participants who know that they are being watched and recorded may change their behavior. Observers can seem less obtrusive by hanging around for several days before actually starting to record their observations. It is important that the novelty of the observer's presence wears off so that behavior can occur naturally.

A classic example of using observations in sport psychology is seen in the seminal work of Smith, Smoll, and Curtis (1979) on the relationship between coaches' behaviors and young athletes' reactions to these behaviors. The first part of the investigation identified what coaches actually do. Observers were trained to watch Little League baseball coaches during practice and games and carefully note what the coaches did. The observers recorded these behaviors over several months. After compiling literally thousands of data points, the researchers attempted to collapse the behaviors into some common categories. The end result of this process categorized coaching behaviors into those initiated by a coach (spontaneous) versus those that were reactions to a player's behavior (reactive).

For example, a coach's yelling at a player who made an error was a reactive behavior. However, a coach's instruction to players on how to slide was considered a spontaneous behavior. Within these categories of reactive and spontaneous behaviors were subcategories, such as positive reinforcement, negative reinforcement, general technical instruction, general encouragement, and mistake-contingent technical instruction. These subcategories of coaching behaviors resulted in the development of an instrument called the Coaching Behavior Assessment System, which allowed the researchers to conduct several studies investigating the relationship between specific coaching behaviors and players' evaluative reactions to these behaviors. For example, in one study, team members who played for coaches who gave predominantly positive (as opposed to negative) reinforcement liked their teammates more, wanted to continue playing next year, and saw their coaches as more knowledgeable and as better teachers than did team members who played for coaches who did not favor positive reinforcement. The cornerstone of the research methodology, however, was collecting qualitative data through the use of observations and using these data to develop the Coaching Behavior Assessment System.

SCALES USED IN SPORT PSYCHOLOGY

Thus far we have provided an overview of sport psychology, the uses and abuses of psychological testing in sport settings, and some basic information on types of scaling procedures. In this section, we highlight the most often used and most popular psychological tests used in sport and physical activity settings. We focus on inventories that have been systematically developed with high standards of reliability and validity and provide examples of how these tests have been used in research and applied settings. For reviews of the scales used in sport psychology research, see Anshel (1987), Ostrow (1996), and Duda (1998).

Competitive Anxiety

One of the most popular topics of study in sport psychology concerns the relationship between anxiety and performance. Athletes, coaches, and researchers generally agree that there is an optimal level of anxiety associated with high levels of performance. Finding that optimal level is not necessarily easy, but the first step is to be able to measure an athlete's anxiety level. As noted earlier, there is a distinction in the general psychology literature between trait anxiety and state anxiety. This distinction is used in developing sport-specific measures of *competitive trait anxiety* and *competitive state anxiety*.

Sport Competition Anxiety Test

One of the most widely used tests in sport psychology is the Sport Competition Anxiety Test (SCAT) developed by Martens (1977). SCAT was developed to provide a reliable and valid measure of competitive trait anxiety. Competitive trait anxiety is a construct that describes individual differences in perception of threat, state anxiety response to perceived threat, or both. *SCAT was developed to provide a measure of how anxious athletes generally feel before competition*. The fact that SCAT is a good predictor of precompetitive state anxiety is important from a practical perspective: It is not always feasible to test athletes right before competition to assess how anxious they feel at that moment (i.e., their competitive state anxiety). SCAT was initially developed for use with youth between the ages of 10 and 15 years, and an adult form of the inventory was constructed shortly thereafter.

Go to the WSG to complete Student Activity 12.7.

Reliability and Validity The internal structure, reliability, and validity of SCAT have been determined independently for the youth and the adult forms on the basis of responses from more than 2500 athletes. SCAT's reliability has been assessed by test–retest and has produced interclass reliability coefficients ranging from .73 to .88, with a mean reliability of .81. Past research findings produced coefficients of internal consistency (intraclass) reliability ranging from .95 to .97 (high values) for both the adult version and the youth version of SCAT. Evidence for the construct validity of SCAT was obtained by demonstrating significant relationships between SCAT and other personality constructs in accordance with theoretical predictions (recall convergent evidence from chapter 6). For example, SCAT is moderately correlated with other general anxiety scales but is not correlated with other general personality scales. Finally, a number of experimental and field studies support the construct validity of SCAT as a valid measure of competitive trait anxiety by providing results in accordance with theoretical predictions. For example, participants scoring high on SCAT exhibit higher levels of precompetitive state anxiety than do low-SCAT participants. SCAT was also found to correlate more strongly with state anxiety in competitive situations than in noncompetitive situations.

Norms and Scoring The adult form of SCAT and norms for high school and college athletes are provided in figure 12.2 and table 12.3. For each item, one of three responses is possible: (a) hardly ever, (b) sometimes, and (c) often. Eight of the test items—2, 3, 5, 8, 9, 12, 14, 15—are scored with the following point values:

Hardly ever = 1

Sometimes = 2

Often = 3

Note that items 6 and 11 are scored in reverse:

Hardly ever = 3

Sometimes = 2

Often = 1

Illinois Competition Questionnaire

Directions: Below are some statements about how people feel when they compete in sports and games. Read each statement and decide if you HARDLY EVER or SOMETIMES or OFTEN feel this way when you compete in sports and games. If your choice is HARDLY EVER, blacken the square labeled A; if your choice is SOMETIMES, blacken the square labeled B; if your choice is OFTEN, blacken the square labeled C. There are no right or wrong answers. Do not spend too much time on one statement. Remember to choose the word that describes how you usually feel when competing in sports and games.

	Hardly ever	Sometimes	Often
1. Competing against others is socially enjoyable.	A ❑	B ❑	C ❑
2. Before I compete I feel uneasy.	A ❑	B ❑	C ❑
3. Before I compete I worry about not performing well.	A ❑	B ❑	C ❑
4. I am a good sport when I compete.	A ❑	B ❑	C ❑
5. When I compete I worry about mistakes.	A ❑	B ❑	C ❑
6. Before I compete I am calm.	A ❑	B ❑	C ❑
7. Setting a goal is important when competing.	A ❑	B ❑	C ❑
8. Before I compete I get a queasy feeling in my stomach.	A ❑	B ❑	C ❑
9. Just before competing I notice my heart beats faster than usual.	A ❑	B ❑	C ❑
10. I like to compete in games that demand considerable physical energy.	A ❑	B ❑	C ❑
11. Before I compete I feel relaxed.	A ❑	B ❑	C ❑
12. Before I compete I am nervous.	A ❑	B ❑	C ❑
13. Team sports are more exciting than individual sports.	A ❑	B ❑	C ❑
14. I get nervous wanting to start the game.	A ❑	B ❑	C ❑
15. Before I compete I usually get uptight.	A ❑	B ❑	C ❑

Figure 12.2 Sport Competition Anxiety Test.

Reprinted, by permission, from R. Martens, 1977, *Sport competition anxiety test* (Champaign, IL: Human Kinetics).

Table 12.3 Sport Competition Anxiety Test: Norms for Male and Female High School and College Athletes

	HIGH SCHOOL (PERCENTILES)		COLLEGE (PERCENTILES)	
Raw score	Males	Females	Males	Females
30	98	98	97	99
29	96	95	93	98
28	93	89	93	96
27	88	83	88	93
26	78	77	82	90
25	65	70	77	86
24	53	60	70	82
23	43	51	63	77
22	33	43	56	69
21	24	34	51	62
20	17	26	43	53
19	12	19	34	43
18	9	14	27	34
17	7	11	21	25
16	6	9	16	18
15	5	6	8	14
14	5	5	4	10
13	4	4	2	6
12	2	2	1	3
11	1	1	0	1
10	0	0	0	0

Reprinted, by permission, from R. Martens, 1977, *Sport competition anxiety test* (Champaign, IL: Human Kinetics).

Items 1, 4, 7, 10, and 13 are not scored; they are included in the inventory as filler items, to direct attention to elements of competition other than anxiety.

Research and Practical Examples Let's look at a study in sport psychology that has used SCAT. Researchers have been interested in studying the differences in perceptions of threat between those who are high and low in competitive trait anxiety. Using male youth soccer players as participants, Passer (1983) found that players high in competitive trait anxiety expected to play less well in the upcoming season and worried more frequently about making mistakes, not playing well, and losing than did players low in competitive trait anxiety. In addition, players high in competitive trait anxiety expected to experience greater emotional upset, shame, and criticism from parents and coaches after failing than did players low in competitive trait anxiety.

These findings demonstrate that athletes who are high and low in competitive trait anxiety significantly differ in their perceptions and reactions to threat and that there are important implications for coaches and parents. Specifically, because young athletes who have high competitive trait anxiety are more sensitive to criticism, failure, and making mistakes, it is important that coaches and parents not overreact when these young athletes do not perform well. Positive reinforcement, encouragement, and support are crucial in helping these young athletes deal with their mistakes and stay involved in sports.

In a study by Bebetsos and Antoniou (2012), a sample of 56 badminton players competing in a national championship completed SCAT shortly before competition. Overall, the athletes were considered calm. This may be expected as badminton is a sport requiring a high level of concentration and composure. However, gender differences emerged as girls reported less competitive anxiety than boys. The authors suggest that these differences may in part be due to gender-based cultural expectancies and competitiveness differences.

A limitation of many questionnaires is a lack of normative data for specific populations. Potgieter (2009) published SCAT norms collected from athletes in a wide variety of less traditionally studied sports (e.g., water polo, squash, netball). A total of 1799 athletes representing 17 sports completed SCAT over approximately 15 years. Potgieter reported that the highest median anxiety scores for individual sports were in track and field, cross country running, and swimming, while the highest scores for team sports were in rugby, netball, and field hockey.

Sport Anxiety Scale-2

Subsequent to the development of SCAT, anxiety research determined that the construct was multidimensional with a cognitive and somatic component. To accommodate this advance, the Sport Anxiety Scale (SAS) was developed to provide a multidimensional assessment of competitive trait anxiety in keeping with contemporary theory. The three subscales that were obtained through confirmatory factor analysis included two factors relating to cognitive anxiety, named concentration disruption and worry, as well as a factor termed somatic anxiety.

Unfortunately, the strong psychometric properties of SAS were not upheld when testing young populations. Data gathered from children and youth did not replicate the three-factor structure of the scale, forcing researchers to use total scores and thereby forfeiting the multidimensional measurement capabilities of the instrument. In addition, although early studies seemed to support the factorial validity of the scale, later studies exposed concerns regarding the Concentration Disruption scale, indicating cross-loading issues with the Worry scale. These limitations of the original scale prompted the development of the Sport Anxiety Scale 2 (SAS-2; Smith, Smoll, Cumming, and Grossbard 2006) to properly assess children and expand its application to older samples.

Since the SAS proved to be ineffective for children, it was hypothesized that item comprehension may have been a limiting factor. As such, the item reading level was reduced to allow greater understanding in younger populations. Using the Flesch–Kincaid reading level analysis, items were set at or below a 4th grade level, with the overall scale set at a reading level of grade 2.4.

Reliability and Validity The psychometric properties of the original SAS proved to be one of its major strengths, and as such, replicating or exceeding this quality in the SAS-2 was a primary concern for the developers. In order to establish reliability, validity, and norms for both children and adults, two separate samples were used. First, 1038 youth athletes (571 males and 467 females) between 9 and 14 years old were selected. A second sample consisted of 593 college students (237 males and 356 females) who were currently participating or had recently participated in sports.

Internal consistency was supported with high Cronbach's alphas for total scores (.91), and somatic (.84), worry (.89), and concentration disruption (.84) subscales. Coefficients were acceptable for all subscales at all age groups. Test–retest reliability (at a 1-week interval) was acceptable with subscale coefficients ranging from .76 to .90. The three-factor structure of the scale was also supported by factor analysis (construct-related validity).

Construct validity was assessed through correlations with similar constructs (e.g., goal orientations, self-esteem) and unrelated constructs (self-perceived competence). As expected, the SAS-2 was highly correlated to the SAS (r = .90). Correlations with other constructs were of the expected direction and significant. For example, significant negative correlations emerged between the SAS-2 and measures of self-esteem, with low correlations to unrelated measures of self-perceived competence. Moreover, discriminant validity was supported with low negative correlations (<–.20) between the SAS-2 and the Children's Social Desirability Scale and the Marlowe–Crowne Social Desirability Scale.

This magnitude of correlation is similar to that of other anxiety scales and indicates that the tendency to present oneself in a positive light is negatively related to SAS scores. In summary, the SAS-2 not only maintained the strong psychometric properties of its parent instrument, but actually improved and expanded upon them.

Norms and Scoring The SAS-2 consists of 15 total items that assess somatic anxiety, worry, and concentration disruption, with five items per subscale. Scores for each subscale range from 5 to 20, while the total anxiety score may range from 15 to 60. All items are scored on a 1 *(not at all)*, 2 *(somewhat)*, 3 *(moderately so)*, and 4 *(very much so)* scale with no reverse scoring. Questions begin with the stem "Before or while I compete in sports…," Example items for each subscale include these: for Somatic Anxiety, "My stomach feels upset"; for Worry, "I worry that I will let others down"; and for Concentration Disruption, "I cannot think clearly during the game." Means for age groups are displayed in table 12.4.

Table 12.4 Means and Standard Deviations of SAS-2 Scores for Child and College-Age Groups

	AGE GROUP			
SAS-2 Scale	9-10 years	11-12 years	13-14 years	College
Somatic anxiety	8.29 (3.14)	7.70 (2.80)	8.34 (3.36)	9.78 (3.61)
Worry	9.05 (3.53)	9.37 (3.54)	10.50 (3.75)	12.12 (3.85)
Concentration disruption	7.54 (2.71)	6.82 (2.28)	7.29 (2.88)	6.93 (2.37)
Total score	24.88 (8.14)	23.88 (7.14)	26.14 (8.40)	28.83 (8.05)

Adapted from R.E. Smith et al., 2006, "Measurement of multidimensional sport performance anxiety in children and adults: The Sport Anxiety Scale-2," *Journal of Sport & Exercise Psychology* 28(4): 479-501.

Research and Practical Examples Let's look at a couple of applications of the SAS. For example, in a study by Smith, Smoll, and Barnett (1995), an intervention designed to train coaches to reduce the stressfulness of the athletic environment by de-emphasizing winning and providing high levels of social support resulted in a significant decrease in the total score on the SAS. Thus, the SAS appears to be sensitive to interventions designed to decrease anxiety. These same results were found in a replication of the original study using the SAS-2 (Smith, Smoll, and Cumming 2007). In another study (Patterson, Smith, Everett, and Ptacek 1998), SAS scores were related to the occurrence of injury under high levels of life stress. Specifically, those scoring high on all SAS subscales were more likely to incur injury when under high levels of life stress. These findings suggest that somatic anxiety, worry, and concentration disruption may all be capable of increasing the risk of anxiety in highly stressed people. Finally, Grossbard and colleagues (2007) used the

SAS-2 to examine the relationship between goal orientations and performance anxiety. Negative correlations were found between task orientation (focus on improvement) and concentration disruption, somatic anxiety, and total SAS-2 scores for males, and with the concentration disruption scale for females. Positive correlations emerged for both genders between ego orientation (focus on winning) and all subscales and overall anxiety scores. Therefore, ego-oriented athletes appear to experience greater performance anxiety than task-oriented athletes, thus promoting the development of task and mastery environments that focus on self-improvement instead of winning.

The interpretation of anxiety is as important (if not more so) than the amount of anxiety experienced. Using the SAS and the Competitive State Anxiety Inventory-2 (CSAI-2), Hanton, Mellalieu, and Hall (2002) found that low scores on the trait concentration disruption were related to more facilitative perceptions of both somatic and cognitive state anxiety symptoms. That is, athletes believed that their feelings of anxiety helped improve their performance. The authors speculate that this may be related to low trait-anxious people being able to selectively attend to only task-relevant information and minimize attending to distracting or threatening information.

Mastery Item 12.1

Download the data from the WSG for Mastery Item 12.1. The scores presented are state anxiety scores from two groups, inexperienced and experienced skydivers. The State Anxiety Scale was administered as subjects boarded the airplane before a jump. The scores presented are from a newly created State Anxiety Scale. If the scale reflects construct validity, as presented in chapter 6, the inexperienced and experienced skydivers should differ on state anxiety. Because there are two groups, you can use an independent group's t-test (as you learned in chapter 5) to see if the groups differ significantly. Use SPSS to confirm that the groups differ significantly (i.e., $t = 6.28$, $p < .001$). These results provide evidence of the construct-related validity of the newly created State Anxiety Scale.

Competitive State Anxiety Inventory–2

The Competitive State Anxiety Inventory-2 (CSAI-2) was developed as a sport-specific inventory of state anxiety and as a revision of the earlier Competitive State Anxiety Inventory (Martens, Vealey, and Burton 1990). In fact, more than 40 published studies have used the CSAI–2 to investigate the relationship between state anxiety and performance (Craft, Magyar, Becker, and Feltz 2003). *The CSAI-2 measures precompetitive state anxiety, which is how anxious an athlete feels at a given time—in this case, right before competition.* The CSAI-2 has three subscales: Somatic Anxiety, Cognitive Anxiety, and Confidence. As noted earlier regarding the SAS, somatic anxiety refers to the physiological component of anxiety and cognitive anxiety to the worry component. These subscales reflect the multidimensional nature of anxiety.

Reliability and Validity Test–retest reliability is inappropriate for state scales because, by definition, scores can change from moment to moment. Thus, the major source of reliability comes from examining the internal consistency of the scale—that is, the degree to which items in the same subscale are homogeneous. Alpha reliability coefficients, estimates of internal consistency (see chapter 6), have ranged from .79 to .90, demonstrating a high degree of internal consistency.

The concurrent validity of the CSAI-2 was examined by investigating the relationship between each of its subscales and eight selected state and trait personality inventories. Results strongly support the concurrent validity of the CSAI-2, because the correlations are highly congruent with hypothesized relationships among the CSAI-2 subscales and other personality constructs. For example, the CSAI-2 Cognitive subscale is highly correlated with the Worry–Emotionality Inventory (Morris, Davis, and Hutchins 1981). The Worry subscale and the CSAI-2 Somatic subscale are highly correlated with the Worry–Emotionality Inventory Emotionality subscale. The construct-related validity of the CSAI-2 was determined by a series of systematic studies demonstrating the relationships between the three subscales and other constructs (e.g., performance, situational variables, individual differences) as predicted by theory.

Cox, Martens, and Russell (2003) later identified potential limitations regarding the factor structure of the CSAI-2 and modified the scale to form the CSAI-2 Revised (CSAI-2R). The authors sought to improve the scale's psychometric robustness and validity through improved investigative methods (e.g., larger sample size, confirmatory factor analysis). As determined through a sample of 506 athletes, it was concluded that 10 items should be deleted (from the 36 presently in the scale) to significantly improve the model fit via confirmatory factor analysis. The new, truncated scale, thus far seems to be an improvement over the CSAI-2. Still, more research is needed to continually test and validate the psychometric properties of the instrument.

Norms and Scoring The CSAI-2 and its norms for high school and college athletes are provided in figure 12.3 and table 12.5, respectively. Scoring for the CSAI-2 is accomplished by computing a separate total for each of the three subscales, with scores ranging from a low of 9 to a high of 36. The higher the score, the greater the cognitive or somatic anxiety or self-confidence.

- The Cognitive Anxiety subscale is scored by adding the responses to items 1, 4, 7, 10, 13, 16, 19, 22, and 25.
- The Somatic State Anxiety subscale is scored by adding the responses to items 2, 5, 8, 11, 14, 17, 20, 23, and 26 (scoring for item 14 must be reversed, i.e., 4–3–2–1).
- The State Self-Confidence subscale is scored by adding up the responses to items 3, 6, 9, 12, 15, 18, 21, 24, and 27.

Research and Practical Examples Using the CSAI-2R, Mesagno and Mullane-Grant (2010) investigated the influence of a preperformance routine on anxiety in relation to choking under pressure. A group of Australian football players completed a kicking task to assess performance outcomes. First, players completed a kick under low-pressure conditions to establish baseline data. After the kick, half of the players were introduced to a preperformance intervention routine to reduce anxiety. Afterward, a high-pressure situation was created via having an audience and being offered monetary incentives. CSAI-2R scores confirmed increases in anxiety in the high-pressure phase. As expected, the group receiving the intervention reported decreased anxiety and increased performance under high pressure. These results are encouraging as performance decrements due to anxiety are especially common in sport, and simple interventions such as the one employed in this study may help to improve performance under these conditions.

A study by Burton (1988) on the relationship between precompetitive state anxiety and performance of collegiate swimmers provides a good illustration of the use of the CSAI-2. Swimmers completed the CSAI-2 right before competing in three separate meets during the season. Results revealed different relationships between each of the subscales and performance in accordance with theoretical predictions. Specifically, cognitive anxiety was negatively related to performance, confidence was positively related to performance, and somatic anxiety exhibited a curvilinear relationship to performance in the shape of an inverted *U* (i.e., there was an optimal level of somatic anxiety, with low and high levels producing decrements in performance).

Illinois Self-Evaluation Questionnaire

Directions: A number of statements that athletes have used to describe their feelings before competition are given below. Read each statement and then circle the appropriate number to the right of the statement to indicate *how you feel right now*—at this moment. There are no right or wrong answers. Do not spend too much time on one statement, but choose the statement which describes your feelings *right now.*

	Not at all	Somewhat	Moderately so	Very much so
1. I am concerned about this competition.	1	2	3	4
2. I feel nervous.	1	2	3	4
3. I feel at ease.	1	2	3	4
4. I have self-doubts.	1	2	3	4
5. I feel jittery.	1	2	3	4
6. I feel comfortable.	1	2	3	4
7. I am concerned that I may not do as well in this competition as I could.	1	2	3	4
8. My body feels tense.	1	2	3	4
9. I feel self-confident.	1	2	3	4
10. I'm concerned about losing.	1	2	3	4
11. I feel tense in my stomach.	1	2	3	4
12. I feel secure.	1	2	3	4
13. I am concerned about choking under pressure.	1	2	3	4
14. My body feels relaxed.	1	2	3	4
15. I'm confident that I can meet the challenge.	1	2	3	4
16. I'm concerned about performing poorly.	1	2	3	4
17. My heart is racing.	1	2	3	4
18. I'm confident about performing well.	1	2	3	4
19. I'm concerned about reaching my goal.	1	2	3	4
20. I feel my stomach sinking.	1	2	3	4
21. I feel mentally relaxed.	1	2	3	4
22. I'm concerned that others will be disappointed with my performance.	1	2	3	4
23. My hands are clammy.	1	2	3	4
24. I'm confident because I mentally picture myself reaching my goal.	1	2	3	4
25. I'm concerned I won't be able to concentrate.	1	2	3	4
26. My body feels tight.	1	2	3	4
27. I'm confident of coming through under pressure.	1	2	3	4

Figure 12.3 Competitive State Anxiety Inventory-2.

Reprinted, by permission, from R. Martens, R. Vealey, and D. Burton, 1990, *Competitive anxiety in sport* (Champaign, IL: Human Kinetics).

Table 12.5 Competitive State Anxiety Inventory-2: Norms for Male and Female High School and College Athletes

	MALE PERCENTILES			FEMALE PERCENTILES		
Raw score	Cognitive	Somatic	Self-confidence	Cognitive	Somatic	Self-confidence
36	99	99	99	99	99	99
35	99	99	96	98	99	98
34	99	99	94	96	99	97
33	99	98	91	94	99	96
32	99	98	87	92	98	94
31	98	97	83	89	98	92
30	98	96	79	87	97	89
29	96	95	76	84	94	86
28	95	94	71	80	91	83
27	93	93	66	76	89	78
26	89	92	60	73	86	73
25	86	89	52	70	83	67
24	83	85	46	65	79	61
23	80	82	39	60	73	55
22	75	79	34	55	67	48
21	68	75	28	49	61	41
20	61	71	22	44	57	35
19	55	63	17	39	51	28
18	48	55	12	33	46	21
17	40	49	7	26	41	14
16	34	42	3	20	35	11
15	28	35	2	16	30	8
14	23	27	1	11	24	5
13	18	21	0	8	18	3
12	12	16	0	5	14	2
11	7	10	0	3	10	1
10	4	5	0	1	6	0
9	1	1	0	0	1	0

Reprinted, by permission, from R. Martens, R. Vealey, and D. Burton, 1990, *Competitive anxiety in sport* (Champaign, IL: Human Kinetics).

This would be useful information for a coach or athlete trying to get an athlete emotionally ready for competition. In particular, it would seem important to reduce worry and fear before competition while building confidence to as high a level as possible. Furthermore, it appears that getting emotionally psyched up and physiologically activated is good up to a point, but that too much arousal decreases performance. Finally, results revealed that athletes reacted differently in terms of the anxiety–performance relationship; thus, group pep talks do not have as much value or sensitivity to each athlete's individual needs (i.e., zone of optimal performance).

One final point relating to anxiety measurement appears appropriate. Over the past 10 years, researchers have been measuring the direction of anxiety as well as the previously discussed intensity. That is, how much anxiety you have would represent the intensity level, but is this anxiety helpful or harmful to performance? This represents the direction of anxiety. Thus, having a high level of anxiety is not necessarily detrimental to performance because it might depend more on how people interpret their levels of anxiety. In fact, research has suggested that the direction of anxiety may be more important than the intensity of anxiety in determining its effects on performance. Thus, measures of direction, which is typically measured in terms of anxiety being facilitative or debilitative, should accompany measures of intensity.

PSYCHOLOGICAL SKILLS

The benefits of mental training in sport are hard to overstate. In fact, many athletes and coaches believe that effective implementation of mental skills constitutes more than 50% of performance (and in some cases, over 90%). With such an appreciation for the psychological aspect of performance, it is not surprising that psychological skills training (PST) has grown in popularity and reputation. In fact, Murphy and Jowdy (1992) reported that over 90% of U.S. and Canadian athletes used a form of mental training (typically imagery) in preparation for their events. Support is not limited to anecdotal sources, however, and in the first review of mental skills training, Greenspan and Feltz (1989) provided early evidence of the efficacy in psychological interventions in sport. A review of 23 interventions included in 19 studies suggested that, overall, educational relaxation training and remedial cognitive restructuring interventions improved performance. Similarly, Weinberg and Comar (1994) reviewed 38 studies, with positive effects reported from 85% of the psychological interventions. Additional reviews (e.g., Fournier, Calmels, Durand-Bush, and Salmela 2005; Brown and Fletcher 2013) employing more sophisticated analyses have further reinforced the efficacy of mental skills training programs.

PST programs are intended to improve performance, and enhance athletes' enjoyment and satisfaction in their sports. Sheard and Golby (2006) provided an example of how a well-designed PST program can improve an athlete's performance as well as psychological health and functioning. They found that not only did PST result in sustained athletic performance increases, but it also improved overall positive psychological development. The participants developed higher levels of self-confidence and attention control, improved goal-setting ability, enhanced perceptions of success, and even elevated mood and positive emotions. Encouragingly, the athletes also reported that these positive psychological developments had permeated their personal lives as well.

Most PST programs include imagery, self-talk, concentration, goal setting, and a variety of other skills. Psychological questionnaires allow a method of measuring the current level of development or use of these skills. An athlete's strengths and weaknesses can then be identified, and ultimately, a customized PST program can be developed for the person or team. Although a wide variety of skills are included in various PST programs, we will focus on the five most common: imagery, self-talk, concentration, confidence, and goal setting.

Imagery

At some point, almost everyone has used imagery (also called visualization, mental rehearsal, or mental practice) to recall a particular scene or picture an imaginary event. For example, you may want to mentally rehearse a presentation before you actually step up to the podium. You might even recall a day you spent on vacation and mentally relive the experience: the warmth of the sun on your skin, the color of the sand, the smell of the ocean, the sound of waves rolling in, perhaps even the mood you were in at the time. Our brains are designed to be able to recall this information remarkably well. Not surprisingly, this ability becomes valuable not only in our day to day lives, but in athletics as well.

Using imagery allows athletes to mentally simulate various skills in great detail, using all of their senses. In much the same way that practicing physical skills enhances an athlete's ability, imagery helps prepare the mind for successful execution of these skills. An athlete can learn to perform the necessary movements in smooth succession—sometimes almost effortlessly—and even optimally regulate emotional responses. Many top athletes have emphatically endorsed the efficacy of imagery, and due to the strong anecdotal and scientific backing supporting its effectiveness, virtually all Olympic athletes report practicing it regularly. In a meta-analysis addressing the efficacy of mental practice (imagery), Driskell, Copper, and Moran (1994) found that, although not as effective as physical practice, mental rehearsal was an effective tool for enhancing performance. The effectiveness, however, was task dependent. In particular, the efficacy of mental rehearsal is greater in tasks that require greater cognitive demands (e.g., archery).

As anecdotal and scientific support continues to validate imagery's efficacy, the clear benefits and inherent capability of each person to practice imagery have led to its widespread popularity and practical applications. The scientific study of this concept has many complexities, and thus a wide variety of questionnaires have been developed to assess these diverse components of imagery. One component that has been related to performance improvements and neurological activation is vividness of imagery. We will consider one questionnaire, which has been used to successfully evaluate vividness of imagery.

Vividness of Movement Imagery Questionnaire-2

An important component of imagery involves the vividness or clarity of the images generated. One instrument designed to measure this aspect is the Vividness of Movement Imagery Questionnaire (VMIQ; Isaac, Marks, and Russell, 1986), revised by Roberts and colleagues to form the Vividness of Movement Imagery Questionnaire-2 (VMIQ-2; 2008). A number of issues in the original instrument prompted its revision: (1) conceptualization issues existed regarding visual and kinesthetic imagery modalities; (2) external visual imagery was assessed by asking the person to imagine *someone else* performing the task, although later research has shown significant improvements in effectiveness when imaging *oneself* in third-person view performing the task; (3) the original VMIQ lacked thorough psychometric testing; and (4) the scoring response procedure generated confusion with many participants and may have led to inaccurate reports. The VMIQ-2 is also shorter in length than the original, which may enhance athletes' willingness to complete it.

The VMIQ-2 is used to measure both kinesthetic and visual imagery of a specific motor task. Unlike other imagery questionnaires (e.g., Movement Imagery Questionnaire), the VMIQ-2 does not require any physical movement. This feature is not necessarily an advantage nor disadvantage, but rather a quality of the instrument, which may make it more or less suitable for specific applications. As discovered through studies, which employed the VMIQ, vividness of imagery has been shown to improve performance, increase neurological activation, and differentiate high-performing from lower-performing athletes. Vividness is particularly important as it relates to an athlete's ability to create or recreate detailed representations in his or her working memory and facilitate neurophysiological activation.

Reliability and Validity

Research has supported the psychometric robustness of the VMIQ-2. Factorial, concurrent, and construct validity were all supported in a series of three studies. Participants completed the Movement Imagery Questionnaire-Revised (MIQ-R; a similar measure of imagery) and the VMIQ-2. Construct validity was assessed from a sample of 777 participants from a range of sports and experience levels, and matched for sport type. Consistent with previous research, all analyses differentiated high-performing from lower-performing athletes, with high-performing athletes reporting significantly more vivid imagery use.

Concurrent validity was assessed through independent t-tests of a sample of 63 athletes from a variety of sports and levels of experience. Results supported concurrent validity with significant correlations between equivalent scales on the VMIQ-2 and the MIQ-R. Reliability has also been supported with high Cronbach's alphas reported for each subscale ranging from .93 to .95. Subsequent studies have also supported the instrument's reliability (Parker and Lovell 2011).

Norms and Scoring

The VMIQ-2 consists of three subscales used to assess vividness of imagery: internal visual imagery (IVI), external visual imagery (EVI), and kinesthetic imagery (KI). The instrument consists of visualizing 12 separate movements (e.g., jumping sideways, swinging on a rope, running downhill). Participants are asked to imagine a specific action and report on the vividness of the image on a 5-point Likert scale from 1 (*perfectly clear and vivid*) to 5 (*no image at all*), with lower total scores representing greater vividness of imagery. Each movement is assessed from three perspectives (IVI, EVI, and KI), for a total of 36 scored items.

Means for high-performing athletes were reported as 23.48 (IVI), 26.53 (EVI), and 23.95 (KI). As predicted, lower-performing athletes' scores reflected significantly less vividness of imagery, with means reported at 27.14 (IVI), 30.22 (EVI), and 28.10 (KI). Norms are also available for specific youth age groups between the ages of 12 and 21 and by total practice volume (see Parker and Lovell 2012 for a review).

Research and Practical Examples

Imagery has been shown to have important implications in motor learning and working memory; therefore, it is not surprising that imagery use in adolescents is a topic of particular interest for many researchers. Parker and Lovell (2011; 2012) have conducted a number of studies using the VMIQ-2 with children and adolescents, and results from these studies suggest that youth athletes are able to generate reasonably vivid images from each measured perspective (IVI, EVI, and KI). It was also found that IVI and KI subscale scores modestly improved with increasing age. The authors suggest that this may in part be due to the increased opportunities for younger athletes to practice imagery in their respective sports. The authors warn, however, that results of imagery questionnaires administered to youth athletes should be interpreted cautiously, since adolescents may have greater difficulty differentiating between the conceptual differences of each subscale.

Numerous studies have shown that elite athletes report more vivid imagery use than nonelite athletes. Therefore, the VMIQ-2 may be useful in helping a coach determine the current imagery capabilities of each athlete, as well as which perspective is most effective for each. As a result, coaches may be able to tailor their feedback to athletes and create effective imagery training protocols to improve each athlete's performance.

Self-Talk

The idea of self-talk as a psychological skill may at first seem strange, since it is something that we all engage in constantly, usually without even trying. Simply engaging in self-talk is not the skill; however, learning and practicing *effective* self-talk strategies is a skill that can be learned and developed. Everyone has experienced the negative inner dialogue that precedes an important performance or follows a failure: *I missed the kick because I'm just not very good; I'm going to fail this test. I never do well on tests.* Negative self-talk often results in significant declines in both psychological states and athletic performance. Sport offers a fast-paced, evaluative environment that may often be greatly affected by an athlete's self-talk. Fortunately, learning how to engage in positive and instructional self-talk can enhance confidence, focus, performance, and psychological health.

Automatic Self-Talk Questionnaire for Sports

Developed by Zourbanos and colleagues (2009), the Automatic Self-Talk Questionnaire for Sports (ASTQS) is one of the sport-specific self-talk scales available. The ASTQS was developed to help athletes and coaches better understand and identify irrational and maladaptive patterns of self-talk. The initial development of the scale resulted in eight subscales measuring both positive and negative self-talk (four subscales each). Subscale examples include worry, confidence, and instruction.

Reliability and Validity

The scale was tested on a sample of 766 athletes from a variety of team and individual sports. Internal consistency was supported with Cronbach's alphas ranging from .79 to .94 for the eight subscales. Comparable reliability coefficients were demonstrated in subsequent studies. Concurrent validity was established through correlations with related scales, including the self-talk subscale on the Test of Performance Strategies (TOPS; a test of psychological skills relevant to sport performance), the SAS (competitive trait anxiety), and the CSAI-2R (revision of the CSAI-2). As predicted, both the positive and negative self-talk subscales from the ASTQS were positively correlated with the equivalent subscales on the TOPS and the confidence subscale on the CSAI-2R. The negative self-talk subscale of the ASTQS was also positively correlated with the anxiety subscales from the SAS, showing that the more negative self-talk, the higher the anxiety level.

Norms and Scoring

Means for each of the eight subscales (from a sample of 766 athletes) are provided in table 12.6. Four positive and four negative subscales are represented by 40 total questions. Total scores vary between subscales. Questions are answered on a 5-point Likert scale ranging from 0 (*never*) to 4 (*very often*).

Table 12.6 Means and Standard Deviations for the Eight Subscales of the ASTQS

Subscale	Psych up	Anxiety control	Confidence	Instruction	Worry	Disen-gagement	Somatic fatigue	Irrelevant thoughts
Mean	3.05	2.39	2.83	2.97	1.05	.65	1.04	.97
SD	.89	.94	.92	.87	.83	.79	.84	.89

Data from Zourbanos et al. 2009.

Research and Practical Examples

Since the development of the ASTQS, Zourbanos and colleagues (2009; 2011; 2014) have conducted a series of studies that demonstrate the utility of measuring self-talk in sport. First, increases in perceived support from a coach predicted comparable increases in athletes' positive self-talk and decreases in negative self-talk. Additionally, other studies have suggested that negative self-talk is even more susceptible to the influence of coaching behaviors. For example, Conroy and Coatsworth (2007) found that athletes who perceived that their coaches blamed them for their mistakes were more likely to internalize this belief and blame themselves more frequently. Finally, task orientation (a form of motivation focused on self-improvement) was positively correlated with each of the four positive self-talk subscales and negatively correlated with negative self-talk. Thus, positive self-talk was related to a focus on improvement, and negative self-talk was associated with a focus on winning.

These results may be particularly important to coaches, because they are largely responsible for determining the motivation of the team and the types of goals each athlete pursues. A coach may therefore want to promote positive self-talk and minimize negative self-talk by fostering an encouraging and supportive task-oriented climate for his or her team.

Concentration

A general measure of concentration which has been used extensively is the Test of Attentional and Interpersonal Style (TAIS), developed by Nideffer (1976). However, the practicality of the TAIS in sport settings had received criticism due to methodological and theoretical concerns. As previously mentioned, scales that have been adapted for specific applications and situations (e.g., sport, academic, business) are typically better able to measure the desired construct and more accurately predict behavior. Consequently, a number of researchers set out to adapt the TAIS for specific sports to improve the scale's psychometric properties and its effectiveness in measuring concentration in sport settings.

The first adaptation of the TAIS came from Van Schoyck and Grasha (1981) who revised the instrument for tennis, resulting in the T-TAIS. As expected, this sport-specific version of the TAIS demonstrated greater reliability and validity than its predecessor. The success of this adaptation inspired the subsequent development of other sport-specific versions of the TAIS, including those for baseball (B-TAIS) and basketball (BB-TAIS), which demonstrated similar psychometric improvements. Reliability and validity statistics and norms are available for each of the specific versions. We will take a deeper look at the baseball adaptation, the B-TAIS.

Baseball Test of Attentional and Interpersonal Style (B-TAIS)

Measurement and interpretation of attentional and interpersonal style are important because these characteristics can influence sport performance. For example, the B-TAIS, developed by Albrecht and Feltz (1987), was not only a successful sport-specific adaptation, but was also skill-specific (batting) to further narrow its focus and promote even greater accuracy in measurement. The B-TAIS showed significant improvements in measures of reliability and validity over its parent instrument.

Reliability and Validity Test–retest coefficients and internal consistency (Cronbach's alpha) coefficients ranged from .72 to .95, and .50 to .85, respectively. Convergent validity was acceptable between the TAIS and B-TAIS, with a moderate correlation between .40 and .60, suggesting that both instruments were measuring the same general phenomena. The B-TAIS was also shown to have greater construct validity than the TAIS since decrements in performance were comparable to related scores on ineffective attentional focus subscales.

Questions were adapted from the TAIS to allow greater construct-related validity and accuracy in measurement. For example, to assess narrow attention the TAIS states, "When I read, it is easy to block out everything but the book"; the B-TAIS wording was revised to "When I bat, it is easy to block out everything but the ball." Although both scales intend to measure the same concept (narrow attention), clearly a person's ability to concentrate may fluctuate greatly depending on the situation. A baseball player may find it easy to concentrate when reading a book, yet struggle to do so when on the field or up to bat.

Therefore, the TAIS would find that an athlete might report having high levels of narrow attention, yet fail to recognize the specific attentional deficits of the athlete in particular situations, such as while batting. The B-TAIS, however, would be more likely to detect an athlete's inability to focus in the specific task of batting, and thus provide much more useful information to both coach and athlete.

Norms and Scoring A total of 59 items from the six attentional and cognitive control subscales of the TAIS were adapted for baseball to form the B-TAIS. Items are rated by their frequency of occurrence on a 5-point scale ranging from 1 (*never*) to 5 (*all the time*). See table 12.7 for means and standard deviations for each subscale.

Table 12.7 Means and Standard Deviations for the B-TAIS

B-TAIS subscales	Mean	SD
Broad-external (BET)	13.77	2.88
Overload-external (OET)	13.48	5.01
Broad-internal (BIT)	16.74	3.77
Overload-internal (OIT)	11.52	3.40
Narrow-attention (NAR)	31.35	6.82
Reduced-attention (RED)	23.00	5.83

Adapted, by permission, from R.R. Albrecht and D.L. Feltz, 1987, "Generality and specificity of attention related to competitive anxiety and sport performance," *Journal of Sport Psychology* 9: 231-248.

Research and Practical Examples To better understand how the B-TAIS offers a unique insight into concentration in baseball, consider this study by Albrecht and Felts (1987). A sample of 29 collegiate baseball players completed the TAIS and B-TAIS, as well as the SCAT and the CSAI-2 (both measures of anxiety). Findings revealed that all ineffective deployment of attention subscales were negatively correlated with performance (as calculated by seasonal contact percentage). Furthermore, the ineffective attention subscales of reduced attention (RED) and overload-internal (OIT) were both correlated to competitive trait anxiety as measured by the SCAT. In summary, more effective use of attentional resources was related to increased performance and decreased anxiety.

Goal Setting

A well-known but often overlooked psychological skill is goal setting. A wealth of knowledge can be gained by understanding what motivates a person and what types of goals a person might set for himself or herself. Effective goal setting can enhance performance and enjoyment, increase participation and interest, and help maintain direction and persistent effort toward achieving a goal. For instance, Duda (1989) found that people who set self-referenced, mastery-based goals reported higher competency levels and greater subjective success.

Although no specific scales exist to directly measure goal setting, a variety of goal-setting strategies have been developed; the most popular is the SMARTS goal method. Using this method involves setting goals that are specific, measurable, attainable, realistic, timely, and self-determined. Besides studying how people use specific goal principles (e.g., SMARTS), researchers have also attempted to measure peoples' goal orientations in terms of how they define success and failure, and what influence this definition process has on their goals. Along these lines, the Task and Ego Orientation in Sport Questionnaire (TEOSQ) was developed to assess these kinds of goal perspectives.

Task and Ego Orientation in Sport Questionnaire

As an adaptation of Nicholls' (1989) academic task and ego orientation questionnaire, the TEOSQ was developed to measure achievement orientations in sport. Task and ego orientations explain how people set goals and define successes and failures. Task mastery involves motivation to improve oneself and strive for subjective success (e.g., break a personal record). Conversely, ego orientation emphasizes external rewards and a desire for objective success (e.g., winning a game or trophy).

Although research often suggests a task orientation, many scholars now note that to be most effective, an athlete needs to score high in both task and ego orientations. This view is due to the fact that a high task orientation may actually buffer the negative outcomes of a high ego orientation, while maintaining the unique benefits of each perspective. As such, many research designs investigate the interactional nature of high and low combinations of task and ego orientations.

Reliability and Validity

The TEOSQ has often emerged as the instrument of choice for measuring achievement orientations. Hanrahan and Biddle (2002) tested four popular achievement measures and determined that the TEOSQ demonstrated superior psychometric properties, making it the preeminent choice for measuring goal orientations in sport.

In the development of the TEOSQ, Duda (1989) reported acceptable Cronbach's alpha reliability coefficients at .62 for task, and .85 for ego. Later studies have supported the instrument's reliability, reporting equal or higher reliability coefficients. Confirmatory factor analysis supported the two-factor structure of the instrument. To support the instrument's validity, the TEOSQ was initially correlated with the Purpose of Sport Questionnaire (PSQ) in which clear relationships emerged in the expected directions. For example, athletes who scored high on task orientation also responded that the purpose of sport should be to enhance mastery, cooperation, and self-esteem.

Norms and Scoring

The TEOSQ includes 13 items (7 for task, 6 for ego) that are scored on a 5-point Likert scale ranging from 1 (*strongly disagree*) to 5 (*strongly agree*). Higher scores represent a stronger orientation. Task orientation is measured through items 2, 5, 7, 8, 10, 12, and 13, while ego orientation is evaluated though items 1, 3, 4, 6, 9, and 11.

Results were determined by a sample of 322 varsity eleventh- and twelfth-grade athletes; means for males were 4.28 (task) and 2.89 (ego) and for females were 4.45 (task) and 2.59 (ego). This initial study reported these gender differences in task and ego as significant; however, subsequent studies have reported mixed results. The TEOSQ is provided in figure 12.4.

Research and Practical Examples

How an athlete defines success and failure can greatly influence his or her performance and participation in sport. Coaches need to recognize what motivates their athletes and understand the impact of different achievement orientations.

For example, in the initial development of the TEOSQ, Duda (1989) found that task-oriented goal perspectives were correlated with prosocial characteristics and values such as cooperation, following the rules, honesty, and participating in sport to enhance one's self-esteem. Conversely, ego-oriented perspectives were related to more maladaptive, antisocial qualities such as emphasizing personal gain, increasing social status, and earning more money.

I feel most successful in sport when . . .

Task orientation	**Ego Orientation**
I learn a new skill and it makes me want to practice more.	I'm the only one who can do the play or skill.
I learn something that is fun to do.	I can do better than my friends.
I learn a new skill by trying hard.	The others can't do as well as me.
I work really hard.	Others mess up and I don't.
Something I learn makes me want to go and practice more.	I score the most points or goals.
A skill I learn really feels right.	I'm the best.
I do my very best.	

Figure 12.4 Question styles for the Task and Ego Orientation in Sport Questionnaire.

Reprinted, by permission, from J.L. Duda, 1992, Motivation in sport settings. In *Motivation in sport and exercise*, edited by Glyn C. Roberts (Champaign, IL: Human Kinetics), 62.

Still, having an ego orientation does not necessarily carry negative psychological and behavioral consequences if paired with a high task orientation. Using the TEOSQ, Xiang and colleagues (2007) provide an example of this buffering effect. Recognizing that as children approach teen years, they are nearing the developmental stage of begin able to differentiate ability from effort, the authors targeted this age group to assess the importance of achievement orientations for members of a physical education running program with a sample of 533 fifth graders. Using task and ego scores, the researchers applied cluster analysis in dividing the sample into four groups: low task–low ego, low task–high ego, high task–low ego, and high task–high ego. The majority of the sample was found to have adaptive motivational patterns (high task–low ego, high task–high ego), with only 20% of the students falling into the low task-low ego group.

The study also demonstrated that the adaptive orientation groups reported greater expectancy beliefs; found running to be more important, interesting, and useful; performed better; and had greater intentions of continuing running. These results indicate once again that the negative consequences often seen with high ego orientations are greatly ameliorated when a concurrent high task orientation is also adopted.

Confidence

Perhaps no psychological factor is more important in sport than confidence (also called self-belief or self-efficacy). So strong is the effect of self-belief that it may, in fact, allow a weaker athlete to overcome a stronger competitor who is low in self-belief. This effect was demonstrated in a study by Nelson and Furst (1972) in which participant confidence levels were manipulated before an arm-wrestling task. Confidence was found to be more predictive of superior performance, to the extent that objectively weaker but highly confident participants almost always outperformed their stronger but less confident competitors. Similarly, by using a test of muscular endurance, Weinberg, Gould, and Jackson (1979) and Weinberg, Gould, Yukelson, and Jackson (1981) showed that participants who were more self-efficacious consistently and significantly outperformed their less confident counterparts.

Having confidence has demonstrated positive results in research and is often considered a critical factor in other constructs. For example, confidence is considered by some researchers to be an essential component of mental toughness (Clough and Strycharczyk 2012; Weinberg, Butt, and Culp 2011).

Confidence (as with other characteristics) may manifest differently in sport than in other environments. That is, a person who is highly confident in his or her ability to perform on the job may lack confidence when participating in a sport. As a result, scales have been developed to assess sport-specific confidence.

Trait and State Sport Confidence Inventories

Vealey (1986) developed the Trait Sport Confidence Inventory (TSCI) and the State Sport Confidence Inventory (SSCI) to measure the benefit or certainty that people have about their ability to be successful in sport. The TSCI indicates the belief an athlete usually has, and the SSCI measures the degree of certainty an athlete has at a particular moment about his or her ability to be successful in sport. The TSCI and SSCI scales are similar to SCAT and the CSAI-2 in that they measure trait and state anxiety, respectively. The two confidence scales were developed based on an interactionalist paradigm in which the individual difference construct of trait self-confidence interacts with the objective sport situation to produce state self-confidence. That is, individual differences in trait self-confidence are predicted to influence how athletes perceive factors within an objective sport situation and predispose them to respond to sport situations with certain levels of state sport confidence.

Reliability and Validity Vealey (1986) established reliability and validity in a series of studies. High school and college athletes took both the SSCI and TSCI inventories twice, either 1 day, 1 week, or 1 month apart. Results revealed test–retest reliability ranging from .63 for the 1-month retest to .89 for the 1-day retest. The internal consistency was high, with alpha coefficients of .93 and .95 for the TSCI and SSCI, respectively. Construct validity was established by analyzing the relationships between the SSCI and TSCI with other personality constructs; all correlations were significant in the predicted direction. For example, both the TSCI and SSCI were positively correlated with other measures of perceived physical ability and self-esteem but negatively correlated with somatic and cognitive state anxiety. Evidence supporting the relationship between the SSCI and TSCI and other constructs in the theoretical model established construct validity. For instance, trait self-confidence was found to be a good predictor of precompetitive and postcompetitive state self-confidence. Figure 12.5 illustrates sample items from the TSCI. Responses to all items are simply added together to get a total score.

Research and Practical Examples As part of the validation of the TSCI and SSCI, Vealey (1986) had elite gymnasts complete the TSCI 24 hours before a national competition and the SSCI approximately 1 hour before the competition. In addition, the gymnasts completed the Competitive Orientation Inventory, which measures whether they are more performance oriented (focusing on improving their own performance) or more outcome oriented (focusing on winning the competition). Gymnasts who were high in trait confidence and also performance orientation exhibited the highest state confidence just before competition. This illustrates the importance of the interaction between an athlete's self-confidence and his or her competitive orientation. *In essence, it appears that athletes who are generally confident and also focus on doing their best instead of winning will be most confident when the time comes to perform.* Because a great deal of research in sport psychology has shown that higher levels of confidence are related to higher levels of performance, it is imperative that coaches and parents foster an orientation that focuses on self-improvement rather than merely on winning.

When you compete, how confident do you generally feel? (Circle number.)

	Low			Medium			High		
1. Compare your confidence in your ability to *execute the skills necessary* to be successful to the most confident athletes you know	1	2	3	4	5	6	7	8	9
2. Compare your confidence in *your ability to make critical decisions during* competition to the most confident athletes you know	1	2	3	4	5	6	7	8	9
3. Compare your confidence in *your ability to perform under pressure* to the most confident athletes you know	1	2	3	4	5	6	7	8	9
4. Compare your confidence in *your ability to execute successful strategy* to the most confident athletes you know	1	2	3	4	5	6	7	8	9
5. Compare your confidence in *your ability to concentrate well enough to be successful* to the most confident athletes you know	1	2	3	4	5	6	7	8	9

Figure 12.5 Sample items from the Trait Sport Confidence Inventory.

Reprinted, by permission, from R.S. Vealey, 1986, "Conceptualization of sport confidence and competitive orientation: preliminary investigation and instrument development," *Journal of Sport Psychology* 8: 221-246.

In seeking to investigate the differences in imagery use and confidence, Abma and colleagues (2002) tested the TSCI in conjunction with two measures of imagery on a sample of track and field athletes. Although confidence was not related to sport experience, athletes higher in confidence reported significantly more use of all types of imagery. However, although highly confident athletes were found to use imagery more often, their skill at using imagery did not differ from that of their less confident counterparts. This suggests that the relationship between imagery and confidence may be more related to frequency of use rather than imagery ability.

Go to the WSG to view Video 12.3.

SCALES USED IN EXERCISE PSYCHOLOGY

Exercise psychology is primarily concerned with studying the psychological and emotional effects of exercise, as well as promoting physical activity and healthy behaviors. Like sport psychology, it is an interdisciplinary field, which integrates knowledge from a diverse body of knowledge. Many sciences have contributed greatly to the discipline of exercise psychology, including biological and social sciences, health psychology, neuroscience, and exercise science. Understandably, measurement in a discipline based on such diversity can be a challenging and multifaceted issue. However, the importance of understanding psychological determinants and consequences has never been more important, because public health concerns have continued to escalate globally, such as those of obesity and lack of adherence to exercise programs.

Questionnaires used in exercise psychology target a range of factors, including self-perceptions, mood (e.g., depression), motivation, attitudes, and emotions regarding exercise. Because exercise psychology is a multidimensional and interdisciplinary field, a combination of questionnaires may provide a more inclusive view of the participants. We will take a brief look at a few questionnaires commonly used in exercise psychology.

Exercise Motivation

With the increased numbers of people who are obese or overweight in society, there has been increased interest in the reasons for participation in physical activity and exercise. There were a number of scales originally developed to assess exercise motivation, including the 1989 Personal Incentives for Exercise Questionnaire and the Exercise Motivation Inventory (Markland and Hardy 1993), although these had some limitations from a psychometric point of view.

Exercise Motivation Scale

The Exercise Motivation Scale (EMS) (Li 1999) was developed to assess the four subtypes of external regulation found on the Self-Determination Theory continuum: external, identified, integrated, and introjected regulation. In addition, Li incorporated Vallerand's (1997) tridimensional view of intrinsic motivation. Vallerand viewed intrinsic motivation as including the intrinsic motivation to know, to move toward accomplishments, and to experience stimulation. Finally, an amotivation subscale was also developed to assess lack of motivation.

Reliability and Validity The EMS (figure 12.6) was administered to 571 college students; participants also completed measures of perceived exercise competence, exercise autonomy, social relatedness, interest, and effort. Internal consistency estimates of reliability ranged from .75 to .90. Model fit statistics from confirmatory factor analyses were below acceptable levels for most indexes. However, Li argued that due to the process of internalization, a simplex model analysis was more appropriate for examining the tenability of the EMS factor.

Analysis of a simplex pattern involves examination of the interrelations among subscales that are expected to form an ordered pattern. Li (1999) noted that the sequence of the relationships provided evidence of factorial validity. Li reported, "Tests on the selected

Responses are scored on a scale of 1 *(strongly disagree)* to 6 *(strongly agree)*.

My reasons for exercising or not exercising include: (Circle number.)

	Strongly disagree			Strongly Agree		
1. Because I think exercise contributes to my health (identified regulation)	1	2	3	4	5	6
2. Because it is consistent with what I value (integrated regulation)	1	2	3	4	5	6
3. Because I feel pressure from others to participate (external regulation)	1	2	3	4	5	6
4. It is not clear to me anymore (amotivation)	1	2	3	4	5	6
5. For the satisfaction it gives me to increase my knowledge about this activity (intrinsic motivation to learn)	1	2	3	4	5	6
6. For the enjoyment that comes from how good it feels to do the activity (intrinsic motivation to experience stimulation)	1	2	3	4	5	6
7. Because I would feel guilty if I did not take the time to do it (introjected regulation)	1	2	3	4	5	6

Figure 12.6 Sample items from the Exercise Motivation Scale (reasons for exercising or not exercising) with corresponding subscales.

Based on Markland and Hardy 1993.

antecedents and consequences of exercise motivation also showed evidence of nomological validity . . ." (p. 111) with a reasonable fit to the data. Li concluded that the EMS is "a theoretically sound and methodologically valid and reliable measure" (p. 112).

Although Li's studies provided initial evidence for the reliability and two forms of validity (factor structure and relationships with related constructs) for the EMS, further testing was conducted by Wininger (2007). Internal consistency reliability estimates using 143 participants ranged from .75 to .90 for the eight subscales.

Research and Practical Examples Wininger (2007) investigated the relationship between the EMS and the stages of change derived from the transtheoretical model (discussed later in this chapter), as it would be expected that peoples' reasons for participation in exercise would be related to their reasons for exercise. Results revealed that, as predicted, the amotivation scale was highest for people in the precontemplation stage, which tends to focus on the negative aspects of exercise. Amotivation scores got progressively lower as high levels of exercise were achieved. Also, according to prediction, scores on the intrinsic subscales of the EMS grew higher as the stages of change model revealed higher levels of exercise. Conversely, the external subscales of the EMS grew higher when exercise levels decreased according to the stages of change. Thus, from a practical perspective, practitioners should attempt to make exercise as interesting and enjoyable as possible, because doing so will typically lead to higher levels of exercise (greater frequency and intensity) over a longer time.

Behavioral Regulation in Exercise Questionnaire

Another exercise-specific measure of motivation that finds its roots in self-determination theory is the Behavioral Regulation in Exercise Questionnaire (BREQ) developed by Mullan, Markland, and Ingledew (1997). To measure the multidimensional components of motivation, the questionnaire consists of subscales for external, identified, introjected, and intrinsic motivation. Based on the principles of self-determination theory, the BREQ proposes that exercise motivation is on a continuum from intrinsic to extrinsic. As such, the subscales are expected to be distinct, but not mutually exclusive.

Reliability and Validity Overall, the BREQ had strong psychometric properties with subscale reliability coefficients consistently exceeding .70. Moreover, the four-factor structure had been supported through confirmatory factor analysis in numerous studies. Convergent validity was established through comparisons of conceptually similar scales, and factorial validity and construct-related validity were also supported. Furthermore, the BREQ demonstrated acceptable internal consistency as alpha coefficients ranged from .76 to .90.

Because the BREQ was seen as a reliable and valid measure, researchers sought to create supplemental subscales for amotivation and integrated regulation while maintaining the original structure of the instrument. Markland and Tobin (2004) revised the BREQ (to the BREQ-2) to include amotivation, which is operationalized as a failure to value exercise. The four original subscales of the BREQ were included in its revision. To determine the psychometric properties of the newly added subscale, as well as to confirm that the original properties were maintained, questionnaires were sent to a sample of 580 exercise participants with a total of 201 completed questionnaires returned. Results were encouraging as factorial validity was supported with the addition of an amotivation scale and the psychometric properties were upheld. Later studies confirmed the reliability with all subscale Cronbach's alphas exceeding .75.

Later, to amend the absence of an integrated regulation subscale, Wilson and colleagues (2006) conducted a series of studies to develop and test a complementary set of items.

Again, the incorporation of the new items did not diminish the original scale's psychometric properties and served as a practical expansion. The new subscale of integrated regulation was determined to be factorially distinct. Reliability was also supported as estimates regularly exceeded .70, with test–retest reliability ranging from .70 to .88 at a 2-week interval. Convergent validity was confirmed through significant correlations with the integrated regulation subscale on the EMS (previously discussed).

Norms and Scoring The BREQ consists of 15 items, with subscales for external (4 items), introjected (3 items), identified (4 items), and intrinsic motivation (4 items). The BREQ-2 expanded this assessment to a total of 19 items with the addition of 4 amotivation items. Similarly, the integrated regulation subscale developed by Wilson et al. (2006) consisted of 4 items.

Regardless of the version, the BREQ is scored on a scale ranging from 0 *(not true for me)* to 4 *(very true for me)*. Questions follow the stem "Why do you exercise?" Means and standard deviations are presented in table 12.8.

Table 12.8 Means and Standard Deviations of the BREQ-2

Subscale	Mean	SD
Intrinsic	2.80	1.02
Identified	3.24	0.87
Introjected	1.74	1.25
External	0.59	0.90
Amotivation	0.30	0.68

Adapted, by permission, from D. Markland and V. Tobin, 2004, "A modification to the Behavioural Regulation in Exercise Questionnaire to include an assessment of motivation," *Journal of Sport & Exercise Psychology* 26(2): 191-196.

Research and Practical Examples In practice, the BREQ has been able to successfully differentiate between physically active and inactive groups, as well as between participants at different stages of change regarding exercise. Edmunds, Ntoumanis, and Duda (2007) used the BREQ-2 to assess influences of adherence to an exercise prescription in a sample of participants who were overweight or obese. A total of 49 participants ranging from 16 to 73 years of age completed the BREQ-2, as well as measures of psychological need satisfaction, exercise behavior, and a variety of other factors. Baseline measurements were collected before participation in the exercise prescription, and again at 1 and 3 months into the program. The findings of the study revealed a wide variety of relationships between behavioral regulation and adherence, commitment, efficacy, and many other measures. For instance, over time, identified regulation (e.g., valuing a goal) and commitment decreased, while introjected regulation (e.g., desire to appease others) increased. These findings suggest that although many people may begin an exercise program with more autonomous motivation and dedication, difficulties following the program may lead to reduced participation or to staying with the program for only a short time from a sense of obligation or guilt. However, those who maintained (or developed) higher levels of autonomous motivation demonstrated greater levels of exercise, commitment, and self-efficacy helpful in overcoming obstacles to exercising.

Stages of Change for Exercise and Physical Activity

Marcus, Selby, Niaura, and Rossi (1992) developed the Stages of Change Scale for Exercise and Physical Activity to measure the specific stage of exercise that participants might be in at a particular time. Stages fall somewhere between states and traits, and although stages may last for considerable lengths of time, they are open to change. This is the nature of most high-risk behaviors: stable over time yet open to change. The stages of change, or transtheoretical model, was developed as a framework to describe the phases involved in the acquisition and maintenance of a behavior (Prochaska and DiClemente 1983; Velicer and Prochaska 1997). Specifically, Marcus and colleagues (1994) suggested that people who engage in a new lifestyle behavior (e.g., physical activity, smoking cessation, condom use, seat-belt wearing) progress in an orderly manner through the following stages:

- ***Precontemplation.*** No intention to change behavior.
- ***Contemplation.*** Intention to change behavior.
- ***Preparation.*** Preparing for action.
- ***Action.*** Involved in behavior change.
- ***Maintenance.*** Sustained behavior change.

People are thought to progress through these stages as they adopt lifestyle behaviors at varying rates; some may move right through each stage, others might get stuck at certain stages, and others may relapse to earlier stages.

Reliability and Validity

Marcus and colleagues (1992) conducted three studies to develop and refine the validity and reliability of the Stages of Change Instrument (SCI). The original SCI was developed by modifying an existing instrument that had previously been developed for smoking cessation. A test–retest reliability estimate of .78 was obtained for the new scale. In addition, concurrent validity was demonstrated by showing that the SCI was significantly related to a 7-day physical activity recall questionnaire (Marcus and Simkin 1993).

The SCI has been used to classify participants into stages of exercise change and thereby to develop specific interventions to help people in each of these stages. A 5-point Likert scale is used to rate each item, ranging from 1 *(strongly agree)* to 5 *(strongly disagree).* Participants are placed into the stage corresponding to the item that they endorsed most strongly (i.e., *agree* or *strongly agree).* Anyone not endorsing any items with *agree* or *strongly agree* is not placed into a stage. Figure 12.7 presents the specific stages and corresponding scale items.

Research and Practical Examples

As noted previously, one of the main benefits of the stages of change model is that it helps practitioners individualize behavioral interventions to increase exercise by identifying the exact stage a person happens to be in at a given time. Using this approach, Marcus and colleagues (1994) tested 610 adults aged 18 to 82 and, using the SCI, classified them into one of five stages of change identified by the model. The researchers devised 6-week interventions using written resource materials and specific exercise opportunities targeted at each participant's specific stage of readiness to adopt or continue an exercise program. Results revealed that 65% of participants in the contemplation stage became active, and 61% of people in the preparation stage became more active. From a practical point of view, practitioners may be able to achieve greater compliance by targeting interventions based on a participant's specific stage of exercise change. In essence, practitioners can devise

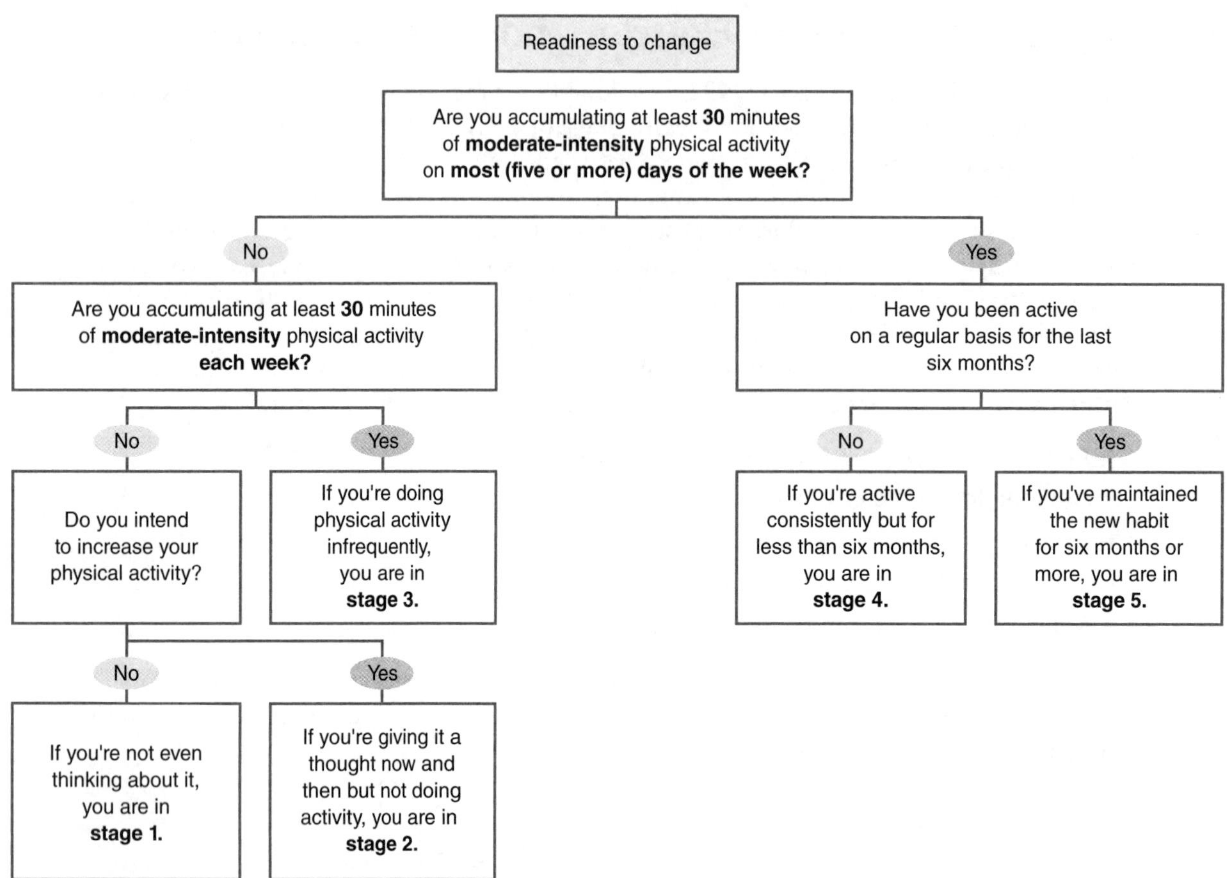

Figure 12.7 Readiness to change flowchart.

Reprinted, by permission, from S.N. Blair et al., 2001, *Active living every day* (Champaign, IL: Human Kinetics), 9.

specific exercise programs and educational materials that would be particularly motivating and relevant for people in certain stages of exercise change. Thus, specific processes can be used to promote transitions from one stage to the next. This, in turn, should enhance both exercise participation and maintenance.

GENERAL PSYCHOLOGICAL SCALES USED IN SPORT AND EXERCISE

Although the trend in sport psychology has been to develop sport-specific versions of psychological inventories, several general psychological inventories have been used regularly in physical activity and competitive sport settings. These inventories have added a great deal to the sport psychology literature, in terms of both performance enhancement and psychological well-being.

Self-Motivation Inventory

The Self-Motivation Inventory (SMI) was designed to measure a person's self-motivation to persist and was originally developed to be used in exercise adherence studies (Dishman and Ickes 1981). Because approximately 50% of all people who start exercise programs drop out in the first 6 months, it is important to know what types of people might be more likely to adhere to or drop out of a formal exercise program.

Reliability and Validity

Internal consistency reliability was established with a sample of 400 undergraduate men and women; the items were found to be highly reliable (r = .91). Test–retest reliability (with a 1-month time interval) was also found to be high (r = .92), thus demonstrating the stability of the construct of **self-motivation**, the desire to take action. Construct-related validity was found to be high: The SMI has consistently discriminated among people who will and will not adhere to exercise programs in athletic and adult fitness environments. In addition, correlations between self-motivation and social desirability, achievement motive, locus of control, and ego strength provided discriminant and convergent evidence for the construct validity of the SMI.

The SMI consists of 40 items in a Likert format that ask participants to describe how characteristic the statement is of them. Responses can range from 1 *(extremely uncharacteristic of me)* to 5 *(extremely characteristic of me)*. There are 21 positively keyed items and 19 negatively keyed items to reduce response bias. Figure 12.8 provides sample items from the SMI.

	Extremely uncharacteristic of me	Somewhat uncharacteristic of me	Neither characteristic nor uncharacteristic	Somewhat characteristic of me	Extremely characteristic of me
1. I'm not very good at committing myself to doing things.	1	2	3	4	5
2. Whenever I get bored with projects I start, I drop them to do something else.	1	2	3	4	5
3. I can persevere at stressful tasks, even when they are physically tiring or painful.	1	2	3	4	5
4. If something gets to be too much of an effort to do, I'm likely to just forget it.	1	2	3	4	5
5. I'm really concerned about developing and maintaining self-discipline.	1	2	3	4	5
6. I'm good at keeping promises, especially the ones I make to myself.	1	2	3	4	5
7. I don't work any harder than I have to.	1	2	3	4	5
8. I seldom work to my full capacity.	1	2	3	4	5

Figure 12.8 Sample items from the Self-Motivation Inventory.

With kind permission from Springer Science+Business Media: *Journal of Behavioral Medicine,* "Self-motivation and adherence to therapeutic exercise," 1981, 4: 421-436, R.K. Dishman and W. Ickes.

Research and Practical Examples

As noted, the SMI was originally devised as a potential dispositional determinant of adherence to exercise programs. Along these lines, Dishman and Ickes (1981) administered the SMI along with a series of other psychological and physiological assessments to a group of people before the start of an exercise program. Rates of adherence to a regularly scheduled exercise program were assessed over a 20-week period. *It was found that only self-motivation and percent body fat were able to predict those people who would adhere to or drop out of the exercise program.* In fact, self-motivation and percent body fat (when taken together) were able to accurately classify participants into actual adherence and dropout groups in approximately 80% of cases.

This result has important implications for you as a health and fitness leader. Specifically, if a person is low in self-motivation, it is more likely that he or she will drop out of the program. Knowing this, you must provide such people with extra reinforcement and encouragement.

Profile of Mood States

McNair, Lorr, and Droppleman (1971) initially developed the Profile of Mood States (POMS) to provide a measure of **mood** states—that is, one's emotional state of mind, feeling, inclination, or disposition. The scale has six subscales, each representing a mood: Vigor, Confusion, Anxiety, Tension, Anger, and Fatigue. The POMS can be used with instructions that ask participants to state how they feel right now or have felt for the last week or the last month. Thus, it can be used as both a state measure and a trait measure.

The use of the POMS in sport and exercise settings has been so extensive that it prompted a special issue in the *Journal of Applied Sport Psychology* (Terry 2000). The POMS has been the major measure of mood, providing a link between physical activity and mental health based on some of Morgan's (1980) groundbreaking research. Although the measure is not without controversy (e.g., Rowley, Landers, Kyllo, and Etnier 1995) in its prediction of performance for successful and less successful athletes, it still remains as one of the most consistent measures of mood in sport and exercise settings.

Reliability and Validity

McNair, Lorr, and Droppleman (1971) found that the reliability and validity of the POMS do not vary for the three time frames. Although such a mood scale would be expected to change over time, test–retest reliability coefficients for the six subscales were found to range from .65 for vigor to .74 for depression. Internal consistency reliability was shown to be consistently high within each subscale, with reliability estimates of approximately .90. Construct-related validity was established by relating the six POMS subscales to other personality measures; results were consistently in the predicted directions for each subscale. The scale consists of 65 items scored on a 5-point Likert scale from *not at all* to *extremely*. Figure 12.9 presents items from the Vigor subscale.

Research and Practical Examples

The POMS has been widely used to study the moods of elite athletes over the course of a season. In a series of studies investigating mood states of elite wrestlers, distance runners, swimmers, and rowers (Morgan and Johnson 1978; Morgan and Pollock 1977), athletes completed the POMS at different times during a competitive season. *One consistent finding was that successful and less successful elite athletes differed in their mood profiles.* The more successful athletes scored high on the Vigor scale (a positive attribute) but low on all the

Below is a list of words that describe feelings people have. Answer *how you feel right now.*

	Not at all	A little	Moderately	Quite a bit	Extremely
1. Lively	0	1	2	3	4
2. Active	0	1	2	3	4
3. Energetic	0	1	2	3	4
4. Alert	0	1	2	3	4
5. Full of pep	0	1	2	3	4
6. Carefree	0	1	2	3	4
7. Vigorous	0	1	2	3	4

Figure 12.9 Vigor subscale from the Profile of Mood States.

Reprinted, by permission, from D.M. McNair, M. Lorr, and L.F. Droppleman, 1971, *EdITS manual for POMS* (San Diego: Educational and Industrial Testing Service).

other scales (negative attributes), whereas the less successful athletes scored higher on all the negative mood states and lower on the positive Vigor scale. In a thorough review of the POMS (Prapavessis 2000) it was found that individual differences really needed to be considered when investigating the relationship between mood (as measured by the POMS) and athletic performance. Specifically, the amount of divergence for each athlete from his or her optimal mood state was related to performance. In essence, the more an athlete's mood deviated from his or her optimal state (whether positive or negative), the poorer the performance. Thus, from a coaching perspective, it is important to individualize athletes' mood states in relationship to what produces the best performance. This might involve several assessments of mood states and performance to determine what types and levels of mood states are associated with peak performance.

Test of Attentional and Interpersonal Style

Nideffer (1976) developed the Test of Attentional and Interpersonal Style (TAIS) to measure a person's attentional and interpersonal characteristics, the appropriateness of one's attentional focus, and the ability or inability to shift from one attentional focus to another. The basis of the TAIS is a theory that seeks to predict behavior based on the interaction between interpersonal and attentional processes and physiological arousal. Attention is hypothesized to vary along two dimensions: width (broad to narrow) and direction (internal to external). It is postulated that both attentional and interpersonal characteristics have state and trait components and that these scales can be examined by manipulating arousal. The test is composed of six attentional subscales—two subscales reflecting behavioral and cognitive control—and nine interpersonal subscales. In sport psychology, most researchers and practitioners have focused on the attentional subscales.

Reliability and Validity

The reliability and validity of the TAIS have been demonstrated in a variety of studies using different samples, although Nideffer (1976) initially developed the test using college students. Test–retest reliability based on a 2-week interval ranged from .60 to .93 with a median of .83. Construct-related validity was examined by correlating TAIS scale scores with scores on other psychological instruments. The overall pattern of results indicated that the TAIS subscales were correlated with conceptually similar scales but were not correlated with scales measuring different constructs. Correlating the TAIS subscales with future behaviors demonstrated predictive validity. For example, swimmers scoring high on the subscale measuring the tendency to make errors of underinclusion were rated by their coaches as choking under pressure, falling apart when they made early performance errors, and becoming worried about one thing and being unable to think about anything else. Poor attentional control and the tendency to make errors of underinclusion were found to be associated with performance deficiencies. Table 12.9 describes the six attentional subscales. Although Nideffer (2007) presents positive evidence of the psychometric properties of the TAIS, other researchers have raised serious concerns regarding its reliability and validity, especially the assumption of subscale independence (e.g., Dewey, Brawley, and Allard 1989; Ford and Summers 1992; Abernethy, Summers, and Ford 1998). In summary, there appears to be some support for the use of the TAIS as a diagnostic tool for helping athletes to identify attentional problems that may be affecting performance (Bond and Sargent 1995). However, "there is little strong empirical support for its use as a research instrument to examine the relationship between attentional abilities and sport performance" (Abernethy, Summers, and Ford 1998).

Table 12.9 Attentional Subscales of the Test of Attentional and Interpersonal Style

Scale	Description
Broad-external	High scores indicate an ability to effectively integrate many external stimuli simultaneously.
External-overload	High scores indicate a tendency to become confused and overloaded with external stimuli.
Broad-internal	High scores indicate an ability to effectively integrate several ideas at one time.
Internal-overload	High scores indicate a tendency to become overloaded by internal stimuli.
Narrow focus	High scores indicate an ability to effectively narrow attention when it is appropriate.
Reduced focus	High scores indicate chronically narrowed attention.

Data from R.M. Nideffer, 1976, "Test of attentional and interpersonal style," *Journal of Personality and Social Psychology* 34: 394-404.

Research and Practical Examples

As noted earlier, the TAIS has been used in practical settings with a number of sports and athletes, providing researchers with a great deal of information on the relationship between attentional processes and performance. In a study conducted by Martin (1983), high school basketball players completed the Narrow Focus and External-Overload subscales of the TAIS before the season. The players who scored high on the Narrow Focus subscale had far superior free throw percentages compared with the players who scored high on External-Overload. Such a finding can be useful for coaches in that it indicates potentially good or poor free throw shooters. A player's high score on External-Overload would indicate to a coach that the player needs help in coping more effectively with the many potential distractions when shooting free throws. Attentional control training designed to focus attention on the relevant cues (the rim) while eliminating irrelevant cues (crowd noises and gestures) could be used at the beginning of the season to help the player develop better attentional skills while shooting free throws.

Dataset Application

The chapter 12 large dataset in the WSG consists of data on 200 boys and girls aged 12 years. The variables include Fitnessgram PACER test results, Healthy Fitness Zone (HFZ) achievement, body mass index (BMI), student responses to 12 Body Satisfaction questions, 6 Endurance Self-Efficacy questions, and a total score for the Body Satisfaction and Endurance Self-Efficacy Scales. Determine the following:

1. What is the alpha coefficient for the 12 Body Satisfaction variables (chapter 6)?
2. What is the alpha coefficient for the 6 Endurance Self-Efficacy variables (chapter 6)?
3. What is the Pearson product-moment correlation between the Body Satisfaction Scale and the Endurance Self-Efficacy Scale (chapters 4 and 12)?
4. What percentage of students are in the Fitnessgram PACER Healthy Fitness Zone (HFZ) (chapters 3, 7, and 10)?
5. What is the Pearson product-moment correlation between the BMI and PACER laps (chapter 4)?
6. Do those achieving the Fitnessgram HFZ differ from those not achieving the HFZ on Body Satisfaction, Endurance Self-Efficacy, BMI, and PACER laps (chapters 3 and 5)?
7. Answer some other interesting questions you might have for this dataset.

MEASUREMENT AND EVALUATION CHALLENGE

After reading this chapter, Coach Bill is more aware of some of the types of psychological inventories at his disposal. He knows about the guidelines for using psychological tests with athletes, and he knows that he can get the best prediction if he uses tests that are sport specific. Armed with this information, Bill decides on the following approach.

1. He will hire a qualified sport psychologist to coordinate the psychological evaluations and to interpret all psychological inventories.
2. He and the test administrator will tell his athletes specifically what the inventories are being used for and will provide feedback about their results.
3. He will use a few trait inventories before the season starts, such as the Sport Competition Anxiety Test (SCAT) and the Trait Sport Confidence Inventory (TSCI), as well as one of the newer multidimensional scales (ACSI-28), to get a better understanding of his athletes' general psychological profiles.
4. He will assess his athletes just before competition using state measures such as the Competitive State Anxiety Inventory-2 (CSAI-2) and the State Sport Confidence Inventory (SSCI) to determine how the athletes are feeling just before games.
5. With this information, Coach Bill, in conjunction with the sport psychologist, will devise a mental training program to help his athletes practice and develop their mental skills.

By combining this psychological testing and training with his work and practice on the physical aspects of training, Coach Bill hopes that his football team will be maximally ready to perform at their optimal level.

SUMMARY

The field of sport and exercise psychology has been expanding rapidly, with the two major areas being performance enhancement and mental health. Part of this expansion has involved developing and refining the measurement of psychological traits and states. These advances have highlighted issues involving the use and abuse of psychological inventories in sport and exercise settings. The American Psychological Association provides guidelines for the use of psychological testing to ensure that athletes are treated in an ethical manner and that test administration and feedback are conducted responsibly.

Although much early work in sport psychology used standardized psychological scales to assess personality and other psychological constructs, more qualitative methods have been developed and used, including in-depth interviews and observation. Along with the increased emphasis on qualitative methods, more accurate and reliable prediction of behaviors in sport and exercise settings has resulted from the development of sport-specific psychological inventories. Although there are still many scales in use without established reliability and validity, more sport-specific psychological inventories are being carefully developed from a psychometric standpoint. Some of the more noteworthy scales are the Sport Competition Anxiety Test (SCAT), the Competitive State Anxiety Inventory-2 (CSAI-2), and the Trait Sport Confidence and State Sport Confidence Inventories. In addition, the Stages of Change Scale for Exercise and Physical Activity questionnaire has been developed to identify a participant's specific stage of exercise behavior so that interventions can be targeted for that particular stage.

Psychological skills training (PST) has also grown rapidly, and with it, a need for precise measurement tools. A wide range of questionnaires have been developed to evaluate imagery, self-talk, concentration, goal setting, and a number of other skills. Scales such as the Vividness of Movement Imagery Questionnaire-2 (VMIQ-2), Automatic Self-Talk Questionnaire for Sports (ASTQS), Baseball Test of Attentional and Interpersonal Style (BB-TAIS), and the Task and Ego Orientation in Sport Questionnaire (TEOSQ) have demonstrated satisfactory psychometric properties and enhanced a practitioner's ability to measure psychological skills and the effectiveness of PST programs. Although the trend is to develop sport-specific tests, several general psychological inventories have also been used extensively in sport and exercise settings; these have added to our understanding of sport-related behavior. Such scales include the Self-Motivation Inventory (SMI), the Test of Attentional and Interpersonal Style (TAIS), and the Profile of Mood States (POMS).

Remember that most instruments consist of several scales. This is because psychological measures need to reflect the multidimensionality of personality, perception, and other psychological factors. Personality and its associated constructs are multifactorial in nature. Subscales help researchers get at these various factors. In fact, there has been a trend in sport and exercise psychology to develop scales that assess a variety of psychological characteristics or skills and are by definition multidimensional in nature. The result is that a number of factors are assessed; however, each factor is typically made up of only a few items and thus may not be as reliable as a scale focusing on that one mental skill (e.g., anxiety, confidence).

Go to the WSG for homework assignments and quizzes that will help you master this chapter's content.

CHAPTER

13

Classroom Grading: A Summative Evaluation

OUTLINE

OBJECTIVES

After studying this chapter, you will be able to

- list the proper criteria for grade assignment;
- illustrate the methods for assigning grades to one test or performance; and
- use the methods for assigning final grades.

You will also understand that there are many uses for grades, including the following:

- Motivating and guiding the learning, educational, and vocational plans of students as well as their personal development
- Communicating to students and parents about student progress
- Condensing information for use in determining college admissions, graduate school admissions, and scholarship recipients
- Communicating a student's strengths and limitations to possible employers
- Helping the school tailor education to student needs and interests and evaluate the effectiveness of teaching methods

 The lecture outline in the WSG will help you identify the major concepts of the chapter.

MEASUREMENT AND EVALUATION CHALLENGE

Tom is getting ready to teach a unit on tennis to his high school physical education class. He wants to weight the cognitive and the psychomotor aspects of the unit equally. Drawing on his experience, Tom plans to give two written tests during the unit to measure cognitive objectives; the first examination will count for 20% of the final grade, and the final examination will count for 30% of the final grade. He plans to use two skills tests, serving and ground strokes, to measure the psychomotor objectives of the unit. The serving test will be weighted as 35% of the grade, and the ground stroke test will be 15% of the final grade. What steps should Tom follow to ensure that the final composite score for each student will reflect these weightings?

Although this chapter is primarily directed to students who plan to become physical education teachers, many of the concepts are relevant to any situation involving the process of obtaining a composite score from multiple measures. The assessment of student performance in public and private schools generally culminates with grade reporting. The grading process can be a challenging one that many beginning, and even veteran, teachers do not look forward to completing. Although teachers and students enjoy assigning and receiving good grades, they do not enjoy assigning and receiving poor grades. However, prospective teachers must realize that assigning grades is a professional obligation and an important component of physical education teaching.

Over the years there have been attempts to devise standard curricula for many public school subjects. Most of these attempts have been at the local or state level, but occasionally some have been at the national level. Although the majority of these efforts have been in classroom-based subjects such as math, English, the sciences, and social studies, a few have attempted to define a national physical education curriculum. The fewer attempts in physical education probably have to do with the large variability in philosophies of what should be taught and required, availability of equipment and facilities, and local customs.

One comprehensive example is P.E. Metrics, produced by the National Association for Sport and Physical Education (NASPE 2008), an Association of the American Alliance for Health, Physical Education, Recreation and Dance (now named SHAPE America). This program is advertised as a standards-based, cognitive, and motor skills assessment package. It attempts to help physical educators identify reliable and valid measuring instruments to be used in the assessment process involved in grading. The importance of these aspects of measurement will become evident throughout this chapter. Along with the need for accurate measuring instruments there are many other nuances to consider in the process of assigning a final grade to each student. These are also considered in the rest of this chapter.

Besides dealing with the many minor difficulties that arise during the process of determining grades, you also need to consider that grades, which are assessments of one person by another, are based on subjective evidence. Sometimes the amount of subjectivity included in a set of grades is large. Grades based on an assessment of a student's answers to an essay examination and participation in class discussions are examples of such situations. Even grades determined through seemingly precise and objective methods often involve more subjectivity than you might think. Objective testing (true–false, multiple-choice, and others) requires somewhat subjective decisions, such as determining what items to include on the test and what response is most correct for each item. Often instructors devise a precise formula for combining certain student achievements, but this formula is generally based

Teachers must work hard to assure reliability of assessments and relevance of topics so that grades truthfully reflect achievement.

on a subjective weighting of the importance of the course objectives, similar to those Tom selected in the measurement and evaluation challenge described at the beginning of this chapter. *Thus, although it is true that some grades involve a greater amount of subjectivity than others, it is also true that all grades involve some degree of subjectivity.*

Generally, the greater the degree of subjectivity involved in the grading system, the greater the unreliability. In other words, a subjective system, if repeated, would probably not assign the same grades to the same students. This lack of objectivity and consequent lowering of reliability lead to a second, more serious, concern about grades: the lack of a common understanding of what a particular grade represents. *Grades are affected by the type of marking system used, the particular instructor who assigns the grade, the makeup of the class, the institution where the grades are earned, and many other related factors.* For example, an E might stand for excellent in one grading system but for failing in another; a grade of B from an instructor who rarely assigns a B has a meaning different from that of a B given by an instructor who seldom gives anything lower; an A obtained in a class of superior students may represent a more significant achievement than an A received in a class with less competition; and the same level of performance might be given an A at one institution but a lower grade at another.

Whereas the use of insufficient objective evidence reduces the *reliability* of grades, the lack of a clear and generally accepted definition of what a particular grade represents affects the *validity* of grades. Grades, which should reflect the degree to which a person has achieved the course objectives or goals, are often contaminated by many other factors. Failure of grades to validly reflect course achievement is important.

No specific rules for grading can be firmly established because any situation in which grades are assigned is different in some respects from any other. Differences exist in areas such as teaching techniques, course objectives, equipment and facilities available, and type of students in the class. However, an effective, consistent grading process allows an instructor and the enrolled students to have confidence in the validity of the assigned grades. If a student understands and accepts the teacher's grading system, he or she may not be happy with a poor grade but can accept it as being fair. Were you ever unhappy when you received a grade different from the one you had expected? Was it easier to accept a poor grade if you felt a teacher's grading method was fair to all students? The challenge of this chapter is to provide you with the skills and knowledge required for developing and using good grading practices in your instructional programs, whether you will use these practices in an academic setting or any similar situation that requires a final assessment of the degree to which objectives have been met.

 Go to the WSG to complete Student Activity 13.1.

EVALUATIONS AND STANDARDS

In chapter 1 we defined several terms important to the measurement and evaluation process. We differentiated between a formative evaluation, conducted during an instruction or training program, and a summative evaluation, a final, comprehensive judgment conducted near the end of an instruction or training program. Evaluations result from a decision-making process that places a judgment of quality on a measurement. In physical activity instruction, teachers make formative evaluations at the beginning of and during the instructional process and use them to detect weaknesses in student achievement and to direct (or redirect) future learning activities. A formative evaluation can be a formal measurement activity, such as a pretest before instruction, or an informal, subjective evaluation given by an instructor (e.g., verbal feedback during tennis practice). This chapter focuses on formal summative evaluations, including the proper steps for conducting such evaluations, which result in a final grade being determined for a student's overall performance for the entire instructional unit or final achievement of the objectives presented.

For a judgment to be made about the quality of performance, the performance must be compared with a standard. In chapters 1, 6, 7, 9, and 10 we discussed norm-referenced and criterion-referenced standards for evaluation. To review briefly, a norm-referenced standard is established by comparing a person's performance to performances of others of the same gender and age or another well-defined group. The establishment of this standard usually requires some type of data analysis. A criterion-referenced standard, on the other hand, is a specific predetermined level of performance that has been established—from past databases or expert opinion—before a participant performs. In a criterion-referenced evaluation, a participant either achieves the standard (passes) or does not achieve the standard (fails). In a norm-referenced evaluation, participants are ranked from excellent to poor based on their position among the comparison participants' scores. Before continuing with this chapter, return to chapters 1, 6, and 7 to review formative and summative evaluations and norm- and criterion-referenced standards.

 Go to the WSG to complete Student Activity 13.2 and view Video 13.1.

PROCESS OF GRADING

On what should students or subjects be graded? As explained in chapter 1, in physical education and human performance, there are three domains of potential objectives:

- ***Psychomotor domain.*** Concerns physical performance or achievement.
- ***Cognitive domain.*** Concerns mental performance or achievement.
- ***Affective domain.*** Concerns attitudes and psychological traits (also called the psychological domain).

Teachers of subjects such as mathematics, science, English, and history have instructional objectives limited mostly to the cognitive domain. Thus, in one sense their evaluation process is simpler than that of a physical education instructor. To evaluate students effectively, an instructor must have a clear understanding of the instructional objectives of the unit he or she is planning to teach. An instructor must select and administer tests and measurements that are relevant to these objectives, compare the resulting test scores to appropriate standards, and finally, determine grades. Figure 13.1 illustrates this continuum.

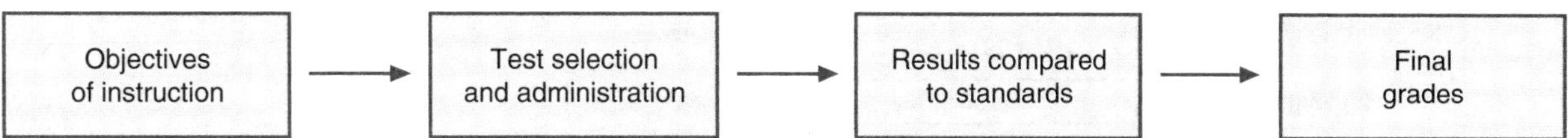

Figure 13.1 The grading process.

An effective and successful grading process requires that students understand the course objectives, know the tests and measurements used for grading, and know the method with which test scores will be combined to determine final grades. Inform your students of these factors at the beginning of instruction, and use formative evaluation techniques to update them on their personal progress throughout the course. Students should always be aware of the nature of the evaluations being conducted in the class and should ultimately not be surprised by the grades assigned because they have been made aware of the grading process, their scores, and the relative weighting of tests and projects.

Go to the WSG to complete Student Activity 13.3.

DETERMINING INSTRUCTIONAL OBJECTIVES

There are three questions to consider when determining whether a potential objective should be part of your instructional unit and thus also a part of your grading process:

1. Is the objective defensible as an important educational outcome?
2. Does every student have an equal chance to demonstrate his or her ability on the objective?
3. Can the objective be measured relatively objectively, reliably, relevantly, and validly?

What Not to Grade On

Often students in physical education are graded on such objectives as attendance, correct uniform, shower taking, leadership, attitude, fair play, participation, team rank, or improvement. Figure 13.2 summarizes the reported attributes used for grading; as you ask the preceding three questions of each of the most graded attributes in figure 13.2, you will see that they are largely inappropriate as bases for students' grades and thus should be eliminated. Rarely is it defensible to grade on one or more of these elements. Research indicates that many of these parameters have been widely overused as objectives in physical education (Hensley and East 1989). Let's briefly analyze why they are inappropriate.

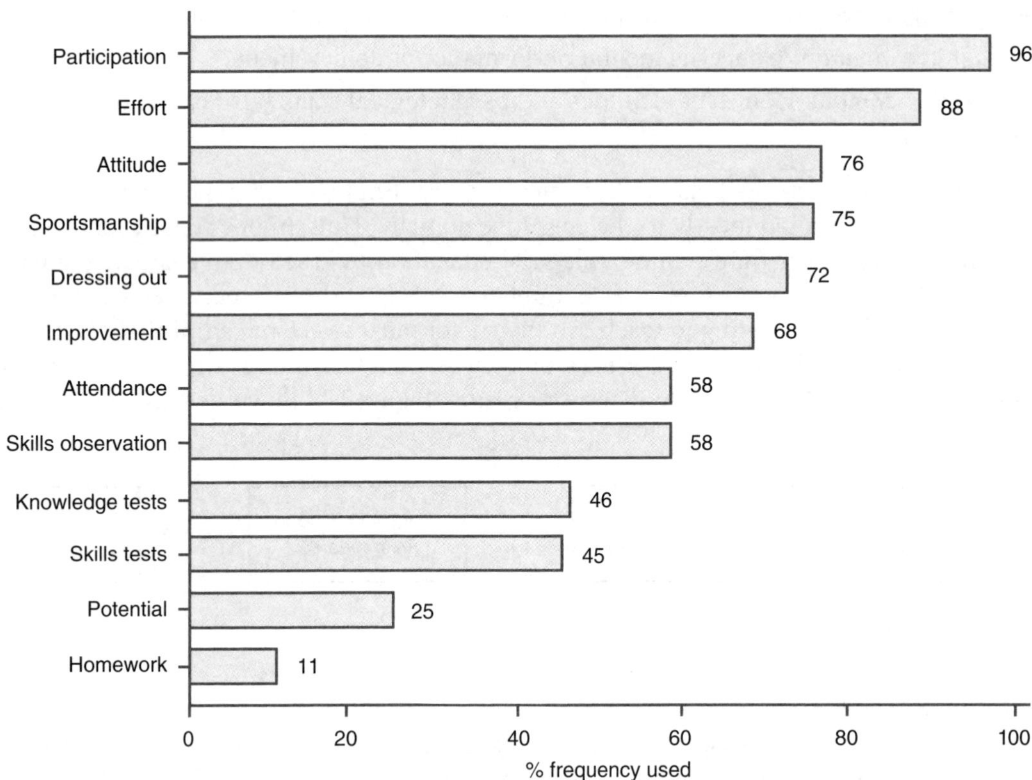

Figure 13.2 Attributes used for grading physical education.

Although attendance, correct uniform, and shower taking are obviously worthwhile and necessary for appropriate physical education instruction, they fail to meet the first criterion: having an important educational outcome. Students failing to meet these requirements should face consequences, but it should not be by a direct lowering of their grades. Students in mathematics may have to attend class and bring their textbooks, but they are seldom graded on these required factors, nor should they be.

Leadership, attitude, fair play, and participation are worthwhile objectives of any physical education or athletic program. However, to grade these factors reliably and validly, an instructor would need to implement a formal and systematic program of measurement and evaluation (third question). Implementing such a program takes time and expertise. Generally, the physical education instructor simply does not have the time or the expertise to perform this task. When these factors are used for grading (unless some of the techniques that are explained in chapter 14 are used), they are usually based on random teacher observations, which tend to be subjective and possibly biased. The result is grades that lack objectivity, reliability, and validity and are thus unfair to students.

Grading on team rank in team sport classes fails to meet the second criterion—provision of an equal opportunity to demonstrate ability. A poor performer might be placed on a good team and receive an A, whereas a good performer could be placed on a poor team and receive a low grade. This strategy is not fair to all students because their grades depend on the performance of other students, over which they have little or no control. *From a measurement and evaluation prospective, grading students in team sport classes may be one of the most problematic tasks a teacher faces.* Sport skills tests are often used to address this problem. However, an individual sport skills test that is isolated from the team game may lack relevance and validity. Sport skills testing is examined in chapter 11. Performance-based assessment (evaluating players as they participate in game conditions, chapter 14) is also used but requires well-developed rating scales and/or rubrics.

Improvement is one of the most attractive objectives to use in grading in physical education. As an instructor you want your students to improve, but grading on improvement presents some difficult problems. One problem is illustrated in figure 13.3. A low beginner (James) improves at a much greater rate than an advanced beginner (Robert). This is a natural phenomenon: The lower beginner has more room for improvement than the advanced beginner. The marathon runner running his or her second marathon will typically show a much more improved time than the experienced runner whose improvement is based on comparing the 20th to the 19th marathon time. Another problem is that improvement scores tend to be less reliable than either the pretest or posttest from which they are computed. A third problem is that some students, knowing that they are to be graded on improvement, may provide false initial performance scores that will inflate their improvement scores. Some teachers attempt to support improvement grading by suggesting that a student's improvement on a final performance be compared with his or her potential. However, the means for validly determining potential are often unreliable or simply don't exist.

Why do you suppose educators, and especially physical educators, like to grade on improvement? It is probably because this practice gives more students a chance to earn high grades, since both those students who score high and those who show great improvement receive high grades. A great number of educators are reluctant to give low grades because they believe low grades discourage effort, which increases the probability of low grades, resulting in a cycle that continues until students dislike the subject. The concern for physical educators is that when students become turned off by the subject, they may be less likely to continue physical activity during their lifetimes.

However, little, if any, evidence exists to support this contention. It is doubtful that students feel rewarded when given high grades for improvement when their actual performance level is lower than most of their peers. For example, poor swimmers know they are poor swimmers (without being told so) and so do their peers; an honest student knows that in the long run it is actual level of achievement that is important, not the rate of improvement. Suppose it is true that low grades cause a student to dislike school, or, more specifically,

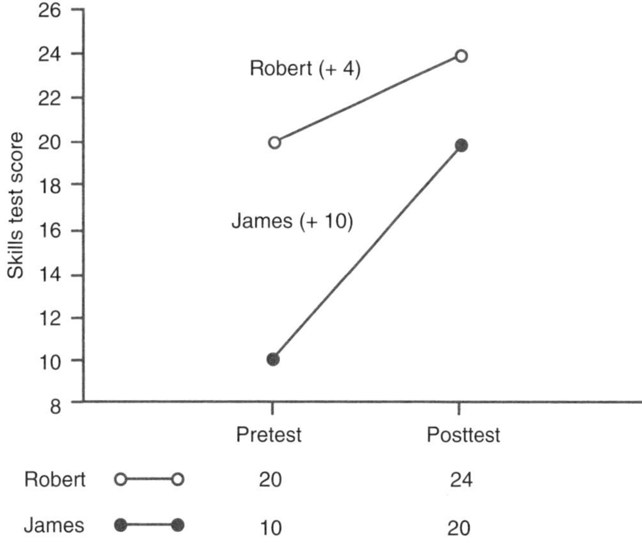

Figure 13.3 The problem with grading on improvement.

that low physical education grades destroy incentive and reduce the probability of a student engaging in physical activity. The solution of not giving low grades is like a physician treating the symptoms rather than the cause of the illness. Instead, instructors need to determine why a student is performing poorly and provide opportunities for success. This might apply to other fitness situations in clinical or community settings as well; you do not want clients to be discouraged from attending fitness classes or programs.

Go to the WSG to complete Student Activity 13.4 and view Video 13.2.

First, an instructor may want to increase the number of opportunities for achieving success by broadening the curriculum from, for example, touch football, basketball, and softball to a larger variety of activities. Because the skills required to succeed in physical education are relatively heterogeneous compared with those required in other subjects, varying the physical education curriculum will probably result in a higher percentage of students achieving success than in the other areas. For example, learning how to swim a particular stroke is not nearly as dependent on knowing how to do a forward roll as, say, learning to derive a square root is dependent on knowing how to divide.

A second means of encouraging success lies in ability grouping and enrolling students in courses suitable to their skill levels. If differences in initial ability are small, grading on improvement becomes unimportant (because it becomes almost the same as grading on achievement). Other subjects, such as mathematics, English, and the sciences, often make use of ability grouping; to adopt this approach, physical educators need to develop valid and reliable measuring devices to classify students by ability.

A third approach involves educating students that a grade is simply an expression of the level of achievement of the course objectives attained by a student; it is not a reward or a punishment. Furthermore, not every student can excel in every course. *One way to help ensure that grades remain a clear indication of a student's performance is to think of grades as measurements rather than as evaluations.* In grading, a measurement is a quantitative description of a student's achievement, whereas an evaluation is a qualitative judgment of a student's achievement. In other words, a grade, as a measurement, represents the degree to which a student has achieved the course objectives. If a grade is considered an evaluation, it should indicate to some degree how adequate a particular student's level of achievement is.

Introducing students to atypical physical education activities is one way to help them discover new activities and perhaps achieve success.

There are several advantages to treating grades as measurements rather than evaluations. Judgments of the adequacy of a student's achievement depend not only on how much he or she achieved but also on the opportunity for achievement and effort. This makes it difficult to report evaluations precisely in a standard grading system. Additionally, it is probably of more value to a future teacher or employer to know that a student was outstanding, average, or poor in a particular course of study than to know that this student did as well as could be expected or that he or she failed to live up to the expectations of a teacher. Given valid and reliable measurement of a student's achievement in various areas, the future teacher or employer can make his or her own evaluations in light of current circumstances, which would probably be more valid than having evaluations made by others under totally different conditions. *The worth of physical education must be conveyed to students through the teaching procedures, personal example, and the curriculum, not through the awarding of high grades regardless of the level of proficiency.*

Remember from the measurement and evaluation challenge described at the beginning of this chapter that Tom has decided to award one half of the total grade based on the cognitive aspects of the tennis unit. Thus, he is broadening the opportunities for students to succeed, and his grading method does not reflect the less defensible objectives of attendance, improvement, and so forth.

What to Grade On

Base grades on reliable and valid tests and measurements that are representative of important instructional objectives. In individual or team sport classes, sport skills tests, expert observations, rating scales, and round-robin tournament performance can be effective criteria for grading. In individual sports, such as bowling, golf, and archery, a student's achievement of objectives can be evaluated through performance. In fitness classes, field tests of physical fitness provide reliable and valid data for evaluation. Because the cognitive domain is important in physical education, knowledge testing should be part of the overall grade. Chapters 8 through 12 provide you with a selection of reliable and valid tests and testing procedures for use in human performance. In summary, use the following procedures to make grading fair, reliable, and valid:

- Carefully determine defensible objectives for each course before it begins.
- Group students according to ability in the physical skills necessary for the course, if possible.
- Inform students about the grading policies, procedures, and expectations.
- Construct tests and measurements as objectively as possible, realizing that all tests are subjective to some degree.
- Remember that no matter how well constructed a test is, no test is perfectly reliable.
- Realize that the distribution of grades for any one class does not necessarily fit any particular curve, but that over the long run physical skills are probably fairly normally distributed.
- Determine grades that reflect only the level of achievement in meeting the course objectives and not other factors.
- Establish grades on the basis of achievement, not improvement.
- Avoid using grades to reward the good effort of a low achiever or to punish the poor effort of a high achiever.
- Consider grades as measurements, not evaluations.

CONSISTENCY IN GRADING

A goal of any grading system should be to achieve consistency in grade determination. Theoretically, a student's grade should not depend on any of the following:

- A particular section of a course. If students are taking PE 100–Beginning Tennis, their grades should not be affected by whether they are in the 9:00 a.m. or the 2:00 p.m. section.
- A particular semester of the class. A student's level of performance should warrant the same grade whether the course is taken in the fall semester or the spring semester.
- Other students in a class. A student's grade depends on his or her performance only and should not be influenced by the performance of other students in the class. This objective suggests that grading on the curve is inappropriate.
- A particular instructor. Instructors X and Y should give the same grade to the same level of performance.
- The method of course delivery. It should not make a difference if the course is delivered online, face-to-face, or blended (assuming each method is an equally effective teaching style).

The goal of consistency is extremely hard to achieve because of student and teacher differences, which are natural phenomena of any instructional setting. As you study further in this chapter, you will discover that lack of consistency in grading is one of the weaknesses of several grading schemes.

Go to the WSG to complete Student Activity 13.5.

GRADING MECHANICS

Four steps are involved in the mechanical process of arriving at a grade to represent each student's level of achievement of course objectives.

1. Determine the course objectives and their relative weights.
2. Measure each student's achievement of the course objectives.
3. Combine measurement results to obtain a composite score for each student.
4. Convert the composite scores to grades.

Each step may be accomplished in a variety of ways, depending on different situations and philosophies.

Step 1: Determine Course Objectives and Their Relative Weights

Determining course objectives and their significance is the most important of the four steps and requires a considerable amount of thought. In fact, doing so is basic to every aspect of teaching a course, not just grading. Determining course objectives precedes planning the sequence of presentation of material, necessary equipment, teaching procedures, assignments, and grading procedures. It should be based on as much knowledge of the abilities of the prospective students as you can obtain; the general objectives of physical education; and such practical considerations as the number of students, the facility limitations, the duration of the course or unit, and the number, frequency, and duration of meeting times.

The objectives that you finally decide on for a physical education course can be classified into the psychomotor, cognitive, and affective (or psychological) domains (see also chapter 1). Stating the objectives in behavioral terms will help make the second step (measuring achievement) easier. In other words, list the actual performance levels, the specific knowledge, and the social conduct expected to be achieved by students. Sample objectives for a badminton unit are as follows:

Sample Badminton Objectives

- Cognitive objective: Know the setting rules when the game becomes tied at certain scores.
- Psychomotor objective: Be able to place at least four out of five low, short serves into the proper portion of the court.
- Affective objective: Be aware of proper etiquette when giving the shuttlecock to an opponent at the end of a rally.

How you weight the course objectives will vary greatly from one situation to another, depending on your philosophy, the age and ability of your students, and so on. For example, you might place less emphasis on the affective domain objectives for a physical fitness unit for tenth-grade boys or girls (single gender) than for a coeducational volleyball unit at the same grade level. *The actual weight you give to each course objective should result in a balanced and defensible list of goals to be achieved by each class member.*

Informing students of what is expected of them at the outset facilitates the planning of student experiences, teaching methods, and grading procedures and should reduce student anxiety.

 Go to the WSG to complete Student Activity 13.6.

Step 2: Measure the Degree of Attainment of Course Objectives

Recall that measurement is the procedure of assigning a number to each member of a group based on some characteristic. In this case, the characteristic involved is the degree of achievement of each course objective. Unlike step 1, steps 2 through 4 occur after the teaching and, presumably, the learning have taken place. This is not to imply that all measuring should be of a summative nature. On the contrary, there is merit in obtaining measures throughout a unit (formative evaluation) because doing so leads to increased awareness on the part of each student, as well as the teacher, of progress toward the course objectives. However, the testing or measuring used for grading students' degree of achievement should obviously not occur before the completion of instruction and learning. Chapters 8 and 11 provide ideas for constructing, evaluating, selecting, and administering tests and other devices to arrive at the numerical value to assign to each student that most accurately represents his or her level of achievement of the course objective being measured.

Step 3: Obtain a Composite Score

Seldom does a course have a single objective. Often, more than one measurement is made to determine the amount of achievement for each objective. *For these reasons, it is usually necessary to combine several scores to arrive at a single value representing a student's overall level of achievement of the course objectives.* This **composite score** is then normally converted into whatever grade format (e.g., A-B-C-D-F; pass/fail) is being used. The correct method to obtain the composite score depends on several factors, the most important of which is the precision of the scores that are to be combined.

In the case of performance scores, it is apparent why a composite score cannot be obtained by simply totaling the various raw scores. Units of measurement often differ: Feet cannot be added to seconds, or the number of exercise repetitions cannot be added to a distance recorded in inches. As was mentioned in chapter 3 under the discussion of standard scores, scores from distributions bearing differing amounts of variability, even if they do have the same units of measure, cannot simply be added together because the variability affects the weight each score contributes to the composite score. The following example illustrates why adding raw scores on tests that actually do have the same units (such as written tests) may not result in the desired composite score: Imagine that the scores from a class of 25 students were distributed as shown in table 13.1 on three tests worth 9 points, 9 points, and 27 points, respectively. This rather extreme example was chosen to make a particular point: that a score from a set of scores having a large variability will have more weight in the composite score than a score from a set of scores with little variability, regardless of the absolute values of the scores. The variability of the scores on the first two tests is greater than the variability for test 3. It would seem that, because the total number of points possible on test 3 is triple that of tests 1 or 2, the score achieved on test 3 would have the most influence on a student's composite score. Notice, however, some possibilities:

Student A was one of the seven students achieving the highest score made on test 3; scores on the first two tests were average. Student A received a composite raw score of 34. The composite raw score for student B, who scored above average on the first two tests but was among the lowest scorers on test 3, is 37, higher than that for student A. Student C scored at the average on two tests and above average on one test, as did student A. However, because student C's above-average performance came on test 1 (on which the scores were more variable than the test on which student A achieved above average, test 3), student C's composite raw score, 37, is also higher than student A's. Finally, even though student D scores as high as anyone in the class on test 3, the low scores made on the first two tests lowered the composite raw score, 30, below the others.

Unless two (or more) sets of scores are similar in variability, summing a student's raw scores to arrive at a composite score may lead to some incorrect conclusions.

Assume that, instead of representing the distributions of scores on three written tests, the values in table 13.1 describe the distributions of the measures of how well cognitive, affective, and psychomotor objectives were met by the students. Further assume that you had decided that achievement of the cognitive objectives, the affective objectives, and the

Table 13.1 Test Score Distributions from a Class of 25 Students

TEST 1 (9 POINTS)		TEST 2 (9 POINTS)		TEST 3 (27 POINTS)	
Score	Frequency	Score	Frequency	Score	Frequency
9	1	9	1	27	
8	2	8	2	26	
7	3	7	3	25	
6	4	6	4	24	7
5	5	5	5	23	11
4	4	4	4	22	7
3	3	3	3	21	
2	2	2	2	20	
1	1	1	1	19	
Student	Score on test 1	Score on test 2	Score on test 3	Total raw score	
A	5	5	24	34	
B	8	7	22	37	
C	9	5	23	37	
D	3	3	24	30	

psychomotor objectives would represent 20%, 20%, and 60% of the final grade, respectively. To achieve this weighting, you simply made the number of points possible for each of the three objectives the desired percentage of the total number of possible points (i.e., 9, 9, and 27). As concluded previously, unless the variability of the three sets of scores is quite similar, the actual weighting will be different from that originally planned.

The solution is to do the weighting after a score for each objective has been obtained rather than attempt to build the weighting factor into the point system, unless you can assume that equal or nearly equal variability among the sets of scores will occur.

If calculating composite scores by combining raw scores is not feasible because of different units of measurement or a lack of equal variability among the sets of scores, how can you form composite scores? As described in chapter 3, convert each set of scores into the same standard distribution so that a common basis is established to make comparing, contrasting, weighting, and summing of scores from several sets of scores possible. There are several methods of doing this (three of which are discussed shortly); selecting the best method depends on the precision of the measurement involved and the assumption of normality.

In the case of the precision of scores, determine whether the scores are on an ordinal scale or an interval or ratio scale. As outlined in chapter 3, when you use an ordinal scale of measurement (such as the rankings from a round-robin tournament), it is only possible to say that A is greater than B. However, with interval and ratio scales (such as the number of free-throws made), it is possible to state how much greater A is than B, because these scales have equal-sized units.

The second consideration involves determining whether the distribution of scores approximates a normal distribution. If it does not, is it because the trait being measured is itself not normally distributed, or is it because, even though the trait is normally distributed, the sample at hand for some reason does not reflect this?

There are five possible situations involving these two considerations (table 13.2). The reason there are not six possible situations is that if the scores are ordinal measures, the distribution of scores—a simple ranking of the students—cannot approximate a normal distribution. Each situation is associated with one of the three methods of converting scores to a standard distribution: rank, normalizing, and standard score. Although statistical tests are available to determine whether a distribution of interval or ratio scores is significantly different from a normal distribution, such tests are beyond the scope of this book. However, a visual inspection of the frequency distribution of the sample is usually sufficient for revealing its closeness to a normal distribution. If you are still uncertain whether a distribution of scores approximates the normal distribution closely enough, you can use the normalizing method.

The end result of the rank method is simply a ranking of the students, whereas the end result of the other two methods is a set of standard scores. *It is not possible to arrive at a*

Table 13.2 Methods of Obtaining a Composite Score Based on Shape of Distribution and Scale of Measurement

	SCALE OF MEASUREMENT	
Shape of distribution of sample	**Ordinal**	**Interval or ratio**
Nonnormal	Rank method	Rank method
Nonnormal but trait is normally distributed	Normalizing method	Normalizing method
Approximately normal	(Not possible)	Standard score method

composite score if some of the sets of scores are converted to ranks and some are converted to standard scores. Therefore, if one of the several scores being summed to obtain a composite score must be expressed as a rank, then all the scores must be expressed as ranks. For this reason, you should plan in advance the type of measuring that you will use during a unit of instruction.

Rank Method

The simplest method, requiring the least precise measurement, is numerical ranking of the performance of the students on each test taken. However, this lack of precision also makes this method less reliable than others, and thus this method should be avoided, if possible. In the rank method, the best performance is given a 1, the second-best performance a 2, and so on until the worst performance is given a rank equal to the number of students being measured. *To obtain a composite score for each student in the numerical ranking system, simply sum the ranks for each student. The lowest total represents the overall best achievement.* In the event that one or more ranks are missing for a student, you may use a mean rank for that student, which is obtained by dividing the sum of the ranks by the number of values contributing to that sum. If you wish to weight the rankings, add in those considered most important more than once to arrive at the total. For example, assume that three rankings were obtained, the first to count 10%, the second 40%, and the third 50% of the final grade. For each student, a composite score would be obtained by summing the first rank, four times the second rank, and five times the third rank. As before, the lowest total would represent the best overall achievement. This may seem to contradict what you studied in chapter 3; however, the robustness of ranks in this situation permits performing mathematical operations on them even though they represent ordinal data.

A variation of the numerical ranking system is to rate each student as belonging to one of a set number of categories. For example, the best five achievements might be rated as 1, the next five as 2, and so on. Another variation goes one step further by giving the categories letter grades rather than numerals. This latter procedure does not require a particular number of students to be classified into each category. It has some merit in that it is more informative to the students than pure numerical rankings, but the principles used to arrive at a composite score are the same. In fact, a composite score is obtained in this system by changing the letter grades to numerals and proceeding in a fashion similar to that described for the numerical ranking system. For example, the categories A+, A, A–, B+, B, B–, C+, C, C–, D+, D, D–, and F are given the values 12, 11, 10, 9, 8, 7, 6, 5, 4, 3, 2, 1, and 0, respectively. (Note that in this method, a higher number is better than a lower one.) A student received an A+ on the affective objectives (10%), a B– on the cognitive objectives (40%), and a C– on the psychomotor objectives (50%). To arrive at a composite score for a student, the letter grades are converted to their numerical equivalents, multiplied by the weight of the corresponding objective, and summed:

$$(12 \times 1) + (7 \times 4) + (4 \times 5) = 12 + 28 + 20 = 60$$

Dividing this sum by 10 (the sum of the weights: 1 + 4 + 5 = 10) and comparing the resulting value to the categories converts this student's performance to a C+ (60 / 10 = 6 = C+).

Mastery Item 13.1

Using the system and weighting scheme just shown, what grade would you assign a student receiving grades of B+, C, and A–, respectively?

Normalizing Method

Use the normalizing method when the scores obtained in measuring a trait that is known or believed to be normally distributed do not appear to result in an approximation of the normal curve. For example, you would expect a distribution of the number of basketball free throws made in a given time period for males in the tenth grade to approach normality as the number of data points increases. If the measurement takes the form of ranks, the normal curve will not be approximated. Other reasons, such as not obtaining a representative sample, may cause a distribution to be other than normal even though the trait being measured should be normally distributed. The normalizing method is much like converting a set of raw scores into a distribution of some standard score such as the T-score. The difference, however, is that in the present case the raw scores are first converted to percentiles, which are converted into a standard score scale (in most cases T-scores are used). A description of the procedures for converting a set of raw scores into percentiles appears in chapter 3, and the resulting percentiles can be converted into T-scores using table 13.3. To illustrate how this procedure converts a nonnormal distribution into a normal distribution, notice that a score, regardless of its raw score standard deviation distance from the mean, that

Table 13.3 Conversion of Percentiles to T-Scores

Percentile	T-score	Percentile	T-score	Percentile	T-score
0.02	15	13.57	39	90.32	63
0.03	16	15.87	40	91.92	64
0.05	17	18.41	41	93.32	65
0.07	18	21.19	42	94.52	66
0.10	19	24.20	43	95.54	67
0.13	20	27.43	44	96.41	68
0.19	21	30.85	45	97.13	69
0.26	22	34.46	46	97.72	70
0.35	23	38.21	47	98.21	71
0.47	24	42.07	48	98.61	72
0.60	25	46.02	49	98.93	73
0.82	26	50.00	50	99.18	74
1.07	27	53.98	51	99.38	75
1.39	28	57.93	52	99.53	76
1.79	29	61.79	53	99.65	77
2.28	30	65.54	54	99.74	78
2.87	31	69.15	55	99.81	79
3.59	32	72.57	56	99.87	80
4.46	33	75.80	57	99.90	81
5.48	34	78.81	58	99.93	82
6.68	35	81.59	59	99.95	83
8.08	36	84.13	60	99.97	84
9.68	37	86.43	61	99.98	85
11.51	38	88.49	62		

Note: Although the T-score scale theoretically extends from 0 to 100, in practice T-scores lower than 15 or higher than 85 are rare and thus are not included in the table.

falls 34.13% above the mean (1 standard deviation above the mean in a normal curve) is equivalent to a T-score of 60, which is 1 standard deviation above the mean in the T-score scale. An example of the normalizing method resulting in a T-score corresponding to each raw score is shown in table 13.4. Once each set of scores is converted to T-scores, the T-scores can be weighted as desired and then combined to arrive at a composite score for each student. These composite scores can then be converted to the appropriate grade of the particular format being used.

Table 13.4 Example of Normalizing Method

Raw score	*F*	*cf*	*cfm*	Percentile	T-score from table 13.3
85	1	6	5.5	91.7	64
74	1	5	4.5	75.0	57
63	1	4	3.5	58.4	52
59	1	3	2.5	41.7	48
53	1	2	1.5	25.0	43
47	1	1	0.5	8.4	36
	6				

Note: In this table, the f column is simply the frequency of occurrence of each score. The cf column is the cumulative frequency associated with each score (beginning with the lowest score). To obtain the values in the cfm (cumulative frequency of the midpoint) column, add one half of the F value for each interval to the cf value from the next lowest score. For example, the value of 3.5 in the cfm column for the interval called 63 is obtained by adding 0.5 (1/2 of the frequency of 1 for this interval) to 3.0 (the cf of the interval below the interval of 63). Finally, to express the values as percentiles, divide 100 by n (in this case, 6) and multiply each cfm value by the resulting quotient.

Standard Score Method

If you judge it safe to assume that the raw score distribution closely approximates a normal distribution, convert the raw scores to a standard score scale such as the T-score scale as described in chapter 3. The net result of this procedure is, as in the normalizing method, a T-score corresponding to each raw score. If you convert all sets of scores to T-scores (or some other standard scale), a common basis exists for comparing scores made on two tests even though the two raw scores were expressed in different units or had differing variability because they all have the same standard deviation (recall that the standard deviation of T-scores is always 10). Furthermore, you can now add T-scores representing various achievements to obtain meaningful composite scores that can be used in determining final grades. As with the rank method, you can also weight the various tests (i.e., the respective objectives on which they are based) by multiplying the T-score by the appropriate weights you determined previously.

Once you obtain a composite score for each student, the final step is to change each score to the appropriate grade of the particular format being used. As in the other three steps, various factors will affect how you do this. The decision of what procedures to follow is based on the form of the composite score, policies of the school system or department, and your individual philosophy.

Step 4: Convert Composite Scores to a Grade

At the end of step 3, the composite scores will be in one of two forms: Each student will have a total or average ranking or a total or average standard score. In effect, both forms are an ordering of the students, although in the rank method the lowest total (or lowest average) usually represents the best achievement, whereas in the normalizing and standard score methods the highest total (or highest average) represents the best achievement. The procedures for converting a set of composite scores to grades are the same regardless of the form of the composite scores and involve answering two related questions:

1. Is the class to be graded below, at, or above average in achievement of the course objectives in comparison to similar classes?
2. What percentage of students should receive each grade?

If tests and measuring devices were absolutely reliable and valid and if course objectives remained constant over time, teachers would not need to answer these questions before converting composite scores to grades. However, measurements are not perfect, objectives change, unexpected or unplanned events occur, facilities and equipment change over time, and several other factors make it impossible to compare the achievement of the current class to previous classes on a strictly objective basis. *Grading is especially difficult for new teachers because they lack experience on which to base their answers to these questions.* After arriving at some subjective answers, you can use several methods to convert the composite scores to grades.

Observation

Observation is one of the simplest methods for determining grades. List the composite scores from best to worst. Examine the list for natural gaps or breaks in the scores. Table 13.5 lists the scores for 15 high school boys. As you see, two gaps appear in the data. Those observed gaps are used for cutoffs for the letter grades A, B, and C.

Table 13.5 Scores Graded by the Observation Method

Scores	Frequency	Grade
150	1	A
140	2	
110	3	B
100	2	
90	2	
80	1	
70	1	
40	1	C
30	1	
20	1	

This method usually works well for a small number of scores, because gaps in the data often appear. However, observation is not useful for a large number of scores, because observable gaps may not be present. *Furthermore, this method does not ensure consistency.* The natural gaps could be quite different in two classes: A grade of A in the fall semester class might fall in the B category in the spring.

Predetermined Percentages

The predetermined-percentages method may be used with composite scores in the form of rankings or standard scores because it is not the value of the score that matters but its position in the distribution. Once you decide on the percentage of students to whom each grade will be assigned, you need only multiply the percentage values by the number of students in the class, as follows:

$$k = N \times (P / 100) \tag{13.1}$$

where k is the number of students to receive a particular grade, N is the total number of students, and P is the percentage of particular grade. The resulting product is the number of each letter grade to be assigned. Table 13.6 shows this procedure for a class of 45 students, which, on the basis of test scores and other evidence, has achieved substantially above the average of other similar classes. Notice that this decision is, in effect, answering the first question under step 4. The teacher decides to allot 15% As, 25% Bs, 45% Cs, 10% Ds, and 5% Fs. If the composite scores are in the form of standard scores or the sum (or average) of several rankings (but not if the composite scores are a single rank in which the best student has a rank of 1, the second-best student has a rank of 2, and so on), you can modify the predetermined-percentages method slightly by using the natural breaks in the distribution of the composite scores. As the last column of the table indicates, the actual number of scores for a grade may be slightly different from the calculated k, but the cutoff points are selected to result in numbers of students close to the predetermined number.

Table 13.6 Grades Determined by Preset Percentages

Grade to be given	Percentage, preset by teacher	Number of students to receive grade (k)	Scores	Frequency	Total actual number of students receiving grade
A	15%	45 × .15 = 6.75 → 7	120 110 100	2 2 2	6
B	25%	45 × .25 = 11.25 → 11	90 80 70	3 4 5	12
C	45%	45 × .45 = 20.25 → 20	60 50 40 30	8 5 5 3	21
D	10%	45 × .10 = 4.5 → 5	20	4	4
F	5%	45 × .05 = 2.25 → 2	10	2	2

As with the observation method, the predetermined-percentages method of grading does not ensure consistency in grade assignment from class to class or semester to semester. Even if the predetermined percentages remain consistent, because scores vary from class to class and semester to semester, the actual cutoff point (in terms of scores) will vary. Thus, a score that translates into an A in the fall semester might be a B in the spring. You can adjust for this, however, by changing the percentages of each grade to be given.

Grading on the Curve

Although *grading on the curve* is a phrase often heard in educational settings, the process is not well understood. Actually, grading on the normal curve is a variation of the predetermined-percentages method in which the assumption is made that the differences in student abilities in a class are normally or at least approximately normally distributed and therefore that the percentages of each grade assigned can be determined through use of the normal curve.

Some confusion about grading on the curve stems from the fact that it is possible to use the normal curve in two ways. The practical limits of the normal curve (±3 standard deviations) can be equally divided into the number of categories of grades to be assigned, or certain standard deviation distances from the mean may simply be selected as the limits for each symbol assigned. Refer to table 13.7 for data used to illustrate each approach. In this example, the composite scores of 65 students have been compiled, and you wish to use the normal curve to determine the cutoff points for assigning the grades into the five categories of A, B, C, D, and F. The mean and standard deviation of the composite scores are 63.4 and 15.09, respectively.

First Approach For practical purposes, the normal curve may be considered to extend ±3 standard deviation units above and below the mean. (Recall that 99.74% of the area under the normal curve is found between these two points.) This total width of 6 standard deviation units is divided equally into the same number of parts as there are categories of grades to be assigned—in this case, five (A, B, C, D, F). Each grade thus encompasses a width of 1.2 standard deviation units (6/5 = 1.2) (figure 13.4). Because the grading format used in this illustration has an uneven number of possible grades, one half of the middle

Table 13.7 Composite Scores for 65 Students

98	78	71	64	60	52	40
93	78	70	64	59	51	38
91	77	69	63	57	50	37
88	76	68	63	57	48	36
86	75	67	63	56	47	26
85	74	67	63	56	47	
83	73	66	62	55	46	
81	73	65	62	55	45	
81	72	65	61	54	44	
79	71	65	61	53	41	

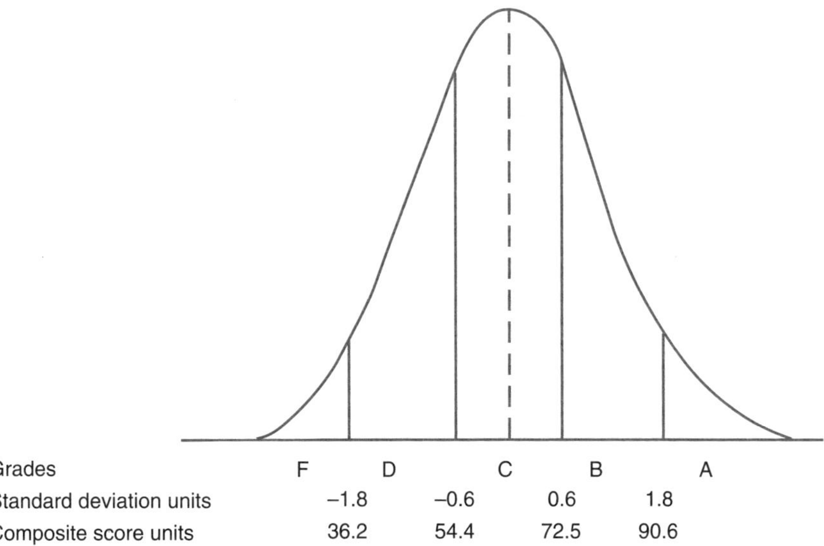

Figure 13.4 Relationship among grades, standard deviation units, and composite score units.

grade (C) falls on either side of the mean. A grade of C will be assigned to those students whose composite scores lie between 0.6 standard deviation units above and 0.6 standard deviation units below the mean. A grade of B will be assigned to students whose composite scores are between 0.6 and 1.8 standard deviation units above the mean; a grade of D will be assigned to those whose composite scores lie between –0.6 and –1.8 standard deviation units below the mean. If this process is continued, it will result in the limits of 1.8 and 3.0 standard deviation units above the mean for grades of A and –1.8 and –3.0 standard deviation units below the mean for grades of F. To account for extreme scores, you can specify that composite scores more than 3.0 standard deviations above the mean also correspond to As and that composite scores more than –3.0 standard deviations below the mean also correspond to Fs.

The final step involves expressing the standard deviation units in terms of the composite score values. Because in this example, 1 standard deviation is equivalent to 15.09 composite score units, the cutoff point between C and B and between C and D must lie 9.05 composite score units above and below the mean, respectively. (The value of 9.05 results from multiplying 15.09 by 0.6.) The two cutoff points are thus 72.5 and 54.4 (63.4 ± 9.05). The cutoff points between B and A and between D and F are obtained by adding to and subtracting from the mean the product of 15.09 and 1.8. The result is 27.16, and thus the cutoff points are 90.56 and 36.24 (see figure 13.4). To simplify, you would probably round the cutoff points; you could also establish a conversion chart for assigning grades to the composite scores (table 13.8).

Table 13.8 Conversion of Composite Scores to Grades: Example of the First Approach

Composite score	Grade
91 and over	A
73-90	B
54-72	C
36-53	D
35 and below	F

If a grading format involving five grades is used in conjunction with the normal curve, the teacher is, in effect, deciding to assign approximately 3.5% of the class As, 24% Bs, 45% Cs, 24% Ds, and 3.5% Fs. (To verify this, review how to use table 3.4.) Applying the conversion chart in table 13.8 to the 65 composite scores results in the assignment of 3 As (4.6% of the class), 15 Bs (23.1%), 30 Cs (46.1%), 14 Ds (21.6%), and 3 Fs (4.6%). Slight differences occur between the normal-curve percentages and the actual percentages because of the rounding in establishing the conversion chart and the fact that the set of 65 scores is not exactly normally distributed.

Second Approach The second approach to grading on the curve involves selecting certain standard deviation distances from the mean as the limits for each grade. In this approach, instead of dividing the 6 standard deviations (the practical limit of the normal curve) into equal units for each grade category, select distances that conform to your notion of the percentage of each grade to be assigned. One possibility, based on the data from table 13.7, is depicted in table 13.9.

Table 13.9 Curve Grading: Example of the Second Approach

Grade	Standard deviation distance for the grade (predetermined by instructor)	Percentage of students to receive grade (use table 3.4)	Number of students to receive the grade (N = 65)	Composite scores corresponding to the grade
A	Above +1.48	7%	.07 × 65 = 5	≥86
B	+0.47 to +1.47	25%	.25 × 65 = 16	71-85
C	−0.47 to +0.46	36%	.36 × 65 = 23	57-70
D	−1.47 to −0.46	25%	.25 × 65 = 16	41-56
F	Below −1.48	7%	.07 × 65 = 5	≤40

Note: The first and last columns would constitute a simple conversion table.

The selection of these particular standard deviation distances would result in assigning approximately 7% As, 25% Bs, 36% Cs, 25% Ds, and 7% Fs to a set of normally distributed scores. As in the first approach, these standard deviations must be converted to the units of the composite scores by multiplying the selected constant by the standard deviation of the composite scores. For the data found in table 13.7 (mean = 63.4, standard deviation = 15.09), the resulting cutoff points are displayed in table 13.9. (As before, slight differences between the normal-curve percentages and the actual percentages occur because of rounding and the fact that the 65 scores are not exactly normally distributed.)

Although any set of standard deviation distances may be chosen, if the assumption regarding normality is generally true, the selected values will necessarily result in symmetrical assignment of grades. If a nonsymmetrical distribution of grades is desired, the second approach to curve grading actually becomes the predetermined-percentages method described previously. *The normal-curve method does not ensure consistency in grading from class to class because the mean and standard deviation may change from class to class.*

Mastery Item 13.2

The mean and standard deviation of a set of composite scores are 38.5 and 6.6, respectively. Calculate the normal-curve grading scale if you are to have a grade distribution of 10% of the students receiving As, 20% Bs, 40% Cs, 20% Ds, and 10% Fs.

Go to the WSG to complete Student Activity 13.7.

Definitional Grading

As mentioned previously, determining relevant course objectives and their weighting should be one of the first tasks in preparing to teach any course. Special care must be taken in assigning weights to various objectives so that a superior performance on the assessment of one objective does not totally offset a poor performance on another objective. As an extreme example, a student could achieve a grade of C by earning an A on the written test on rules and strategy and nearly failing an assessment of the physical performance required if each of the two objectives has an equal weighting of 50% of the grade.

One method to ensure this does not occur is through the use of what is often called definitional grading. Briefly, in this grading system, an instructor sets a minimum performance required for each objective. *The level of achievement on each objective determines the grade for each objective, and the final grade is equal to the lowest grade achieved on any objective.*

Table 13.10 illustrates a simplified possible set of minimums for grades of objectives in the three domains of a tennis unit. In this class the instructor plans to measure the cognitive objective with a 50-item written test on tennis rules and strategies, the psychomotor objective by counting the number of good serves made out of 10 attempts, and the affective objective with a 5-point rating scale applied while observing students' etiquette during tennis matches.

Table 13.10 Illustration of Definitional Grading System

	Cognitive	Psychomotor	Affective
A	45+ ↑	8+ ↑	5
B	40-44	7	4
C	35-39	6	3
D	30-34	5	2
F	25-29	4	1

To achieve an A in this class, a student must answer at least 45 questions correctly on the written test, make 8 or more good serves on the serving test, and be given a rating of 5 on the observation of etiquette. Although tables can be constructed to allow various weightings of the objectives to determine the final grade, one option is to have the final grade for a student be determined by the lowest level of performance on any of the three objectives. For example, a student receiving a 37 on the written test, a 9 on the serving test, and a 4 on the etiquette observation would receive a grade of C. Looney (2003) provides a discussion of this system, how these tables can be constructed, and other nuances in using this grading system.

Norms

Norms are derived from a large number of scores on tests and measurements from a specifically defined population. The scores are statistically analyzed to produce a descriptive statistical analysis, which allows the production of percentile norms or standard score norms. Norms are available for many physical fitness and sport skills tests. The U.S. government performs large-scale surveys of various health-related variables such as cholesterol and blood pressure. These normative data are used to develop national health profiles, and the variables are related to morbidity and mortality. The National Children and Youth Fitness Studies I and II produced percentile norms on various physical fitness tests for American youth (Pate, Ross, Dotson, and Gilbert 1985; Ross et al. 1987). Although it is not recommended that you grade on these characteristics, the norms and profiles provide examples of large-scale normative data. For example, if you were teaching a volleyball unit, it would be helpful if you had normative data for various volleyball skills available. Having such data would help you interpret student performance, assign performance grades, and provide students with information for goal setting.

 Go to the WSG to complete Student Activity 13.8.

Norms can be used both for assigning grades on individual test items and for making your own norms for composite scores to arrive at final grades. For example, table 13.11 provides the 75th and 25th percentile norms for 1-mile (1.6 km) walk test scores for 400 males and 426 females aged 18 to 30 years. You might assign an A to the top quartile (above P_{75}), a B to the two middle quartiles (P_{25}- P_{75}), and a C to the last quartile (below P_{25}). *National*

and other published norms can be used to establish grading scales. However, the fairest norms for grade determination are local norms. To be used in grading, norms must be representative of the students' gender, age, training, and instructional situation. Norms developed at the instructional location on students similar to those who are to be graded will ensure representativeness.

Table 13.11 Percentile Norms for the 1-Mile (1.6 km) Walk Test (min:sec)

Percentile	Males	Females
75	11:42	12:49
25	13:38	14:12

Data from Jackson, Solomon, and Stusek 1992.

You can develop local norms for composite scores at your facility by following these steps:

1. Determine the objectives of your instructional or training program.
2. Select the most accurate possible tests and measurements to assess these objectives.
3. Administer the same tests with standardized procedures for several years.
4. Collect enough data to have at least 200 scores for each gender and age you intend to evaluate.
5. Conduct a statistical analysis of the data and establish percentile or standard score norms.

If you follow these steps, you will be able to establish norms that are representative of your students and their learning situation. The grading standards will provide consistency in grade assignment between classes, semesters, and instructors if all instructors are involved in the development and use of the norms. However, until you have sufficient data to establish norms, you will have to use some other grading technique. *Ultimately, you should work toward establishing local norms, because this method reduces most of the inconsistencies in grading.*

Dataset Application

Use the chapter 13 large dataset to accomplish three things related to grading. Assume that the 500 scores there represent total points earned in a class across several semesters and you are interested in determining what the distribution of points looks like. Do the following:

1. Create a histogram of the TotalPoints to see if it is normally distributed (chapters 2 and 3).
2. Compare the distributions for the females and males in your classes by creating two histograms (chapter 3) using Graphs → Legacy Dialogs → Histogram and enter TotalPoints in the Variable box and Gender in the Panel by Rows area. Do the distributions of scores appear similar for females and males?
3. Compare the gender means to see if they are significantly different (chapter 5). Refer to the procedures you used for independent t-test in chapter 5. Confirm that the mean for the females is 496.83 and that the mean for males is 503.46. Are they significantly different?
4. Calculate the mean and standard deviation with SPSS. Use the *z*-table in chapter 3 to determine cut points for various grades that you might assign based on the percentage of letter grades.

Arbitrary Standards

The previous techniques for converting scores to grades have involved data analysis of the observed measurements. Using arbitrary standards for grade assignment does not require data analysis of test scores. With this process, criterion-referenced standards are established for each grade. Table 13.12 provides an example of such standards on a 100-point knowledge test.

Table 13.12 Arbitrary Standards for Grading a 100-Point Knowledge Test

Point range	Grade
100-90	A
89-80	B
79-70	C
69-60	D
59-0	F

The biggest advantage of this type of grading system is that it provides consistency in grading. A 90 on the knowledge test will be an A in whatever class or semester a student takes the test. The system is simple and easy to understand. However, setting the standards in physical performance tests without knowledge of expected student performance levels results in a best guess process that may result in an undesirable grade distribution. If the standards you use are accurate reflections of student achievement levels, the arbitrary system can be used to good effect, but, as with the norms method, it relies on a relatively long period of data accumulation.

A specific version of using arbitrary standards in evaluation is the pass/fail assignment. Student performance is compared with a specific criterion-referenced standard that represents a minimum level of acceptable performance or ability. However, the setting of these minimum pass/fail standards is not an easy task and often takes years of experimentation and data collection.

Dataset Application

Return to the chapter 13 large dataset in the WSG and see if you can determine grades for the 500 students there based on some of the methods that have been presented. For example, can you set cut points based on the distribution for grades of A, B, C, D, and F?

Go to the WSG to complete Student Activity 13.9.

Mastery Item 13.3

Consider a course you might teach. Develop grading criteria and standards.

MEASUREMENT AND EVALUATION CHALLENGE

After thinking through what objectives he wants his students to achieve, Tom decides to weight two cognitive tests and two psychomotor skills tests to make up certain percentages of the final tennis unit grade for each student. To accomplish this he has to convert the scores on all the tests to a standard score format (e.g., T-scores) and adjust these converted scores in accordance with the percentages he has chosen. The four adjusted standard scores are then added together to arrive at a total composite score representing each student's class achievement. Tom then can select one of the methods described in the last section of this chapter to convert the composite scores to final grades.

SUMMARY

Determining and issuing grades to students is difficult. However, because grades are vital to a great number of important decisions, you must take them seriously and determine them fairly and in a meaningful way. This chapter identifies many of the issues involved in the process of grading and suggests methods for you to consider when faced with this task. Determine your own answers to some of the philosophical issues of assessing the performance of others, and use the mechanical tools provided here for measuring, weighting, and combining various scores, as well as determining the final grade.

Go to the WSG for homework assignments and quizzes that will help you master this chapter's content.

CHAPTER 14

Performance-Based Assessment: Alternative Ways to Assess Student Learning

Jacalyn L. Lund, Georgia State University

The chapter and book authors acknowledge the contributions of Dr. Larry D. Hensley (University of Northern Iowa) to earlier versions of the text in this chapter.

OUTLINE

OBJECTIVES

After studying this chapter, you will be able to

- define performance-based assessment and distinguish it from traditional, standardized testing;
- discuss current trends in assessment practices in education;
- identify various types of performance-based assessments;
- identify criteria for judging the quality of performance when using performance-based assessments;
- create a performance-based assessment, complete with scoring criteria;
- explain the advantages and disadvantages of performance-based assessment; and
- identify guidelines for developing and using performance-based assessment.

 The lecture outline in the WSG will help you identify the major concepts of the chapter.

MEASUREMENT AND EVALUATION CHALLENGE

Mariko is a middle school physical education teacher who is planning to teach a soccer unit. After completing the unit, her students will have the knowledge and skills to play a small-sided (four vs. four players) soccer game. When evaluating game play, she wants to also assess student affective domain behaviors. During the unit, Mariko wants to use several formative assessments to increase student learning since they will help her diagnose student problems and offer feedback to students about how to improve. Because she is anxious to encourage students to use higher-level thinking skills, Mariko will use several types of assessment. Her goal is to have her students play soccer to demonstrate competence on several standards from the *National Standards for K-12 Physical Education* (SHAPE America 2014). Knowing that one of the characteristics of performance-based assessment is that students must know the criteria by which they will be assessed before the start of instruction, Mariko plans to develop her assessments before beginning the unit. What assessments should Mariko use to assess each student's understanding and performance of the various skills, knowledge, and attitudes needed to play soccer?

Assessment and accountability are two of the most frequently discussed topics in education. Since the mid-1980s there has been increasing interest in using assessment as a way to enhance student learning. This interest coincides with the educational reform movement that led to the development of standards and standards-based instruction. Increased interest in assessment and having students use higher-level thinking skills has led to a focus on performance-based assessment techniques that align with outcome-based content standards.

The term **assessment** has been popularized by educators during the last two decades and now is used more often in education than either of the terms measurement or evaluation. *Evaluation is the judgment made about the quality of a product. Assessment is "the gathering of evidence about a student's level of achievement and making inferences on student progress based on that evidence"* (SHAPE America 2014, p. 90).

But what is this new type of assessment, and how does it affect assessment in physical education? How can it be used to assess learning in all domains and thus be used to determine whether a student is physically educated? The next section will address these and other related questions.

IMPETUS FOR DEVELOPING A NEW TYPE OF ASSESSMENT

Since the mid-1980s there has been widespread concern among parents, business leaders, government officials, and educators about school effectiveness. The educational reform movement that began in the mid-1980s called for a dramatic shift in the way schools delivered education. The reform movement led to the development of standards that were written in terms of what students should know and be able to do.

Content standards specify what should be learned in various subject areas. Several disciplines (e.g., math, science, language arts, social science) have national standards defining essential skills and knowledge for that content area. The National Association for Sport and Physical Education (NASPE) first announced content standards for physical education in 1995 in the publication *Moving Into the Future—National Standards for Physical Education:*

A Guide to Content and Assessment. The standards were revised slightly and updated in 2004. In 2014, new national standards were published along with a curriculum framework for kindergarten through twelfth grade physical education. The current SHAPE America National Standards for K-12 Physical Education (SHAPE America 2014, p. 12) are as follows:

> *Standard One: The physically literate individual demonstrates competency in a variety of motor skills and movement patterns.*
>
> *Standard Two: The physically literate individual applies knowledge of concepts, principles, strategies, and tactics related to movement and performance.*
>
> *Standard Three: The physically literate individual demonstrates the knowledge and skills to achieve and maintain a health-enhancing level of physical activity and fitness.*
>
> *Standard Four: The physically literate individual exhibits responsible personal and social behavior that respects self and others.*
>
> *Standard Five: The physically literate individual recognizes the value of physical activity for health, enjoyment, challenge, self-expression, and/or social interaction.*

Assessment was seen as a critical part of the reform movement, and standardized tests were deemed insufficient for measuring student learning. Selected response questions (e.g., multiple-choice, true–false) found on many physical education tests are used to measure knowledge, which, according to Bloom's taxonomy, is one of the lowest forms of learning. The reform movement wanted to put increased demands on student learning, requiring students to demonstrate higher levels of learning and mastery of content. Measuring students' abilities to analyze, synthesize, and evaluate knowledge, thus demonstrating the ability to use the information rather than just acquire it, required a different approach to testing.

Increased demands for accountability and the heightened emphasis on assessment occurred at a time of growing dissatisfaction with the traditional forms of assessment (e.g., selected response, machine-scored tests). Wiggins (1989) called for a new type of assessment that would assess higher levels of student thinking and learning and be part of the learning process. The use of the word *assessment* instead of *testing* signaled a shift from focusing on a test at the end of instruction to the use of assessment to enhance student instruction. The word *assessment* comes from the French word *assidere*, which means "to sit beside." The shift implied that teachers were no longer gatekeepers who judged whether students had learned, but rather coaches who were responsible for increasing student learning. *The gotcha mentality was replaced by the belief that assessment methods should facilitate teaching, enhance learning, and result in greater student achievement.*

Wiggins' work resonated with the educational community, and the move toward a different type of assessment gained momentum. Along with the change in the purpose for assessment, Wiggins also advocated for assessments that would have meaning for students. He argued that students should demonstrate learning in more ways than by penciling in answers on a standardized test. Feuer and Fulton (1993) identified several performance-based assessments appropriate for use in education. Those assessments appropriate for physical education include essays, **exhibitions**, portfolios, and oral discourse. These assessments were complex and represented meaningful activities that might be done by professionals in the field.

Although most teacher education programs required majors to complete a course in tests and measurement, the information from this class was not fully used by teachers once they began teaching. Traditional tests in physical education include written tests, standardized sport skills tests, and physical fitness tests. Kneer (1986) reported a gap between assessment theory and teacher use of assessments in the field. Several studies have indicated that teachers were more likely to hold students accountable for managerial concerns in physical education than for achievement (Imwald, Rider, and Johnson 1982; Hensley, Lambert, Baumgartner, and Stillwell 1987). In 1988 a study by Veal showed little evidence of teachers using traditional forms of assessment to evaluate students. Participants in her study indicated a strong desire to include elements such as effort and improvement to determine student grades. Additionally, teachers wanted assessments that could be used for purposes other than grading (e.g., remediation, improvement of performance) while being efficient (i.e., not taking up too much class time) and easy to administer. In short, the teachers in this study were calling for assessments that could be linked with instruction but, when administered, would not take vast amounts of time away from instruction. Wood (1996) suggested, "Traditional psychometric assessment devices (e.g., multiple choice tests, sport skills tests) may no longer be sufficient for assessment in the quickly changing educational landscape characterized by emphasis on learning outcomes, higher order cognitive skills, and integrated learning" (p. 213). Furthermore, traditional assessment instruments and techniques tend to measure narrowly defined characteristics, do not facilitate integration of skills or processes, and are frequently artificial in nature. Various calls for change have resulted in a shift to increase the use of other forms of assessment and thus validate techniques that many physical educators have used for years. Clearly, the landscape of educational assessment in general and specifically in kinesiology and physical education is undergoing change (Wood 2003).

Clarification of Terminology

As people embraced the concept of doing things differently, they added adjectives to the word *assessment* to ensure that people (educators as well as the general public) understood that this way of measuring student learning was indeed different. Originally Wiggins had used the term **authentic assessment**. The terms **performance-based assessment** and **alternative assessment** also started appearing in the literature. Some people like to distinguish between the terms while others (Herman, Aschbacher, and Winters 1992; Lund and Kirk 2010) use the terms interchangeably.

Using any adjective is problematic if it causes confusion among those using the term. Arguments have occurred between teachers (regular classroom and physical educators) about whether an assessment was authentic. For example, in the early days of this new assessment movement, one physical educator advocated that a tennis forehand test delivered with a ball machine was not authentic, but that if a player was rallying with another player, it would be an authentic assessment. In fact, while both are perfectly good skill tests, neither is an example of the complex assessment to which performance-based assessment alludes. With the switch to this new type of assessment, the authenticity of the assessment becomes important only because it tends to increase the assessment's meaningfulness and relevance to students. *The more authentic one can make an assessment (it represents something associated with the learning that a professional in the field might do or value), the greater chance that students will see the relevance of the assessment and get excited about completing it.*

For the purpose of this chapter, the term *performance-based assessment* will be used to indicate the complex assessments used to determine student learning. The term performance-based assessment will refer to "the variety of tasks and situations in which students are given opportunities to demonstrate their understanding and to thoughtfully apply knowledge, skills, and habits of mind in a variety of contexts" (Marzano, Pickering, and McTighe 1993, p. 13).

In physical education, the term performance-based assessment should not be confused with tests of performance such as skills tests and fitness tests. Skills and fitness tests are not complex and meaningful as entities (e.g., Wimbledon involves people playing tennis games and not players hitting forehand shots back and forth over the net), and so are not considered performance-based assessments.

Performance-based assessment uses nontraditional techniques to ensure students learn and retain physical activity knowledge.

Assessment was a cornerstone of the educational reform movement—not only to provide data for accountability purposes but to serve as an integral part of the instructional process. Diagnostic assessments are used before the start of a unit to inform teachers about prior student knowledge and to ensure that teachers don't assume that all students in the class are beginners with little or no previous knowledge of the unit (Lund and Veal, 2013). **Formative assessment** is the term used to describe assessment done during the learning process. The results do not contribute to student grades but rather are used by teachers to provide feedback to students to correct or improve the final performance. Formative assessments can be self- or peer assessments or administered by a teacher. Students can use results from formative assessments to set goals and monitor their progress toward the final learning goal or objective. Formative assessments also provide teachers with data to use for planning future instruction. Although performance-based assessments are used as formative assessments as part of the coaching and teaching process to enhance student learning, traditional assessments such as skills tests can be used as formative assessments as well. Formative assessments are useful for goal setting by teachers or with students. Teachers can establish goals for student learning to map progress toward final learning goals. **Summative assessments** are those given at the conclusion of instruction. They represent the final level of student achievement and are associated with grading, evaluation, or both. Summative assessments are administered by a teacher or someone with authority for the class.

The terms formative and summative refer to the purpose of the assessment rather than a certain format or type of assessment. Formative assessments might be used throughout a unit for feedback and teacher planning but then become a summative assessment if used at the conclusion of a unit as a final assessment of student learning. Although students can improve their performance with formative assessments, there is no opportunity to improve a grade or score on a summative assessment.

Mastery Item 14.1

How does the learning measured by a traditional test (selected response) differ from that measured by performance-based assessments?

How does performance-based assessment differ from traditional assessment? Performance-based assessment is considered different from the conventional forms of standardized testing (i.e., multiple-choice or true–false tests, sport skills tests, or physical fitness tests) in several ways. Performance-based assessments involve a wide variety of nontraditional techniques for assessing student achievement. Herman, Aschbacher, and Winters (1992) identified several characteristics to describe performance-based assessments. These characteristics help explain what makes performance-based assessments different from other traditional testing formats.

- ***Students are required to perform, create, produce, or solve something.*** Several types of assessments allow students to demonstrate their knowledge of a topic or their ability to do something through the creation of a product or performance. Creating a jump rope routine or a dance or playing lacrosse requires a different type of learning that traditional tests would not measure. With performance-based assessments students can do the work individually or as part of a group. The resulting work typically takes several days to complete, and the process used to complete it is as important as the final product or performance. Note the similarity in student expectations with the highest level of Anderson and Krathwohl's (2001) revision of Bloom's cognitive taxonomy, where the highest level of the taxonomy is that of creating.
- ***Students are required to use higher-level thinking and problem-solving skills.*** It is difficult to measure student ability to analyze, synthesize, or evaluate with a test consisting of selected response questions (e.g., true–false, multiple-choice, matching). When measuring higher levels of thinking, teachers can give students a problem and have them solve it. In physical education classes, written tests often ask questions about the dimensions of a court, when a sport was first played, who invented the sport, or how to define terms associated with the activity. These are all examples of knowledge or comprehension according to Bloom's taxonomy. Assessments of game play or the ability to choreograph a dance require students to use higher levels of thinking and problem-solving skills, thus making them more challenging for students. Additionally, these assessments often reveal true levels of student understanding or misunderstanding.
- ***Tasks are used that represent meaningful instructional activities.*** Students rarely get excited about the chance to take a written test. Performance-based assessments provide students the opportunity to demonstrate learning in a variety of ways. For example, instead of taking a written test on the rules, students could demonstrate knowledge of the rules while officiating a game or serving as the game announcer. These latter activities may be far more meaningful and relevant than responding to questions on a written test. Teachers might ask students in a dance class to write a critique of a dance performance to demonstrate their knowledge of dance choreography. Playing a game and performing a dance routine are meaningful tasks that allow students to demonstrate what they have learned.
- ***When possible, real-world applications are used.*** The preceding examples are all things that occur in the real world of sport or other types of physical activity. Because students regularly see these events occurring in the world around them, they are more likely to see the relevance of the task. Some assessment books recommend that student performance should have an audience when possible. Playing a game for others or distributing a publication gives the assessment an additional level of accountability simply because someone other than a teacher will have an opportunity to view it.
- ***People use professional judgment to score the assessment results.*** Performance-based assessments are administered by a person who is assessing the performance or product using a list of criteria previously identified as significant for the assessment. Those scoring the assessment are professionals who have training and experience in the subject or

content being evaluated. When an external panel (i.e., not a class teacher) is used to do the scoring, training on using the scoring rubrics is typically done to ensure that the assessors understand the criteria and the expectations for the assessment and that assessments are scored with consistency.

- ***Students are coached throughout the learning process.*** This is one of the most exciting aspects of performance-based learning. As students work to complete the performance-based assessment, they are provided with guidance and feedback by their teachers so that the final performance or product is the best that it can be. Students can receive feedback about their performance during a game in the same way that a coach provides feedback to members of a team. Because students know the criteria in advance, they are capable of comparing their performance to the criteria (self-assessment for formative assessment) or can assist other students by providing peer feedback. Since the final goal is outcome based (e.g., students can play soccer), various people can provide knowledge and assistance to help each student achieve that goal.

TYPES OF PERFORMANCE-BASED ASSESSMENT

As previously mentioned, there are many types of performance-based assessments. Each type of assessment brings with it strengths and deficiencies relative to providing credible and dependable information about student learning. Because it is virtually impossible for a single assessment tool to adequately assess all aspects of student performance, the real challenge comes in selecting or developing performance-based assessments that complement both each other and more traditional assessments to equitably assess students in physical education and human performance.

The goal for assessment is to accurately determine whether students have learned the materials or information taught and reveal whether they have complete mastery of the content with no misunderstandings. Just as researchers use multiple data sources to determine the truthfulness of the results, teachers can use multiple types of assessment to evaluate student learning. Because assessments involve the gathering of data or information, some type of product, performance, or recording sheet must be generated. The following are some examples of various types of performance-based assessments used in physical education.

Using Observation in the Assessment Process

Human performance provides many opportunities for students to exhibit behaviors that may be directly observed by others, a unique advantage of working in the psychomotor domain. Wiggins (1998) used physical activity when providing examples to illustrate complex assessment concepts, because they are easier to visualize than would be the case with a cognitive example. *The nature of performing a motor skill makes assessment through observational analysis a logical choice for many physical education teachers.* In fact, investigations of measurement practices of physical educators have consistently shown a reliance on observation and related assessment methods (Hensley and East 1989; Matanin and Tannehill 1994; Mintah 2003).

Observation is a skill used during instruction by physical education teachers to provide students with feedback to improve performance. However, without some way to record results, feedback to a student is not an assessment. Peers can assess each other using observation. They might use a checklist of critical elements or some type of event-recording scheme to tally the number of times a behavior occurred. Keeping game play statistics is an example of recording data using event recording techniques. Students can self-analyze their own performances and record their performances using criteria provided on a checklist or a game play rubric. Table 14.1 is an example of a recording form that could be used for peer assessment.

Table 14.1 Peer Assessment: Checklist for Softball Batting

Behavior	Observed	Not observed
Feet shoulder-width apart		
Bat in air ready to swing		
Draws or cocks the bat backward in preparation to swing		
Steps, putting total body behind the ball		
Swings through the ball		
Rolls wrists to generate power		
Follows through across body		

When using peer assessment, it is best to have the assessor do only the assessment. When the person recording assessment results is also expected to take part in the assessment (e.g., tossing the ball to the person being assessed), he or she cannot both toss and do an accurate observation. In the case of large classes, teachers might even use groups of four, in which one student is being evaluated, a second student is feeding the ball, a third student is doing the observation, and a fourth student is recording the results that the third student provides orally.

 Go to the WSG to complete Student Activity 14.1 and view Videos 14.1 and 14.2.

Mastery Item 14.2

Why must an observation include some type of written record to be considered an assessment?

Individual or Group Projects

Projects have long been used in education to assess a student's understanding of a subject or a particular topic. Projects typically require students to apply their knowledge and skills while completing the prescribed task, which often calls for creativity, critical thinking, analysis, and synthesis. Examples of student projects used in physical education include the following: demonstrating knowledge of invasion game strategies by designing a new game; demonstrating knowledge of how to become an active participant in the community by doing research on obesity and then developing a brochure for people in the community that presents ideas for developing a physically active lifestyle; demonstrating knowledge of fitness components and how to stay fit by designing one's own fitness program using personal fitness test results; demonstrating knowledge of how to create a dance with a video recording of a dance that members of the group choreographed; and doing research on childhood games and teaching children from a local elementary school how to play them. Criteria for assessing the projects are developed and the results of the project are recorded.

Group projects involve a number of students working together on a complex problem that requires planning, research, internal discussion, and presentation. Group projects should include a component that each student completes individually to avoid having a student receive credit for work that others did. Another way to avoid this issue is for each member of the group to award "paychecks" to the other group members (e.g., split a $10,000 check) and provide justifications for the amount given to each person. To encourage reflections on the contributions of others, students are not allowed to give an equal amount to everyone. These "checks," along with the rationales for allocation of funds to members of the group, are confidential and submitted directly to the teacher in an envelope so that others in the group do not see the amount that other group members "paid" them.

The following example of a project designed for middle school, high school, or college students involves a research component, analysis and synthesis of information, problem solving, and effective communication.

DANCE PROJECT

The purpose of this assessment is for students to demonstrate their ability to read and understand the directions for an aerobic, folk, or other type of structured dance and then teach that dance to peers. To do so they must demonstrate the ability to teach the steps, count to the music, and coach students to use the music and steps in harmony to perform the dance. Students are evaluated based on their ability to read a set of dance directions and interpret them, teach the dance to others, and perform with competence the dances that they are taught by other students in the group.

Directions to students: *Working in groups of six, you will demonstrate the ability to perform six structured dances to music. Each student will be required to read the directions for a dance, learn those steps, and then perform the dance to the music. If you do not know how to do the step that the routine calls for, do some research (e.g., Internet, books) to learn how to do the step. After successfully doing the first step, each student will teach the dance to the other five students in the group. All dances will be performed by the group and videotaped. Additionally, each student will be videotaped while teaching others the dance steps. Each teacher must demonstrate the ability to explain the steps to others, count to the music, and coach others to eliminate any errors on the dance.*

Mastery Item 14.3

Develop an idea for a simple project and identify the type of learning this project would document.

Portfolios

***Portfolios** are systematic, purposeful, and meaningful collections of a student's work designed to document learning over time.* Since a portfolio provides evidence of student learning, the knowledge and skills that a teacher desires to have students document guides the structure of the portfolio. The type of portfolio, its format, and the general contents are usually prescribed by the teacher. Portfolio artifacts may also include input provided by teachers, parents, peers, administrators, or others.

The guidelines used to format a portfolio will be based on the type of learning that the portfolio is used to document. Teachers often have students use working portfolios as repositories of documents or artifacts accumulated over a certain period. Other types of process information could also be included, such as drafts of student work, records of student achievement, or documentation of progress over time.

A showcase portfolio contains the artifacts submitted for assessment. It consists of work samples selected by a student that are examples of that student's best work. Typically teachers specify three to five documents or artifacts that all students must submit and then allow students to choose two or three optional artifacts. For the self-selected items, each student needs to consciously evaluate his or her work and choose those products that best represent the type of learning identified for this assessment. Each artifact included in a showcase portfolio is accompanied by a reflection, in which the student whose portfolio it is explains the significance of the item and the type of learning it represents.

It's a good idea to limit the portfolio to a certain number of artifacts to prevent the portfolio from becoming a scrapbook that has little meaning to a student and to avoid giving teachers a monumental task when scoring them. This also requires students to exercise some judgment about which artifacts best fulfill the requirements of the portfolio task and exemplify their level of achievement. The portfolio itself is usually a file or folder that contains a student's collected work. The contents could include items such as a training log, student journal or diary, written reports, photographs or sketches, letters, charts or graphs, maps, copies of certificates, computer disks or computer-generated products, completed rating scales, fitness test results, game statistics, training plans, report of dietary analyses, and even video or audio recordings. Collectively, the artifacts selected provide evidence of student growth and learning over time as well as current levels of achievement. The potential items that could become portfolio artifacts are almost limitless. The following list of possible portfolio artifacts may be useful for physical activity settings (Kirk, 1997). A teacher would never require that a portfolio contain all of these items. The list is offered as a way to generate ideas for possible artifacts.

POTENTIAL PORTFOLIO PROJECTS FOR PHYSICAL ACTIVITY

- Write a self-assessment of current skill level and playing ability, with individual goals for improvement.
- Conduct ongoing self- and peer assessments of skill performance and playing performance (process–product checklists, rating scales, criterion-referenced tasks, task sheets, game play statistics).
- Prepare a graph or chart that shows and explains performance of particular skills or strategies across time.
- Analyze your game-playing performance (application of skills and strategies) by collecting and studying your individual game statistics (e.g., shooting percentage, assists, successful passes, tackles, steals).
- Create and perform an aerobic dance, step aerobic, or gymnastic routine, with application knowledge and skills (evidence: a routine script or videotape of the performance).
- Document participation in practice, informal game play, or organized competition outside of class.
- Keep a daily physical activity journal in which you set daily goals; record successes, setbacks, and progress; and analyze situations to make recommendations for present and future work.
- Using a self-analysis or diagnostic assessment, select or design an appropriate practice program and complete schedule. Record results.
- Set up, conduct, and participate in a class tournament. Keep group and individual records and statistics (as an individual or as a part of a group).
- Write a newspaper article as if you were a sports journalist reporting on the class tournament or a game (must demonstrate knowledge of the game).
- Develop and edit a class sport or fitness magazine.

- Complete and record a play-by-play and color commentary of a class tournament game as if you were a radio (audiorecord) or television (videorecord) announcer.
- Interview a successful competitor about his or her process of development as an athlete and his or her current training techniques and schedule (audiorecord or videorecord).
- Interview an athlete with a disability about his or her experience of overcoming adversity. Apply what you have learned to your situation (audiorecord, videorecord, or article).
- Write an essay on the subject *What I learned and accomplished during gymnastics [or any activity unit] and what I learned about myself in the process.*

A rubric (scoring tool) should be used to assess portfolios in much the same manner as any other product or performance. Providing a rubric to students along with the assignment allows them to self-assess their work and thus be more likely to produce a portfolio of high quality. Portfolios, since they are designed to show growth and improvement in student learning, are assessed holistically. The reflections that describe the artifact and why the artifact was selected for inclusion in the portfolio provide insights into levels of student learning and achievement. Teachers should remember that format is less important than content and that the rubric should be weighted to reflect this. Table 14.2 illustrates a qualitative analytic rubric for assessing a portfolio along three dimensions.

Table 14.2 Qualitative Scoring Rubric for Complete Portfolio

	Format and design	Content items for each section	Reflection on portfolio tasks
Bull's-eye	Follows prescribed format with no errors. Design is attractive. Shows creativity. Well organized. Strong alignment with instructional goals associated with the portfolio.	A variety of artifacts shows both breadth and depth of learning. Items are selected that document growth and evidence of student learning and achievement. Comprehensive and complete coverage of desired student learning.	Reflections are thoughtful and document depth of understanding. Reflection provides insight about student learning and mastery of the desired goals for learning.
On target	Follows prescribed format with few errors. Design is neat with some creativity. Format is organized and easy to follow. Effective instructional alignment with learning goals.	Artifacts selected demonstrate competence in the subject. Some breadth of understanding and few if any misconceptions. Student is able to demonstrate growth and understanding of the subject.	Reflections provide a clear rationale for including the artifact. The intent of the reflection is easy to determine. Clearly written and easy to understand.
Getting close	Follows prescribed format for most part. Little evidence of creativity or imagination. Some grammar and spelling errors. Format may be messy or lack organization in some parts.	Basic information is presented so that a student is able to present an acceptable level of understanding. Artifacts help document student learning and growth. One or more artifacts may not align with instructional goals.	Reflections are to the point and direct, conveying reasons that the artifacts demonstrate learning. Some are short and fail to capture a depth of understanding.
Missed the mark	Does not follow prescribed format. Numerous errors. Inappropriate design. Disorganized and not aligned with the goals of learning. Gives the sense that it was completed at the last minute.	Many inappropriate or unrelated items included, indicating a lack of basic knowledge. Poor-quality work. Little variety or depth.	Little evidence of personal reflection on the tasks. Vague and repetitive. Little thought or rationale for including the various artifacts.

For additional information about portfolio assessments, see Lund and Kirk's (2010) chapter on developing portfolio assessments. An article published as part of a *JOPERD* feature presents a suggested scoring scale for a portfolio (Kirk 1997). Melograno's Assessment Series publication (2000) on portfolios also contains helpful information.

EXAMPLE OF A PORTFOLIO PROJECT

The following is an example of a portfolio project for a middle school or high school activity unit. This portfolio project could be used with a variety of sports, such as volleyball, basketball, tennis, badminton, or soccer.

Directions to students: To demonstrate your growth in the ability to play a sport, you will submit up to eight artifacts that were completed during one of your sport education units this year. Using one of the units taught in this class, you must demonstrate your ability to apply your knowledge of the skills, rules, and strategies of a sport and your ability to analyze the game. You must also document your contributions as a member of the team. The following three items are required:

1. An article reporting the class tournament game that you observed, including game statistics and an analysis of the game based on team or individual performances.
2. The results from a game play assessment done on your team that demonstrates your skill and competence and ability to play the sport.
3. An example of how you fulfilled a role on a duty team that also documents your knowledge of the game. This might include an example of your ability to officiate, announce, or keep game statistics. (Note: "Duty team" members are not playing games but are the referees, announcers, scorekeepers, line judges, etc.)

Select two to five additional artifacts of your own choosing. Each artifact must be accompanied by a reflection that explains the type of knowledge demonstrated by the artifact and a justification as to why you selected this item for inclusion in your portfolio.

Mastery Item 14.4

What if your instructor for this class required you to submit a portfolio that represented your learning in the class? Make a list of the important things that you have learned in this class (the learning that you wish to document) and then identify possible artifacts that you could use to document this growth. Some of your initial work might contain several errors (remember, you are trying to document your growth) while other artifacts represent high levels of achievement. Identify three or four key artifacts that your instructor might require from everyone (key assignments). Write a reflection for one of the assignments that explains why you chose it. (Note: if this were an actual portfolio assignment, reflections for each artifact would be written.) Describe how this portfolio would allow you to show your growth and competence in measurement and evaluation.

Mastery Item 14.5

How would the learning demonstrated in a portfolio differ from that demonstrated on a final exam (e.g., selected response, short-answer essay questions)?

Performances

Student performances can be used as culminating assessments at the completion of an instructional unit. Teachers might organize a gymnastics or track and field meet at the conclusion of one of those units to allow students to demonstrate the skills and knowledge that they gained during instruction. Game play during a tournament is also considered a student performance. Rubrics for game play can be written so that students are assessed on all three learning domains (psychomotor, cognitive, and affective). Students might demonstrate their skills and learning in one of the following ways:

- Performing an aerobic dance routine for a school assembly
- Organizing and performing a jump rope show at the half-time of a basketball game
- Performing in a folk dance festival at the county fair
- Demonstrating wu shu (a Chinese martial art) at the local shopping mall
- Training for and participating in a local road race or cycling competition

Although performances do not produce a written product, there are several ways to gather data to use for assessment purposes. A score sheet can be used to record student performance using the criteria from a game play rubric. Game play statistics are another example of a way to document performance. Performances can also be videotaped to provide evidence of learning.

In some cases teachers might want to shorten the time used to gather evidence of learning from a performance. *Event tasks are performances that are completed in a single class period.* Students might demonstrate their knowledge of net or wall game tactics by playing a scripted game as it is being videotaped. The ability to create movement sequences or a dance that uses different levels, effort, or relationships could be demonstrated during a single class period with an event task. Many adventure education activities that demonstrate affective domain attributes can be assessed using event tasks.

Mastery Item 14.6

After graduation, many exercise science students teach their clients as personal trainers, physical therapists, and so on. How could you document your ability to teach using a performance or an exhibition?

Student Logs

Documenting student participation in physical activity (National Standard 3, SHAPE America 2014) is often difficult. Teachers can assess participation in an activity or skill practice trials completed outside of class using logs. Practice trials during class that demonstrate student effort can also be documented with logs. A log records behavior over time (see figure 14.1). Often the information recorded shows changes in behavior, trends in performance, results of participation, progress, or the regularity of physical activity. A student log is an excellent artifact for use in a portfolio. Because logs are usually self-recorded documents, they are not used for summative assessments unless as artifacts in a portfolio or for a project. If teachers wanted to increase the importance placed on a log, a method of verification by an adult or someone in authority should be added.

Name ______________________ Teacher ______________________
Class ______________________

Activity	Date	Length of time participated

Figure 14.1 Activity log.

Mastery Item 14.7

What type of learning could you document with a log?

Journals

Journals can be used to record student feelings, thoughts, perceptions, or reflections about actual events or results. The entries in journals often report social or psychological perspectives, both positive and negative, and may be used to document the personal meaning associated with one's participation (National Standard 5, SHAPE America 2014). Journal entries would not be an appropriate summative assessment by themselves, but might be included as an artifact in a portfolio. Journal entries are excellent ways for teachers to take the pulse of a class and determine whether students are valuing the content of the class. Teachers must be careful not to assess affective domain journal entries for the actual content, because doing so may cause students to write what teachers want to hear (or give credit for) instead of true and genuine feelings. Teachers could hold students accountable for completing journal entries. Some teachers use journals as a way to log participation over time.

Mastery Item 14.8

Which type of performance-based assessment would you use to best represent your knowledge gained from the class about measurement and evaluation?

ESTABLISHING CRITERIA FOR PERFORMANCE-BASED ASSESSMENTS

One of the characteristics of a performance-based assessment is that the criteria used for scoring it are provided when the assessment task is first presented to students. When developing assessments, some teachers make the mistake of not determining the criteria. Without an effective scoring standard, the assessment task is essentially nothing more than an instructional activity and should not be called an assessment. Teachers must always clarify their expectations to students regarding assessments, and these expectations are the criteria that accompany the assessment.

Setting criteria for performance-based assessments is different from setting criteria for traditional assessments because a numerical score isn't always used to judge the performance (e.g., performance levels on a rubric might be termed adequate, excellent), but the fundamental principle of informing students about the expected level of performance remains the same. When students are given the criteria in advance, they no longer need to play the guessing game of *What's on the test?* The concept of giving the criteria with the assessment should not be new to a physical education major. In many physical education tests, the criteria used for the assessment are known. For example, Fitnessgram provides a range of scores needed to reach the Healthy Fitness Zone. Similarly, teachers often tell students the score they must achieve on a skills test to receive a certain grade or will require students to reach a certain level of expertise before allowing them to play a game. With both fitness and skills tests, teachers can inform students of the criteria that they must strive to reach.

There are several types of criteria that teachers can use to denote the standards for performance on an assessment. **Process criteria** look at the quality of performance for students completing the assessment or how students complete a task. In physical education, the term *process criteria* is used to denote the critical elements needed for a student to demonstrate correct form. Table 14.1 provides an example of the critical elements that are important for softball batting. Process criteria are critical for students when they are first learning a skill, because focusing on correct form is important. In many assessments for elementary school children, process criteria are typically used to assess student performance. For regular classroom teachers, the term *process criteria* is used in reference to class participation, doing homework, or the effort expended by a student.

When the criteria are stated in terms of the outcome of student performance, they are called **product criteria**. Examples of product criteria are the number of times a student must serve the volleyball or the number of consecutive forward rolls students must do in a tumbling sequence. A rule of thumb is that teachers should not specify product criteria without also requiring correct form.

Guskey and Bailey (2001) report the use of **progress criteria** by teachers when they want to measure student improvement. Progress criteria do not look at levels of achievement, but instead measure the amount of student improvement. Some teachers like progress criteria, especially with students who have little or no experience with a sport or activity. A problem with using progress criteria is that they create a moving target, meaning that some students may be required to reach a higher level of performance than others to receive a similar grade (e.g., there is no established score that everyone must reach—it changes from one student to another). For example, Taylor had never played tennis, and on the pretest she was unable to do a forehand rally and received a score of zero. Chris had played tennis before and was able to hit three consecutive forehands during the test. Using the data from the pretest, the teacher determined that Taylor would need to get six hits on the continuous rally test while Chris would need to get 10 in order to receive an A. At the conclusion of the unit, both students were able to do six hits on the continuous rally test. Taylor received an A for the test while Chris did not. An additional problem with using progress criteria is that some students might sandbag on the pretest, meaning that they perform at a level below what they are capable of so that they can demonstrate improvement with relatively little personal skill development.

Simple Scoring Guides

The simplest type of scoring guide for performance-based assessments is a checklist. With checklists, no decision is made about the degree to which the characteristic is present or about the quality of the performance: The trait is either demonstrated or it is not. With a checklist, all traits or characteristics are of equal worth. Checklists are often used by teachers to evaluate process criteria for sport skills (e.g., forearm pass, overhead pass, spike) or fundamental motor skills (e.g., hop, skip, leap). When using a checklist, teachers indicate whether the element is present or not by marking a recording sheet. Some teachers use a single blank in front of the element while others have two columns of blanks so that the evaluator can check either Yes or No to indicate whether the element was observed. The latter method is preferred when checklists are used as peer or self-assessments to ensure that students have observed each item on the list. Checklists are relatively simple to create and are useful as a written product to document an observation. When checklists are used, teachers must not make them too long, especially for younger children. It is difficult for even an experienced teacher to observe six or seven things simultaneously. Younger children should look at no more than two or three items, depending on their age. When using checklists containing multiple characteristics for peer assessments, teachers should require students to perform a sufficient number of trials so that the observer has an adequate opportunity to accurately observe the skill (see figure 14.2). Peers must be taught the purpose of checklists and how to use them if they are to be effective peer assessments.

Volleyball forearm pass	Trial 1		Trial 2		Trial 3	
Hands clasped, knees bent	Yes	No	Yes	No	Yes	No
Contact on flat surface of the arms	Yes	No	Yes	No	Yes	No
Uses legs to provide force without swinging arms	Yes	No	Yes	No	Yes	No
Platform directed to target on follow through	Yes	No	Yes	No	Yes	No

Figure 14.2 Repeated trials of volleyball assessment.

The point system scoring guide is another type of performance record with a listing of traits or characteristics considered important for the performance. It differs from a checklist in that it has points listed for each characteristic so that those items considered to have greater value or importance are given more points. As with checklists, when using a point system performance list, the observer makes no judgment of quality—the trait is either present or is not. Performance lists can be used by teachers for grading purposes. After the presence or absence of the traits or elements has been determined, the number of points can be added and a final or total score assigned. Teachers should not award partial credit for an item that is partially present, because this practice can affect the reliability of the rubric. If an element can be broken down into subelements and if teachers want to give credit for the various subelements demonstrated, then each of the subelements should be assigned a point value so that partial credit for the category is awarded with consistency. Figure 14.3 is an example of a point system scoring guide.

SPORT EDUCATION NOTEBOOK

Name ______________________ Final score (out of 100 points) __________

Contributions to team points (15 points)

____Attendance, dress, and participation logs (2 points)

____Player profile page (5 points)

____Duty team role contributions (5 points)

- *Explanation of contributions based on evaluation (3 points)*
- *Documentation of duty team evaluation (2 points)*

____Explanation of team role contributions (3 points)

Documentation of cognitive learning (25 points)

____Certifying test for officials (5 points)

____Rubric for officiating (10 points)

- *Preliminary round game (5 points)*
- *Tournament game (5 points)*

____Completed stats sheet (10 points)

- *Preliminary round game (5 points)*
- *Tournament game (5 points)*

Documentation of skill (40 points)

____Practice (10 points)

- *Logs (6 points)*
- *Peer evaluation checklist (4 points)*

____Graphs of data showing change of performance (5 points)

____Statistics sheet from game (10 points)

____Game play assessment rubric results (15 points)

- *Preliminary round (5 points)*
- *Tournament game (10 points)*

Fair play and team contributions (5 points)

____Fair play checklist completed by an opponent (2 points)

____Fair play checklist completed for an opponent (3 points)

Organization and presentation of notebook (15 points)

____Table of contents (2 points)

____Reflections on documents that involved student choice on selection (10 points)

____Neatness, grammar, spelling (3 points)

Figure 14.3 Point system scoring example.

Rubrics

Rubrics are used to assess complex assessments (e.g., game play or dances, a student project, portfolio). When using a rubric, assessors are required to make a judgment about the level of quality of student performance. Because making a judgment about the quality of the performance requires more time than simply deciding whether a characteristic is present, typically fewer traits or characteristics are evaluated on rubrics. There are two types of rubrics—analytic and holistic. **Analytic rubrics** are used to evaluate individual traits or characteristics of the performance separately. Analytic rubrics are useful for formative assessments so students can see how well they performed or how they were scored on the various traits or descriptors and use the results to make improvements.

There are two types of analytic rubrics. **Quantitative rubrics** are useful for assessing sports with sequential skills where consistency of form affects the quality of performance. Target games such as archery and bowling and individualized activities such as diving and the field events in track and field are some examples of where qualitative rubrics might be used. The numerical scores given are based on whether a student performs the various phases in the performance sequence consistently (e.g., six out of six) or just occasionally (e.g., two out of six), or somewhere between those two levels. When developing a quantitative rubric, a teacher might identify four or five phases of the performance and then assess each of these for consistency (see table 14.3). A **qualitative rubric** has written explanations that describe the quality of the performance for the various levels of the traits or characteristics being evaluated. These rubrics are used to assess complex performances that involve multiple domains or dimensions of cognitive learning. Game play or dance performances are two examples of complex performances involving more than one domain. Some qualitative rubrics identify errors commonly seen at certain levels of performance. These descriptions help communicate what the evaluator should or should not be looking for with regard to the trait. Table 14.4 is an analytic qualitative rubric for softball game play.

Holistic rubrics are used to evaluate the entire performance with a single score. With an analytic rubric, each of the characteristics is given a separate or individual score. A holistic rubric describes all the characteristics of a level of performance in a single paragraph. Holistic rubrics are most commonly used with summative assessments when students are not reading teacher feedback and trying to improve their product or performance based on that feedback. Table 14.5 is an example of a holistic rubric.

An issue with using holistic rubrics can occur because student performance for every characteristic is rarely at the same level; students may have some items at two levels. This situation can be addressed in one of two ways. If a student demonstrates some characteristics at one level and the rest of the characteristics at another level, then the lowest level at which the characteristics were demonstrated is the score assigned to that student's work. A second way to address multiple levels of performance is to weight the characteristics listed on the rubric and score the work according to how the items were rated. Those items of highest importance determine the score given to the performance or product. The decision about how to apply the rubric must be decided before scoring, and all raters must use the same method for consistent results. When using holistic scoring, evaluators are often given examples of past student work called *exemplars*. Because a holistic score is an overall impression of the total work, these exemplars help evaluators calibrate the intent of the rubric. When scoring student work, it is helpful to compare items to the exemplars to maintain high levels of reliability. Exemplars for written work are fairly easy to provide. When holistic rubrics are used for student performances, the use of exemplars is much more difficult since video recordings are needed.

Developmental rubrics are used for students at all levels of proficiency, from beginning to advanced players. Players can see where they are on a continuum of skill development—where they currently are with regard to skillfulness and how much they still need to master. Table 14.6 is an example of a simple developmental rubric for catching and throwing. Sports such as gymnastics or diving use a developmental scoring system; all performers regardless of age are judged using the same criteria.

Table 14.3 Shot Put Quantitative Rubric: Class Recording Sheet

Steps	Starting Position	Start of glide	Turn	Explosion	Landing
Critical elements	Shot is cradled against the neck. Holds the shot on finger pads. Body is leaning forward. Weight is on throwing side foot and knees are bent.	Thinker position. Kicks backward with non-supporting leg as the supporting leg straightens. Back is facing the direction of the throw. Supporting foot starts the turn.	Plants non-supporting leg and opens hips. Supporting leg starts to straighten. Body starts to straighten.	Push shot, extending arm. Weight shifts to non-throwing side leg. Releases the shot at a 45° angle as jumps into the air.	Lands on throwing side leg with side of body facing the target. Avoids falling out of the circle when releasing the shot. Exits by the back of the ring.
Name	**Score**	**Score**	**Score**	**Score**	**Score**

Directions: Observe each athlete and rate each of the steps on a scale using the following definitions:

4 = exhibits the behaviors consistently or almost always (3 × or more in a row)

3 = usually exhibits the behaviors (4 out of 6 trials)

2 = exhibits behaviors sometimes or about half the time

1 = exhibits behaviors occasionally (2 × or fewer out of 6 trials)

0 = does not complete the assignment or does not participate

When writing rubrics, teachers can write a rubric specific to a single sport or activity or one that can be used for multiple sports or activities. **Task-specific rubrics** are written for a single sport or activity as in table 14.4, which was developed for softball. **Generalized rubrics** can be used for several related sports or activities. Consider level 5 from the tennis rubric found in table 14.7. If the terminology concerning the serve were changed slightly to read "Service is strong and usually lands in bounds," then the rubric could be used for serving with several net games. Removing the phrase about spin would be necessary for badminton games. These are examples of how a task-specific rubric could be modified to a generalized rubric. In many instances, games within a category (e.g., invasion, net or wall, target, field) or activities that are alike (e.g., activities involving dance and rhythm such as folk and square dance, hip hop, jazz, social dance) could be assessed with a single rubric that would capture the essence of learning for a variety of sports and activities within the classification.

When creating generalized rubrics, some people fail to include any meaningful benchmarks, which results in a rubric that leaves much room for subjective decisions. Table 14.8 is an example of a hypergeneral rubric that provides little guidance to those using it. Teachers should avoid using hypergeneral rubrics for assessment.

Table 14.4 Softball Game Play Analytic Qualitative Rubric

Characteristic	Rookie	Junior varsity	Varsity	All star
Base running	Ignores base coach while running. Runs on every hit regardless of the situation. Jogs slowly around the bases.	Hustles to reach the base. Watches the base coach to keep track of whether to stay or advance.	Knows own abilities and does not overestimate the ability to safely advance. Rounds the bases using the left foot for greatest efficiency. Occasionally slides to avoid a tag.	Stretches a hit to extra bases when possible. Slides to avoid being tagged or to slow the approach to the bag. Follows the base coach's advice but is capable of making accurate decisions.
Batting	Chops when swinging the bat. Swings at every pitch or fails to swing. Often makes mistakes about whether the pitch is good enough to swing at.	Good form. Swings through the ball; knows which pitch is good enough to swing at.	Typically makes contact with the ball when batting. Good judge of the strike zone.	Consistently hits the ball even when it is thrown with speed. Rarely strikes out. Is able to place the ball depending on the situation.
Fielding and throwing	Fails to judge many of the balls, which causes errors; overruns the ball. Waits for ground balls to come to the player.	Can make a play and throw a runner out. Struggles to throw the ball for a distance longer than 40 ft (12.2 m). Ball is thrown in general direction of the target—usually throws a catchable ball.	Can throw from third to first to make the play. Knows when to make the play on the hop but usually tries to catch the ball on the fly. Charges ground balls.	Judges a batted ball accurately. Charges ground balls and can make an accurate throw on the run. Is able to leap and make difficult catches. Covers the field and plays the position well.
Knowledge of rules	As a base runner, does not know whether to advance or stay when the ball is hit. Knows basic rules (e.g., 3 strikes is an out; 4 balls is a walk; 3 outs per inning).	Knows the difference between a force out and a tag out. Knows enough of the rules to keep the scorebook.	Good knowledge of the rules. Knows more complicated rules such as infield fly rule. Can coach players while playing defense.	Can accurately officiate a game. Can answer most questions about rules correctly.
Strategies	Knows some strategies but applies them inconsistently during a game. Does not back up other players.	Knows what to do with runners on various bases with various outs.	Backs up the play. Uses offensive and defensive strategies to improve chances to win. Changes positions on the field depending on the game situation (runners and outs).	Reads the players and moves around position according to the game situation (runners and outs) and who is up to bat.
Fair play	May berate others when they make a mistake. Does not compliment others. Cares about own play with little regard to team success.	Ignores the mistakes of others and congratulates good play of others on the team.	Talks to teammates when they make a mistake to calm them down. Enthusiastic about the play of others.	Congratulates others on good play, even players on the opposing team. Is willing to sacrifice hit for the team. Will sit the bench if it benefits the team.

Table 14.5 Holistic Softball Game Play Rubric

Level	Description
Level 4	Stretches a hit to extra bases when possible. Slides to avoid being tagged or to slow the approach to the bag. Follows the base coach's advice but is capable of making accurate decisions. Consistently hits the ball even when it is thrown with speed. Rarely strikes out. Is able to place the ball depending on the situation. Judges a batted ball accurately. Charges ground balls and can make an accurate throw on the run. Is able to leap and make difficult catches. Covers the field and plays the position well. Can accurately officiate a game. Can answer most questions about rules correctly. Reads the players and moves around position according to the game situation (runners and outs) and who is up to bat. Congratulates others on good play, even players on the opposing team. Is willing to sacrifice hit for the team. Will sit the bench if it benefits the team.
Level 3	Knows own abilities and does not overestimate the ability to safely advance. Rounds the bases using the left foot for greatest efficiency. Occasionally slides to avoid a tag. Typically makes contact with the ball when batting. Good judge of the strike zone. Can throw from third to first to make the play. Knows when to make the play on the hop but usually tries to catch the ball on the fly. Charges ground balls. Good knowledge of the rules. Knows more complicated rules such as infield fly rule. Can coach players while playing defense. Backs up the play. Uses offensive and defensive strategies to improve chances to win. Changes positions on the field depending on the game situation (runners and outs). Talks to teammates when they make a mistake to calm them down. Enthusiastic about the play of others.
Level 2	Hustles to reach the base. Watches the base coach to keep track of whether to stay or advance. Good form. Swings through the ball; knows which pitch is good enough to swing at. Can make a play and throw a runn er out. Struggles to throw the ball for a distance longer than 40 ft (12.2 m). Ball is thrown in general direction of the target—usually throws a catchable ball. Knows the difference between a force out and a tag out. Knows enough of the rules to keep the scorebook. Knows what to do with runners on various bases with various outs. Ignores the mistakes of others and congratulates good play of others on the team.
Level 1	Ignores base coach while running. Runs on every hit regardless of the situation. Jogs slowly around the bases. Chops when swinging the bat. Swings at every pitch or fails to swing. Often makes mistakes about whether the pitch is good enough to swing at. In the field, fails to judge many of the balls, which causes errors; overruns the ball. Waits for ground balls to come to the player. As a base runner, does not know whether to advance or stay when the ball is hit. Knows basic rules (e.g., 3 strikes is an out; 4 balls is a walk; 3 outs per inning). Knows some strategies but applies them inconsistently during a game. Does not back up other players. May berate others when they make a mistake. Does not compliment others. Cares about own play with little regard to team success.

Table 14.6 Developmental Catching and Throwing Rubric for Prekindergarten Through Age 9

Developmental level	Catching	Throwing
6	Can catch an object thrown with increased velocity or catch an object while moving; can sequence catching and throwing skills smoothly	Can throw with velocity and accuracy; is able to hit a target from a distance of 25 ft (7.6 m)
5	Can transfer catching skills to a game situation	Can transfer throwing skills to a game situation
4	Can catch a variety of objects at different levels with a partner	Shows trunk rotation and accuracy
3	Can catch a variety of self-tossed objects	Follows through toward target
2	Can catch a bounced ball from a partner	Shows opposition when stepping
1	Extends arms toward thrower, shows avoidance reaction; traps the ball rather than catching it	Shows limited body movement; arm dominates the throw

Adapted from the Wichita Public Schools, Kansas.

Table 14.7 Sample Scoring Rubric for Holistic Assessment of Tennis Playing Ability

5—Excellent	Consistently executes all strokes with little or no conscious effort, resulting in few unforced errors. Uses spin to add difficulty to shots and for strategic advantage. Uses the rules to gain an advantage. Anticipates opponent's shots and can control shots to employ effective strategies that lead to winning the point. First serve is strong and usually lands in bounds; difficult to return. Calls shots fairly and accurately.
4—Good	Demonstrates competency and ability to perform basic tennis skills while making few errors. Demonstrates an understanding of the rules by self-officiating a game correctly. When asked, provides a strategy appropriate for an upcoming game. Usually selects appropriate strategy and shot selection for situation and generally displays consistent performance. Is able to serve consistently and accurately.
3—Satisfactory	Displays basic understanding of tennis and is able to perform fundamental skills adequately to play a game. Performance is steady, relying mostly on ground strokes with some net play. Understands basic strategies but lacks ability to effectively employ them. Often relies on second serve during a game. Knows the rules well enough to play a basic game of tennis.
2—Fair	Primarily uses the forehand and some backhand shots to keep the game going. May run around his or her backhand to avoid using the shot. Has difficulty placing shots away from an opponent. Uses little strategy and frequently returns the ball to an opponent without making the player(s) move. Is able to keep score but gets confused and cannot answer questions. Serve is inconsistent.
1—Poor	Rarely, if ever, performs skills well enough to sustain a rally. Fails to return most serves. Is able to get the serve in play less than half of the time. Makes mistakes on basic rules such as where to serve, line calls, etc. Is not skilled enough to adjust to the performance of the opponent.

Table 14.8 Example of a Hypergeneral Rubric

Level	Knowledge
4	Demonstrates a thorough understanding of the important concepts or generalizations and provides new insights into some aspect of that information
3	Displays a complete and accurate understanding of the important concepts or generalizations
2	Displays an incomplete understanding of the important concepts and generalizations and has notable misconceptions
1	Demonstrates little understanding of the concepts and generalizations and has several misconceptions

Adapted from the Mid-continent Research for Education and Learning (McREL), Aurora, CO.

Developing a Rubric

The rubric for an assessment often determines whether the person using the assessment is able to make a valid inference about student learning. Most performance-based assessment tasks have strong content or face validity—they represent what a teacher wants students to know and be able to do. For example, when teaching a badminton unit, the ultimate goal is to teach students to play the game of badminton at a reasonable level of competence. Requiring students to play a game allows the assessment of that ability. Therefore, the game is a strong indicator of student competence. However, to assess this ability, the rubric must include the right items or an evaluator will not be able to make an assessment of each student's ability to play. Knowing the rules is an important part of being able to play a game. If the rubric failed to include knowledge of rules as a characteristic that was evaluated, there would be no way to determine whether students had learned them. A competent player will use a variety of shots to move the opponent around the court and try to place the shuttlecock where the opponent cannot return it. This trait should be an item included on the rubric. Since the assessment must provide information on which a teacher makes a determination of ability, if the rubric fails to allow a teacher to gather adequate information about students' ability to play a badminton game, the decision about game play competence will not be based on good data. Selecting the right characteristics to use on a rubric is an essential part of the process used to develop an assessment.

Lund and Veal (2013) identified 7 basic steps to follow when writing qualitative rubrics:

1. Identify what needs to be assessed.
2. Brainstorm ideas for the behaviors representative of what you are trying to achieve.
3. Identify the descriptors.
4. Decide how many levels are needed
5. Write levels for the quantitative rubric.
6. Pilot the assessment.
7. Revise the rubric.

Writing a qualitative rubric is difficult unless one has a product or performance to use when creating the written descriptions for the various levels. Doing an actual assessment of a performance or product reveals levels of student misunderstandings and common errors, as well as identifying essential elements for the various levels of performance. A holistic rubric is fairly easy to develop from a qualitative rubric since the descriptors for a given level are combined into a single paragraph.

As stated earlier, quantitative rubrics are often selected for sports containing a single skill for which consistency of performance is critical to success. These sports are typically performed in a fairly closed environment. Quantitative rubrics are excellent for assessing target games such as bowling and archery. The example rubric in table 14.3 was written for the shot put. When developing a quantitative rubric,

1. identify the content,
2. decide on the phases or steps important to consistent performance,
3. identify the critical elements for each phase (no more than two or three),
4. determine how many levels that you wish to use,
5. define the levels by indicating the level of consistency related to that level,
6. develop a recording sheet,
7. pilot the rubric, and
8. revise the assessment.

See Lund and Veal (2013) for a detailed discussion of how to develop rubrics.

Mastery Item 14.9

Develop a rubric that could be used with a game play assessment for a sport or a rubric for a physical activity (e.g., dance, gymnastics, in-line skating).

SUBJECTIVITY: A CRITICISM OF PERFORMANCE-BASED ASSESSMENTS

Performance-based assessments are often criticized because of the subjectivity associated with the scoring process. As Danielson (1997) points out, there is a degree of subjectivity with all testing and assessment. The decision about what knowledge to assess on a test is quite subjective. Think about classes in which professors use two tests to evaluate whether students had mastered all the material covered in a course lasting an entire semester. Did instructors use questions from the book, lectures, or a mixture of the two? Some instructors choose to use questions from the instructor's guide that were written by someone who had never met the instructor or attended a single class. Have you ever had a teacher who asked questions about the captions under the pictures in the textbook? Selection of the content for the questions was done subjectively, by the test writer. Some multiple-choice questions actually have two plausible answers (depending on how one interprets the stem) and only one opportunity to select the right response. In this case, subjectivity occurred when the person selected the response that was considered correct. Many of these issues are mitigated when tests are developed following the guidelines provided in chapter 8.

The point of this discussion is to recognize that there is subjectivity in most types of testing and assessment, with the difference being when it occurs. With traditional written tests, the chance for subjectivity occurs with the selection of content, the writing of questions, and the selection of the correct response. With performance-based assessments, the chance for subjectivity occurs when the assessment is evaluated. Danielson notes that most teachers who use performance-based assessments are competent professionals. Instead of referring to their judgments of quality using the rubrics as subjective, she states that teachers are using professional judgment to make decisions about student learning. A physical educator or coach who is knowledgeable about a sport can observe players and assess a level of performance with a high degree of accuracy. Performance-based assessments formalize this process and let students know the criteria used to assess the performance while at the same time allowing teachers to focus on those elements identified on the rubric as being important. Performance-based assessments allow teachers to coach students and thus maximize student achievement.

SELECTING APPROPRIATE PERFORMANCE-BASED ASSESSMENTS

One of the dangers of introducing new ideas in education is that teachers (and administrators) will jump on the bandwagon of an idea that seems good on the surface and not think comprehensively through all the details. *Generally, the overall purpose of assessment is to provide good information for making valid decisions about learning.* Good assessment information provides an accurate indication of a student's performance and enables a teacher, coach, or other decision maker to make appropriate decisions. But what constitutes good assessment information? What determines the quality of an assessment? Herman, Aschbacher, and Winters (1992) suggest that when talking about the quality of an assessment, we are really asking the following:

1. Does an assessment provide accurate information for decision making?
2. Do the results permit accurate and fair conclusions about student or athlete performance?

3. Does using the results contribute to sound decisions?

To be able to answer yes to these questions, three criteria must be met: reliability, validity, and fairness.

Reliability: Accurate and Consistent Information

As previously defined, reliability relates to the consistency of results. Assessments that are scored inconsistently are essentially useless because they do not provide meaningful information to the user. Inasmuch as performance-based assessment depends heavily on the professional judgment of teachers (or other assessors) to score and interpret the performance or product of the assessment task, one needs to be particularly concerned about interrater reliability or objectivity. Similar to the concept presented in chapter 6, this means that if more than one person is doing an evaluation, a student's score should be the same regardless of who did the evaluation. Intrarater reliability is also important. A teacher evaluating students in the morning must apply the criteria in the same way when using the assessment in the afternoon. When teachers apply criteria differently as the evaluation process continues, they *drift* from the original intent of the assessment. In some instances, teachers discover that a rubric is insufficient and begin to mentally add criteria to the rubric to clarify it, thus affecting reliability. People doing the evaluation must use the rubric to complete the assessment and not add other criteria. Extraneous variables such as the personality of a student (e.g., a teacher likes Student A but thinks that Student B is obnoxious) also must not enter into the decision. In short, an evaluator must strive to minimize inconsistency (i.e., error) in scoring to have confidence that the judgment is a result of the actual performance.

A good performance-list scoring guide or rubric is essential for consistent scoring. When using a rubric, an evaluator must be able to distinguish between levels, and the words used to describe the levels must create a clear picture of what the performance or product should demonstrate. When developing performance lists, teachers must select the most important traits or characteristics. Judging a performance is difficult when too many characteristics are identified. To ensure that performance lists or rubrics are thorough and accurate but

It's important that an instructor create assessment guidelines that are reliable, consistent, and fair to all students, regardless of their gender, ability, or economic status.

not unwieldy, always pilot test them before using them for an important assessment (e.g., when they mean a lot to a grade, whether a player makes a team). When multiple people will be using the same assessment and rubric (as with a state- or districtwide common assessment), training is essential. With training, scorers can reach high levels of reliability (Herman, Aschbacher, and Williams 1992). The use of exemplars helps to ensure that the scoring stays consistent.

When developing rubrics, the more levels one uses, the lower the reliability (Guskey and Bailey 2001). It is easier to make distinctions between 2 levels (acceptable and unacceptable) of performance than to make distinctions among 10 levels. Common sense tells one that as the distinctions between the levels become smaller, agreement between evaluators will decrease. The purpose of the assessment should determine the number of levels used. When developing assessments and rubrics, teachers should avoid making the rubric unwieldy.

Validity

As you know from chapter 6, validity is an indication of how well an assessment actually measures what it is supposed to measure. Although reliability is necessary, it is not a sufficient condition for validity. *An assessment could be perfectly reliable but not relevant to the decision for which it is intended.* If the assessment result is not related to the characteristic reportedly being measured, it may jeopardize accurate conclusions about a person's performance and subsequent decisions. Broadly speaking, validity relates to the meaning and consequences attached to the test scores (Messick 1995). Popham (2003) suggests that the question is not whether an assessment is valid, but rather whether the assessment allows users to make valid judgments about the results gathered from the process.

The primary form of validity reported for performance-based assessments, particularly those developed by teachers for classroom use, is face (content-related) validity, based on the assumed relationship between instruction and assessment. Although this linkage is important, content-related (face) validity alone should not be accepted as sufficient evidence for the use of an assessment method. Logically, the most accomplished students should receive the best scores. Performance-based assessments will be more believable if there is corroborating evidence that performance-based assessments allow teachers to reach valid conclusions. However, inasmuch as most performance-based assessments are substantially different from standardized testing, traditional validation procedures may be inappropriate (Miller and Legg 1993). Measurement and evaluation experts must work with teachers to resolve this issue and develop ways to ensure validity for performance-based assessments.

To make a valid judgment about student learning, teachers must use assessment tasks that have good face validity (have the potential to measure the type of learning that is targeted), use rubrics with characteristics that identify those behaviors and appropriate levels of student performance, and use multiple data sources to gather evidence about student learning.

Fairness

Although fairness is not a psychometric property in the same sense as reliability and validity, it is critically important in all forms of assessment, whether traditional or alternative. **Fairness** simply means that an assessment allows all students, regardless of gender, ethnicity, or background, equal opportunity to do well. Although there is tremendous diversity within our society and students do not come to school with the same background, exposure, motivation, or values, all students should have an equal opportunity to demonstrate the skills and knowledge being assessed. *Fairness should be evident in the development or selection of the assessment task as well as in the criteria used for judging the performance or product.*

Is the assessment task fair and free from bias? Does the assessment task favor boys or girls, students from a particular ethnic group, students who have lived in a particular area, or those whose families have greater financial resources? To be fair, the task should reflect knowledge, values, and experiences that are familiar to and appropriate for all students, and should seek to measure the knowledge and skills that all students have had adequate time to acquire. Scoring procedures and criteria for assessing a performance or the product created must be free from bias. Making sure this occurs helps to ensure that the ratings of a performance reflect an examinee's true capability and are not a function of the biases and perceptions of the person judging the performance. See chapter 13 for more detail about assigning student grades.

Some people believe that traditional evaluation procedures tend to demonstrate greater reliability, objectivity, and validity than performance-based assessment, largely because of the more objective nature of the traditional assessment. However, many of the objective, valid, and reliable assessments provide only indirect measures of student achievement. If teachers use only traditional testing formats, they are left with evaluating student ability to play a game by measuring knowledge of rules through a written test and the various skills with skill assessments. Students' abilities to apply knowledge (e.g., rules, strategy), to use skills in an open environment as well as demonstrate knowledge of when to use the skill, and to play fairly simply cannot be captured with traditional assessments. Performance-based assessments allow a **holistic assessment** of student learning. The challenge for practitioners who use these assessments is to carefully consider the issues of validity, reliability, and fairness to ensure that any assessment used leads to appropriate decisions about student learning.

ISSUES TO CONSIDER WHEN DEVELOPING PERFORMANCE-BASED ASSESSMENTS

It should be obvious by now that the nature of performance-based assessments precludes the publication of assessments that have universal appeal and usefulness. Clearly, one size does not fit all when it comes to performance-based assessments. As a result, most professionals generate their own assessments. This does not mean that one cannot adapt and enhance ideas obtained from other sources, but in many instances teachers will need to create their own assessments that meet their specific needs. *When developing any type of assessment, remember that it consists of both a task and performance criteria.* That is, to create an assessment, one needs both to create a task and to identify the criteria used to evaluate it.

To be most useful in an instructional setting, a performance-based assessment should consider both context (situation–task) and performance (construct–skill). That is, the assessment task should represent a completed performance having contextualized meaning that is directly related to the eventual use of the skill (Siedentop 1996). Skill assessments are useful for allowing students to practice skills in a closed environment. They also provide an opportunity to guarantee that teachers will be able to observe students performing a skill, which is something that doesn't always happen during game play. Because basic individual skills may serve as the foundation for future activity or performance in a game, assessing both the individual skills and the application of these skills in game performance is necessary. Too often students are able to perform the skills on a test but then unable to use them in a game. Performance-based assessments provide teachers with a way to assess the application of skills in a meaningful context.

The following are some things to consider when developing assessments.

Determining Purpose

The critical first step in the process of developing assessments is determining the purpose of the assessment. For instance, is the purpose of the assessment to diagnose initial skill levels to know where to start instruction? Perhaps the purpose is to determine deficiencies in a student's performance or the product he or she has created (formative assessment) and give feedback to the student about what has been learned. Or is the purpose to ascertain individual achievement of specified objectives to award a grade (summative assessment)? Or is the purpose of the assessment to obtain information for evaluating a physical education program or to comply with state reporting requirements (curricular assessment)? The purpose of doing the assessment will determine the types of assessments that are selected.

Deciding What You Are Willing to Accept as Evidence for Student Learning

What type of evidence will be used to document student learning? Some teachers might want to document fitness improvement by requiring students to complete logs of their activities. Others might simply pre- and posttest students using heart rate monitors and the information from the Fitnessgram PACER item. If a student is a beginner learning a skill, then process variables (correct form) are important to emphasize. Adventure education activities allow students to demonstrate several elements of National Standard 4, exhibiting responsible personal and social behavior, (SHAPE America 2014), but how does one capture those behaviors with an assessment? Self-assessments are good for documenting the learning process but are not appropriate for use as a summative assessment.

Selecting the Appropriate Assessment Task

An assessment task must allow a teacher to gather the evidence needed to make the correct decision about student learning. If the task does not provide the information needed to assess student learning, then it serves no useful purpose for assessment, and the results are not meaningful for decision making. One must ask if the assessment task elicits the desired performance or work. Have students had the opportunity to acquire the knowledge and skills needed for the task? If not, consider assessing something else. Also keep in mind that not all targets or desired outcomes are best measured using performance-based assessments.

The following performance-based assessments are useful for assessing process or facilitating student learning:

- Logs
- Journals
- Self-evaluations
- Peer assessments
- Interviews
- Performance checklists

For assessing product or outcome measures, the following performance-based assessments could be used:

- Projects
- Performances (e.g., game play, event tasks, performances)
- Portfolios

Setting Performance Criteria

Previous parts of this chapter have discussed the need for and importance of **performance criteria.** *A task alone does not constitute an assessment—criteria are necessary for judging the performance or product and telling you how good is good enough.* Some teachers and professionals, particularly those just starting out with performance-based assessment, tend to focus only on the task, to the exclusion of the criteria. How will you know if a student has accomplished the learning goal or target that was set? How will you know when the performance is good enough? How will you know if the portfolio is acceptable? How will you be able to diagnose a student's performance and provide meaningful feedback? Each of these questions is answered by the criteria, not the task.

Time

Time is a critical factor in selecting assessments. Assessments should not be a burden to teachers. Performance-based assessments can be time consuming and labor intensive. Teachers should select assessments that will allow them to accurately evaluate student learning. However, assessments should be carefully scheduled so that teachers can realistically manage them.

In large classes, teachers struggle to assess students. Performance-based assessments that are part of the learning process can help bridge the gap between wanting to assess student learning and lack of time to do assessments. If a teacher is checking for student learning, it is sufficient to sample students with a range of abilities and not assess each student in the class. Peer and self-assessments can improve cognitive learning and provide feedback as formative assessments. Teachers must also realize that adequate time must be allocated to teach a unit. Some teachers allocate 1 to 2 weeks per unit or activity. This is an insufficient amount of time to adequately teach a skill. If students don't have the opportunity to learn the skills and how to participate in the activity, there really is nothing to assess. Typically with large classes, more time is spent on managerial issues and less time on instruction. Therefore, more time must be allocated if student learning is to be optimal and meaningful.

Go to the WSG to view Video 14.3.

IMPROVING ASSESSMENT PRACTICES IN PHYSICAL EDUCATION SETTINGS

The purpose of this chapter is not to convince people that performance-based assessments are the solution to the assessment dilemmas that teachers in kindergarten through grade 12 schools face. Performance-based assessments do expand the possible ways in which teachers can assess their students and are probably best used along with other traditional forms of assessment. For example, skills tests are a good way to look at skills in a closed situation. They are excellent formative assessments that can provide feedback to students and teachers about current levels of student ability. Consider the following example that uses both types of assessment:

Mrs. Gaylor is an experienced teacher who has selected pickleball to teach net and wall tactics. She wants her students to be able to play pickleball at the conclusion of the unit and has decided that she will use a rubric to assess game play to determine students' overall abilities to play the game during a class doubles tournament. The rubric used for assessing game play will require students to use correct form when executing shots, to strategically place the ball away from the opponent, to work with a partner, to demonstrate positive sport behaviors toward the partner and opponent, to know the rules, and to use the serve to gain an offensive advantage (e.g., not just get it over the net, but put spin on it and place it away from the opponent if possible). She will not assess individual skills during the game because often form is sacrificed as players make an attempt at an errant ball. Instead, skills tests will be used for evaluating the volley shot, the serve, and a continuous rally. All of the skills tests are done against a wall so that a student doing the skills test does not need to depend on another student to demonstrate his or her own skillfulness. Although knowledge of rules will be one of the categories on the game play rubric, an additional written test (selected response, short answer essay) will be given so that lower-skilled students who might know the rules but not be able to demonstrate them during game play will have the opportunity to demonstrate that knowledge.

On the first day of the unit, Mrs. Gaylor informs students of her expectations, demonstrates the skills tests, and provides students with the game play rubric, which is also posted on the wall so that students can refer to it when needed. During the unit, Mrs. Gaylor incorporates the skills tests into her teaching progressions. Students are allowed to take the skills tests before and after the instructional part of class and during the class tournament when they are not playing a game. Students can take the skills tests multiple times—the goal is to reach the criterion scores that Mrs. Gaylor gave when explaining the tests. At the start of class, students pick up a 3-inch × 5-inch (7.6 cm × 12.7 cm) note card that has a place to record practice trials and results (this is a participation log). Students are allowed 10 minutes at the beginning of class to dress for class and practice their skills. Those students demonstrating more effort change their clothes quickly and have more opportunity to test, practice for the skills tests, or both. By looking at the practice logs (the note cards) Mrs. Gaylor can see which skills need more work and use this information while planning her lessons. Additionally, Mrs. Gaylor is keeping track of those students passing their skills tests and knows which skills need additional instruction, the students who need more assistance, and the students who need additional challenges because they have achieved the basic level of competence.

The tasks used for instruction are designed to teach students the skills and tactics needed to play pickleball. The content of the lessons is guided by the information (data) that Mrs. Gaylor is getting from her formative assessments. Before the start of game play, a rules test is given to ensure that all students have cognitive knowledge of the rules. When students start to play games, she will observe them multiple times using the game play rubric. The areas of lower performance will be addressed in future lessons. Classes conclude with students completing exit slips that require them to answer questions about the content of the day's lesson. On the exit slip, students also have an opportunity to ask any questions about things from the class that they didn't understand, request additional instruction on an area that is proving difficult for them to learn, or ask for challenges such as additional skills or game play tactics to help them continue to improve.

Mastery Item 14.10

Identify the types of student learning that Mrs. Gaylor can document using the procedure just explained. How do traditional skill assessments work with performance-based assessments to enhance student learning?

As physical educators and instructors strive to improve the quality of their programs, many will heed the reform initiatives of education experts who propose that standards-based education and performance-based assessment offer great promise for enhancing the education system. To improve the preparation of physical education teachers and other physical activity specialists so that they are able to conduct meaningful assessments, our profession needs to embrace a new way of thinking about assessment. *Although this chapter proposes that physical education teachers include performance-based assessment techniques in their repertoire of assessment methods, it is not suggesting that teachers completely abandon traditional, standardized testing techniques.* As shown in the example, there is a need for both, depending on the purpose of the assessment. Regardless of the approach taken, teachers should use meaningful assessment in physical education class or other physical activity settings. The design and incorporation of clear, developmentally appropriate, and explicitly defined scoring rubrics are essential to ensure valid inferences about learning, consistency, and fairness.

Stiggins (1987) suggested that the most important element in designing performance-based assessments is the explicit definition of the performance criteria. Moreover, Herman, Aschbacher, and Winters (1992) stated that criteria for judging student performance lie at the heart of performance-based assessment. *If performance-based assessments are to realize their promise and live up to expectations, then it is essential that high-quality assessments be accompanied by clear, meaningful, and credible scoring criteria.*

The following guidelines (adapted from Gronlund 1993) provide ways to improve the credibility and usefulness of performance-based assessment in physical education.

1. Ensure that assessments are congruent with the intended outcomes and instructional practices of the class.
2. Recognize that, together, observation and informed judgment with written results compose a legitimate and meaningful method of assessment.
3. Use an assessment procedure that will provide the information needed to make a judgment about the intended student learning.
4. Use authentic tasks in a realistic setting, thus providing contextualized meaning to the assessment.
5. Design and incorporate clear, explicitly defined scoring rubrics with the assessment.
6. Provide scoring rubrics and evaluative criteria to students and other interested persons.
7. Be as objective as possible in observing, judging, and recording performance.
8. Record assessment results during the observation.
9. Use multiple assessments whenever possible.
10. Use assessment to enhance student learning.

A balanced approach to assessment is the prudent path to follow. The issue is not whether one form of assessment is intrinsically better than another. No assessment model is suited for every purpose. The real issue is determining what type of performance indicator best serves the purpose of the assessment and then choosing an appropriate assessment method that is suitable for providing this type of information.

Dataset Application

Use the chapter 14 large dataset, which is based on figure 14.2. (That dataset is in the WSG.) Confirm the total scores for each of the three trials (chapter 3). Determine the alpha reliability coefficient (chapter 6) for the reliability *across* total scores for the three trials. Comment on the reliability. Calculate the correlation coefficients (chapter 4) among the four measures taken during the first trial. Comment on these correlation coefficients. Determine whether boys and girls differ significantly (chapter 5) for the sum across all four measures across all three trials.

MEASUREMENT AND EVALUATION CHALLENGE

Let's return to Mariko from the opening assessment challenge. She has decided to use soccer to teach invasion games concepts and that her primary purpose is for her students to be able to play a game of soccer. Psychomotor assessments will be emphasized in the unit. Mariko has decided to develop a game play rubric for soccer that includes characteristics about being able to maintain possession of the ball and advance the ball toward the goal (dribbling and passing), scoring (shooting or kicking to target), and preventing scoring (defense). She also wants to observe student knowledge of the rules during game play and whether her students know the tactics that are important in an invasion game. Her rubric also includes elements of fair play, since National Standard 4 is important to her. Skills tests for dribbling, passing, and shooting are used as formative assessments. Mariko has also developed some log sheets that allow students to document the effort that they demonstrate by practicing skills outside of class. By looking at whether students have improved their skills, Mariko has an indirect way to determine whether students are practicing as indicated. To document knowledge about tactics, she has created a project that requires students to develop a playbook with tactics and strategies for playing. During the class tournament, students would be expected to demonstrate knowledge of rules as they officiate the small-sided soccer games. A rubric was also created to assess officiating knowledge.

SUMMARY

Performance-based assessment conducted in authentic settings has become a foundation of current educational reform taking place in schools. This approach to assessment is believed to enhance learning and achievement. Performance-based assessments take many forms, but those typically used in physical education generally include individual or group projects, portfolios, performances, and student logs or journals. A method for recording learning is also essential so that teachers can use the information as they plan future lessons. A guiding principle for the development and use of any type of performance-based assessment is that it consists of both task and performance criteria. Furthermore, we have emphasized that consistency, validity, and fairness are just as important with performance-based assessment as they are with traditional, standardized testing. For those who are interested in more information about performance-based assessments, resource texts are provided by Lund and Kirk (2010) and Lund and Veal (2013). In addition, the several volumes published in the NASPE Physical Education Assessment Series are designed to provide a collection of current, appropriate, and realistic assessment tools for physical educators.

Go to the WSG for homework assignments and quizzes that will help you master this chapter's content.

APPENDIX

Microsoft Excel Applications

As we indicated in the textbook's preface, many of the procedures used to conduct the analyses associated with human performance measurement and evaluation decisions can be rather time consuming. However, the use of computers greatly facilitates conducting these analyses. We have used SPSS (IBM SPSS Statistics) throughout the book to illustrate step-by-step procedures for analyzing data. SPSS is a large, sophisticated software package that is used for research, business, education, and personal decisions. Less expensive student versions of SPSS are available, but they are generally limited in the types of procedures that can be completed, in the number of participants or variables (or both) that can be included, or in both of these ways. SPSS is widely available at universities, businesses, education and governmental agencies, and research centers. However, your textbook authors are aware that some students may have difficulty accessing SPSS.

Microsoft Excel is a widely available spreadsheet that can be used to conduct all of the procedures that are presented in your textbook. Excel was not created to be an analytical package, as was SPSS, but it can serve this purpose. Many of the analyses that may be performed with Excel involve more steps than SPSS. Essentially, SPSS has point and click or point and drag functionality that makes analyses quite easy. Excel, because it was not created as a statistical analysis package, is a bit more cumbersome to use when completing analyses related to reliability, validity, measurement, and evaluation decisions. Its advantages are its universal availability and relatively low cost.

All of the data tables and large datasets in each chapter are included with the web study guide (WSG) associated with your textbook; you may access the WSG by going to www.HumanKinetics.com/MeasurementAndEvaluationInHumanPerformance. These tables are presented in both SPSS and Excel formats.

In this appendix, we provide you with directions and screen captures (from PC computers) that illustrate Excel use. We illustrate the procedures presented in chapters 2 through 7. Chapters 8 through 14 use the procedures introduced in chapters 2 through 7. It is not our intent to teach you how to become an Excel expert. Rather, as with SPSS, it is our intent to illustrate the specific tools that you will encounter when making measurement and evaluation decisions in kinesiology and human performance. As with SPSS, once you have learned the procedures in chapters 2 through 7, you are expected to be able to generalize these analyses to the remaining text chapters.

Some of the Excel procedures for analyses used in chapters 5 and 7 are quite cumbersome. Thus, we have included templates for these analyses in chapters 5 and 7 of the WSG. In particular, the chi-square (χ^2) analysis introduced in chapter 5 and used again in chapter 7 has a template. Similarly, a template for the epidemiologic analyses presented in chapter 7 is included in the WSG. With the templates, you simply enter values into the cells provided and the results are calculated for you. A PowerPoint presentation illustrating Excel is in the WSG in chapter 2.

Many of the procedures listed in this appendix use Excel's Analysis ToolPak. The following descriptions are for activating the PC Analysis ToolPak if it is not already activated in your PC version of Excel 2010.

1. Open the Excel File Menu and click on Options.

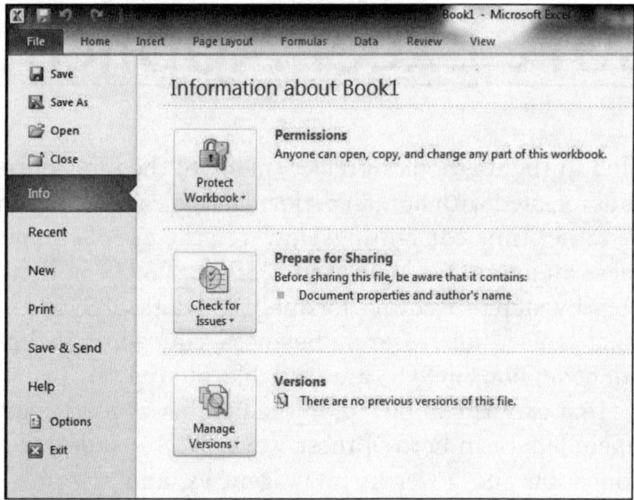

2. Under Excel Options click on Add Ins.

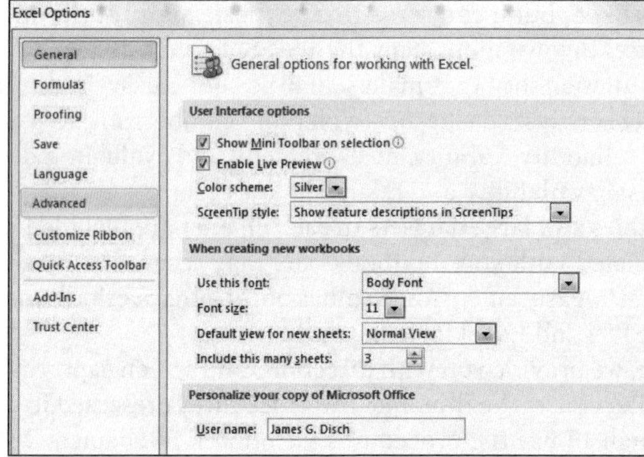

3. Highlight Analysis ToolPak – VBA and click Go at the bottom of the screen.

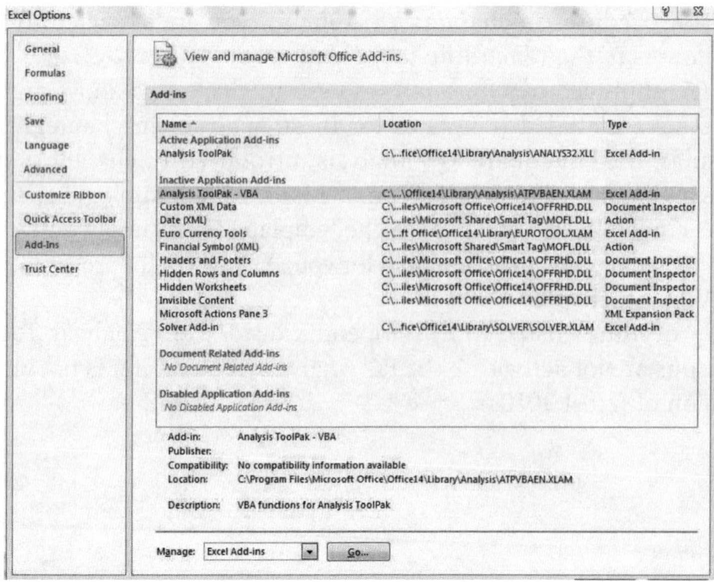

4. Check Analysis ToolPak and then click OK. Data Analysis should appear on the far right of your Data menu.

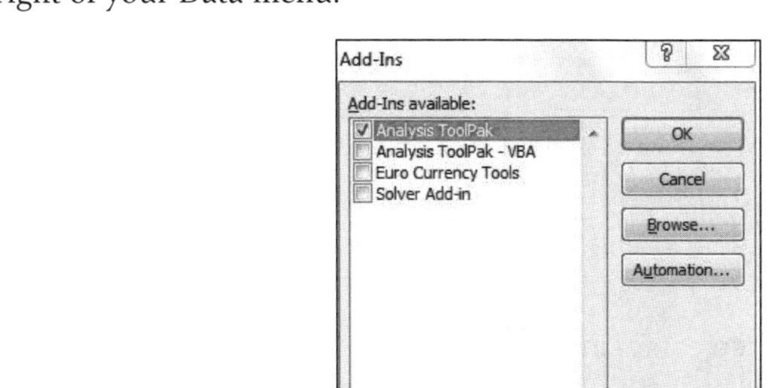

The Analysis ToolPak that allows you to perform many of the analyses is no longer available for recent versions of Excel for the Mac. The program StatPlus:mac LE is available at no cost from AnalystSoft. It may be downloaded from www.analystsoft.com/en/products/statplusmacle. Excellent documentation for Mac uses is available along with this program. For more information about using Excel for statistical analysis please see *Excel Statistics: A Quick Guide*, 2nd ed. (Salkind 2013).

CHAPTER 2: USING TECHNOLOGY IN MEASUREMENT AND EVALUATION

The commands presented are from Excel 2010 (PC version). In the following tasks, you'll work with table 2.1, presented again here, as an introduction to working in Excel.

Sample Database (Data Matrix)

id	gender	age	weightkg	heightcm	stepsperday
1	0	20	50	165	5000
2	0	24	51	160	6000
3	0	21	62	173	7000
4	0	19	59	178	6500
5	0	23	43	145	4500
6	1	22	86	193	4800
7	1	25	65	183	4000
8	1	24	61	178	4200
9	1	28	75	173	3900
10	1	20	70	178	3500
Mean		22.6	62.2	172.6	4940
Stdevs		2.76	12.71	13.29	1183.4

Task 1

Input and verify the data.

Task 2

Compute BMI—steps.

1. Type "BMI" in column G, row 1.
2. Create BMI formula:
   a. Type = in cell G2.
   b. Click on cell D2 and then enter / in cell G2.
   c. Enter (and click on cell E2.
   d. Enter * and click on cell E2 again.
   e. Enter /10000).
   f. Press Return.
3. Use Excel to compute BMI for all 10 people.
   a. Place your cursor in cell G2. Click and hold the little box in the lower right corner of the cell.
   b. Drag the cursor down to cell G11 to provide BMIs for all participants.
   c. Use the Number box on the Home tab to round to two or three decimal places to make the display more aesthetically pleasing.

Task 3

Compute selected descriptive statistics (using the functions menu).

1. Type "Mean" in cell A12.
2. Move your cursor to cell C12.
3. Go to the Formulas menu, click on More Functions and then Statistical, and then find Average.
   a. Click on Average.
   b. Place your cursor in cell C2 and scroll in all of the data for "Age." Note: Excel will read blanks as missing values!
   c. Press Return.
4. To calculate all other means keep your cursor in cell C12.
   a. Place crosshairs over the small rectangle in lower right corner of cell C12.
   b. Drag the small rectangle with your cursor to cell G12—doing this should calculate all means.
   c. Round as you like!
5. To calculate standard deviation, type "Stdev.s" in cell A13.
   a. Move your cursor to cell C13.
   b. Go to the Formulas menu and then More Functions and then Statistical and click on stdev.s in the Statistical Functions menu.
   c. Scroll down to enter the numbers from C2 through C11. Do NOT include C12 as that is the mean. Click OK.
   d. Place your cursor over the small rectangle in the lower right corner of the box and drag the cursor to cell G13 as you did in step 4.b. for the means—this should calculate all standard deviations.
   e. Round to one decimal using the Number box on the Home tab.

6. Note that we did not include gender when calculating the mean and standard deviation because gender is not a continuous variable.

Tasks 4 and 5

Compute maximum (use MAX) and minimum (use MIN) as just explained by using the Statistical Functions menu as well. Place the labels in cells A14 and A15, respectively. Compute in the same manner as for mean (average) or standard deviation (stdev.s). Be sure to include only data from row 2 through row 11 each time.

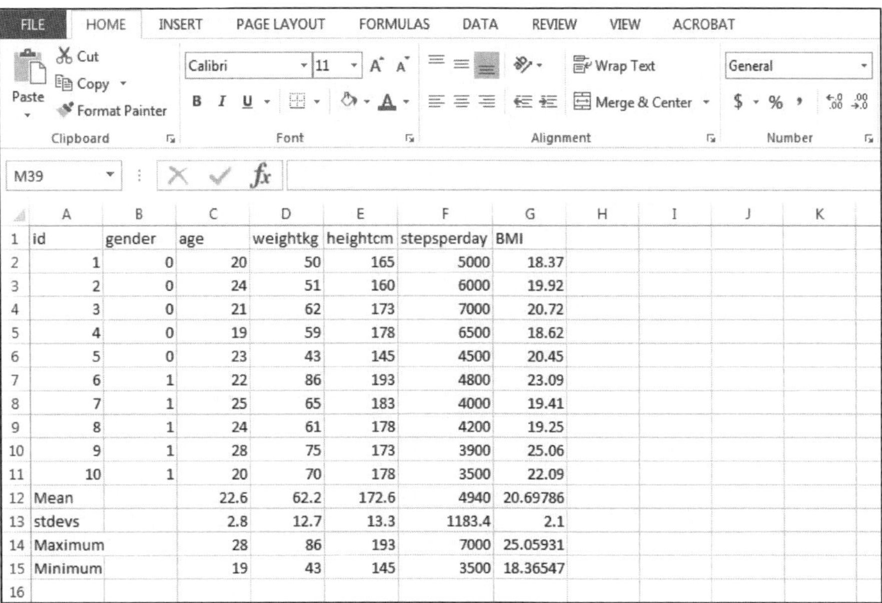

	A	B	C	D	E	F	G
1	id	gender	age	weightkg	heightcm	stepsperday	BMI
2	1	0	20	50	165	5000	18.37
3	2	0	24	51	160	6000	19.92
4	3	0	21	62	173	7000	20.72
5	4	0	19	59	178	6500	18.62
6	5	0	23	43	145	4500	20.45
7	6	1	22	86	193	4800	23.09
8	7	1	25	65	183	4000	19.41
9	8	1	24	61	178	4200	19.25
10	9	1	28	75	173	3900	25.06
11	10	1	20	70	178	3500	22.09
12	Mean		22.6	62.2	172.6	4940	20.69786
13	stdevs		2.8	12.7	13.3	1183.4	2.1
14	Maximum		28	86	193	7000	25.05931
15	Minimum		19	43	145	3500	18.36547
16							

CHAPTER 3: DESCRIPTIVE STATISTICS AND THE NORMAL DISTRIBUTION

In the following steps, you'll use the Data Analysis ToolPak to compute descriptive statistics for the data in table 2.1.

1. Go to the Data menu in Excel and open Data Analysis.
2. You should get the following screen:

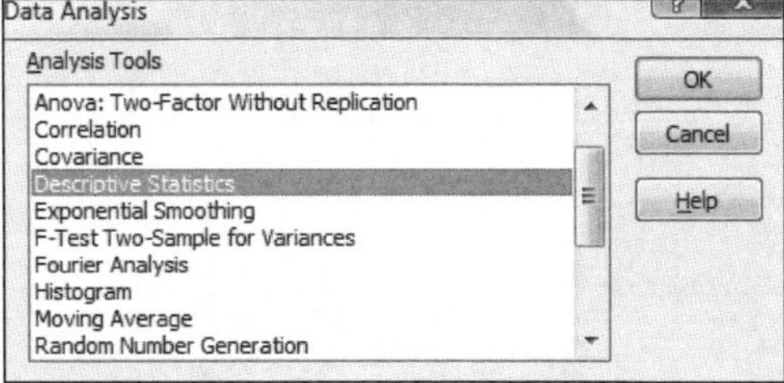

3. Select Descriptive Statistics and follow these instructions:
   a. Scroll in all of the data (ONLY the data and NOT the mean, stdev.s, maximum and minimum), including variable names.
   b. Click on Labels in first row.
   c. Leave Output options selected for New Worksheet Ply.
   d. Click on Summary statistics.

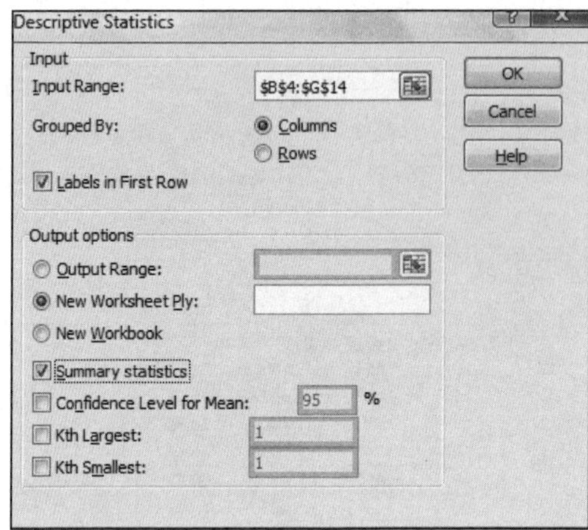

4. Click OK, and your output should look like this:

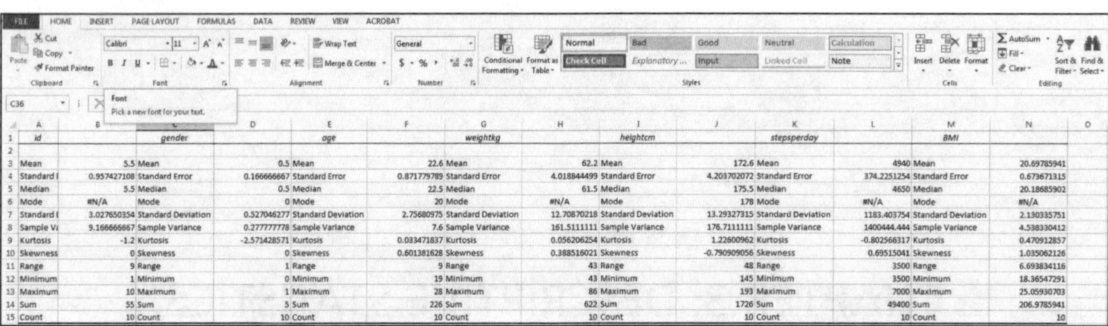

	A	B	C	D	E	F	G	H	I	J	K	L	M	N
1	id		gender		age		weightkg		heightcm		stepsperday		BMI	
2														
3	Mean	5.5	Mean	0.5	Mean	22.6	Mean	62.2	Mean	172.6	Mean	4940	Mean	20.69785941
4	Standard I	0.957427108	Standard Error	0.166666667	Standard Error	0.871779789	Standard Error	4.018844499	Standard Error	4.203702072	Standard Error	374.2251254	Standard Error	0.673671315
5	Median	5.5	Median	0.5	Median	22.5	Median	61.5	Median	175.5	Median	4650	Median	20.18685902
6	Mode	#N/A	Mode	0	Mode	20	Mode	#N/A	Mode	178	Mode	#N/A	Mode	#N/A
7	Standard I	3.027650354	Standard Deviation	0.527046277	Standard Deviation	2.75680975	Standard Deviation	12.70870218	Standard Deviation	13.29327315	Standard Deviation	1183.403754	Standard Deviation	2.130335751
8	Sample Va	9.166666667	Sample Variance	0.277777778	Sample Variance	7.6	Sample Variance	161.5111111	Sample Variance	176.7111111	Sample Variance	1400444.444	Sample Variance	4.538330412
9	Kurtosis	-1.2	Kurtosis	-2.571428571	Kurtosis	0.033471837	Kurtosis	0.056206254	Kurtosis	1.22600962	Kurtosis	-0.802566317	Kurtosis	0.470912857
10	Skewness	0	Skewness	0	Skewness	0.601381628	Skewness	0.388516021	Skewness	-0.790909056	Skewness	0.69515041	Skewness	1.035062126
11	Range	9	Range	1	Range	9	Range	43	Range	48	Range	3500	Range	6.693834116
12	Minimum	1	Minimum	0	Minimum	19	Minimum	43	Minimum	145	Minimum	3500	Minimum	18.36547291
13	Maximum	10	Maximum	1	Maximum	28	Maximum	86	Maximum	193	Maximum	7000	Maximum	25.05930703
14	Sum	55	Sum	5	Sum	226	Sum	622	Sum	1726	Sum	49400	Sum	206.9785941
15	Count	10	Count	10	Count	10	Count	10	Count	10	Count	10	Count	10

5. Clean up the output by removing the redundant columns, rounding entries, and so on.
6. The cleaned-up output should look like this:

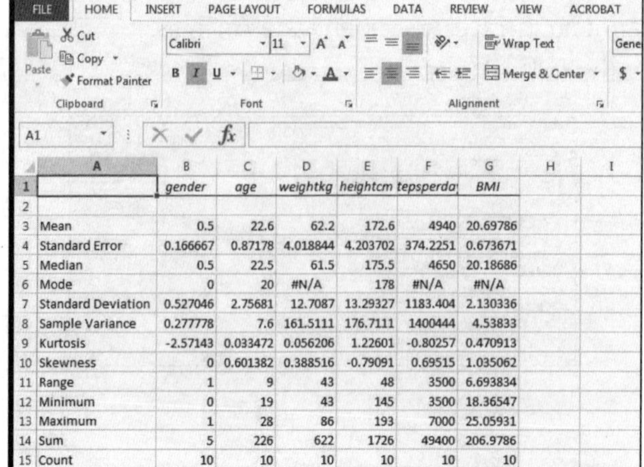

	A	B	C	D	E	F	G
1		gender	age	weightkg	heightcm	tepsperda	BMI
2							
3	Mean	0.5	22.6	62.2	172.6	4940	20.69786
4	Standard Error	0.166667	0.87178	4.018844	4.203702	374.2251	0.673671
5	Median	0.5	22.5	61.5	175.5	4650	20.18686
6	Mode	0	20	#N/A	178	#N/A	#N/A
7	Standard Deviation	0.527046	2.75681	12.7087	13.29327	1183.404	2.130336
8	Sample Variance	0.277778	7.6	161.5111	176.7111	1400444	4.53833
9	Kurtosis	-2.57143	0.033472	0.056206	1.22601	-0.80257	0.470913
10	Skewness	0	0.601382	0.388516	-0.79091	0.69515	1.035062
11	Range	1	9	43	48	3500	6.693834
12	Minimum	0	19	43	145	3500	18.36547
13	Maximum	1	28	86	193	7000	25.05931
14	Sum	5	226	622	1726	49400	206.9786
15	Count	10	10	10	10	10	10

To create a histogram for the data in table 3.1, use the following steps:

1. Go to the Home menu and perform a descending sort (the commands will vary based on the version of Excel and the type of computer you are using, but you want to perform a descending sort). An example is shown in the next screen capture.

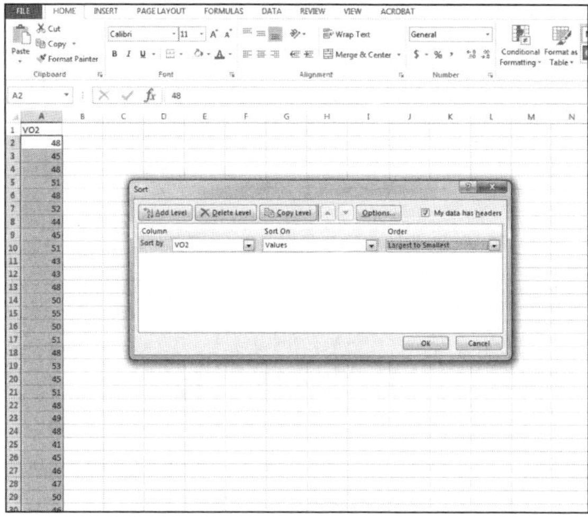

2. Scroll in all of your data.
3. Open the Sort function and perform a descending sort on VO2.
4. Be sure to click Descending and Header rows.
5. Click OK to sort your scores.
6. To create a histogram, you now have to create what Excel calls Bins.
7. The Bins are used to set up the frequency distribution for the histogram.
   a. Type the word "Bins" in the cell next to the label VO2 for your sorted data (cell B1).
   b. Type a column of numbers in the Bins column that corresponds to the range of the scores from highest to lowest (in this case, from 55 to 41).
   c. Now open the histogram program under Data and the Data Analysis. Scroll in your labels and scores in the Input Range and then scroll your Bins in where it asks for Bin Range.
   d. Be sure to click Labels and Chart Output.
   e. Use the New Worksheet Ply option for output (the next screen shows how it should look to this point).
   f. Click OK.

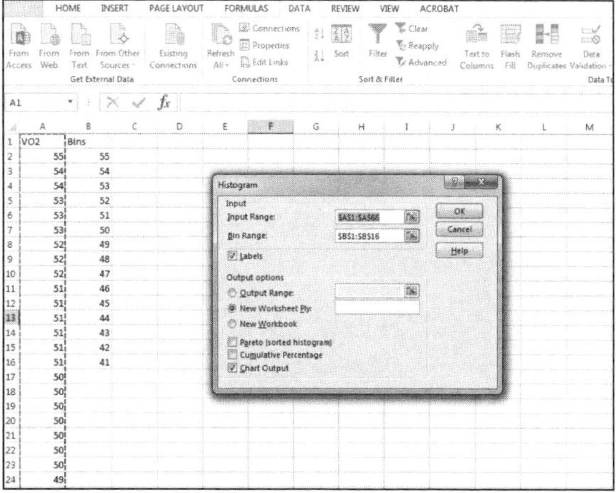

8. To clean up your output, click on the histogram chart (which usually comes out too flat).
   a. A box will appear around it.
   b. Place your cursor over the lower-right corner and drag the corner down and to the right to create an aesthetically pleasing graph.

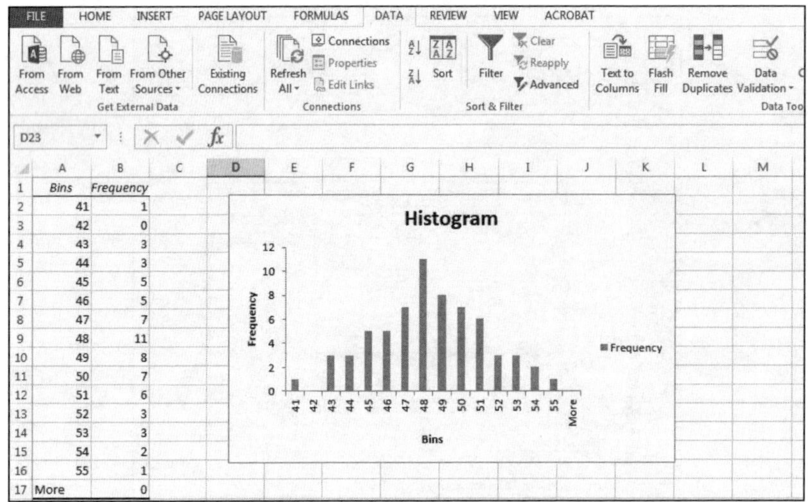

Bins	Frequency
41	1
42	0
43	3
44	3
45	5
46	5
47	7
48	11
49	8
50	7
51	6
52	3
53	3
54	2
55	1
More	0

To calculate percentiles with Excel, use the Rank and Percentile program. Be on the Excel sheet where your data are located.

1. Select the Rank and Percentile program from the Data Analysis menu.
2. Input your data by placing your cursor over the variable name (VO2) and scroll in all of your data.
3. Be sure to click on Labels in the first-row button.

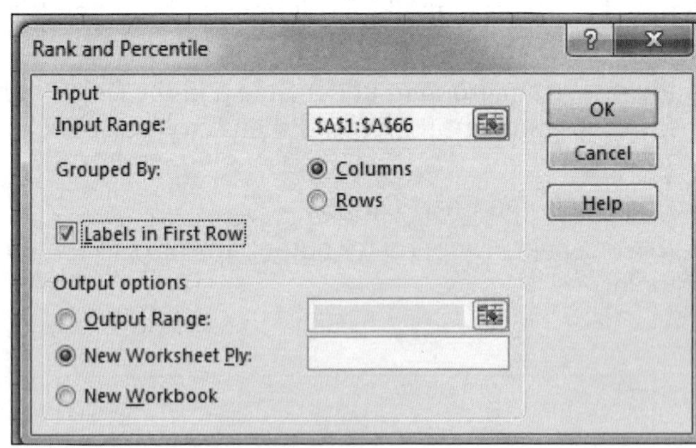

4. Output your data to a New Worksheet Ply.
5. Click OK, and your output should look like the output shown here. Note that the ranks use the highest rank for tied ranks and the percent column is interpreted as percentiles, similar to SPSS.

	A	B	C	D
1	*Point*	*VO2*	*Rank*	*Percent*
2	14	55	1	100.00%
3	62	54	2	96.80%
4	64	54	2	96.80%
5	18	53	4	92.10%
6	35	53	4	92.10%
7	61	53	4	92.10%
8	6	52	7	87.50%
9	55	52	7	87.50%
10	58	52	7	87.50%
11	4	51	10	78.10%
12	9	51	10	78.10%
13	16	51	10	78.10%
14	20	51	10	78.10%
15	45	51	10	78.10%
16	56	51	10	78.10%
17	13	50	16	67.10%
18	15	50	16	67.10%
19	28	50	16	67.10%
20	54	50	16	67.10%
21	60	50	16	67.10%
22	63	50	16	67.10%
23	65	50	16	67.10%
24	22	49	23	54.60%
25	30	49	23	54.60%
26	33	49	23	54.60%
27	38	49	23	54.60%
28	48	49	23	54.60%
29	50	49	23	54.60%
30	57	49	23	54.60%
31	59	49	23	54.60%
32	1	48	31	37.50%
33	3	48	31	37.50%
34	5	48	31	37.50%
35	12	48	31	37.50%
36	17	48	31	37.50%
37	21	48	31	37.50%
38	23	48	31	37.50%
39	31	48	31	37.50%
40	32	48	31	37.50%
41	40	48	31	37.50%
42	52	48	31	37.50%
43	27	47	42	26.50%
44	36	47	42	26.50%
45	39	47	42	26.50%
46	41	47	42	26.50%
47	42	47	42	26.50%
48	44	47	42	26.50%
49	53	47	42	26.50%
50	26	46	49	18.70%

CHAPTER 4: CORRELATION AND PREDICTION

Excel allows you to calculate pairwise correlations or an entire correlation matrix. To calculate the individual correlations (pairwise), use the Formulas menu. Go to More Functions and Statistical to get to Correl. It does not matter which variable you put in-—array 1 or array 2. Be *certain* to remember what you did and label it correctly. The correlation will appear in the cell where you have placed your cursor.

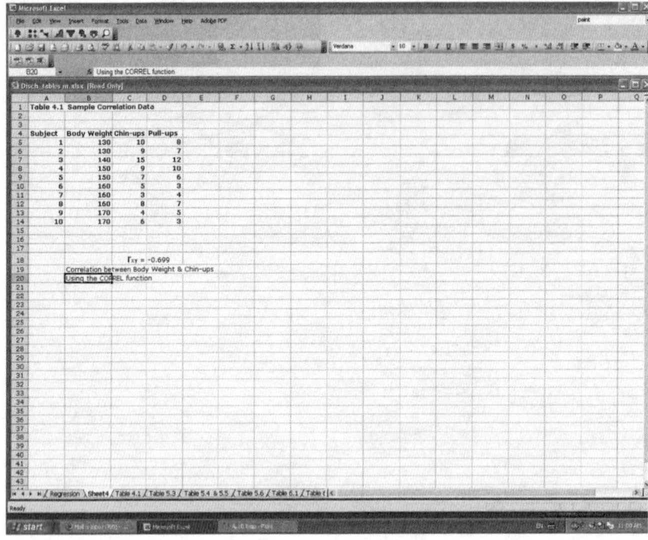

As an example, the correlation between body weight and chin-ups in table 4.1 is shown in the next figure. (Remember: No labels, and always round correlations to three places with the formatting palette.) For practice, calculate the correlations for body weight and pull-ups and for chin-ups and pull-ups.

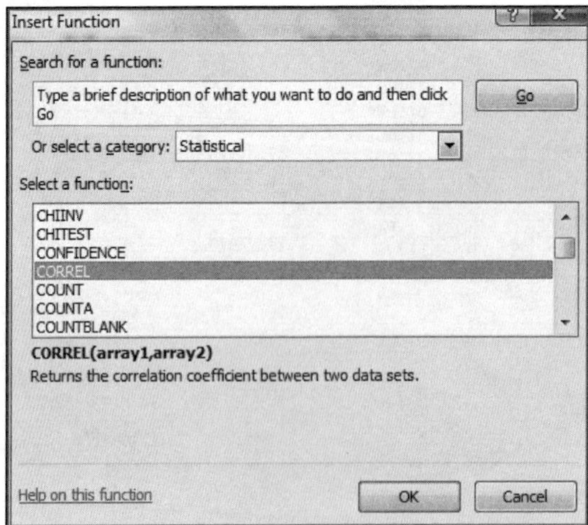

Using Data Analysis, you can create an entire correlation matrix for the three variables.

1. Select the Correlation program.
2. Scroll in all of the data, including variable names.
3. Click on Labels.
4. Select Output to New Worksheet Ply. The screen should look like this:

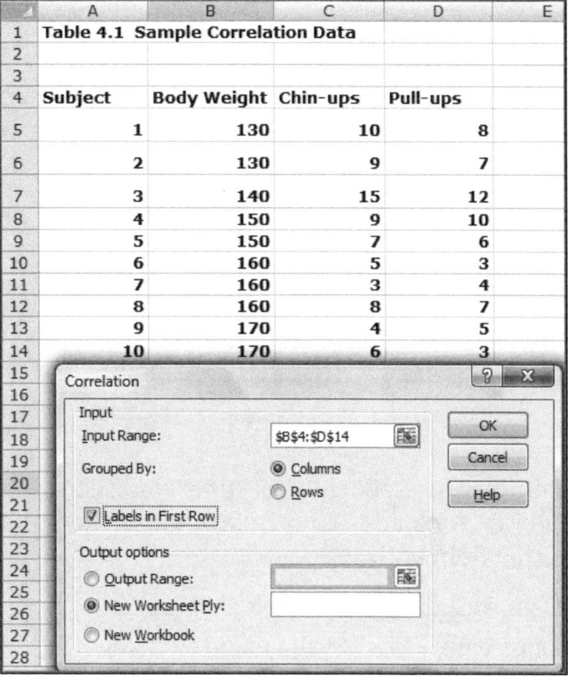

Table 4.1 Sample Correlation Data

Subject	Body Weight	Chin-ups	Pull-ups
1	130	10	8
2	130	9	7
3	140	15	12
4	150	9	10
5	150	7	6
6	160	5	3
7	160	3	4
8	160	8	7
9	170	4	5
10	170	6	3

5. Click OK.

You will get the following output:

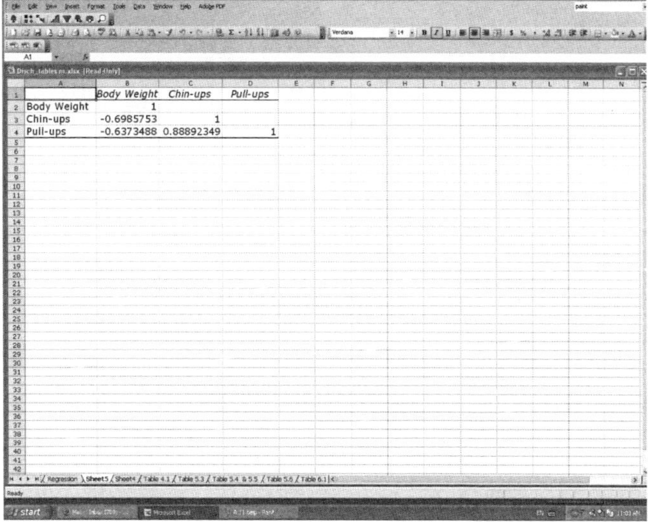

	Body Weight	Chin-ups	Pull-ups
Body Weight	1		
Chin-ups	-0.6985753	1	
Pull-ups	-0.6373488	0.88892349	1

6. To clean up, scroll in all of your output values and reduce the number of decimals to three with the formatting palette. The final matrix should look like this:

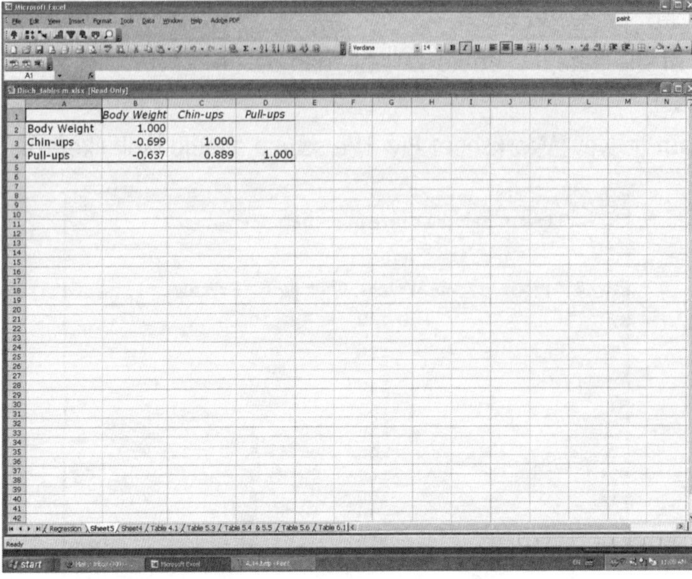

Regression

The same data in table 4.1 may be used to demonstrate prediction (called regression in Excel). Single-variable prediction is computed using Data Analysis: Regression. For this example, calculate pull-ups predicted from body weight.

1. Scroll in Pull-Up Data, including the name for Y Input.
2. Scroll in Body-Weight Data, including the name for X Input.
3. Be sure to click Labels.
4. Leave Output option at New Worksheet Ply.

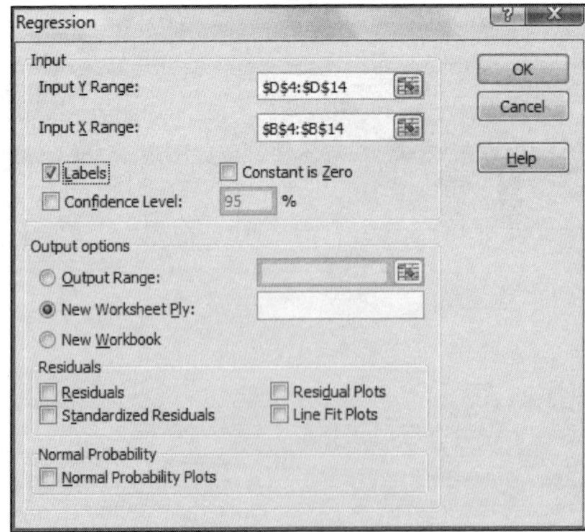

5. Click OK.
6. On Summary Output, reduce the number of decimal places to three with the formatting palette.

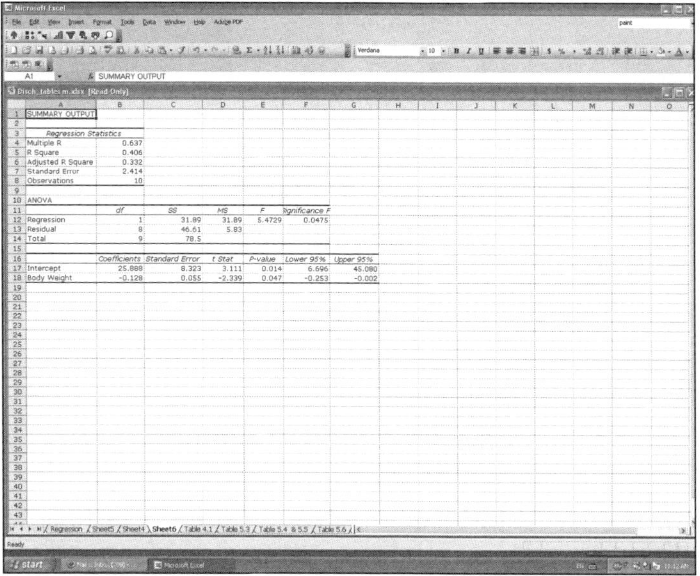

CHAPTER 5: INFERENTIAL STATISTICS

Excel will calculate all of the statistics included in chapter 5. Some of the options available in SPSS are not available, however. Also, Excel has some other limitations associated with it. They will be explained as they are presented.

Table 5.3

The calculation of chi-square (χ^2) is quite cumbersome in Excel. We have provided you with a template for calculating χ^2 in the WSG associated with your textbook. You simply enter values into the cells provided, and χ^2 and associated statistics are calculated for you. The screen capture for χ^2, Kappa, Phi, and the proportion of agreement is presented here:

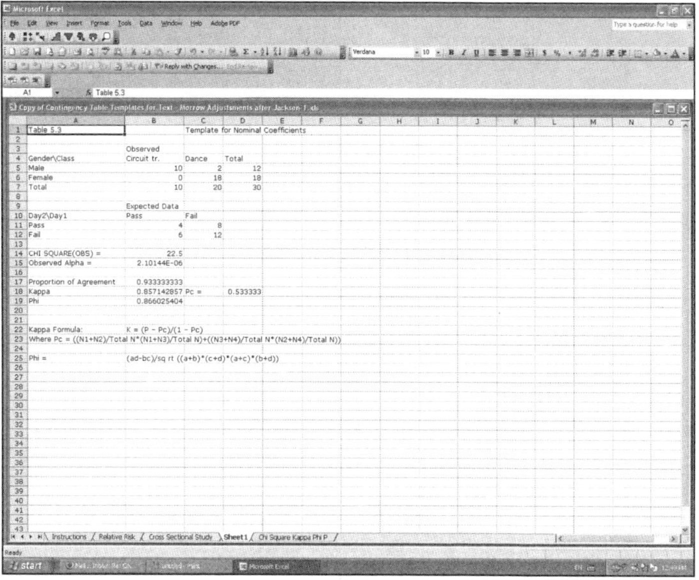

Table 5.4

1. From the Data Analysis programs, select t-test: Two-Sample Assuming Equal Variance.

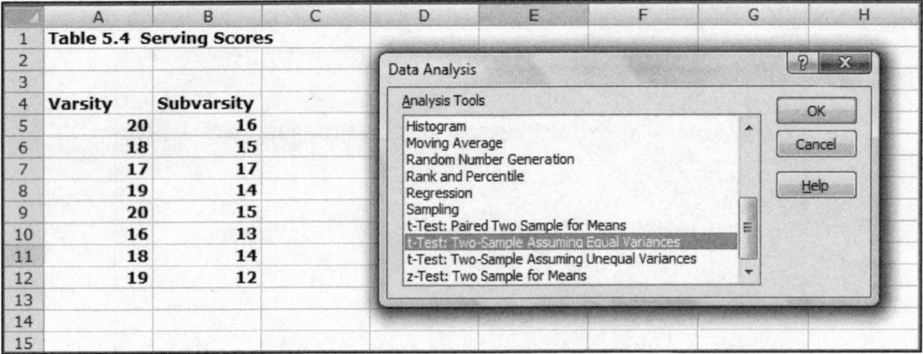

Table 5.4 Serving Scores	
Varsity	Subvarsity
20	16
18	15
17	17
19	14
20	15
16	13
18	14
19	12

2. Scroll Varsity Data in for Variable 1.
3. Scroll Subvarsity Data in for Variable 2.
4. Select Output to New Worksheet Ply.

Note: The direction of t-test will be based on which variable is entered first.

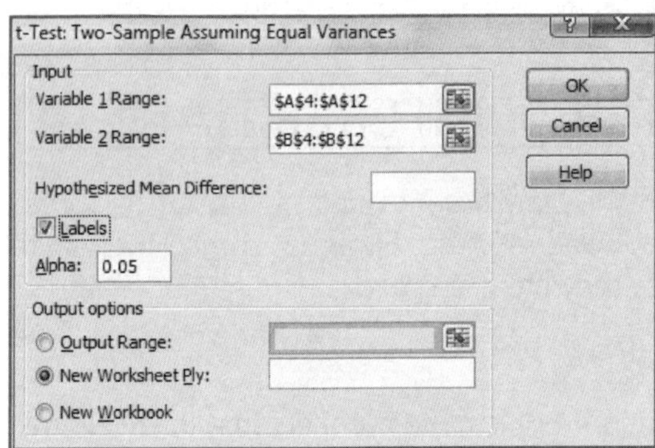

5. Click OK.

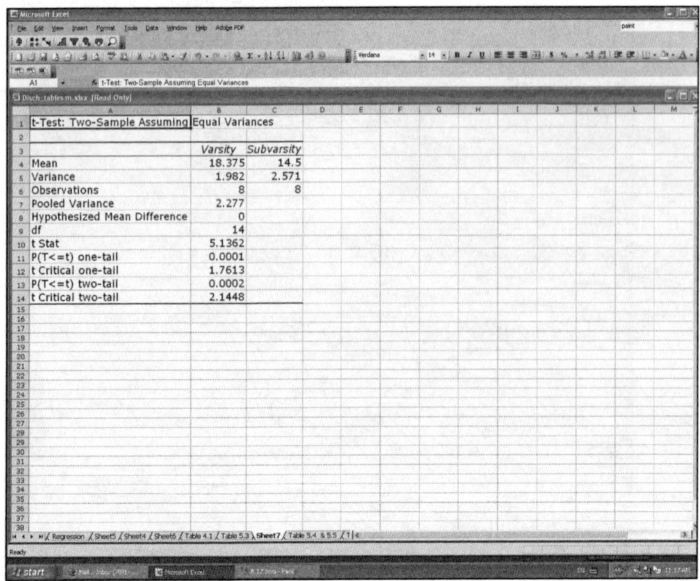

t-Test: Two-Sample Assuming Equal Variances		
	Varsity	Subvarsity
Mean	18.375	14.5
Variance	1.982	2.571
Observations	8	8
Pooled Variance	2.277	
Hypothesized Mean Difference	0	
df	14	
t Stat	5.1362	
P(T<=t) one-tail	0.0001	
t Critical one-tail	1.7613	
P(T<=t) two-tail	0.0002	
t Critical two-tail	2.1448	

Table 5.5

1. From the Data Analysis programs, select t-test: Paired Two-Sample for Means.

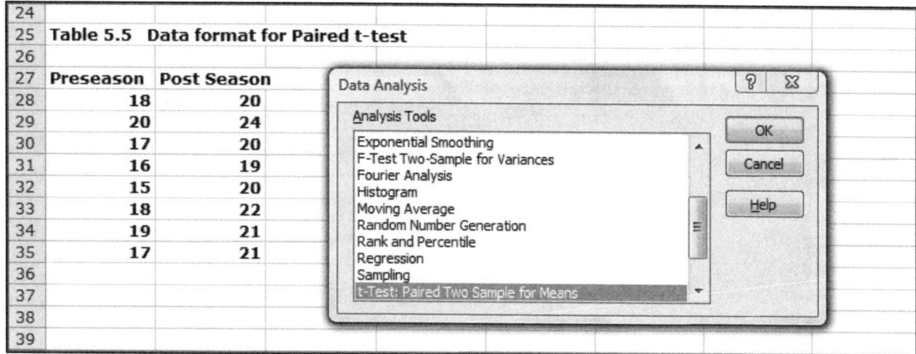

2. Scroll Postseason Data in for Variable 1.
3. Scroll Preseason Data in for Variable 2.
4. Click on Labels.
5. Select Output to New Worksheet Ply.

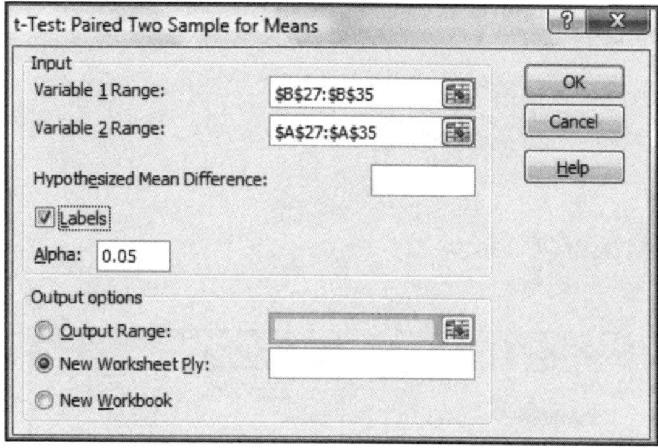

6. Click OK.

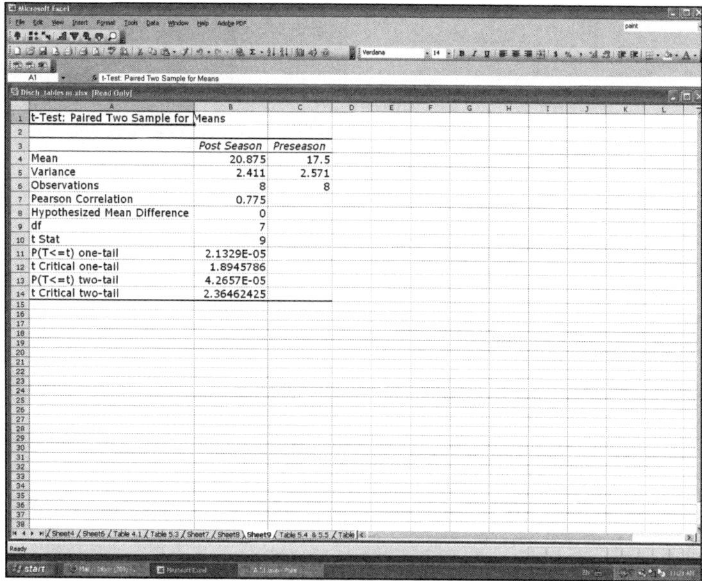

t-Test: Paired Two Sample for Means		
	Post Season	Preseason
Mean	20.875	17.5
Variance	2.411	2.571
Observations	8	8
Pearson Correlation	0.775	
Hypothesized Mean Difference	0	
df	7	
t Stat	9	
P(T<=t) one-tail	2.1329E-05	
t Critical one-tail	1.8945786	
P(T<=t) two-tail	4.2657E-05	
t Critical two-tail	2.36462425	

Table 5.6

To use Excel for one-way ANOVA, you need to have equal *N* and you need to input data as shown here. Note the data are formatted differently than with SPSS.

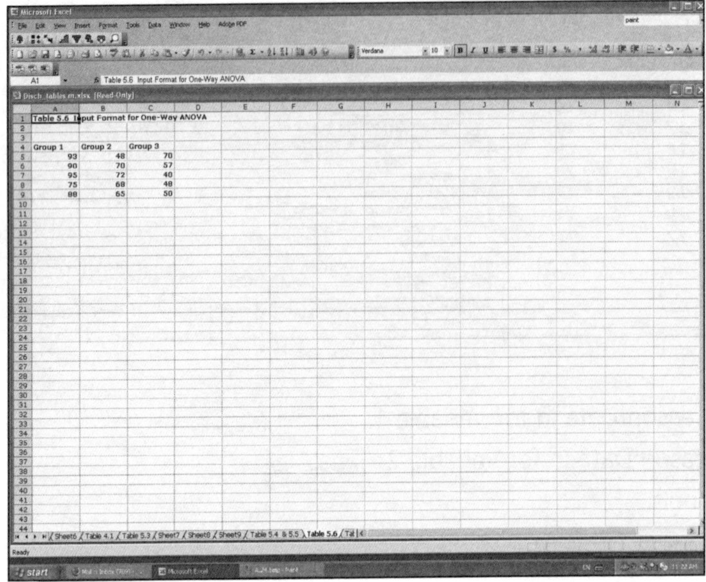

1. From the Data Analysis programs, select ANOVA Single Factor.
2. Scroll in all of the data, including group labels.
3. Click Labels.
4. Select Output to New Worksheet Ply.

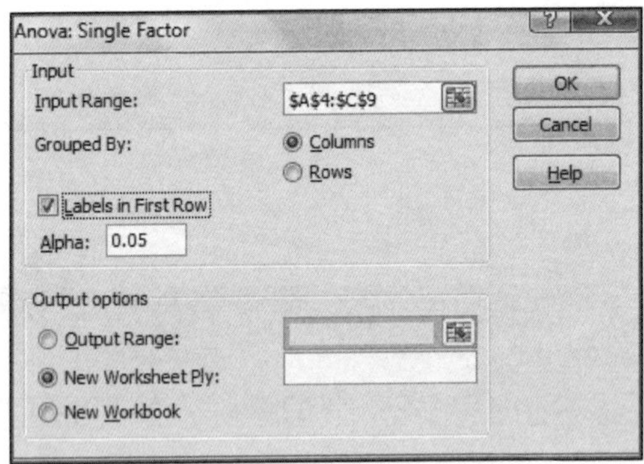

5. Click OK.

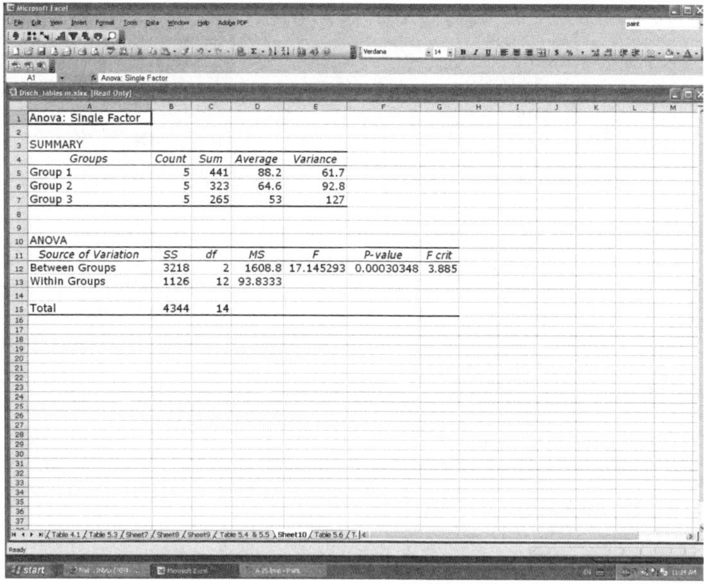

Table 5.7

To use Excel for two-way ANOVA, you need to have equal *N* and input data as shown here. Note the data are formatted differently than with SPSS.

Sample Database

	Pitchers	Position
Scrimmages	29	34
	30	34
	24	31
	23	30
	28	32
	29	28
Drill	29	28
	34	34
	34	31
	38	26
	28	27
	29	32

1. From the Data Analysis programs, select ANOVA: Two-Factor With Replication

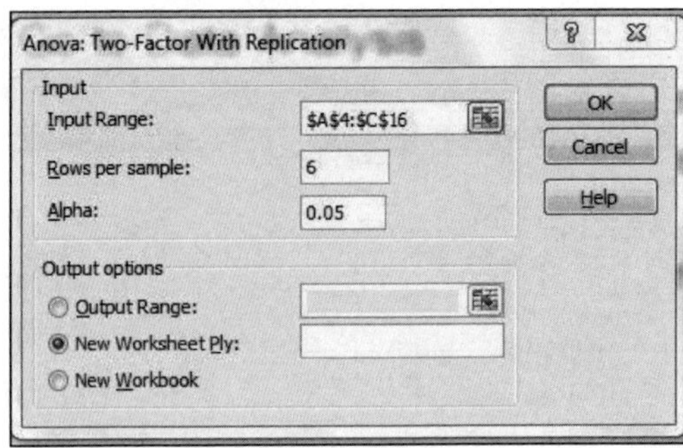

2. Scroll in all of the data, including group labels.
3. Identify rows per sample. In our example, this is 6 (there are 6 people in the scrimmage group and 6 people in the drills group). Note: Each cell must have equal *n*.
4. Designate output location.
5. Click OK.

Anova: Two-Factor With Replication

SUMMARY	Pitchers	Position	Total
Scrimmages			
Count	6	6	12
Sum	163	189	352
Average	27.17	31.50	29.33
Variance	8.57	5.50	11.52
Drills			
Count	6	6	12
Sum	192	178	370
Average	32.00	29.67	30.83
Variance	15.60	9.87	13.06
Total			
Count	12	12	
Sum	355	367	
Average	29.58	30.58	
Variance	17.36	7.90	

ANOVA

Source of Variation	*SS*	*df*	*MS*	*F*	*P-value*	*F crit*
Sample (Method)	13.5	1	13.5	1.366	0.256	4.351
Columns (P/Pos)	6	1	6	0.607	0.445	4.351
Interaction	66.67	1	66.67	6.745	0.017	4.351
Within	197.67	20	9.88			
Total	283.83	23				

Note: In the Excel ANOVA Summary table the sources of variation will simply be labeled Sample and Columns. You will need to go in and type the labels for the independent variables associated with each source of variation.

CHAPTER 6: NORM-REFERENCED RELIABILITY AND VALIDITY

There is no program in Excel to compute the alpha coefficient directly. It can be computed using the Functions menu and the formula. The steps presented here use table 6.6 as an example.

As an alternative, you could create a total score and then calculate all of the variances associated with the trials and the total; you would then enter these into the alpha coefficient formula and complete the calculations by hand.

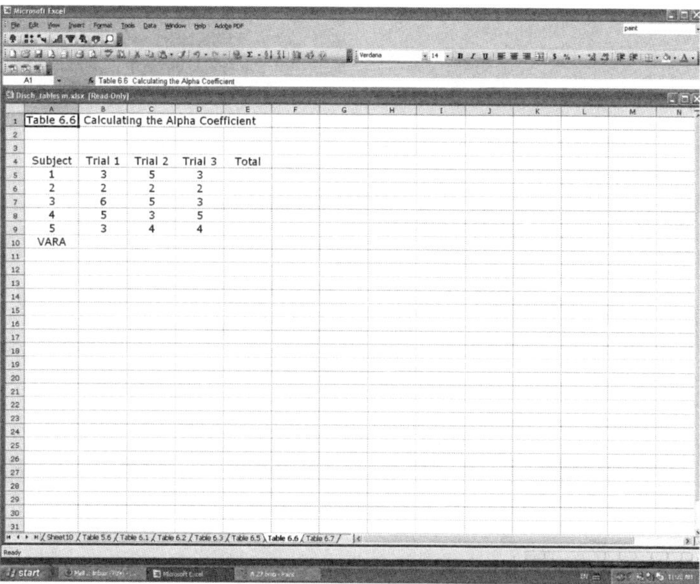

Table 6.6 Calculating the Alpha Coefficient

Subject	Trial 1	Trial 2	Trial 3	Total
1	3	5	3	
2	2	2	2	
3	6	5	3	
4	5	3	5	
5	3	4	4	
VARA				

1. First, compute a total score for all of the trials for each participant.
2. Use Excel's Sum function to get a total score for each person. The sum for person 1 is 11. Note: The sum will include that person's ID (subject) by default. You will need to scroll in the data for the trials only.
3. Fill in a Total Score for each participant.

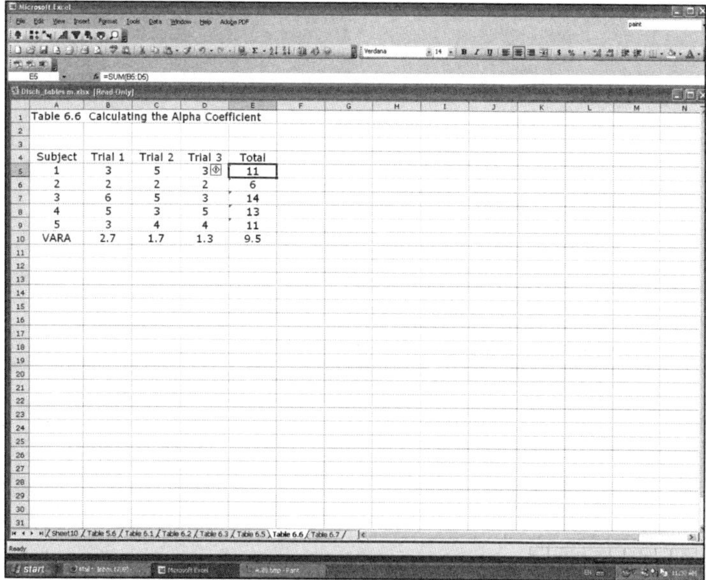

Table 6.6 Calculating the Alpha Coefficient

Subject	Trial 1	Trial 2	Trial 3	Total
1	3	5	3	11
2	2	2	2	6
3	6	5	3	14
4	5	3	5	13
5	3	4	4	11
VARA	2.7	1.7	1.3	9.5

4. Then position your cursor in the cell below the first trial,
5. Go to Formulas and then More Functions and then Statistical and find VARA.
6. Since you have no missing data, Excel will compute the sample variance for the five participants.
7. Now place your cursor in the lower-right corner of the variance for trial 1 and drag the cursor across—that will produce variances for each trial as well as the total variance.

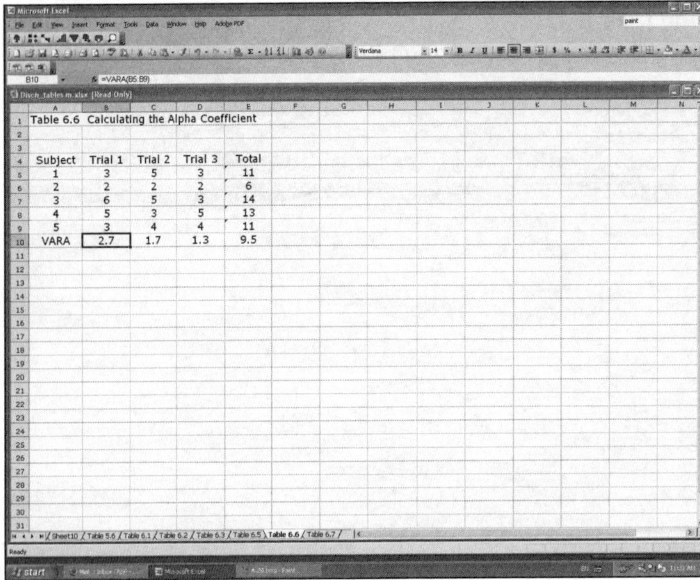

8. Once you have the variances for the trials and total, you can easily calculate alpha by hand. Notice we have included the alpha coefficient formula in cell B14.

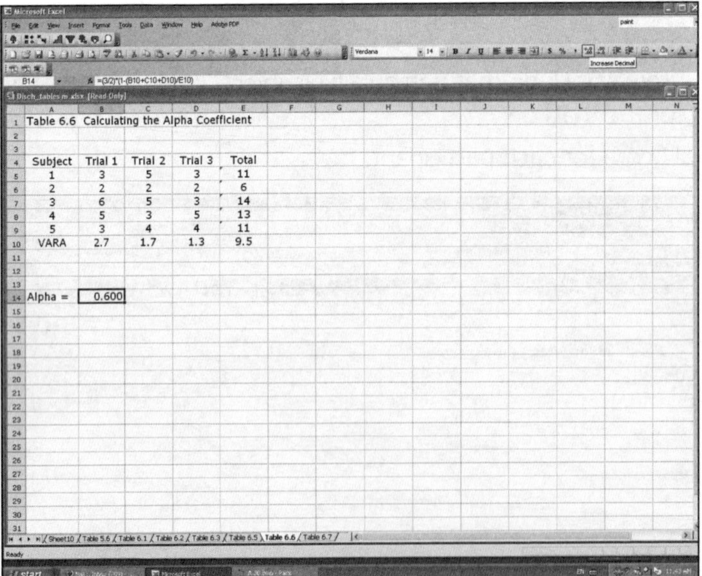

CHAPTER 7: CRITERION-REFERENCED RELIABILITY AND VALIDITY

The calculated statistics associated with criterion-referenced reliability and validity are based on the chi-square (χ^2) examples presented in chapter 5. Calculation of these statistics is quite cumbersome in Excel. We have provided you with a template for calculating χ^2 in the WSG for chapter 7. You simply enter values into the cells provided, and χ^2 and associated statistics are calculated for you. The following screen captures illustrate the templates. (The Excel template in the WSG calculates these statistics for you.)

Chi-square, Kappa, Phi, and the Proportion of Agreement

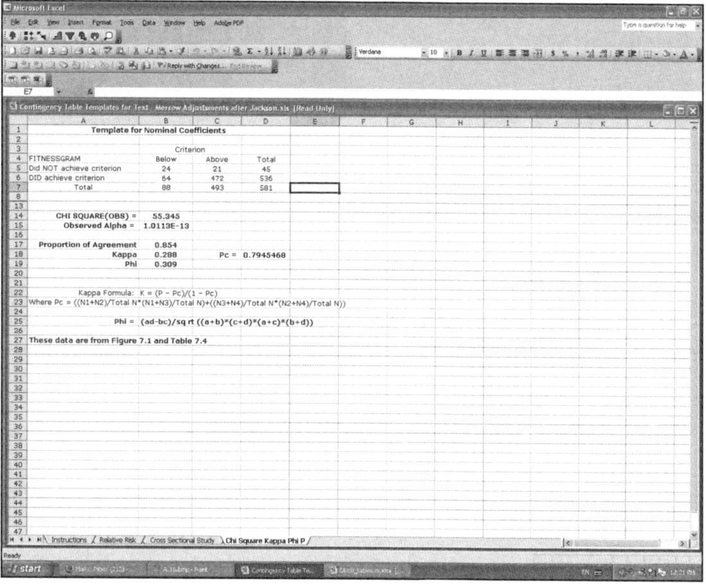

Relative Risk

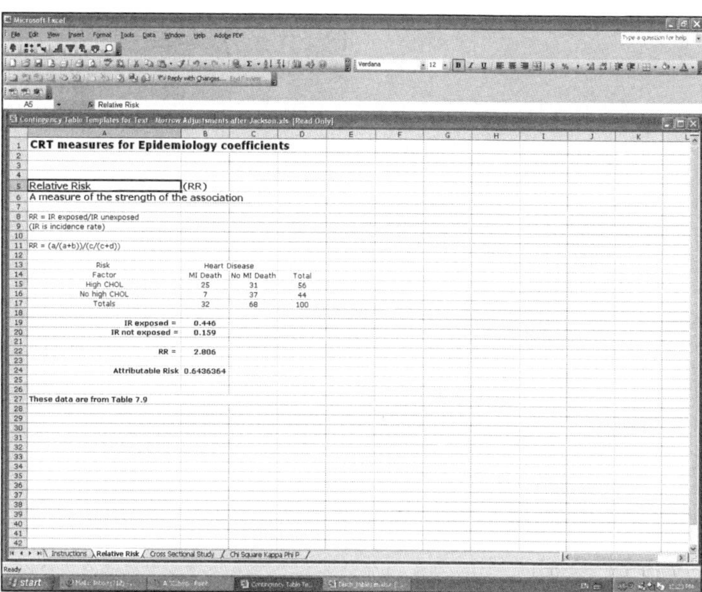

Cross-Sectional Study

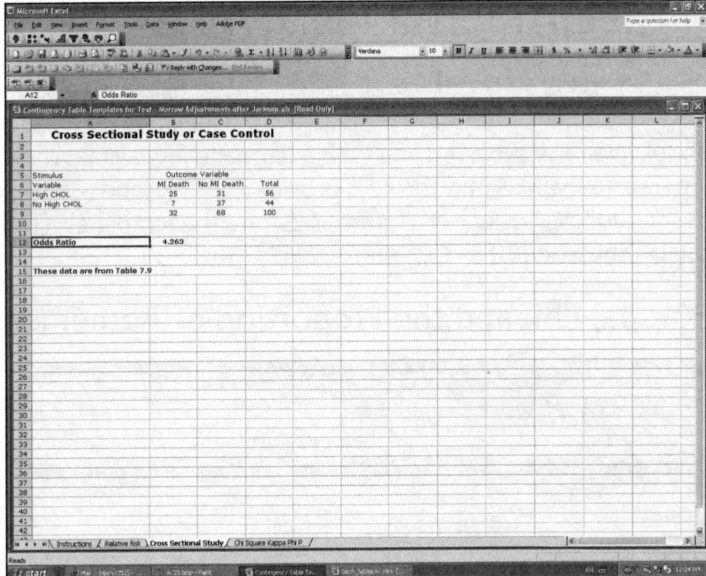

CHAPTERS 8 THROUGH 14

The data, examples, and measurement and evaluation analyses presented in chapters 8 through 14 use the procedures that you learned in chapters 2 through 7. When you are analyzing the large datasets or working through problems in the remaining chapters, simply return to the analytical procedures learned earlier and use the appropriate SPSS or Excel procedure.

Glossary

ability—A general, innate psychomotor trait.

absolute endurance—A measure of repetitive performance against a fixed resistance (e.g., number of repetitions of a 100-pound [45.4 kg] bench press).

absolute rating—Evaluating performance on a fixed scale.

accuracy-based skills test—A skills test that assesses one's ability to serve or project an object (e.g., ball, shuttlecock) into a prescribed area or for distance and accuracy.

achievement test—In the cognitive domain, a test designed to measure the extent to which an examinee comprehends some body of knowledge.

adapted physical education—Physical education adjusted to accommodate children and youth with physical or mental disabilities.

aerobic power—The body's ability to supply oxygen to the working muscles during physical activity. Often called $\dot{V}O_2$max.

affective domain—Involves attitudes, perceptions, and psychological characteristics.

alpha level—The probability of falsely rejecting a null hypothesis (i.e., claiming that there is a relation between the variables when, in fact, there is not). Also called the significance level or the probability of making a type I error.

alpha reliability—See *intraclass reliability*.

alternative assessment—An assessment technique that is different from traditional, standardized testing. Also called *authentic assessment*.

analytic rubrics—A way to score an assessment that lists the characteristics or traits important to the completion or performance of the task and then provides the opportunity for a rater to evaluate the degree to which these characteristics or traits were met.

analytic scoring—A method of scoring responses to essay questions that involves identifying specific facts, points, or ideas in the answer and awarding credit for each one located.

anxiety—Distress and tension caused by apprehension.

assessment—The process of collecting information and judging what it means.

athletic fitness or motor fitness—Physical fitness related to sport performance (e.g., 50-yard [45.7 m] dash, speed, agility).

authentic assessment—Assessment designed to take place in a real-life setting that provides authenticity and contextualized meaning.

behavioral objectives—Goals with specific measurable steps for their achievement.

body composition—The physical makeup of the body, including weight, lean weight, and fat weight.

cardiorespiratory endurance—The body's ability to extract and use oxygen in a manner that permits continuous exercise, physical work, or physical activities (e.g., jogging).

central tendency—Statistics that are located near the center of a set of scores.

central-tendency error—The type of rating scale error that results from the hesitancy of raters to assign extreme ratings.

checklist—A typically dichotomous rating of a trait.

coefficient of determination—A measure of variation in common between two variables. Interpreted as a percent, it is the square of the correlation (r^2).

cognitive domain—Involves knowledge and mental achievement.

composite score—A total score developed from the scores of a set of separate tests or performances.

concurrent validity—The relationship between a test (a surrogate measure) and a criterion when the two measures are taken relatively close together in time. It is based on the Pearson product-moment (PPM) correlation coefficient between the test and criterion.

construct-related validity—The highest form of validity; it combines both logical and statistical evidence of validity through the gathering of a variety of statistical information that, when viewed collectively, adds evidence for the existence of the theoretical construct being measured.

content objectives—The specific course objectives determined by the instructor.

content-related validity—Evidence of truthfulness based on logical decision making and interpretation. Also called *face validity* or *logical validity*.

contingency table—A table used for cross-referencing two nominal variables.

correlation—A measure of the relationship between two variables.

correlation coefficient (r)—An index of the linear relationship between two variables, indicating the magnitude, or amount of relationship, and the direction of the relationship.

criterion-referenced standard—A specific, predetermined level of achievement.

criterion-referenced test (CRT)—A test with specific, predetermined performance or health standards.

criterion-related validity—Evidence that a test has a statistical relationship with the trait being measured; also called *statistical validity* and *correlational validity.*

curvilinear relationship—An association among the variables that can best be depicted by a curve.

cutoff scores—Scores that establish identifiable groups or levels of performance; typically used with criterion-referenced testing.

dependent variable—The variable used as a criterion or that you are trying to predict *(Y). Also called outcome variable.*

depression—A mental condition of gloom, sadness, or dejection.

descriptive statistics—Mathematics used to organize, summarize, and describe data.

developmental rubric—A system used to judge performance across all levels from that of a beginner to that of an expert.

direct relationship—A positive relationship between two variables such that higher scores on one variable are associated with higher scores on the other variable. Similarly, lower scores on one variable are associated with lower scores on the other variable.

distance or power performance test—A skills test that assesses one's ability to project an object for maximum displacement or force.

distractor—An incorrect response to a multiple-choice test item.

educational objectives—General educational goals defined by various pedagogical experts.

epidemiology—The study of the incidence, distribution, and frequency of diseases (e.g., the study of the effects of physical inactivity on coronary heart disease).

error score—An unobservable but theoretically existent score that contributes to an inaccurate estimate of individual differences.

evaluation—A dynamic decision-making process that places a value judgment on the quality of what has been measured (e.g., a test score, physical performance).

exhibition—A public presentation or performance in which a person displays his or her knowledge and/or skills.

extrinsic ambiguity—The characteristic of a test item that appears ambiguous to a respondent who does not understand the concept that the item is testing.

factor analysis—A statistical method using correlational techniques to describe the underlying unobserved dimensions (called factors) from measured variables.

fairness—A characteristic of an assessment, meaning the absence of bias, that allows all participants an equal opportunity to perform to the best of their capabilities.

flexibility—The range of motion of a joint or group of joints.

foil—An incorrect response to a multiple-choice test item. Also called distractor.

formative assessment—Assessment that takes place during instruction and provides feedback to teachers and students so that students can make adjustments to improve their performance and enhance learning.

formative evaluation—An evaluation conducted during (as opposed to at the end of) an instructional or training program.

frequency distribution—A list of the observed scores and their frequency of occurrence.

functional capacity—The ability to perform the normal activities of daily living.

general motor ability (GMA)—The overall proficiency to perform a variety of psychomotor tasks.

generalized rubric—An evaluation system that can be used for multiple assessments that typically are related conceptually or by content.

global rating—Evaluating overall performance rather than assessing individual components of the action.

global scoring—A method of assessing an answer to an essay question that involves converting the general impression into a score obtained from reading the answer.

halo effect—The tendency to elevate a person's score because of positive bias. May also work in reverse (reduce a score because of negative bias).

health-related physical fitness—Attributes of fitness related to health outcomes; cardiorespiratory, body composition, and muscular fitness.

histogram—A graph using vertical bars to present the frequency distribution of the observed scores.

holistic assessment—Analysis based on the overall quality of a performance or product.

holistic rubric—A way to evaluate an assessment that provides paragraph descriptions of the desired levels of performance for the various characteristics or traits considered important. Typically used with summative assessments.

hypothesis—A statement of relation between at least two variables.

independent variable—The variable related to a dependent variable. Often used as a predictor *(X).*

index of difficulty—A mathematical expression used in item analysis to estimate the percentage of examinees answering a test item correctly.

index of discrimination—A mathematical expression used in item analysis to estimate how well a test item discriminates among examinees who have been categorized by some criterion (typically the total score on the examination).

index of reliability—The theoretical correlation between observed score and true score; the square root of the reliability coefficient.

indirect relationship—See *inverse relationship*.

inferential statistics—Statistics used to test a hypothesis within a small group (sample) to infer to a larger group (population).

interclass reliability—A reliability coefficient computed with the Pearson product-moment correlation coefficient.

Internet—A network of computers for high-speed transmission of information.

intraclass reliability—A type of internal consistency reliability, based on ANOVA, where there may be unlimited trials. Alpha, KR_{20}, and KR_{21} are intraclass reliability estimates.

intrinsic ambiguity—The characteristic of a test item that is truly ambiguous, even to the respondent who understands the concept that the item is testing.

inverse relationship—A negative relationship between two variables such that higher scores on one variable are associated with lower scores on the other variable. Similarly, lower scores on one variable are associated with higher scores on the other variable.

item analysis—A prescribed process used to examine the quality (e.g., the difficulty and the discrimination) of individual items on a written test.

kappa (K)—A measure of agreement or association between categorical variables that is adjusted for chance.

keyed response—The correct response to a multiple-choice test item.

kurtosis—An indication of the shape of a distribution that specifies how flat or peaked the distribution is.

linear relationship—An association between two variables that can best be depicted by a straight line.

marginal—The sum of observations across a specific row or column of a contingency table.

mastery test—In the cognitive domain, a test designed to measure whether an examinee has achieved enough knowledge to satisfy a prescribed standard or criterion. Typically used with criterion-referenced testing.

maximal exercise test—An aerobic fitness test that requires the subject to exercise until reaching exhaustion (e.g., treadmill stress test).

maximal oxygen consumption ($\dot{V}O_2max$)—The criterion measure of aerobic capacity. Also called maximal oxygen uptake.

measurement—The act of assessing (e.g., assessing a knowledge or psychomotor test score or one's attitude toward physical activity).

mental disabilities—Having mental or psychological limitations (e.g., autism).

Microsoft Excel—A spreadsheet product of Microsoft that results in rows of subjects and columns of variables.

mood—An emotional state of mind, feeling, inclination, or disposition.

motor educability—The ability to learn motor skills.

multiple correlation—The relationship between one outcome (dependent) variable and multiple predictor (independent) variables.

muscular endurance—The physical ability to perform continuous muscular work.

muscular strength—The force that can be generated by the musculature that is contracting.

negative correlation—A correlation in which high scores on one variable are associated with low scores on another variable and low scores are associated with high scores; also called *indirect* or *inverse correlation*.

Net D—An index of discrimination for written test items that indicates the proportion of good discriminations remaining after neutral and bad discriminations are removed.

normal distribution—A bell-shaped, symmetric probability distribution.

norm-referenced fitness standard—A level of achievement relative to a clearly defined subgroup, such as all males or females your age.

null hypothesis—A statement indicating there is no relationship between variables X (independent) and Y (dependent).

objectivity—The degree of interrater reliability; the ability of two or more raters to equivalently score a test.

observed score—One's score on a test. The observed score is the sum of a person's true score and error score.

parameter—A measure in the population of interest (e.g., the population mean).

Pearson product-moment correlation coefficient (PPM) (r)—See *correlation coefficient*. The calculated Pearson Product-moment must be between -1.00 and + 1.00.

perceived exertion—The mental perception of the intensity of physical work.

percentile—The percentage of observations occurring at or below a given score.

performance-based assessment—Testing method that requires a participant to create a product or performance that demonstrates his or her knowledge or skills.

performance criteria—Standards for judging a performance or product.

performance ratio—Dividing a performance score by another measure to better compare performance between subjects (e.g., weight, speed).

personality—The totality of a person's distinctive psychological traits.

phi coefficient—The Pearson product-moment correlation between two dichotomously scored variables, each of which can take on the value of 0 or 1.

physical activity—The act of bodily movement that requires the contraction of muscles and the expenditure of energy.

physical disabilities—Having physical or organic limitations (e.g., cerebral palsy).

physical fitness—A set of attributes that people have or achieve that relates to the ability to perform physical activity.

point system scoring guide—A list of characteristics used to judge a performance or product with points assigned to each item so that the characteristics are weighted in importance.

population—The target group of people or observations for which research findings are to be inferred.

portfolio—A systematic, purposeful, and meaningful collection of one's work that is assembled over time.

power—The amount of work performed in a fixed amount of time.

prediction—The ability to estimate the value of one variable from one or more other variables.

predictive validity—The relationship between a test (a surrogate measure) and a criterion when the criterion is measured in the future. It is based on the PPM correlation coefficient between the test and the criterion.

process criteria—Standards used to evaluate the quality of performance or how a student completes a task.

product criteria—Standards used to measure the outcome of a performance.

progress criteria—Standards used to measure student improvement during the course of instruction.

proportion of agreement (P)—Percentage of agreement between two measures.

psychomotor domain—Involves physiological and physical performance.

qualitative—Perceptive measurement, typically textual in nature.

qualitative rubric—Written descriptions that provide qualitative information about the performance being assessed.

quantitative—Precise measurement, typically numerical in nature.

quantitative rubric—Numbers (quantities) are assigned to indicate quality of the performance being assessed.

range—A measure of variability obtained by subtracting the lowest score from the highest score.

relationship—The statistical association between two or more variables.

relative endurance—A measurement of repetitive performance related to maximum strength (e.g., number of repetitions completed at 50% of 1RM).

relative risk—The risk of mortality (death) or morbidity (disease) associated with one group compared with another (e.g., smokers vs. nonsmokers; physically active vs. not active).

relative scale—Evaluating performance relative to others in a specific group.

relative scoring—A method of scoring responses to essay questions that involves reading all the answers to one question and arranging the papers in order according to degree of adequacy.

relevance—The degree to which a test pertains to the objectives of the measurement.

reliability—The degree to which repeated measurements of the same trait are reproducible under the same conditions; consistency.

repetition maximum (RM)—The maximum number of repeated movements that can be conducted (e.g., 1RM is the maximum weight that can be lifted one time).

repetitive-performance test—A skills test that involves continuous performance of an activity for a specified period of time (e.g., volleying).

research hypothesis—The researcher's hypothesis of the relationship between the independent variable (X) and the dependent variable (Y).

residual volume—The air left in the lungs after maximal forced expiration of air.

sample—A subgroup of the population with which scientific research is conducted.

scatterplot—A graphic representation of the relationship or correlation between two variables.

scientific method—A method of inquiry that requires the development of a hypothesis and the subsequent statistical testing of its plausibility.

self-motivation—The desire to take action.

significance—A probability of rejecting the null hypothesis when it is true (alpha). See *type I error*.

simple linear prediction—Using one variable (*X*) to predict the criterion (*Y*).

skewness—An indication of the shape of a distribution that specifies a lack of symmetry.

skill—A learned trait based on the abilities a person has.

specific determiner—A word or a phrase in a written test item that provides an unintended clue to the correct answer.

specificity—Refers to motor abilities or skills that are unique to individual psychomotor tasks.

SPSS (IBM SPSS Statistics)—Data analysis software. From 2009 to 2010 the same software was referred to as PASW (Predictive Analytics Software).

stability reliability—Consistency of measures across time.

standard deviation—A linear measure of variability that takes into account every score in the distribution; the square root of the variance.

standard error—The type of rating scale error that results from differences in the standards of evaluation applied by raters of the same performance. Results from different raters having different standards.

standard error of estimate (SEE)—An indication of the amount of error when predicting *Y* from *X*; the standard deviation of the errors of prediction. Also called *standard error (SE)* or *standard error of prediction (SEP)*. A validity statistic.

standard error of measurement (SEM)—A value reflecting the degree to which a person's observed score fluctuates as a result of errors of measurement; it is interpreted in the same way as a standard deviation. A reliability statistic.

standard score—A score resulting from the conversion of observed values into a score with a given mean and a standard deviation. z-scores and T-scores are standard scores.

state—A psychological attribute that is related to situational changes.

statistic—A numerical value calculated in the sample to estimate a population parameter (e.g., a sample mean).

subjective rating—A value that an instructor places on a skill or performance based on personal observation.

submaximal exercise test—A fitness test that requires the subject to put forth less than maximal effort (e.g., using submaximal cycle ergometry to assess aerobic capacity).

summative assessment—Assessment that occurs at the conclusion of instruction and is designed to measure how much a student learned.

summative evaluation—A final, comprehensive evaluation conducted near or at the end of an instruction or training program.

surrogate measure—A test used to estimate the criterion (e.g., skinfold measurement is a surrogate measure of the criterion percent body fat that is obtained by DXA or underwater weighing).

table of specifications—A written test blueprint that indicates the proportion of test items dealing with each combination of content objective and educational objective.

target weight—The body weight required to achieve a target percent body fat.

task-specific rubric—A rubric written specifically for a single assessment task. Criteria are written to specifically measure only that assessment.

taxonomy—A classification system based on common characteristics (e.g., the cognitive, affective, and psychomotor domains).

test—A measuring instrument (e.g., written test, performance test, or a wide variety of other instruments) used to make a particular measurement.

theory of basic abilities—Fleishman's theory that skills are based on a number of basic psychomotor abilities that are combined to perform the skill.

torque—A force producing rotation about an axis.

total body movement test—A test that assesses the speed at which a performer completes a task that involves movement of the whole body in a restricted area (e.g., basketball defensive shuffle test).

trait—A relatively stable, common, consistent psychological attribute.

trials-to-criterion testing—The performance of a skill until a standard of achievement is reached.

true score—An unobservable but theoretically existent score that contributes to one's observed test score; it contributes to an accurate estimate of individual differences.

type I error—A false rejection of the null hypothesis. Deciding there is a relation between variables when there actually is none. Expressed as a probability (α).

type II error—A false rejection of the alternative or research hypothesis. Failing to discern a relation between variables when a relation actually exists. Expressed as a probability (β).

validity—The degree of truthfulness in a test.

variability—The spread, or dispersion, of scores in a set of data, the result of the fact that not all scores are exactly the same.

variance (s^2)—A measure of variability; a measure of the spread of a set of scores based on the average squared deviation of each score from the mean.

waist–hip girth ratio—The waist circumference divided by the hip circumference. This measure provides an estimate of body fat distribution, which is a risk factor for cardiovascular disease.

work—The result of the physical effort that is performed; the product of the amount of force applied and the distance over which it is applied.

youth fitness battery—A group of fitness tests to provide an overall assessment of physical fitness (e.g., Fitnessgram; President's Challenge).

zero correlation ($r = 0$)—An indication that there is no linear relationship between two variables.

References

Abernethy, B., J. Summers, and S. Ford. 1998. Issues in the measurement of attention. In *Advances in sport and exercise psychology measurement,* ed. J. Duda. Morgantown, WV: Fitness Information Technology.

Abma, C.L., M.D. Fry, Y.Y. Li, and G.G. Relyea. 2002. Differences in imagery content and imagery ability between high and low confident track and field athletes. *Journal of Applied Sport Psychology* 14(2):67-75.

Ainsworth, B., W. Haskell, A. Leon, D. Jacobs, H. Montoye, J. Sallis, and R. Paffenbarger. 1993. Compendium of physical activities: Classification of energy costs of human physical activities. *Medicine & Science in Sports & Exercise* 25:71-80.

Ainsworth, B.E., W.L. Haskell, M.C. Whitt, M.L. Irwin, A.M. Swartz, S.J. Strath, W.L. O'Brien, D.R. Bassett Jr., K.H. Schmitz, P.O. Emplaincourt, D.R. Jacobs Jr., and A.S. Leon. 2000. Compendium of physical activities: An update of activity codes and MET intensities. *Medicine & Science in Sports & Exercise* 32(Suppl.):498-504.

Ainsworth, B.E., and C.E. Matthews. 2001. Descriptive research in physical activity epidemiology. In *Research methods in physical activity,* 4th ed., ed. J.R. Thomas and J.K. Nelson, 291-308. Champaign, IL: Human Kinetics.

Alamar, B. 2013. *Sports analytics: A guide for coaches, managers, and other decision makers.* New York: Columbia University Press.

Albrecht, R.R., and D.L. Feltz. 1987. Generality and specificity of attention related to competitive anxiety and sport performance. *Journal of Sport Psychology* 9, 231-248.

American Alliance for Health, Physical Education, Recreation and Dance (AAHPERD). 1976. *AAHPERD youth fitness test manual* (Rev.1976 edition). Reston, VA: AAHPERD.

———. 1980. *Health-related physical fitness test manual.* Reston, VA: AAHPERD.

———. 1985. *Norms for college students: Health-related physical fitness test.* Reston, VA: AAHPERD.

———. 1988. *Physical best.* Reston, VA: AAHPERD.

American College of Sports Medicine (ACSM). 2010. *ACSM's health-related physical fitness assessment manual.* 3rd ed. Philadelphia: Lippincott, Williams & Wilkins.

———. 2014a. *ACSM's guidelines for exercise testing and prescription.* 9th ed. Philadelphia: Lippincott, Williams & Wilkins.

———. 2014b. *ACSM's resource manual for guidelines for exercise testing and prescription.* 7th ed. Philadelphia: Lippincott, Williams & Wilkins.

American Heart Association. 1994. *Heart and stroke facts.* Dallas: American Heart Association.

American Psychological Association (APA). 1999. *Standards for educational and psychological testing.* Washington, DC: American Psychological Association.

American Red Cross. 2009. *American Red Cross water safety instructor's manual.* St. Louis: Mosby Lifeline.

Anderson, L.W., and D.R. Krathwohl, eds. 2001. *A taxonomy for learning, teaching and assessing: A revision of Bloom's taxonomy of educational objectives.* Complete ed. New York: Longman.

Anshel, M. 1987. Psychological inventories used in sport psychology research. *Sport Psychologist* 1:331-349.

Åstrand, P., and I. Rhyming. 1954. A nomogram for calculation of aerobic capacity (physical fitness) for pulse rate during submaximal work. *Journal of Applied Physiology* 7:218-221.

Barlow, D.A. 1970. Relation between power and selected variables in the vertical jump. In *Selected topics on biomechanics,* ed. J.M. Cooper, 233-241. Chicago: Athletic Institute.

Barrow, H.M. 1954. Test of motor ability for college men. *Research Quarterly* 25:253-260.

Bartlett, J., L. Smith, K. Davis, and J. Peel. 1991. Development of a valid volleyball skills test battery. *Journal of Physical Education and Dance* 62(2):19-21.

Bass, R.I. 1939. An analysis of the components of tests of semicircular canal function and static and dynamic balance. *Research Quarterly* 2:33-52.

Battinelli, T. 1984. From motor ability to motor learning: The generality-specificity connection. *Physical Educator* 41(3):108-113.

Baumgartner, T.A., A.S. Jackson, M.T. Mahar, and D.A. Rowe. 2016. *Measurement for evaluation in kinesiology.* 9th ed. Burlington, MA: Jones & Bartlett Learning.

Bebetsos, E., and P. Antoniou. 2012. Competitive state anxiety and gender differences among youth Greek badminton players. *Journal of Physical Education and Sport* 12(1):107-110.

Berger, R.A. 1966. Relationship of chinning strength to total dynamic strength. *Research Quarterly* 37:431-432.

Blair, S. 1992. Are American children and youth fit? The need for better data. *Research Quarterly for Exercise and Sport* 63:120-123.

Blair, S., W. Kannel, H. Kohl, and N. Goodyear. 1989. Surrogate measures of physical activity and physical fitness: Evidence for sedentary traits of resting tachycardia, obesity, and low vital capacity. *American Journal of Epidemiology* 129:1145-1156.

Blair, S., H. Kohl, R. Paffenbarger, D. Clark, K. Cooper, and L. Gibbons. 1989. Physical fitness and all-cause mortality: A prospective study of healthy men and women. *Journal of the American Medical Association* 262:2395-2401.

Blair, S.N., J.B. Kampert, H.W. Kohl, III, C.E. Barlow, C.A. Macera, R.S. Paffenbarger Jr., and L.W. Gibbons. 1996. Influence of cardiorespiratory fitness and other precursors on cardiovascular disease and all-cause mortality in men and women. *Journal of the American Medical Association* 276:205-210.

Bland, J.M., and D.G. Altman. 1986. Statistical methods for assessing agreement between two methods of clinical measurement. *Lancet* 327:307-310.

Bloom, B.S., ed. 1956. *Taxonomy of educational objectives: Cognitive domain.* New York: McKay.

Bloom, G., D. Stevens, and T. Wickwire. 2003. Expert coaches' perceptions of team building. *Journal of Applied Sport Psychology* 15:129-143.

Bond, J., and G. Sargent. 1995. Concentration skills in sport: An applied perspective. In *Sport psychology: Theory, applications and issues,* ed. T. Morris and J. Summers, 386-419. Brisbane: Wiley.

Booth, M.L., A. Okely, T. Chey, and A. Bauman. 2002. The reliability and validity of the Adolescent Physical Activity Recall Questionnaire. *Medicine & Science in Sports & Exercise* 34:1986-1995.

Borg, G. 1962. *Physical performance and perceived exertion.* Lund, Sweden: Gleerup.

———. 1998. *Borg's perceived exertion and pain scales.* Champaign, IL: Human Kinetics.

Brace, D.K. 1927. *Measuring motor ability.* New York: Barnes.

Brewer, B., J. Vose, J. Raalte, and A. Petitpas. 2011. Metaqualitative reflections in sport and exercise psychology. *Qualitative Research in Sport, Exercise and Health* 3(3):329-334.

Briere, N., R. Vallerand, M. Blais, M., and L. Pelletier. 1995. Development and validation of the French form of the Sport Motivation Scale. *International Journal of Sport Psychology* 26:465-489.

Brose, A., U. Lindenberger, and F. Schmiedek. 2013. Affective states contribute to trait reports of affective well-being. *Emotion* 13(5): 940-948.

Brown, D., and D. Fletcher. 2013. *Does sport psychology work? A systematic and meta-analytic review of the effects of psychosocial interventions on sport performance.* Paper presented at the Annual Meeting of the Association for Applied Sport Psychology, New Orleans, LA.

Brown, L.E., ed. 2000. *Isokinetics in human performance.* Champaign, IL: Human Kinetics.

Bull, S.J., C.J. Shambrook, W. James, and J. Brooks. 2005. Towards an understanding of mental toughness in elite English cricketers. *Journal of Applied Sport Psychology* 17(3):209-227.

Bungum, T.J., D.L. Peaslee, A.W. Jackson, and M.A. Perez. 2000. Exercise during pregnancy and type of delivery in nulliparae. *Journal of Obstetric, Gynecologic, and Neonatal Nursing* 29:258-264.

Burton, D. 1988. Do anxious swimmers swim slower? Reexamining the elusive anxiety-performance relationship. *Journal of Sport & Exercise Psychology* 10:45-61.

Canadian association of sports sciences, fitness appraisal certification and accreditation program: Fitness Canada. 1987. Ottawa: Government of Canada, Fitness and Amateur Sport.

Canadian standardized test of fitness (CSTF): Operations manual (3rd. ed.). 1986. Ottawa: Government of Canada, Fitness and Amateur Sport.

Caspersen, C. 1989. Physical activity epidemiology: Concepts, methods, and applications to exercise science. In *Exercise and sport science reviews,* ed. K. Pandolph, 423-473. Baltimore: Williams & Wilkins.

Chi, L., and J.L. Duda. 1995. Multi-sample confirmatory factor analysis of the task and ego orientation in sport questionnaire. *Research Quarterly for Exercise and Sport* 66:91-98.

Clark, B.K., E. Winkler, G.N. Healy, P.G. Gardiner, D.W. Dunstan, N. Owen, and M.M. Reeves. 2013. Adults' past-day recall of sedentary time: Reliability, validity, and responsiveness. *Medicine & Science in Sports & Exercise* 45:1198-1207.

Clarke, H.H., and H.A. Bonesteel. 1935. Equalizing the ability of intramural teams at a small high school. *Research Quarterly Supplement* 6(March):193-196.

Clarke, H.H., and R. Munroe. 1970. *Test manual: Oregon cable tension strength test batteries for boys and girls from fourth grade through college.* Eugene, OR: Microcard Publications in Health, Physical Education, and Recreation.

Clough, P., and D. Strycharczyk. 2012. *Developing mental toughness: Improving performance wellbeing and positive behaviour in others.* Philadelphia, PA: Kogan Page

Coleman, R., S. Wilkie, L. Viscio, S. O'Hanley, J. Porcari, G. Kline, B. Keller, S. Hsieh, P. Freedson, and J. Rippe. 1987. Validation of 1-mile walk test for estimating $\dot{V}O_2$max in 20-29 year olds [Abstract]. *Medicine & Science in Sports & Exercise* 19(Suppl. 2):S29.

Collins, D.R., and P.B. Hodges. 2001. *A comprehensive guide to sports skills tests and measurement.* 2nd ed. Lanham, MD: Rowman & Littlefield.

Conroy, D. E., and J.D. Coatsworth. 2007. Coaching behaviors associated with changes in fear of failure: Changes in self-talk and need satisfaction as potential mechanisms. *Journal of Personality* 75:384-419.

Considine, W.J. 1970. A validity analysis of selected leg power tests utilizing a force platform. In *Selected topics on biomechanics,* ed. J.M. Cooper, 243-250. Chicago: Athletic Institute.

Cooper, K. 1968. A means for assessing maximal oxygen intake. *Journal of the American Medical Association* 203:201-204.

Cooper Institute. 1999. *Fitnessgram test administration manual* 2nd ed. Champaign, IL: Human Kinetics.

———. 2004. *Fitnessgram & Activitygram test administration manual* 3rd ed. Champaign, IL: Human Kinetics.

———. 2010. *Fitnessgram & Activitygram test administration manual* 4th ed. Champaign, IL: Human Kinetics.

Cooper Institute for Aerobics Research. 1987. *Fitnessgram.* Dallas, TX: Cooper Institute for Aerobics Research.

———. 1992. *Fitnessgram.* Dallas, TX: Cooper Institute for Aerobics Research.

Corbin, C.B., and R. Lindsey. *Fitness for Life.* 5th ed. Champaign, IL: Human Kinetics; 2005.

Corbin, C.B., and R.P. Pangrazi. 1992. Are American children and youth fit? *Research Quarterly for Exercise and Sport* 63:96-106.

Cox, J., and K. Cox. 2008. *Your opinion, please! How to build the best questionnaires in the field of education.* Thousand Oaks, CA: Sage.

Cox, R.H., M.P. Martens, and W.D. Russell. 2003. Measuring anxiety in athletics: The Revised Competitive State Anxiety Inventory-2. *Journal of Sport & Exercise Psychology* 25:519-533.

Craft, L., M. Magyar, B. Becker, and D. Feltz. 2003. The relationship between the Competitive State Anxiety Inventory-2 and sport performance: A meta-analysis. *Journal of Sport & Exercise Psychology* 25:44-65.

Craig, C.L., C. Cameron, and C. Tudor-Locke. 2013. CANPLAY pedometer normative reference data for 21,271 children and 12,956 adolescents. *Medicine & Science in Sports & Exercise* 45:123-129.

Culver, D., W. Gilbert, and P. Trudel. 2003. A decade of qualitative research in sport psychology journals 1990-1999. *Sport Psychologist* 17:1-15.

Cureton, K.J., and G.L. Warren. 1990. Criterion-referenced standards for youth health-related fitness tests: A tutorial. *Research Quarterly for Exercise and Sport* 61:7-19.

Danielson, C. 1997. Designing successful performance tasks and rubrics. Audio cassette tape #297072. Recorded live at the 52nd Annual Conference ASCD, Baltimore, March 22-25, 1997. Alexandria, VA: Association for Supervision and Curriculum Development.

Dennison, B., J.H. Straus, D. Mellits, and E. Charney. 1988. Childhood physical fitness tests: Predictor of adult physical activity levels? *Pediatrics* 82:324-330.

Dewey, D., L. Brawley, and F. Allard. 1989. Do the TAIS attentional style scales predict how visual information is processed? *Journal of Sport & Exercise Psychology* 11:171-186.

Disch, C.F., and J.G. Disch. 1979. Predictive analysis of a battery of anthropometric and motor performance tests for classifications of male volleyball players. *Volleyball Technical Journal* 4:93-98.

Disch, J.G. 1978. The construction and analysis of a test battery related to volleyball playing capacity in females. Report No. ED 148815. Washington, DC: ERIC Clearinghouse in Teacher Education.

———. 1979. A factor analysis of selected tests for speed of body movement. *Journal of Human Movement Studies* 5:141-151.

Disch, J.G., and S.C. Disch. 2005. Performance testing athletes. *Olympic Coach* 17(3):17-21.

Dishman, R.K., and W. Ickes. 1981. Self-motivation and adherence to therapeutic exercise. *Journal of Behavioral Medicine* 4:421-436.

Docherty, D., ed. 1996. *Measurement in pediatric exercise science.* Champaign, IL: Human Kinetics.

Doyle, J., and G. Parfitt. 1996. Performance profiling and predictive validity. *Journal of Applied Sports Psychology* 8:160-170.

Driskell, J.E., C. Copper, and A. Moran. 1994. Does mental practice enhance performance? *Journal of Applied Psychology* 79:481-492.

Duda, J. 1998. *Advances in sport and exercise psychology measurement.* Morgantown, WV: Fitness Information Technology.

Duda, J.L. 1989. Relationship between task and ego orientation and the perceived purpose of sport among high school athletes. *Journal of Sport & Exercise Psychology* 11:318-335

Dunn, J.H., J. Causgrove Dunn, and D.G. Syrotuik. 2002. Relationship between multidimensional perfectionism and goal orientations in sport. *Journal of Sport & Exercise Psychology* 24:376-395.

Durand-Bush, N., Salmela, J., Green-Demers, I (2001). The Ottawa Mental Skills Assessment (OMSAT-3). *The Sport Psychologist*, 15, 1-19.

Ebel, R. 1965. *Measuring educational achievement.* Englewood Cliffs, NJ: Prentice-Hall.

Edmunds, J., N. Ntoumanis, and J.L. Duda. 2007. Adherence and well-being in overweight and obese patients referred to an exercise on prescription scheme: A self-determination theory perspective. *Psychology of Sport and Exercise* 8:722-740.

Eklund, L., W. Haskell, J. Johnson, F. Whaley, M. Criqui, and D. Sheps. 1988. Physical fitness as a predictor of cardiovascular mortality in asymptomatic North American men. *New England Journal of Medicine* 319:1379-1384.

Ellenbrand, D.A. 1973. Gymnastics skills tests for college women. Unpublished master's thesis, Indiana University, Bloomington.

Engelman, M.E., and J.R. Morrow Jr. 1991. Reliability and skinfold correlates for traditional and modified pull-ups in children grades 3-5. *Research Quarterly for Exercise and Sport* 62:88-91.

Eysenck, H.J., and S.B.G. Eysenck. 1968. *Eysenck Personality Inventory manual.* London: University of London Press.

Feuer, M., and K. Fulton. 1993. The many faces of performance assessment. *Phi Delta Kappan* 74:478.

Fiatarone, M.A., E.F. O'Neill, N.D. Ryan, K.M. Clements, G.R. Solares, M.E. Nelson, S.B. Roberts, J.J. Kehayias, L.A. Lipsitz, and W.J. Evans. 1994. Exercise training and nutritional supplementation for physical frailty in very elderly people. *New England Journal of Medicine* 330:1769-1775.

FitzGerald, S.J., C.E. Barlow, J.B. Kampert, J.R. Morrow Jr., A.W. Jackson, and S.N. Blair. 2004. Muscular fitness and all-cause mortality: Prospective observations. *Journal of Physical Activity & Health* 1:7-18.

Fleishman, E.A. 1964. *The structure and measurement of physical fitness.* Englewood Cliffs, NJ: Prentice-Hall.

Fleishman, E.A., and M.K. Quaintance. 1984. *Taxonomies of human performance.* New York: Academic Press.

Ford, S., and J. Summers. 1992. The factorial validity of the TAIS attentional style subscales. *Journal of Sport & Exercise Psychology* 14:283-297.

Fournier, J.F., C. Calmels, N. Durand-Bush, and J.H. Salmela. 2005. Effects of a season-long PST program on gymnastic performance and on psychological skill development. *International Journal of Sport & Exercise Psychology* 3:59-78.

Fox, K., and C. Corbin. 1989. The Physical Self-Perception Profile: Development and preliminary validation. *Journal of Sport & Exercise Psychology* 11:408-430.

Glaser, R., and D.J. Klaus. 1962. Proficiency measurement: Assessing human performance. In *Psychological principles in systems development,* ed. R.M. Gagne, 419-474. New York: Holt, Rinehart & Winston.

Glass, G.V, and K.D. Hopkins. 1996. *Statistical methods in education and psychology.* 3rd ed. Englewood Cliffs, NJ: Prentice-Hall.

Getchell, L.H., D. Kirkendall, and G. Robbins. 1977. Prediction of maximal oxygen uptake in young adult women joggers. *Research Quarterly* 48:61-67.

Golding, L. 2000. *YMCA fitness testing and assessment manual* 4th ed. Champaign, IL: Human Kinetics.

Golding, L., C. Myers, and W. Sinning. 1989. *Y's way to physical fitness.* Champaign, IL: Human Kinetics.

Goldsmith, W. 2005. Testing: How, why, who, what, and when (and how to make sense of it). *Olympic Coach* 17(3):13-16.

Gotwals, J.K., and J.H. Dunn. 2009. A multi-method multi-analytic approach to establishing internal construct validity evidence: The Sport Multidimensional Perfectionism Scale 2. *Measurement in Physical Education and Exercise Science* 13:71-92.

Gotwals, J.K., J.H. Dunn, J. Causgrove Dunn, and V. Gamache. 2010. Establishing validity evidence for the Sport Multidimensional Perfectionism Scale-2 in intercollegiate sport. *Psychology of Sport and Exercise* 11:423-432.

Gould, D., K. Dieffenbach, and A. Moffett. 2002. Psychological characteristics and their development in Olympic champions. *Journal of Applied Sport Psychology* 14:172-204.

Gould, D., R. Horn, and J. Spreeman. 1984. Competitive anxiety in junior elite wrestlers. *Journal of Sport Psychology* 5:58-71.

Graves, J., M. Pollock, D. Carpenter, S. Leggett, A. Jones, M. MacMillan, and M. Fulton. 1990. Quantitative assessment of full range-of-motion isometric lumbar extension strength. *Spine* 15:289-294.

Green, K.N., W.B. East, and L.D. Hensley. 1987. A golf skills test battery for college males and females. *Research Quarterly for Exercise and Sport* 58:72-76.

Greenspan, M.J., and D.L. Feltz. 1989. Psychological interventions with athletes in competitive situations: A review. *Sport Psychologist* 3:219-236.

Gronlund, N.E. 1993. *Assessment of student achievement.* 6th ed. Boston: Allyn and Bacon.

Grossbard, J.R., S.P. Cumming, M. Standage, R.E. Smith, and F.L. Smoll. 2007. Social desirability and relations between goal orientations and competitive trait anxiety in young athletes. *Psychology of Sport and Exercise* 8: 491-505.

Grove, J.R. 2001. Practical screening tests for talent identification in baseball. *Applied Research in Coaching and Athletics Annual* 16:63-77.

Gucciardi, D., S. Gordon, and J. Dimmock. 2008. Toward an understanding of mental toughness in Australian football. *Journal of Applied Sport Psychology* 20:261-281.

Guskey, T., and J. Bailey. 2001. *Developing grading and reporting systems for student learning.* Thousand Oaks, CA: Corwin Press.

Hagberg, J.M., J.E. Graves, M. Limacher, D.R. Woods, S.H. Leggett, C. Cononie, J. Gruber, and M.L. Pollock. 1989. Cardiovascular responses of 70-79 year old men and women to exercise training. *Journal of Applied Physiology* 66:2589-2594.

Hales, D., A.E. Vaughn, S. Mazzucca, M.J. Bryant, R.G. Tabak, C. McWilliamsJ. Stevens, D.S. Ward. 2013. Development of HomeSTEAD's physical activity and screen time physical environment inventory. *International Journal of Behavioral Nutrition and Physical Activity* 10:132.

Hall, C., D. Mack, A. Paivio, and H.A. Hausenblas. 1998. Imagery use by athletes: Development of the Sport Imagery Questionnaire. *International Journal of Sport Psychology* 29:73-89.

Hall, G., R. Hetzler, D. Perrin, and A. Weltman. 1992. Relationship of timed sit-up tests to isokinetic abdominal strength. *Research Quarterly for Exercise and Sport* 63:80-84.

Hanrahan, S.J., and S.H. Biddle. 2002. Measurement of achievement orientations: Psychometric measures, gender, and sport differences. *European Journal of Sport Science* 2(5):1-12.

Hanton, S.S., S.D. Mellalieu, and R.R. Hall. 2002. Re-examining the competitive anxiety trait-state relationship. *Personality and Individual Differences* 33:1125-1136.

Harris, M.L. 1969. A factor analytic study of flexibility. *Research Quarterly* 40:62-70.

Harris, C.D., K.B. Watson, S.A. Carlson, J.E. Fulton and J.M. Dorn. May 3, 2013. Adult participation in aerobic and muscle-strengthening physical activities - United States, 2011. MMWR 62(17):326-330.

Harrow, A.J. 1972. *A taxonomy of the psychomotor domain*. New York: McKay.

Henry, F.M. 1956. Coordination and motor learning. In *59th Proceedings of the Annual College Physical Education Association*, 68-75. Washington, DC.

———. 1958. Specificity vs. generality in learning motor skills. In *61st Annual Proceedings of the College Physical Education Association*, 126-128. Washington, DC.

Hensley, L.D., ed. 1989. *Tennis for boys and girls: Skills test manual*. Reston, VA: AAHPERD.

Hensley, L., L. Lambert, T. Baumgartner, and J. Stillwell. 1987. Is evaluation worth the effort? *Journal of Physical Education, Recreation and Dance* 58(6):59-62.

Hensley, L.D., and W.B. East. 1989. Testing and grading in the psychomotor domain. In *Measurement concepts in physical education and exercise science*, ed. M.J. Safrit and T.M. Wood, 297-321. Champaign, IL: Human Kinetics.

Hensley, L.D., W.B. East, and J.L. Stillwell. 1979. A racquetball skills test. *Research Quarterly* 50:114-118.

Herman, J.L., P.R. Aschbacher, and L. Winters. 1992. *A practical guide to alternative assessment*. Alexandria, VA: Association for Supervision and Curriculum Development.

Holt, N., and A. Sparkes. 2001. An ethnographic study of cohesiveness in a college soccer team over a season. *Sport Psychologist* 15:237-259.

Hopkins, D.R., J. Schick, and J.J. Plack. 1984. *Basketball for boys and girls: Skills test manual*. Reston, VA: AAHPERD.

Howley, E.T., and B.D. Franks. 2007. *Health fitness instructor's handbook*. 5th ed. Champaign, IL: Human Kinetics.

Imwold, C., R. Rider, and D. Johnson. 1982. The use of evaluation in public school physical education programs. *Journal of Teaching in Physical Education* 2:13-18.

Isaac, A., D.F. Marks and D.G. Russell. 1986. An instrument for assessing imagery of movement: The Vividness of Movement Imagery Questionnaire (VMIQ). *Journal of Mental Imagery* 10(4):23-30.

Jackson, A.S., S.N. Blair, M.T. Mahar, L.T. Wier, R.M. Ross, and J.E. Stuteville. 1990. Prediction of functional aerobic capacity without exercise testing. *Medicine & Science in Sports & Exercise* 22:863-870.

Jackson, A.S., and R.J. Frankiewicz. 1975. Factorial expressions of muscular strength. *Research Quarterly* 46:206-217.

Jackson, A.S., and M. Pollock. 1978. Generalized equations for predicting body density of men. *British Journal of Nutrition* 40:497-504.

Jackson, A.S., M. Pollock, and A. Ward. 1980. Generalized equations for predicting body density of women. *Medicine & Science in Sports & Exercise* 12:175-182.

Jackson, A.W., and A. Baker. 1986. The relationship of the sit and reach test to criterion measures of hamstring and back flexibility in young females. *Research Quarterly for Exercise and Sport* 57:183-186.

Jackson, A.W., A.S. Jackson, and J. Bell. 1980. A comparison of alpha and the intraclass reliability coefficients. *Research Quarterly for Exercise and Sport* 51:568-571.

Jackson, A.W., and N. Langford. 1989. The criterion-related validity of the sit and reach test: Replication and extension of previous findings. *Research Quarterly for Exercise and Sport* 60:384-387.

Jackson, A.W., D.C. Lee, X. Sui, J.R. Morrow Jr., T.S. Church, A.L. Maslow, and S.N. Blair. 2010. Muscular strength is related to prevalence and incidence of obesity in adult men. *Obesity* 18:1988-1995.

Jackson, A.W., J.R. Morrow Jr., H.R. Bowles, S.J. FitzGerald, and S.N. Blair. 2007. Construct validity evidence for single-response items to estimate physical activity levels in large sample studies. *Research Quarterly for Exercise and Sport* 78:24-31.

Jackson, A.W., J.R. Morrow Jr., P.A. Brill, H.W. Kohl III, N.F. Gordon, and S.N. Blair. 1998. Relations of sit-up and sit-and-reach tests to low back pain in adults. *Journal of Orthopaedic & Sports Physical Therapy* 27(1):22-26.

Jackson, A.W., J. Solomon, and M. Stusek. 1992. One-mile walk test: Reliability, validity, and norms for young adults [Abstract]. *Research Quarterly for Exercise and Sport* 63:A52.

Jackson, A.W., M. Watkins, and R. Patton. 1980. A factor analysis of twelve selected maximal isotonic strength performances on the universal gym. *Medicine & Science in Sports & Exercise* 12:274-277.

Jacobs, D., B. Ainsworth, T. Hartman, and A. Leon. 1993. A simultaneous evaluation of 10 commonly used physical activity questionnaires. *Medicine & Science in Sports & Exercise* 25:81-91.

Jensen, C., and C. Hirst. 1980. *Measurement in physical education and athletics.* New York: Macmillan.

Johnson, B., and J. Nelson. 1979. *Practical measurements for evaluation in physical education.* 3rd ed. Minneapolis: Burgess.

Jones, G.G., S.S. Hanton, and D.D. Connaughton. 2002. What is this thing called mental toughness? An investigation of elite sports performers. *Journal of Applied Sport Psychology* 14:205-218.

Jurca, R., A.S. Jackson, M.L. LaMonte, J.R. Morrow Jr., S.N. Blair, N.J. Wareham, W.L. Haskell, M.W. van Mechelen, T.S. Church, J.M. Jakicic, and R. Laukkanen. 2005. Assessing cardiorespiratory fitness without performing exercise testing. *American Journal of Preventive Medicine* 29:185-193.

Kann, L., S. Kinchen, S.L. Shanklin, K.H. Flint, J. Hawkins, W.A. Harris, R. Lowry, E.O. Olsen, T. McManus, D. Chyen, L. Whittle, E. Taylor, Z. Demissie, N. Brener, J. Thornton, J. Moore, and S. Zaza. 2014. Youth risk behavior surveillance-United States, 2013. MMWR 63(No. 4).

Kelly, G.A. 1955. *The psychology of personal constructs.* New York: Norton.

Kenyon, G.S. 1968a. A conceptual model for characterizing physical activity. *Research Quarterly* 39:96-105.

———. 1968b. Six scales for assessing attitude toward physical activity. *Research Quarterly* 39:566-574.

Kirby, R.F., ed. 1991. *Kirby's guide to fitness and motor performance tests.* Cape Girardeau, MO: Ben Oak.

Kirk, M.F. 1997. Using portfolios to enhance student learning and assessment. *Journal of Physical Education, Recreation and Dance* 68(7):29-33.

Kline, G., J. Porcari, R. Hintermeister, P. Freedson, A. Ward, R. McCarron, J. Ross, and J. Rippe. 1987. Estimation of $\dot{V}O_2$max from a one-mile track walk, gender, age, and body weight. *Medicine & Science in Sports & Exercise* 19:253-259.

Kneer, M. 1986. Description of physical education instruction theory/practice gap in selected secondary schools. *Journal of Teaching in Physical Education* 5:91-106.

Krathwohl, D.R., B.S. Bloom, and B.A. Masia. 1964. *Taxonomy of educational objectives: Handbook II: The affective domain.* New York: McKay.

LaPorte, R., H. Montoye, and C. Caspersen. 1985. Assessment of physical activity in epidemiologic research: Problems and prospects. *Public Health Reports* 100:131-146.

Larson, L.A. 1941. A factor analysis of motor ability variables and tests, with test for college men. *Research Quarterly* 12:499-517.

Last, J. 1992. *Dictionary of epidemiology.* 2nd ed. New York: Oxford University.

LeBlanc, A.G.W., and I. Janssen. 2010. Difference between self-reported and accelerometer measured moderate-to-vigorous physical activity in youth. *Pediatric Exercise Science* 22:523-534.

Leone, M., G. Lariviere, and A.S. Comtois. 2002. Discriminant analysis of anthropometric and biomotor variables among elite adolescent female athletes in four sports. *Journal of Sport Sciences* 20:443-449.

Lewis, M. 2004. *Moneyball: The art of winning an unfair game.* New York: W.W. Norton.

Li, F. 1999. The Exercise Motivation Scale: Its multifaceted structure and construct validity. *Journal of Applied Sport Psychology* 11:97-115.

Lohman, T. 1989. Assessment of body composition in children. *Pediatric Exercise Science* 1:19-30.

Looney, M., and S. Plowman. 1990. Passing rates of American children and youth on the FITNESSGRAM criterion-referenced physical fitness standards. *Research Quarterly for Exercise and Sport* 61:215-223.

Looney, M.A. 2003. Facilitate learning with a definitional grading system. *Measurement in Physical Education and Exercise Science* 7:269-275.

Lund, J., and M. Kirk. 2010. *Performance-based assessment for middle and high school physical education.* Champaign, IL: Human Kinetics.

Lund, J., and M.L. Veal. 2013. *Assessment-driven instruction in physical education: A standards-based approach to promoting and documenting learning.* Champaign, IL: Human Kinetics.

MacDougall, J.D., and H.A. Wenger. 1991. The purpose of physiological testing. In *Physiological testing of the high performance athlete,* 2nd ed., ed. J.D. MacDougall, H.A. Wenger, and H.J. Green, 1-5. Champaign, IL: Human Kinetics.

Mahar, M.T., and D.A. Rowe. 2002. Construct validity in physical activity research. In *Physical activity assessments for health-related research*, ed. G.J. Welk, 51-72. Champaign, IL: Human Kinetics.

———. 2014. A brief exploration of measurement and evaluation in kinesiology. *Kinesiology Review* 3:80-91.

Mahar, M.T., D.A. Rowe, C.R. Parker, F.J. Mahar, D.M. Dawson, and J.E. Holt. 1997. Criterion-referenced and norm-referenced agreement between the mile run/walk and PACER. *Measurement in Physical Education and Exercise Science* 1:245-258.

Mahoney, M., T. Gabriel, and T. Perkins. 1987. Psychological skills and exceptional athletic performance. *Sport Psychologist* 1:181-199.

Marcus, B.H., S.W. Banspach, R.C. Lefebvre, J.S. Rossi, R.A. Carleton, and D.B. Abrams. 1994. Using the stages of change model to increase the adoption of physical activity among community participants. *American Journal of Health Promotion* 6:424-429.

Marcus, B.H., V.C. Selby, R.S. Niaura, and J.S. Rossi. 1992. Self-efficacy and the stages of exercise behavior change. *Research Quarterly for Exercise and Sport* 63:60-66.

Marcus, B.H., and L. Simkin. 1993. The stages of exercise behavior. *Journal of Sports Medicine and Physical Fitness* 33:83-88.

Markland, D., and L. Hardy. 1993. The Exercise Motivation Inventory: Preliminary development and validity of a measure of individuals' reasons for participation in physical activity. *Personality and Individual Differences* 15:289-296.

Markland, D., and V. Tobin. 2004. A modification to the Behavioural Regulation in Exercise Questionnaire to include an assessment of motivation. *Journal of Sport & Exercise Psychology* 26:191-196.

Martens, R. 1977. *Sport competition anxiety test.* Champaign, IL: Human Kinetics.

Martens, R., R. Vealey, and D. Burton. 1990. *Competitive anxiety in sport.* Champaign, IL: Human Kinetics.

Martin, R.H. 1983. Effectiveness of attentional focus and basketball free-throw percentage: An attempt at prediction. Unpublished master's thesis, California State University, Fullerton.

Marzano, R.J., D. Pickering, and J. McTighe. 1993. *Assessing student outcomes: Performance assessment using the dimensions of learning model.* Alexandria, VA: Association for Curriculum and Development.

Matanin, M., and D. Tannehill. 1994. Assessment and grading in physical education. *Journal of Teaching in Physical Education* 13:395-405.

Mayhew, T., and J. Rothstein. 1985. Measurement of muscle performance with instruments. In *Measurement in physical therapy,* ed. J. Rothstein, 57-102. New York: Churchill Livingstone.

McCloy, C.H. 1932. *The measurement of athletic power.* New York: Barnes.

McKenzie, T.L., D.A. Cohen, A. Sehgal, S. Williamson, and D. Golinelli. 2006. System for Observing Play and Leisure Activity in Communities (SOPARC): Reliability and feasibility measures. *Journal of Physical Activity & Health* 1:S203-217.

McKenzie, T.L, S.J. Marshall, J.F. Sallis, and T.L. Conway. 2000. Leisure-time physical activity in school environments: An observational study using SOPLAY. *Preventive Medicine* 30:70-77.

McKenzie, T.L, J.F. Sallis, and P.R. Nader. 1991. SOFIT: System for Observing Fitness Instruction Time. *Journal of Teaching in Physical Education* 11:195-205.

McKenzie, T.L., J.F. Sallis, T.L. Patterson, J.P. Elder, C.C. Berry, J.W. Rupp, C.J. Atkins, M.J. Buono, and P.R. Nader. 1991. BEACHES: An observational system for assessing children's eating and physical activity behaviors and associated events. *Journal of Applied Behavior Analysis* 24:141-151.

McNair, D.M., M. Lorr, and L.F. Droppleman. 1971. *EDITS manual for POMS.* San Diego: Educational and Industrial Testing Service.

Melograno, V. 2000. *Portfolio assessment for K-12 physical education.* Reston, VA: National Association for Sport and Physical Education Publications.

Mesagno, C., and T. Mullane-Grant. 2010. A comparison of different pre-performance routines as possible choking interventions. *Journal of Applied Sport Psychology* 22:343-360.

Messick, S. 1995. Standards of validity and the validity of standards in performance assessment. *Educational Measurement: Issues and Practice* 14(1):5-8.

Miller, M.D., and S.M. Legg. 1993. Alternative assessment in a high-stakes environment. *Educational Measurement: Issues and Practice* 12(3):9-15.

Miller, P. 1985. Assessment of joint motion. In *Measurement in physical therapy,* ed. J. Rothstein, 103-136. New York: Churchill Livingstone.

Mintah, J.K. 2003. Authentic assessment in physical education: Prevalence of use and perceived impact on students' self-concept, motivation, and skill achievement. *Measurement in Physical Education and Exercise Science* 7:161-174.

Montoye, H.J., H.C. Kemper, W.H.M. Saris, and R.A. Washburn. 1996. *Measuring physical activity and energy expenditure.* Champaign, IL: Human Kinetics.

Mood, D.P., Jackson, A.W., and Morrow, J.R., Jr. 2007. Measurement of physical fitness and physical activity: Fifty years of change. *Measurement in Physical Education and Exercise Science* 11:217-227.

Morgan, W.P. 1980. Test of champions: The iceberg profile. *Psychology Today* (July):92-93, 97-99, 102, 108.

Morgan, W.P., and R.W. Johnson. 1978. Psychological characteristics of successful and unsuccessful oarsmen. *International Journal of Sport Psychology* 11:38-49.

Morgan, W.P., and M.L. Pollock. 1977. Psychologic characterization of the elite distance runner. *Annals of the New York Academy of Science* 301:382-403.

Morris, L., D. Davis, and C. Hutchins. 1981. Cognitive and emotional components of anxiety: Literature review and revised worry-emotionality scale. *Journal of Education Psychology* 73:541-555.

Morrow, J.R. Jr. 2005. Are American children and youth fit? It's time we learned. *Research Quarterly for Exercise and Sport* 76:377-388.

Morrow, J.R. Jr., and A. Ede. 2009. Statewide physical fitness testing: A BIG waist or a BIG waste? *Research Quarterly for Exercise and Sport* 80:696-701

Morrow, J.R. Jr., T. Fridye, and S. Monaghen. 1986. Generalizability of the AAHPERD health-related skinfold test. *Research Quarterly for Exercise and Sport* 57:187-195.

Morrow, J.R. Jr., S.B. Going, and G.J. Welk, eds. 2011. Fitnessgram development of criterion-referenced standards for aerobic capacity and body composition. *American Journal of Preventive Medicine* Suppl. 2 41(4):S63-S144.

Morrow, J.R. Jr., A. Jackson, P. Bradley, and H. Hartung. 1986. Accuracy of measured and predicted residual lung volume on body density measurement. *Medicine & Science in Sports & Exercise* 18:647-652.

Morrow, J.R. Jr., S.B. Martin, and A.W. Jackson. 2010. Reliability and validity of the Fitnessgram: Quality of teacher-collected health-related fitness surveillance data. *Research Quarterly for Exercise and Sport* 81(Suppl.):S24-S30.

Morrow, J.R. Jr., J.S. Tucker, A.W. Jackson, S.B. Martin, C.A. Greenleaf, and T.A. Petrie. Meeting physical activity guidelines and health-related fitness in youth. *American Journal of Preventive Medicine* 44: 439-444.

Morrow, J.R. Jr., W. Zhu, B.D. Franks, M.D. Meredith, and C. Spain. 2009. 1958-2008: 50 years of youth fitness tests in the United States. *Research Quarterly for Exercise and Sport* 80:1-11.

Morrow, J.R. Jr., W. Zhu, and M.T. Mahar. 2013. Physical fitness standards for children. In *Fitnessgram/Activitygram Reference Guide*. 4th ed. (Internet Resource). In S.A. Plowman and M.D. Meredith, eds., chapter 4, 2-12. Dallas, TX: Cooper Institute.

Mullan, E., D. Markland, and D.K. Ingledew. 1997. A graded conceptualisation of self-determination in the regulation of exercise behaviour: Development of a measure using confirmatory factor analytic procedures. *Personality and Individual Differences* 23:745-752.

Murphy, S.M., and D.P. Jowdy. 1992. *Imagery and mental practice*. Champaign, IL: Human Kinetics.

Murray, T.D., J.L. Walker, A.S. Jackson, J.R. Morrow Jr., J.A. Eldridge, and D.L. Rainey. 1993. Validation of a 20-minute steady-state jog as an estimate of peak oxygen uptake in adolescents. *Research Quarterly for Exercise and Sport* 64:75-82.

National Association for Sport and Physical Education (NASPE). 2004. *Moving into the future: National standards for physical education*. Reston, VA: NASPE.

———. 2008. *PE metrics: Assessing the national standards*. Reston, VA: NASPE.

National Center for Health Statistics. 2013. *Health, United States, 2012: With special feature on emergency care*. Hyattsville, MD: National Center for Health Statistics.

Neilson, N.P., and F.W. Cozens. 1934. *Achievement scales in physical education activities for boys and girls in elementary and junior high schools*. New York: Barnes.

Nelson, J.K., S.H. Yoon, and K.R. Nelson. 1991. A field test for upper body strength and endurance. *Research Quarterly for Exercise and Sport* 62:436-441.

Nelson, L.R., and M.L. Furst. 1972. An objective study of the effects of expectation on competitive performance. *Journal of Psychology* 81:69-72.

Nicholls, J.G. 1989. *The competitive ethos and democratic education*. Cambridge, MA: Harvard University Press.

Nideffer, R.M. 1976. Test of attentional and interpersonal style. *Journal of Personality and Social Psychology* 34:394-404.

———. 2007. Reliability and validity of the Attentional and Interpersonal Style (TAIS) inventory concentration scales. In *Essential readings in sport and exercise psychology*, ed. D. Smith and M. Bar-Eli, 265-277. Champaign, IL: Human Kinetics.

Nieman, D.C. 1995. *Fitness and sports medicine: A health-related approach*. 3rd ed. Mountain View, CA: Mayfield.

Nitko, A.J. 1984. Defining "criterion-referenced test." In *A guide to criterion-referenced test construction*, ed. R.A. Berk, 8-28. Baltimore: Johns Hopkins University Press.

Odom, L.R., and Morrow, J.R. Jr. 2006. What is this r? A correlational approach to explaining validity, reliability, and objectivity coefficients. *Measurement in Physical Education and Exercise Science*, 10:137-145.

Ostrow, A. 1996. *Directory of psychological tests in the sport and exercise sciences*. 2nd ed. Morgantown, WV: Fitness Information Technology.

Parker, J.K., and G. Lovell. 2011. The influence of experience upon imagery perspectives in adolescent sport performers. *Journal of Imagery Research in Sport and Physical Activity* 6(1). doi:10.2202/1932-0191.1048

Parker, J.K., and G.P. Lovell. 2012. Age differences in the vividness of youth sport performers' imagery ability. *Journal of Imagery Research in Sport and Physical Activity* 7(1). doi:10.1515/1932-0191.1069

Passer, M.W. 1983. Fear of failure, fear of evaluation, perceived competence, and self-esteem in competitive-trait-anxious children. *Journal of Sport Psychology* 5:172-188.

Pate, R. 1988. The evolving definition of physical fitness. *Quest* 40:174-179.

Pate, R., J. Ross, C. Dotson, and G. Gilbert. 1985. The new norms: A comparison with the 1980 AAHPERD norms. *Journal of Physical Education, Recreation and Dance* 56(1):70-72.

Pate R.R., C.Y. Wang, M. Dowda, S.W. Farrell, and J.R. O'Neill. 2006. Cardiorespiratory fitness levels among US youth 12 to 19 years of age: Findings from the 1999-2002 National Health and Nutrition Examination Study. *Archives of Pediatric Adolescent Medicine* 16:1005-1012.

Patterson, E.L., R.E. Smith, J.J., Everett, and J.T. Ptacek. 1998. Psychosocial factors as predictors of ballet injuries: Interactive effects of life stress and social support. *Journal of Sport Behavior* 21:101-112.

Patton, M. 1990. *Qualitative evaluation and research methods*. 2nd ed. Newbury Park, CA: Sage.

Perusse, A., C. Tremblay, C. Leblanc, and C. Bouchard. 1989. Genetic and environmental influences on level of habitual physical activity and exercise participation. *American Journal of Epidemiology* 129:1012-1022.

Pettee-Gabriel, K.K, J.R. Morrow Jr., and A.L. Woolsey. 2012. Framework for physical activity as a complex and multidimensional behavior. *Journal of Physical Activity & Health* 9(Suppl. 1): S11-S18.

Pew Internet and American Life Project. Generations online in 2009. 2009 Pew Research Center.

Plowman, S.A. 1992. Criterion referenced standards for neuromuscular physical fitness tests. *Pediatric Exercise Science* 4:10-19.

Pollock, M., R. Bohannon, K. Cooper, J. Ayres, A. Ward, S. White, and A. Linnerud. 1976. A comparative analysis of four protocols for maximal treadmill stress testing. *American Heart Journal* 92:39-46.

Popham, W.J. 2003. *Test better, teach better*. Alexandria, VA: Association for Supervision and Curriculum Development.

Potgieter, J.R. 2009. Norms for the Sport Competition Anxiety Test (SCAT). *South African Journal for Research in Sport, Physical Education and Recreation* 31(1):69-79.

Powell K.E., A.M. Roberts, J.G. Ross, M.A. Phillips, D.A. Ujamaa, and M. Zhou. 2009. Low physical fitness among fifth- and seventh-grade students, Georgia, 2006. *American Journal of Preventive Medicine* 36:304-310.

Prapavessis, H. 2000. The POMS and sports performance: A review. *Journal of Applied Sport Psychology* 12:34-48.

President's Council on Physical Fitness and Sports (PCPFS). 1999. *The Presidential Physical Fitness Award program*. Washington, DC: PCPFS.

Prochaska, J.O., and C.C. DiClemente. 1983. Stages and processes of self-change in smoking: Toward an integrative model of change. *Journal of Consulting and Clinical Psychology* 51:390-395.

Reiff, G., W. Dixon, D. Jacoby, G. Ye, C. Spain, and P. Hunsicker. 1985. *The President's Council on Physical Fitness and Sports 1985 National School Population Fitness Survey*. Washington, DC: President's Council on Physical Fitness and Sports.

Ridgers, N.D., G. Stratton, and T.L. McKenzie. 2010. Reliability and validity of the System for Observing Children's Activity and Relationships During Play (SOCARP). *Journal of Physical Activity & Health*, 7:17-25.

Rikli, R.E. 1991. *Softball for boys and girls: Skills test manual*. Reston, VA: AAHPERD.

Rikli, R.E., and C.J. Jones. 1999a. Development and validation of a functional fitness test for community-residing older adults. *Journal of Aging and Physical Activity* 7:129-161.

———. 1999b. Functional fitness normative scores for community-residing older adults, ages 60-94. *Journal of Aging and Physical Activity* 7:162-181.

———. 2013a. Development and validation of criterion-referenced clinically relevant fitness standards for maintaining physical independence in later years. *Gerontologist* 53:255-267.

———. 2013b. *Senior fitness test manual*. 2nd ed. Champaign, IL: Human Kinetics.

Rikli, R.E, C. Petray, and T.A. Baumgartner. 1992. The reliability of distance run tests for children in grades K-4. *Research Quarterly for Exercise and Sport* 63:270-276.

Roberts, R., N. Callow, L. Hardy, D. Markland, and J. Bringer. 2008. Movement imagery ability: Development and assessment of a Revised Version of the Vividness of Movement Imagery Questionnaire. *Journal of Sport & Exercise Psychology* 30:200-221.

Robertson, L., and H. Magnusdottir. 1987. Evaluation of criteria associated with abdominal fitness testing. *Research Quarterly for Exercise and Sport* 58:355-359.

Rodriguez, M.C. 2005. Three options are optimal for multiple-choice items: A meta-analysis of 80 years of research. *Educational Measurement* 24:3-13.

Ross, J., C. Dotson, G. Gilbert, and S. Katz. 1985. New standards for fitness measurement. *Journal of Physical Education, Recreation and Dance* 56(1):62-66.

Ross, J., R. Pate, L. Delby, R. Gold, and M. Svilar. 1987. New health-related fitness norms. *Journal of Physical Education, Recreation and Dance* 58(9):66-70.

Rotter, J.B. 1966. Generalized expectancies for internal versus external control of reinforcement. *Psychological Monographs* 80 (No. 609).

Rowe, D.A., and M.T. Mahar. 2006. Validity. In *Measurement theory and practice in kinesiology,* ed. T.M. Wood and W. Zhu, 9-26. Champaign, IL: Human Kinetics.

Rowland, T. 1990. *Exercise and children's health.* Champaign, IL: Human Kinetics.

Rowley, A., D. Landers, B. Kyllo, and J. Etnier. 1995. Does the iceberg profile discriminate between successful and less successful athletes? A meta-analysis. *Journal of Sport & Exercise Psychology* 17:185-199.

Ruiz, J.R., X. Sui, F. Lobelo, J.R. Morrow Jr., A.W. Jackson, M., Sjostrom, S.N, and Blair. 2008. Association between muscular strength and mortality in men: Prospective cohort study. *British Medicine Journal* 337:a439. doi:10.1136/bmj.a439

Safrit, M. 1986. *Introduction to measurement in physical education and exercise science.* St. Louis: Mosby.

Safrit, M., L. Hooper, S. Ehlert, M. Costa, and P. Patterson. 1988. The validity generalization of distance run tests. *Canadian Journal of Sport Sciences* 13:188-196.

Safrit, M., and M. Looney. 1992. Should the punishment fit the crime? A measurement dilemma. *Research Quarterly for Exercise and Sport* 63:124-127.

Safrit, M., and C. Pemberton. 1995. *Complete guide to youth fitness testing.* Champaign, IL: Human Kinetics.

Safrit, M.J., T.A. Baumgartner, A.S. Jackson, and C.L. Stamm. 1980. Issues in setting motor performance standards. *Quest* 32:152-162.

Salkind, N.J. 2013. *Excel statistics: A quick guide.* 2nd ed. Washington, D.C.: Sage.

Sallis, J.F., M.J. Buono, J.J. Roby, F.G. Micale, and J.A. Nelson. 1993. Seven-day recall and other physical activity self-reports in children and adolescents. *Medicine & Science in Sports & Exercise* 25:99-108.

Sarason, I.G. 1975. Test anxiety and the self-disclosing coping model. *Journal of Consulting and Clinical Psychology* 43:148-153.

Sargent, D.A. 1921. The physical test of man. *American Physical Education Review* 26(April):188-194.

Sax, G. 1980. *Principles of educational and psychology al measurement and evaluation.* Belmont, CA: Wadsworth.

Schick, J., and N.G. Berg. 1983. Indoor golf skill test for junior high boys. *Research Quarterly for Exercise and Sport* 54:75-78.

Seaman, J., and K. DePauw. 1989. *The new adapted physical education: A developmental approach.* Mountain View, CA: Mayfield.

SHAPE America. 2014. *National standards & grade-level outcomes for K-12 physical education.* Champaign, IL: Human Kinetics.

Sheard, M., and J. Golby. 2006. Effect of a psychological skills training program on swimming performance and positive psychological development. *International Journal of Sport & Exercise Psychology* 4:149-169.

Shephard, R. 1990. *Fitness in special populations.* Champaign, IL: Human Kinetics.

Shifflett, B., and B.J. Shuman. 1982. A criterion-referenced test for archery. *Research Quarterly for Exercise and Sport* 53:330-335.

Siedentop, D. 1996. Physical education and education reform: The case for sport education. In *Student learning in physical education: Applying research to enhance instruction,* ed. S.J. Silverman and C.D. Ennis, 247-267. Champaign, IL: Human Kinetics.

Sierer, S., C. Battaglini, J. Mihalik, E. Shields, and N. Tomasini. 2008. The National Football League combine: Performance differences between drafted and nondrafted players entering the 2004 and 2005 drafts. *Journal of Strength and Conditioning Research* 22:6-12.

Singer, R.N. 1968. *Motor learning and human performance.* New York: Macmillan.

Siri, W. 1956. Gross composition of the body. In *Advances in biological and medical physics,* ed. J. Lawrence, 239-280. New York: Academic Press.

Smith, R., R. Schutz, F. Smoll, and J. Ptacek. 1995. Development and validation of a multidimensional measure of sport-specific psychological skills: The Athletic Coping Skills Inventory-28. *Journal of Sport & Exercise Psychology* 17:379-398.

Smith, R., F.L. Smoll, and B. Curtis. 1979. Coach effectiveness training: A cognitive behavioral approach to enhancing relationship skills in youth and sport coaches. *Journal of Sport Psychology* 1:59-75.

Smith, R.E., F.L. Smoll, and N.P. Barnett. 1995. Reduction of children's sport performance anxiety through social support and stress-reduction training for coaches. *Journal of Applied Developmental Psychology* 16:125-142.

Smith, R.E., F.L. Smoll, and S.P. Cumming. 2007. Effects of a motivational climate intervention for coaches on young athletes' sport performance anxiety. *Journal of Sport & Exercise Psychology* 29:39-59.

Smith, R.E., F.L. Smoll, and R.W. Schutz. 1990. Measurement and correlates of sport-specific cognitive and somatic trait anxiety: The Sport Anxiety Scale. *Anxiety-Research* 2:263-280.

Smith, R.E., F.L. Smoll, S.P. Cumming, and J.R. Grossbard. 2006. Measurement of multidimensional sport performance anxiety in children and adults: The Sport Anxiety Scale-2. *Journal of Sport & Exercise Psychology* 28:479-501.

Society of Health and Physical Educators (SHAPE America). 2014. *National standards & grade-level outcomes for K-12 physical education.* Champaign, IL: Human Kinetics.

Sorrentino, R.M., and B.H. Sheppard. 1978. Effects of affiliation-related motives on swimmers in individual versus group competition: A field experiment. *Journal of Personality and Social Psychology* 36:704-714.

Sparkes, A. 1998. Validity in qualitative inquiry and the problem of criteria: Implications for sport psychology. *Sport Psychologist* 12:333-345.

Spielberger, C.D., R.L. Gorsuch, and R.F. Lushene. 1970. *Manual for the State–Trait Anxiety Inventory.* Palo Alto, CA: Consulting Psychologists Press.

Spink, K.S., and A.V. Carron. 1994. Group cohesion effects in exercise classes. *Small Group Research* 25:26-42.

Stiggins, R. 1987. Design and development of performance assessment. *Educational Measurement* 6(3):33-42.

Stone, D.B., W.R. Armstrong, D.M. Macrina, and J.W. Pankau. 1996. *Introduction to epidemiology.* Madison, WI: Brown and Benchmark.

Strand, B., and R. Wilson. 1993. *Assessing sports skills.* Champaign, IL: Human Kinetics.

Strean, W. 1998. Possibilities for qualitative research in sport psychology. *Sport Psychologist* 12:333-345.

Stuart, M. 2003. Sources of subjective task value in sport: An examination of adolescents with high or low value for sport. *Journal of Applied Sport Psychology* 15:239-255.

Sui, X., M.J. LaMonte, and S.N. Blair. 2007. Cardiorespiratory fitness as a predictor of nonfatal cardiovascular events in asymptomatic women and men. *American Journal of Epidemiology* 165:1413-1423.

Suzuki, N., and S. Endo. 1983. A quantitative study of trunk muscle strength and fatigability in the low-back pain syndrome. *Spine* 8:69-74.

Tarter, B.C., L. Kirisci, R.E. Tarter, S. Weatherbee, V. Jamnik, E.J. McGuire, and N. Gledhill. 2009. Use of aggregate fitness indicators to predict transition into the National Hockey League. *Journal of Strength and Conditioning Research* 23:1828-1832.

Taylor, M., D. Gould, and C. Roio. 2008. Performance strategies of US Olympians in practice and competition. *High Ability Studies* 19:19-36.

Telama, R., X. Yang, L. Laasko, and J. Viikari. 1997. Physical activity in childhood and adolescence as predictor of physical activity in young adulthood. *American Journal of Preventive Medicine* 13:317-323.

Tenenbaum, G., R.C. Eklund, and A. Kamata, eds. 2012. *Measurement in sport and exercise psychology.* Champaign, IL: Human Kinetics.

Terry, P. 2000. Perspectives on mood in sport and exercise. *Journal of Applied Sport Psychology* 12:1-4.

Thelwell, R., N. Weston, and I. Greenlees. 2005. Defining and understanding mental toughness within soccer. *Journal of Applied Sport Psychology* 17:326-332.

Thomas, J.R., J.K. Nelson, and S.J. Silverman. 2015. *Research methods in physical activity.* 7th ed. Champaign, IL: Human Kinetics.

Thomas, P., S. Murphy, and L. Hardy. 1999. Test of Performance Strategies: Development and preliminary validation of a comprehensive measure of athletes' psychological skills. *Journal of Sport Sciences* 17:691-711.

Troiano, R.P., D. Berrigan, K.W. Dodd, L.C. Masse, T. Tilert, and M. McDowell. 2008. Physical activity in the United States measured by accelerometer. *Medicine & Science in Sports & Exercise* 40:181-188

Trudelle-Jackson, E., A.W. Jackson, C.M. Frankowski, K.M. Long, and N.B. Meske. 1994. Interdevice reliability and validity assessment of the Nicholas Hand-Held Dynamometer. *Journal of Orthopaedic & Sports Physical Therapy* 20:302-306.

Tudor-Locke, C., and D.R. Bassett Jr. 2004. How many steps/day are enough? Preliminary pedometer indices for public health. *Sports Medicine* 34:1-8.

U.S. Department of Health and Human Services (USDHHS). 1996. *Physical activity and health: A report of the Surgeon General.* Atlanta: USDHHS, Centers for Disease Control and Prevention, National Center for Chronic Disease Prevention and Health Promotion.

———. 2008. *2008 physical activity guidelines for Americans.* Washington, DC: U.S. Department of Health and Human Services.

Vallerand, R.J. 1997. Toward a hierarchial model of intrinsic and extrinsic motivation. *Advances in experimental social psychology* 29:271-360.

Van Schoyck, R.S., and A.F. Grasha. 1981. Attentional style variations and athletic ability: The advantage of a sport-specific test. *Journal of Sport Psychology* 2:149-165.

Veal, M.L. 1988. Pupil assessment perceptions and practices of secondary teachers. *Journal of Teaching in Physical Education* 7:327-342.

Vealey, R.S. 1986. Conceptualization of sport-confidence and competitive orientation: Preliminary investigation and instrument development. *Journal of Sport Psychology* 8:221-246.

Velicer, W.F., and J.O. Prochaska. 1997. Introduction: The transtheoretical model. *American Journal of Health Promotion* 12:6-7.

Vema, J.P., A.S. Sajwan, and M. Debnath. 2009/2. A study on estimating $\dot{V}O_2$max from different techniques in field situation. *International Quarterly of Sport Science* 2:42-47.

Verducci, F.M. 1980. *Measurement concepts in physical education.* St. Louis: Mosby.

Viera, A.J., and J.M. Garrett. 2005. Understanding interobserver agreement: The kappa statistic. *Family Medicine* 37:360-363.

Vingren, J.L., A.T. Woolsey, and J.R. Morrow, Jr. 2014. Assessing physical activity, fitness, and progress in older adults. In *ACSM's exercise for older adults*, ed. W. J. Chodzko-Zajko. Champaign, IL: Human Kinetics.

Weinberg, R., J. Butt, and B. Culp. 2011. Coaches' views of mental toughness and how it is built. *International Journal of Sport and Exercise Psychology* 9:156-172.

Weinberg, R., D. Gould, and A. Jackson. 1979. Expectations and performance: An empirical test of Bandura's Self-efficacy Theory. *Journal of Sport Psychology* 1:320-331.

Weinberg, R.S., and W.W. Comar. 1994. The effectiveness of psychological interventions in competitive sport. *Sports Medicine* 18:406-418.

Weinberg, R.S., D. Gould, D. Yukelson, and A. Jackson. 1981. The effect of preexisting and manipulated self-efficacy on a competitive muscular endurance task. *Journal of Sport Psychology* 3:345-354.

Welk, G.J., ed. 2002. *Physical activity assessments for health-related research*. Champaign. IL: Human Kinetics.

Weston, A.T., R. Petosa, and R.R. Pate. 1997. Validation of an instrument for measurement of physical activity in youth. *Medicine & Science in Sports & Exercise* 29:138-143.

Widmeyer, W.N., L.R. Brawley, and A.V. Carron. 1985. *The measurement of cohesion in sport teams: The group environment questionnaire*. (Available from Sports Dynamics, 11 Ravenglass Crescent, London, ON, Canada N6G 3X7)

Wier, L.T., A.S. Jackson, G.W. Ayers, and B. Arenare. 2006. Nonexercise models for estimating $\dot{V}O_2$max with waist girth, percent fat, or BMI. *Medicine & Science in Sports & Exercise* 38:555-561.

Wiggins, G. 1989. A true test: Toward more authentic and equitable assessment. *Phi Delta Kappan* 69:703-713.

Wiggins, G. 1998. *Educative assessment: Designing assessments to inform and improve student performance*. San Francisco: Jossey-Bass.

Wilson, P.M., W.M. Rodgers, C.C. Loitz, and G. Scime. 2006. "It's who I am . . . really!" The importance of integrated regulation in exercise contexts. *Journal of Applied Biobehavioral Research* 11(2):79-104.

Wininger, S. 2007. Self-determination theory and exercise behavior: An examination of the psychometric properties of the Exercise Motivation Scale. *Journal of Applied Sport Psychology* 19:471-486.

Winnick, J.P., and F.X. Short. 1999. *The Brockport physical fitness test manual*. Champaign, IL: Human Kinetics.

Winston, W. 2009. *Mathletics: How gamblers, managers, and sports enthusiasts use mathematics in baseball, basketball, and football*. Princeton, Princeton University Press.

Wood, T.M. 1996. Evaluation and testing: The road less traveled. In *Student learning in physical education: Applying research to enhance instruction*, ed. S.J. Silverman and C.D. Ennis, 199-219. Champaign, IL: Human Kinetics.

———. 2003. Assessment in physical education: The future is now. In *Student learning in physical education: Applying research to enhance instruction*, 2nd ed., ed. S.J. Silverman and C.D. Ennis, 187-203. Champaign, IL: Human Kinetics.

Xiang, P., R.E. McBride, A. Bruene, and Y. Liu. 2007. Achievement goal orientation patterns and fifth graders' motivation in physical education running programs. *Pediatric Exercise Science, 19*(2), 179–191.

Yukelson, D., R. Weinberg, and A. Jackson. 1984. A multidimensional group cohesion instrument for intercollegiate basketball teams. *Journal of Sport Psychology* 6: 103–117.

Zhu, W., M.J. Safrit, and A. Cohen. 1999. *FitSmart test user manual: High school edition*. Champaign, IL: Human Kinetics.

Zourbanos, N., A. Hatzigeorgiadis, S. Chroni, Y. Theodorakis, and A. Papaioannou. 2009. Automatic Self-Talk Questionnaire for Sports (ASTQS): Development and preliminary validation of a measure identifying the structure of athletes' self-talk. *Sport Psychologist* 23: 233–251.

Zourbanos, N., A. Hatzigeorgiadis, M. Goudas, A. Papaioannou, S. Chroni, and Y. Theodorakis. 2011. The social side of self-talk: Relationships between perceptions of support received from the coach and athletes' self-talk. *Psychology of Sport and Exercise* 12: 407–414.

Zourbanos, N., A. Papaioannou, E. Argyropoulou, and A. Hatzigeorgiadis. 2014. Achievement goals and self-talk in physical education: The moderating role of perceived competence. *Motivation and Emotion* 38: 235–251. doi:10.1007/s11031-013-9378-x

Index

Note: Page numbers followed by an italicized *f* or *t* refer to the figure or table on that page, respectively.

About the Authors

Photo by Jonathan Reynolds, University of North Texas.

James R. Morrow, Jr., PhD, is a regents professor emeritus in the department of kinesiology, health promotion, and recreation at the University of North Texas at Denton. Morrow regularly teaches courses in measurement and evaluation in human performance. He has authored more than 150 articles and chapters on measurement and evaluation, physical fitness, physical activity, and computer use and has made approximately 300 professional presentations. He has also conducted significant research using the techniques presented in the text.

Morrow served as president of the National Academy of Kinesiology and as chair of the President's Council on Physical Fitness and Sports Science Board. He has received research funding from the United States Olympic Committee, the U.S. Centers for Disease Control and Prevention, the National Institutes of Health, and the Cooper Institute. He is a fellow of the American College of Sports Medicine (ACSM); the National Academy of Kinesiology (NAK); and the North American Society of Health, Physical Education, Recreation, Sport and Dance Professionals. He is also a research fellow of SHAPE America. Morrow has chaired the AAHPERD Measurement and Evaluation Council and is a recipient of that council's Honor Award. He has produced four fitness-testing software packages, including the AAHPERD Health-Related Physical Fitness test, and was editor in chief of *Research Quarterly for Exercise and Sport* from 1989 to 1993. He was the founding coeditor of the *Journal of Physical Activity and Health*. He enjoys playing golf, reading, traveling, and spending time with his grandchildren.

Courtesy of Dale Mood

Dale P. Mood, PhD, is a professor emeritus and former associate dean of arts and sciences at the University of Colorado at Boulder. Mood has taught measurement and evaluation, statistics, and research methods courses since 1970 and has published extensively in the field, including 47 articles and 6 books. He has been a consultant to five NFL football teams and chair of the Measurement and Evaluation Council of AAHPERD, and he is a former president of AAALF. He is a reviewer for *Medicine and Science in Sports and Exercise, Measurement in Physical Education and Exercise Science*, and *Research Quarterly for Exercise and Sport*. In his leisure time, Mood enjoys reading, officiating summer league swimming meets, traveling, following the activities of his 17 grandchildren, and participating in a variety of physical activities.

Courtesy of James Disch

James G. Disch, PED, is an associate professor in the sport management department at Rice University. From 1986 to 1991 he was master of Richardson College at Rice. From 1995 to 2001 he was chair of the kinesiology department. Disch has authored numerous articles, chapters, manuals, and texts in the areas of applied measurement, prediction in sport, and applied sport science. A member of AAHPERD (now SHAPE America) since 1974, he has been chair, secretary, and advisory board member of the measurement and evaluation council of AAHPERD. He was vice president of the college division of TAHPERD and had numerous section chair appointments. Disch is also a reviewer for *Research Quarterly for Exercise and Sport* and *Medicine and Science in Sports and Exercise.*

Disch has coordinated several workshops and symposia on measurement and evaluation. He was a major contributor to the development of AAHPERD health-related fitness norms in 1980 and has worked as a consultant and advisor for Olympic and professional teams. In 1999 he received the National Measurement and Evaluation Council Honor Award, and in 2011 he was named a Sport Ethics Fellow of the Year by the International Institute of Sport and the Positive Coaching Alliance. He was named the TAHPERD Scholar for 2012.

Dr. Disch served on the board of the RBI Foundation (Recycled Baseball Items) from 2009 until 2014 and is the current chair of the Houston chapter of the Positive Coaching Alliance. He is also on the local planning committee for the Joe Niekro Foundation Knuckleball.

Disch earned his PED in biomechanics and measurement from Indiana University in 1973.

Courtesy of Minsoo Kang

Minsoo Kang, PhD, is a professor in the department of health and human performance at Middle Tennessee State University. He received his PhD in kinesmetrics (measurement and evaluation) in kinesiology with emphasis in IRT, Rasch, and psychometrics from the University of Illinois at Urbana-Champaign. Kang's research has focused on measurement and statistical methods and their applications to assessments of physical activity and sedentary behavior. He has published more than 70 refereed journal articles, made 9 book contributions, and presented more than 200 research projects. He teaches courses on data analysis, research methods, meta-analysis, research seminar, and current measurement issues in human performance.

Kang is a fellow of the American College of Sports Medicine (ACSM) and a research fellow of SHAPE America. He has chaired the AAHPERD Measurement and Evaluation Council and is a recipient of that council's Honor Award. Kang received the Distinguished Research Award at Middle Tennessee State University. He currently is an associate editor of the *Research Quarterly for Exercise and Sports,* a section editor of *Measurement in Physical Education and Exercise Science,* and a member of the editorial board for those journals. He enjoys playing badminton, golf, and tennis.

About the Contributors

Allen W. Jackson, EdD, is a regents professor in the department of kinesiology, health promotion, and recreation at the University of North Texas, where he has taught kinesiology with research, statistics, and computer applications since 1978. He has published extensively in measurement and evaluation, including more than 100 articles, and has presented more than 200 scientific papers. He has received research funding from the Centers for Disease Control and Prevention, the National Institutes of Health, the Robert Wood Johnson Foundation, and the Cooper Institute. He has reviewed articles for *Measurement in Physical Education and Exercise Science, Medicine and Science in Sport and Exercise, Research Quarterly for Exercise and Sport,* and the *Journal of Pediatrics.*

Dr. Jackson earned his EdD in 1978 at the University of Houston. He is an ACSM fellow, a NAK fellow, a former member of the Science Board for the President's Council on Physical Fitness, Sport & Nutrition, and a member of SHAPE America. Dr. Jackson's favorite leisure activities are cardiorespiratory fitness conditioning, walking, and resistance training.

Joseph F. Kerns is a graduate student at Miami University in Oxford, Ohio, working as a graduate assistant with Dr. Robert S. Weinberg. He is an alumna of Washington State University, and a former personal trainer and nutrition consultant. He has also consulted with the National Academy of Sports Medicine as a Subject Matter Expert. After completing his M.S. he is planning on working with the military as a performance psychology specialist.

Jacalyn Lea Lund, PhD, is a professor and chair in the department of kinesiology and health at Georgia State University in Atlanta. She taught for 16 years in public schools before entering the doctoral program at Ohio State University to prepare for a second career in teacher education. She received her PhD in 1990. Dr. Lund has presented on assessment at numerous workshops and has taught numerous classes on assessment in physical education.

Dr. Lund has been a member of the National Association for Sport and Physical Education (NASPE) for more than 40 years. She was on the committee that developed the 1995 NASPE content standards for physical education and has served as NASPE president. In 2009 she received a Service Award from the National Association of Kinesiology and Physical Education in Higher Education and was inducted into the NASPE Hall of Fame in 2013. In March of 2015, Dr. Lund was elected President-Elect of SHAPE America. She loves spending time with her family, dancing, reading, and, as she puts it, "having her dogs take her for a walk."

Robert S. Weinberg, PhD, is a professor in Kinesiology and Health at Miami University in Oxford, Ohio. Weinberg has more than 30 years of experience in both the scholarly and applied aspects of sport psychology. He has written numerous research articles, including more than 150 refereed articles in scholarly journals, as well as books, book chapters, and applied articles for coaches, athletes, and exercisers. His text (with co-author Dan Gould) entitled, "Foundations of Sport and Exercise Psychology" is the most popular text in the field of sport psychology in the world.

Weinberg was voted one of the top 10 sport psychologists in North America by his peers. He is a past president of the North American Society for Psychology of Sport and Physical Activity (NASPSPA) and of the Association for Applied Sport Psychology (AASP). He is a Fellow in the Association for Applied Sport Psychology, American Psychological Association (Division 47), and National Academy of Kinesiology. He is also a certified AASP consultant, where he consults with athletes of all sports and ages.

Weinberg was named a Distinguished Scholar in Sport Psychology at Miami University in 2005. In addition, he was the editor of the Journal of Applied Sport Psychology and was voted outstanding faculty member in the School of Education and Allied Professions at Miami University in 1998. In his leisure time, he enjoys playing tennis, traveling, and gardening.